Optical pattern recognition

Pattern recognition is an important application of modern optical technology. Many systems exist in which objects or patterns are imaged optically and then analyzed with sophisticated computer algorithms. This book provides a comprehensive review of optical pattern recognition, covering theoretical aspects as well as details of practical implementations and signal-processing techniques.

The first chapter is devoted to pattern recognition performed with optical correlators. New approaches based on neural networks, wavelet transforms, and the fractional Fourier transform are then discussed. Nonlinear filter methods are also covered, as are optical-electronic hybrid systems. The final part of the book deals with the devices and materials employed in modern systems, such as photorefractive crystals, microlasers, and liquid-crystal spatial light modulators.

This book gives many examples of working systems that integrate optics, electronics, and computers, and covers a range of new developments from mathematical theories to novel optical materials. It will be of great interest to graduate students and researchers in optical engineering and machine vision.

Francis T. S. Yu is a Professor of Electrical Engineering at the Pennsylvania State University, University Park. The author or coauthor of seven books, he is a Fellow of the Optical Society of America, SPIE – The International Society for Optical Engineering, and the Institute of Electrical and Electronics Engineers. Dr. Yu received the Ph.D. degree (1964) in electrical engineering from the University of Michigan, Ann Arbor.

Suganda Jutamulia is an Associate Professor of Biomedical Engineering at the University of Northern California and the President of In-Harmony Technology Corporation, Petaluma, California. He is a senior member of the Institute of Electrical and Electronics Engineers and a member of the Optical Society of America, SPIE – The International Society for Optical Engineering and the Japan Applied Physics Society. Dr. Jutamulia received the Ph.D. degree (1985) in electronic engineering from Hokkaido University, Sapporo, Japan.

Optical pattern recognition

Edited by

FRANCIS T. S. YU

and

SUGANDA JUTAMULIA

CAMBRIDGE
UNIVERSITY PRESS

CAMBRIDGE UNIVERSITY PRESS
Cambridge, New York, Melbourne, Madrid, Cape Town, Singapore, São Paulo, Delhi

Cambridge University Press
The Edinburgh Building, Cambridge CB2 8RU, UK

Published in the United States of America by Cambridge University Press, New York

www.cambridge.org
Information on this title: www.cambridge.org/9780521465175

First published 1998
This digitally printed version 2008

A catalogue record for this publication is available from the British Library

Library of Congress Cataloguing in Publication data

Optical pattern recognition / edited by Francis T. S. Yu, Suganda
Jutamulia.
p.   cm.
ISBN 0-521-46517-6
1. Optical pattern recognition.   I. Yu, Francis T. S., 1932– .
II. Jutamulia, Suganda.
TA1650.0655   1998   97-36651
621.39′9 – dc21   CIP

ISBN 978-0-521-46517-5 hardback
ISBN 978-0-521-08862-6 paperback

Contents

Contributors

Dr. Tien-Hsin Chao
Jet Propulsion Lab
4800 Oak Grove Drive
Pasadena, CA 91109-8099

Prof. Robert W. Cohn
Department of Electrical Engineering
University of Louisville
Louisville, KY 40292

Dr. Gregory Gheen
KLA Instrument Corporation
160 Rio Robles
San Jose, CA 95161

Prof. Don A. Gregory
Physics Department
University of Alabama in Huntsville
Huntsville, AL 35899

Prof. Laurence G. Hassebrook
Department of Electrical Engineering
University of Kentucky
Lexington, KY 40506

Dr. Suganda Jutamulia
In-Harmony Technology Corp.
101 South Antonio Road
Petaluma, CA 94952

Dr. Yao Li
NEC Research Institute
Physical Science Division
Princeton, NJ 08540

Dr. Guowen Lu
Applied Research laboratory
The Pennsylvania State University
University Park, PA 16802

Dr. Taiwei Lu
In-Harmony Technology Corp.
101 South Antonio Road
Petaluma, CA 94954

Dr. Mingzhe Lu
Institute of Modern Optics
Nankai University
Tianjin, 300071, P.R. China

Prof. David Mendlovic
Faculty of Engineering
Tel-Aviv University
Tel-Aviv, Israel 69978

Prof. Guoguang Mu
Institute of Modern Optics
Nankai University
Tianjin, 300071, P.R. China

Prof. Haldun M. Ozaktas
Faculty of Electrical Engineering
Bilkent University
Ankara, Turkey

Dr. Eung-Gi Paek
National Institute of Standards and
Technology
Gaithersburg, MD 20899

Prof. Joseph Shamir
Department of Electrical
Engineering
Technion-Israel Institute of
Technology
Technion City
Haifa, Israel 32000

Prof. Yunlong Sheng
Physics Department
Laval University
Sainte-Foy 61K7P4, Canada

Prof. Q. Wang Song
Department of Electrical and Computer
Engineering
Syracuse University
Syracuse, NY 13244-1240

Prof. Yin Sun
Institute of Modern Optics
Nankai University
Tianjin, 300071, P.R. China

Dr. Aris Tanone
JAYA Corporation
University Square
Huntsville, AL 35816

Prof. Xiangyang Yang
Department of Electrical
Engineering
University of New Orleans
New Orleans, LA 70148

Prof. Leonid Yaroslavsky
Department of Interdisciplinary Studies
Engineering Faculty
Tel-Aviv University
Tel-Aviv 69978, Israel

Prof. Shizhuo Yin
Department of Electrical Engineering
The Pennsylvania State University
University Park, PA 16802

Prof. Francis T. S. Yu
The Pennsylvania State University
University Park, PA 16802

Dr. Zeev Zalevsky
Faculty of Engineering
Tel-Aviv University
Tel-Aviv, Israel 69978

Dr. Yu-He Zhang
Department of Electrical and Computer
Engineering
Syracuse University
Syracuse, NY 13244-1240

Preface

Smart automatic machines that may reduce our working load and minimize risk in work have been sought for a long time. This suggests supplementing rather than supplanting, and extending rather than denying, our human capabilities to work. To be smart, the machine must understand a scene as a human does. In other words, the machine must be able to recognize the scene by comparing the present scene and the past scenes. The machine operates based more on the scene it views, and it is less controlled by a set of thousands of instructions. For these smart machines, a picture is, indeed, worth a thousand words. Pattern recognition is the primary task of any smart automatic machine. Because the pattern to be recognized is often received optically, it is perhaps most natural and straightforward to recognize an optical pattern by using optics. This book reviews in depth the recent progress in optical pattern recognition, although no attempt has been made to give an encyclopedic presentation.

The book was designed to incorporate multiple facets and approaches essential for today's optical pattern recognition research enterprise to proceed. To this end, we have brought together leading researchers worldwide, all focusing their efforts on selected aspects of optical pattern recognition issues in 15 chapters. The first chapter overviews pattern recognition that is performed mainly with an optical correlator. The following four chapters describe new approaches to optical pattern recognition: neural networks, wavelet transform, fractional Fourier transform, and mathematical morphology. Nonlinear methods are discussed in the following three chapters. One method employs a nonlinear device in a Fourier plane. The other two methods apply nonlinear-quadratic and composite filters, respectively. These nonlinear filters are implemented in a conventional linear optical system. The next two chapters describe optoelectronic hybrid systems that use an optical system and a digital computer in recognizing an input pattern. The remaining five chapters present devices and materials used in an optical pattern recognition system: photorefractive crystals, microlasers, bacteriorhodopsin, liquid-crystal spatial light modulators, and complex-function spatial light modulators.

We thank all the authors for their excellent contributions, and we are honored to have had an opportunity to work with them. It is early yet to estimate the magnitude of the contribution optical pattern recognition will make to the extension of human capabilities and the corresponding convenience in human life, and it would be more than a little reckless to rank it now along with optical disk and fiber communication. But the contribution will be in that class and will indeed be profound.

Francis T. S. Yu
Suganda Jutamulia

1

Pattern recognition with optics

Francis T. S. Yu and Don A. Gregory

1.1 Introduction

The roots of optical pattern recognition can be traced back to Abbé's work in 1873 [1], when he developed a method that led to the discovery of spatial filtering to improve the resolution of microscopes. However, optical pattern recognition was not actually appreciated until the complex spatial filtering work of VanderLugt in 1964 [2]. Since then, techniques, architectures, and algorithms have been developed to construct efficient optical correlators for pattern recognition application.

Our objective in this chapter is to discuss the optical architectures and techniques as applied to recent advances in pattern recognition. Basically there are two approaches in the optical implementation of pattern recognition, namely, the correlation approach and the neural net approach. In the correlation approach, there are two frequently used architectures: the VanderLugt correlator (VLC) and the joint transform correlator (JTC). The first JTC architecture was demonstrated by Weaver and Goodman in 1966 [3] and independently by Rau [4]. Because of a lack of interface devices, the JTC was virtually stagnant until 1984, when a real-time programmable JTC was reported by Yu and Lu [5]. Since then the JTC has assumed a major role in various processing applications.

Aside from correlation applications to pattern recognition, artificial neural networks (NN's) are also well suited. The first optical NN is attributed to Psaltis and Farhat in 1985 [6], who showed that pattern retrieval can be achieved with a lenslet array interconnection network.

In this chapter, advances in this rapidly growing field are reviewed and the basic optical architectures and techniques addressed. The pros and cons of each approach are discussed. Because of recent technical advances in interface devices [such as electronically addressable spatial light modulators (SLM's), nonlinear optical devices, etc.], new philosophies and new algorithms have been developed for the design of better pattern recognition systems.

1.2 Optical correlators

The optical implementation of pattern recognition can be accomplished with either Fourier-domain complex matched filtering or spatial-domain filtering. Correlators that use Fourier-domain matched filtering are commonly known as VLC's [2], and an example of spatial-domain filtering is the JTC [3–5]. The basic distinctions between them are that the VLC depends on Fourier-domain spatial filter synthesis (e.g., Fourier hologram), whereas the JTC depends on spatial-domain (impulse-response) filter synthesis. In other words,

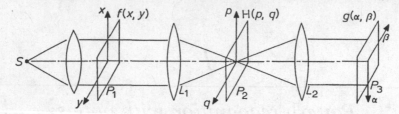

Fig. 1.1. VLC.

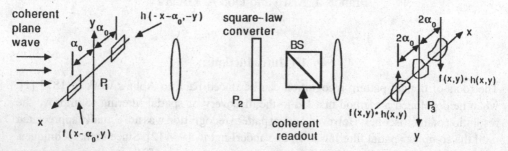

Fig. 1.2. JTC.

the complex spatial detection of the VanderLugt arrangement is input scene *independent*, whereas the joint transform method is input scene *dependent* [7]. The basic optical set-ups of these two types of correlator are depicted in Figs. 1.1 and 1.2. A prefabricated Fourier-domain matched filter $H(p, q)$ is needed in the VLC, whereas a matched filter is not required in the JTC but a spatial-domain impulse response $h(x, y)$ is needed. Although the JTC avoids spatial filter synthesis problems, it generally suffers lower detection efficiency, particularly when applied to multitarget recognition or targets embedded in intense background noise [8]. Nonetheless, the JTC has many merits, particularly when interfaced with electronically addressable SLM's.

The JTC has other advantages, such as higher space–bandwidth products, lower carrier frequency, higher index modulation, and suitability for real-time implementation. Disadvantages include inefficient use of illuminating light, a large transform lens, stringent spatial coherence requirements, and the overall small size of the joint transform spectrum. A quasi-Fourier-transform JTC (QFJTC) that can alleviate some of these limitations is shown in Fig. 1.3 [9, 10]. The depth of focus is given by [10]

$$\delta \leq 2\lambda(f/b)^2. \tag{1.1}$$

To illustrate the shift-invariant property of the QFJTC, an input object such as that shown in Fig. 1.4(a) is used. The joint transform power spectrum (JTPS) is recorded as a photographic transparency, which could be thought of as a joint transform hologram (JTH). When the recorded JTH is simply illuminated with coherent light, the cross-correlation distribution can be viewed in the ouput plane [shown in Fig. 1.4(b)], where autocorrelation peaks indicating the location of the input character, G, are detected.

Representative experimental results were obtained with the QFJTC with the input object and the reference functions of Fig. 1.5(a). Three JTHs – for $\delta = 0$, $\delta = f/10 = 50$ mm, and $\delta = f/5 = 100$ mm – are shown in Fig. 1.5(b), which shows that the size of the JTPS enlarges as δ increases. Figure 1.6 illustrates that the correlation peak intensity increases as δ

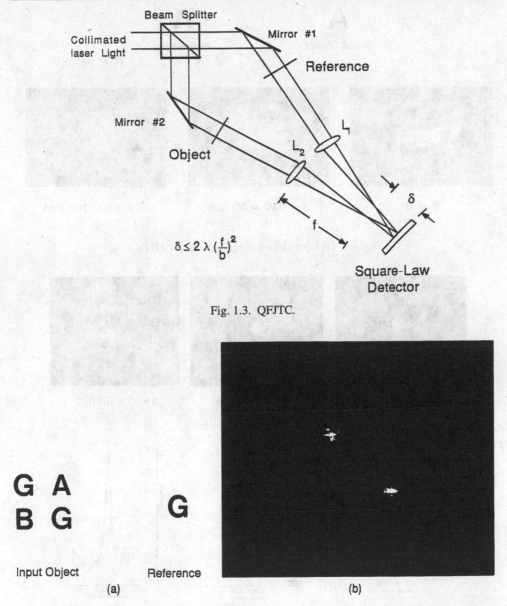

Fig. 1.3. QFJTC.

Fig. 1.4. (a) Input object and reference function, (b) output correlator distribution.

increases, whereas the size of the correlation spot decreases as δ increases. Thus the QFJTC architecture can improve the signal-to-noise ratio (SNR) and the accuracy of detection.

1.3 Hybrid optical correlators

It is apparent that a purely optical correlator has drawbacks that make certain tasks difficult or impossible to perform. The first problem is that optical systems are difficult to program, in the sense of programming general-purpose digital–electronic computers. A purely optical system can be designed to perform specific tasks (analogous to a hard-wired electronic computer), but it cannot be used when more flexibility is required. A second problem is that

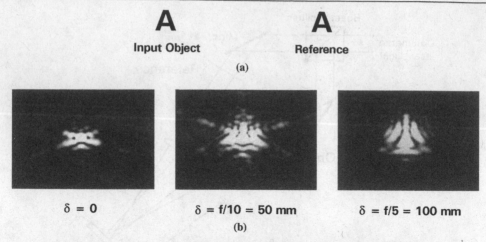

Fig. 1.5. (a) Input and reference objects, (b) JTH's.

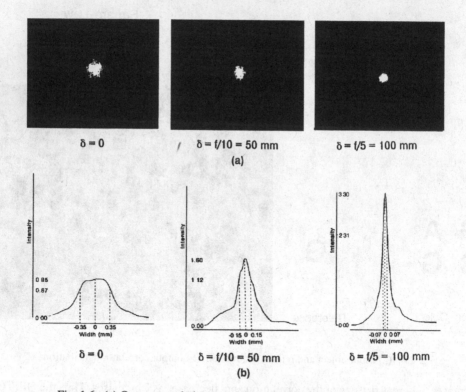

Fig. 1.6. (a) Output correlation spots, (b) output correlation distributions.

a system based on Fourier optics is naturally analog, which often makes accuracy difficult to achieve. A third problem is that optical systems by themselves cannot easily be used to make decisions. Even the simplest type of decision making is based on the comparison of an output with a stored value. Such an operation cannot be performed optically without the intervention of electronics.

Many deficiencies of optical systems happen to be strong points in their electronic counterparts. For example, accuracy, controllability, and programmability are some obvious

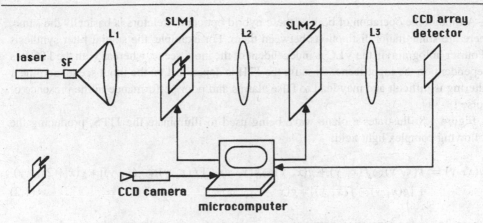

Fig. 1.7. Hybrid optical VLC. SF, spatial filter; L's, lenses.

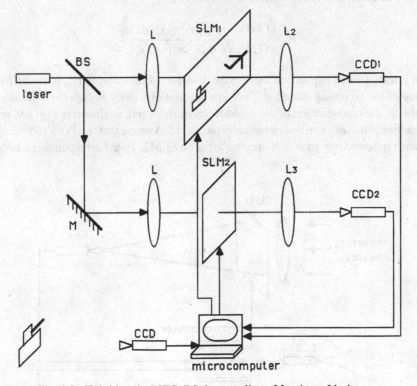

Fig. 1.8. Hybrid optical JTC. BS, beam splitter; M, mirror; L's, lenses.

traits of digital computers. Thus the idea of combining an optical system with its electronic counterpart is rather natural as a means of applying the rapid processing speed and parallelism of optics to a wider range of applications. In this section, this concept is applied to general-purpose microcomputer-based optical correlators. Examples are shown in Figs. 1.7 and 1.8, in which SLM's are used for object and spatial filter devices. One of the important aspects of these hybrid optical architectures is that decision making can be done by the computer.

Although the operation of both of these hybrid optical correlators is basically the same, there are some major distinctions between them: For example, the spatial filter synthesis (Fourier hologram) in the VLC is independent of the input scene, whereas with the JTC it is dependent on the input scene. Because the JTH is dependent on the input scene, nonlinear filtering is difficult and may lead to false alarms and poor performance in the presence of noise [8, 11].

Figure 1.8 illustrates a plane wave being used to illuminate the JTPS, producing the following complex light field:

$$u(x, y) = f(x, y) \circledast f(x, y) + g(x, y) \circledast g(x, y) + [f(x, y) \circledast g(x, y)] * g(x + 2x_0, y)$$
$$+ [g(x, y) \circledast f(x, y)] * g(x - 2x_0, y), \tag{1.2}$$

where $2x_0$ is the separation between $f(x, y)$ and $g(x, y)$, $\circledast$ denotes the correlation operation, and $*$ denotes convolution. Because the JTPS represents a small region of the square-law detector, the readout efficiency is [12, 13]

$$\varepsilon = \varepsilon_m \frac{\iint |f(x, y) \circledast g(x, y)|^2 \, dx \, dy}{(1/R) \iint |u(x, y)|^2 \, dx \, dy}, \tag{1.3}$$

where R is the ratio of the spectrum's size to the detector's dimension and ε_m is the diffraction efficiency of the recording material. Thus low readout efficiency is expected. To alleviate this problem, the configuration of the readout can be modified, as shown in Fig. 1.9, so that the diffraction efficiency can be dramatically increased. Assume that the JTPS is detected by CCD1 and replicated into an $N \times N$ spectral array on SLM2. Then corresponding amplitude

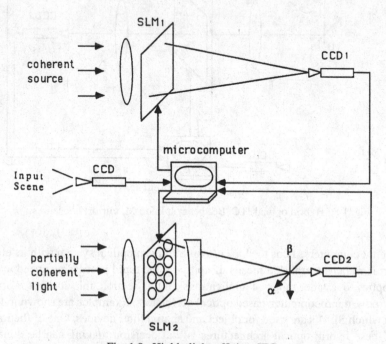

Fig. 1.9. Highly light-efficient JTC.

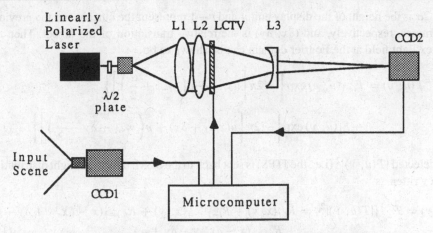

Fig. 1.10. Single-SLM JTC. L's, lenses; LCTV, liquid-crystal TV.

transmittance is given by

$$(p, q) = 2 \sum_{n=1}^{N} \sum_{m=1}^{N} |F(p - nd, q - md)|^2 \{1 + \cos[2x_0(p - nd)]\}, \qquad (1.4)$$

where (p, q) is the spatial frequency coordinate and d is the bandwidth of the Fourier spectrum. Thus, with normal coherent readout, the overall correlation diffraction efficiency can be written as [12]

$$\varepsilon = N^2 \varepsilon_m \frac{\iint [f(x, y) \circledast f(x, y)]^2 \, dx \, dy}{(1/R) \iint |u(x, y)|^2 \, dx \, dy}, \qquad (1.5)$$

which is an increase of $\sim N^2$ times, compared with Eq. (1.3).

The above-mentioned hybrid optical JTC can be modified by use of a single-SLM architecture, shown in Fig. 1.10, which was first demonstrated by Yu *et al.* [14] and later by Tam *et al.* [15]. Obviously the price paid must be are operating duty cycle time that is twice that of the two-SLM JTC of Fig. 1.8.

1.4 Autonomous target tracking

In this section autonomous target tracking with a JTC [15, 16] is discussed. The idea is to correlate the object in the current frame with the object in the previous frame. This makes the hybrid optical JTC adaptive by constantly updating the reference object with the dynamic input scene.

Two sequential scenes of a moving object are displayed with a video framegrabber on the SLM of Fig. 1.10 with the previous and the current frames positioned in the upper and the lower halves of the SLM. This is represented mathematically by

$$f_{t-1}\left(x - x_{t-1}, y - y_{t-1} - \frac{\alpha}{2}\right),$$

$$f_t\left(x - x_{t-1} - \delta x, y - y_{t-1} - \delta y + \frac{\alpha}{2}\right), \qquad (1.6)$$

where 2α is the height of the display unit, t and $t-1$ represent the current and the previous time frames, respectively, and $(\delta x, \delta y)$ is the relative translation of the object. Then the complex light field at the Fourier domain can be shown to be

$$T(u, v) = F_{t-1}(u, v)\exp\left\{-i2\pi\left[ux_{t-1} + v\left(y_{t-1} + \frac{\alpha}{2}\right)\right]\right\}$$

$$+ F_1(u, v)\exp\left\{-i2\pi\left[u(x_{t-1} + \delta x) + v\left(y_{t-1} + \delta y - \frac{\alpha}{2}\right)\right]\right\}. \quad (1.7)$$

If the detected $|T(u, v)|^2$ (i.e., the JTPS) is sent back to the SLM, the output light distribution can be written as

$$C(x, y) = \mathcal{F}^{-1}\{|T(u, v)|^2\} = R_{t,t}(x, y) + R_{t-1,t-1}(x, y) + R_{t,t-1}(x - \delta x, y + \delta y - \alpha)$$
$$+ R_{t-1,t}(x - \delta x, y - \delta y + \alpha), \quad (1.8)$$

where

$$R_{m,n}(x, y) = \mathcal{F}^{-1}\left[F_m(u, v)F_n^*(u, v)\right]$$

$$= \int_{-\infty}^{+\infty}\int_{-\infty}^{+\infty} f_m(u, v)f_n^*(u - x, v - y)\,\mathrm{d}u\,\mathrm{d}v$$

represents the correlations of f_m and f_n, which are diffracted at $x_1 = \delta x$, $y_1 = (\delta y - \alpha)$ and $x_2 = \delta x$, $y_2 = (-\delta y + \alpha)$, respectively. If the angular and the scale tolerances of the JTC are approximately $\pm 5°$ and $\pm 10\%$, respectively [17, 18], and the motion of the target is relatively slow compared with that of the processing cycle of the correlator, then f_{t-1} correlates strongly with f_t. Two high-intensity correlation peaks are diffracted into the output plane to locations given by

$$x_t = (x_{t-1} + x_1), \qquad y_t = (y_{t-1} + y_1 + \alpha). \quad (1.9)$$

A simple C language program can then be used to evaluate the target position. A block diagram of the system configuration is shown in Fig. 1.11. In spite of hardware limitations, the system constructed ran at approximately 1.2 s/cycle. It is possible for the hybrid system to run at half the video frame frequency, which is $1/2 \times 30 = 15$ cycles/s, if specialized support hardware is used.

To test the tracking performance, a tank-shaped object was used as the test target. The tank was set to revolve in an elliptical path, as shown in Fig. 1.12(a). Figures 1.12(b) and 1.12(c) show the locations of the tracked tank. Excellent tracking of the object's location is observed at the end of the first revolution. However, after four revolutions, some deviations of the tracked points from the correct loci are apparent. These deviations reveal that detection errors accumulate during each correlation cycle. This is a general problem with object tracking based on the correlation between images in sequential frames.

A major benefit of this technique is its adaptivity. Suppose a situation in which a camera mounted on a moving space vehicle is focused on a fixed target on the ground for automatic navigation. As the space vehicle approaches the target, the detected scene changes continuously, the target size appears larger, and its orientation and shape change because of the motion of the vehicle. With computer-aided-design graphics, a three-dimensional treelike model was created as a simulated target on the ground. Three of the nine image sequences,

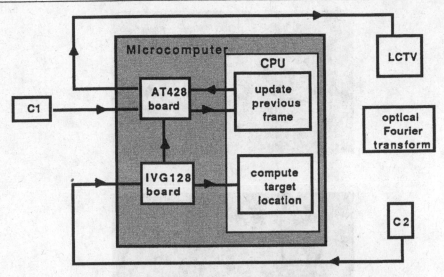

Fig. 1.11. Optical–digital interface diagram. C's, CCD.

simulating the changing scene viewed by a moving space vehicle, are shown in Fig. 1.13. The JTC tracking system has little difficulty in correlating targets from different frames, even though the targets in the first frame and the last frame look different. Figure 1.14 shows the tracked locations of the target as seen from the vehicle.

1.5 Optical-disk-based joint transform correlator

Recent improvements in many real-time addressable SLM's have made pattern recognition by use of optical correlation techniques much more realistic. A single input image, however, might require correlation with a huge library of reference images before a match is found; therefore the lack of large-capacity optimal information storage devices that can provide high-speed readout is still a major obstacle to practical optical correlation systems. Optical disks (OD's), developed in recent years as mass-storage media for many consumer products, are excellent candidates for this task. Psaltis *et al.* [19] and Lu *et al.* [20] have recently proposed the use of OD's in correlation and NN systems, respectively. In this section, an OD-based JTC [21] is discussed as part of the continuing effort to develop a practical optical correlator. The system employs an electrically addressable SLM to display the input image and an OD to provide the reference image in the JTC architecture shown in Fig. 1.15. A target captured by a video camera is first displayed on SLM1. A beam expansion–reduction system is then used to reduce the size of the input image m times. The reference image stored on the optical disk is read out in parallel and magnified m times by another set of expansion lenses so that the reference image and the input target are the same size. The joint transform is done by transform lens FL1, and the JTPS is recorded on the write side of optically addressed SLM2. After FL2 Fourier transforms the JTPS, the resulting correlation is recorded by a high-speed CCD camera located in the output plane. The OD advances to the next reference image and so on until a match is found.

Because of the limited resolution of the optically addressed modulator, SLM2, an upper cutoff spatial frequency is imposed on the JTPS. This can be alleviated either by the insertion

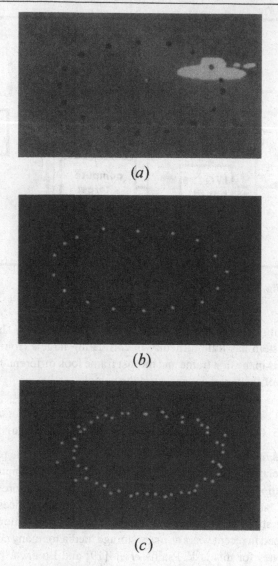

(a)

(b)

(c)

Fig. 1.12. Autonomous object tracking: (a) a revolving object in an elliptical path, (b) tracked positions in one revolution, (c) after four revolutions.

Fig. 1.13. Sequence of nine images are recorded that simulate the exact scenes as captured by a camera mounted on a moving space vehicle. Out of nine images, only frames 1, 5, and 9 are shown here.

Fig. 1.14. Tracked positions of the ground target as seen from the vehicle's coordinate frame.

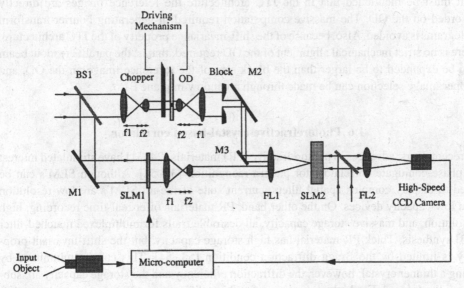

Fig. 1.15. An OD-based JTC. BS, beam splitter; M's, mirrors; FL's, Fourier-transform lenses.

of an imaging lens in front of SLM2 to magnify the JTPS or by the addition of a negative lens behind the transform lenses so that the effective focal length of these lenses is significantly increased. This in turn decreases the effective spatial frequency in the Fourier plane.

It has been proposed that the transparent write-once, read-many-times OD be used in this architecture [21]. To record a library of reference images on the OD, a laser beam is focused to an approximately 0.5-μm spot on the photosensitive layer of the OD to record each pixel of the image sequentially. This method results in the reference images' being recorded in binary form [19]. Images with gray scale can be recorded by area modulation of the spots on the OD and read out by the differential interferometric technique proposed by Rilum and Tanguay [22].

If binary images of 200 × 200 pixels are to be recorded, each image will occupy a 0.2 mm × 0.2 mm block on the OD. If a 0.01-mm spacing is assumed between adjacent blocks, then more than 27,000 reference images can be recorded on a single 120-mm-diameter OD. Because the block size is on the order of a fraction of a millimeter, the minute phase variation caused by a nonuniform thickness in the plastic cover of the OD can be neglected. OD's with optical-quality covers could be made if necessary.

The light source should be gated at a frequency sychronized with the image access time of the OD so as to freeze the image on the rotating OD. To estimate the sequential access time for a block of 0.2 mm $\times$ 0.2 mm on a standard OD, assume that the average revolution speed is 1122 rev/min, the average radius is 40 mm, and all the images are scanned sequentially on each consecutive band of the OD. The minimum access time is then approximately 40 μs. To estimate the operating speed of this system further, assume that the write side of an optically addressed ferroelectric liquid-crystal SLM is used as the square-law detector, the response time of which is between 10 and 155 μs [23]. Completing one correlation process (with the response time of the slower device) should take 40–155 μs. This is the same as performing more than 6400 correlations per second, a number that can hardly be matched by current electronic technology.

It must be mentioned that in the JTC architecture the reference images are directly recorded on the OD. The massive computation required in generating Fourier-transform holograms is avoided. Also, because of the shift-invariance property of the JTC architecture, there is no strict mechanical alignment of the OD required, that is, the parallel readout beam can be expanded to be larger than the block size of the reference image on the OD, and proper image selection can be made through a window in plane P1.

1.6 Photorefractive-crystal-based correlator

In recent years, advances in photorefractive (PR) materials [24–31] have stimulated interest in phase-conjugate correlators for pattern recognition [32–52]. Although SLM's can be used to display complex spatial filters, current state-of-the-art SLM's are low-resolution and low-capacity devices. On the other hand, PR materials offer real-time recording, high resolution, and massive storage capacity, all desirable traits for multiplexed matched-filter [53] synthesis. Thick PR material has high storage capacity, but the shift-invariant property is limited by the Bragg diffraction condition [53, 54]. This can be minimized by using a thinner crystal; however, the diffraction efficiency and the storage capacity are substantially reduced. For high storage capacity, high diffraction efficiency, and a large shift invariance, a reflection-type wavelength-multiplexed PR matched filter can be used [55]. A reflection-type matched-filter correlator is shown in Fig. 1.16. A z-cut PR crystal is used for

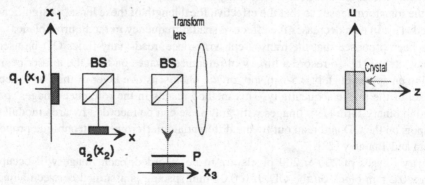

Fig. 1.16. Wavelength-multiplexed reflection-type matched-filter correlator. BS's, beam splitters; P, polarizer.

matched-filter synthesis. The matched filter is recorded by the combination of the Fourier spectrum of an object beam $q_1(x_1)$ with a reference plane wave from the opposite direction. Correlation can be done by the insertion of an input object $q_2(x_2)$ at plane x. The output correlation distribution can then be observed at plane x_3. To separate the read beam $q_2(x_2)$ from the write beam, the read beam can be made orthogonally polarized to the write beam by a polarized beam splitter. The insertion of polarizer P at front plane x_3 prevents the polarized writing beams from reaching the output plane. To analyze the correlation performance of the PR-based correlator, consider the Bragg diffraction wave vectors given by

$$\mathbf{k}_0 = -\frac{n}{\lambda}\hat{z},$$

$$\mathbf{k}_1 = -\frac{x_1}{\lambda f}\hat{u} + \frac{n}{\lambda}\left(1 - \frac{x_1^2}{2n^2 f^2}\right)\hat{z},$$

$$\mathbf{k}_2 = -\frac{x_2}{\lambda f}\hat{u} + \frac{n}{\lambda}\left(1 - \frac{x_2^2}{2n^2 f^2}\right)\hat{z},$$

$$\mathbf{k}_3 = \frac{x_3}{\lambda f}\hat{u}\frac{1}{\lambda}(x_1 - x_2 + x_3)\hat{u} - \frac{n}{\lambda}\left(1 - \frac{x_3^2}{2n^2 f^2}\right)\hat{z}, \tag{1.10}$$

where $\mathbf{k}_0$ and $\mathbf{k}_1$ are the writing vectors, $\mathbf{k}_2$ and $\mathbf{k}_3$ are the reading and the diffracted wave vectors, respectively, and n is the refractive index. Momentum conservation requires the deviation of the transverse and the longitudinal waves to be

$$\Delta\mathbf{k}_u = \frac{1}{\lambda}(x_1 - x_2 + x_3)\hat{u} = 0, \tag{1.11}$$

$$\Delta\mathbf{k}_z = \frac{x_3(x_1 - x_3)}{n\lambda f^2}\hat{u}. \tag{1.12}$$

The output correlation peak intensity as a function of shift-variable S can be shown to be [54]

$$R(S) = \left| \int q_1^*(x_1)q_1(x_1) \, \mathrm{sinc}\left[-\frac{\pi D}{n\lambda}\frac{S(x_1 - S)}{f^2}\right] \mathrm{d}x_1 \right|^2, \tag{1.13}$$

where D is the thickness of the crystal.

Because the shift-variable S can be either positive or negative, a figure of merit (FOM) (shift tolerance) can be defined as

$$(\mathrm{FOM})_{\mathrm{RC}} = X S_{\max} = \frac{4n\lambda}{D}f^2 - S_{\max}^2, \tag{1.14}$$

where $S_{\max}$ denotes the maximum allowable shift. Plots of Eq. (1.14) along with FOM's for the VLC and the JTC are shown in Fig. 1.17, which shows that the reflection-type PR correlator performs better. In other words, the reflection-type matched-filter correlator has a higher shift tolerance compared with that of the JTC or the VLC, ~1 order more tolerant than the VLC. The wavelength-multiplexed hologram also offers a higher and more uniform wavelength selectivity compared with that of the angular-multiplexed technique. The wavelength-multiplexed reflection-type matched filter should then be the best choice [55].

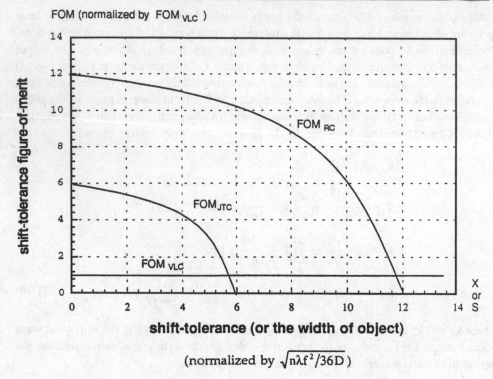

Fig. 1.17. FOM's for different types of correlators.

1.7 Optical neural networks

Electronic computers can solve some classes of computational problems thousands of times faster and more accurately than the human brain. However, for cognitive tasks, such as pattern recognition, understanding and speaking a language, etc., the human brain is much more efficient. In fact, these tasks are still beyond the reach of modern electronic computers. A human brain consists of millions of neurons, which are massively interconnected by synapses. Techniques for simulating artificial NN's are basically drawn from cognitive psychology and biological models.

A NN consists of a collection of processing elements, called neurons. Each neuron has many input signals, but only one output signal, which is fanned out to many pathways connected to other neurons. These pathways interconnect with other neurons to form a network. The operation of a neuron is determined by a transfer function that defines the neuron's output as a function of the input signals. Every connection entering a neuron has an adaptive coefficient, called a *weight*, assigned to it. The weight determines the interconnection strength between neurons, and it can be changed through a learning rule that modifies the strengths in response to input signals and the transfer function. The learning rule allows the response of the neuron to change with time, depending on the nature of the input signals. This means that the network *adapts* itself to the environment and organizes the information within itself, a type of learning. A score of NN models have been developed in the past several decades [56–61].

Generally speaking, a one-layer NN of N neurons has N^2 interconnections. The transfer function of a neuron can be described by a nonlinear relationship such as a step function,

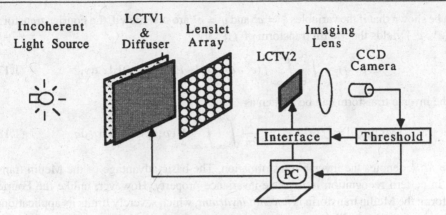

Fig. 1.18. LCTV optical NN.

making the output of a neuron either 0 or 1 (binary), or a sigmoid function, which gives rise to analog values. The state of the ith neuron in the network can be represented by a *retrieval equation*, given by

$$u_i = f\left(\sum_{j=1}^{N} T_{ij}u_j - \theta_i\right),$$ (1.15)

where u_i is the activation potential of the ith neuron, T_{ij} is the *interconnection weight* matrix (IWM) (associative memory) between the jth and the ith neuron, and f is a nonlinear processing operator.

Light beams propagating in space will not interfere with each other, and optical systems generally have large space–bandwidth products. These are the primary features that prompted the optical implementation of NN's. The first optical implementation of NN's was proposed by Psaltis and Farhat [62]. Since then, a score of optical NN architectures have been proposed [63–69]. A typical hybrid optical NN with a liquid-crystal TV (LCTV) SLM is shown in Fig. 1.18. The lenslet array provides the interconnection between the IWM and the input pattern. The transmitted light field after LCTV2 is collected by an imaging lens, focusing at the lenslet array and imaging onto a CCD array detector. The array of detected signals is sent to a thresholding circuit, and the final pattern can be viewed at the monitor or sent back for the next iteration. The data flow is controlled primarily by the microcomputer. The LCTV-based neural net just described is an *adaptive* optical NN.

1.8 Scale- and rotational-invariant correlation

Pattern recognition by complex spatial filtering is sensitive to orientation and scale change in the target. Methods have been proposed to overcome scaling and rotation problems, but most of them fall short. One technique that handles misscaling is the Mellin transform, which is given by

$$M(ip, iq) = \int_0^\infty \int_0^\infty f(\xi, \eta)\xi^{-(ip+1)}\eta^{-(iq+1)} \, d\xi \, d\eta.$$ (1.16)

It can be shown that if the variables $\xi = e^x$ and $\eta = e^y$ are substituted, the Fourier-transform of $f(e^x, e^y)$ yields the Mellin transform of $f(\xi, \eta)$:

$$M(p, q) = \int_{-\infty}^{\infty} \int_{-\infty}^{\infty} f(e^x, e^y) \exp[-i(px + qy)] \, dx \, dy, \qquad (1.17)$$

and the inverse transform can be written as

$$M^{-1}[M(p, q)] = f(e^x, e^y) = \frac{1}{2\pi} \int_{-\infty}^{\infty} \int_{-\infty}^{\infty} M(p, q) \xi^{ip} \eta^{iq} \, dp \, dq, \qquad (1.18)$$

where M^{-1} denotes the inverse transformation. The basic advantage of the Mellin transform in pattern recognition is its scale-invariance property. However, unlike the Fourier transform, the Mellin transform is *not shift invariant*, which severely limits its applications. The implementation of Mellin transforms in optical correlators is in practical applications to pattern recognition. Methods for implementing Mellin transforms in real time with an electronically addressed device have been demonstrated by Casasent and Psaltis [70, 71] and later by Sheng and Arsenault [72].

Rotational invariance in pattern recognition can be accomplished with circular harmonic expansion techniques, as advocated by Hsu and Arsenault [73] and later applied to a JTC by Yu *et al.* [74].

If a function $f(r, \theta)$ is continuous and integrable over the region $(0, 2\pi)$, it can be expanded in a Fourier series as

$$f(r, \theta) = \sum_{m=-\infty}^{+\infty} F_m(r) \exp(im\theta), \qquad (1.19)$$

where

$$F_m(r) = \frac{1}{2\pi} \int_0^{2\pi} f(r, \theta) \exp(-im\theta) \, d\theta \qquad (1.20)$$

and $F_m(r, \theta) = F_m(r) \exp(im\theta)$ is the mth-order *circular harmonic*. If the object is rotated by an angle α, we have

$$f(r, \theta + \alpha) = \sum_{m=-\infty}^{\infty} F_m(r) \exp(im\alpha) \exp(im\theta). \qquad (1.21)$$

Denote $f(x, y)$ and $f_\alpha(x, y)$ as object functions $f(r, \theta)$ and $f(r, \theta + \alpha)$, respectively. From the classic matched-filtering theory, when a rotated object $f_\alpha(r, \theta + \alpha)$ addresses an optical correlator, the output light field is given by

$$g_\alpha(x, y) = \int_{-\infty}^{+\infty} \int_{-\infty}^{+\infty} f_\alpha(\xi, \eta) f^*(\xi - x, \eta - y) \, d\xi \, d\eta. \qquad (1.22)$$

It is apparent that if $\alpha = 0$, the autocorrelation peak appears at $x = y = 0$. It is evident that, if one of the circular harmonic functions is used as the reference function, such as

$$f_{\text{ref}}(r, \theta) = F_m(r) \exp(im\theta),$$

then the correlation between a rotated target and the reference function occurs independently of the object orientation. The value of the circular harmonic expansion lies in its role in practical application to optical correlators.

1.9 Wavelet transform filtering

Wavelet transform (WT) analysis [75–80] is a feasible alternative to Fourier transforms for optical recognition. The WT has been utilized for multiresolution image analysis and in optical implementations by several investigators [81–93]. The two-dimensional WT of a signal $s(x, y)$ is given by

$$w(a_x, a_y, b_x, b_y) = WT[s(x, y)] = \iint s(x, y) h^*_{ab}(x, y)\, dx\, dy, \qquad (1.23)$$

where the asterisk represents the complex conjugate of the wavelet, which is defined as

$$h_{ab}(x, y) = \frac{1}{(a_x a_y)^{1/2}} h\left(\frac{x - b_x}{a_x}, \frac{y - b_y}{a_y}\right). \qquad (1.24)$$

This function is obtained by translating and dilating the analyzing wavelet function $h(x, y)$. Thus the WT of Eq. (23) can be written as

$$w(a_x, a_y, b_x, b_y) = \frac{1}{(a_x a_y)^{1/2}} \int_{-\infty}^{\infty} \int_{-\infty}^{\infty} s(x + b_x, y + b_y) h^*\left(\frac{x}{a_x}, \frac{y}{a_y}\right) dx\, dy$$

$$= \frac{1}{(a_x a_y)^{1/2}} s(x, y) \circledast h^*\left(\frac{x}{a_x}, \frac{y}{a_y}\right), \qquad (1.25)$$

where $\circledast$ denotes the correlation operation. This is essentially the cross-correlation between the signal and the dilated analyzing wavelet. Furthermore, $h(x/a_x, y/a_y)$ can be interpreted as a bandpass filter governed by the (a_x, a_y) dilation. Thus both the dominant frequency and the bandwidth of the filter can be adjusted when the dilation of $h(x/a_x, y/a_y)$ is changed. In other words, the WT is essentially a filtered version of the input signal $s_i(x, y)$. The correlation of two WT's with respect to the input signal is an estimation of the similarity between the two signals. Because of the inherent local feature selection characteristic of the WT [94], the wavelet matched filter generally offers a higher discriminability than the conventional matched filter.

The implementation of the WT in matched-filter correlators depends on the practical application. Methods of implementing the wavelet matched filter in a VLC have been reported by Li and Zhang [83] and others [84–93] and applied to a JTC by Lu et al. [94].

1.10 Discriminant filtering

Another way to achieve distortion-invariant pattern recognition, which can include out-of-plane rotation, is with the synthetic discriminant function (SDF) filter [95–97], which is described briefly here.

Suppose a set of images $\{f_n(x, y)\}$, which represents N perspective views of an object $O_t(x, y, z)$. Then the SDF of this set of images is a linear combination of N training images, which is given by

$$h(x, y) = \sum_{n=1}^{N} a_n f_n(x, y). \qquad (1.26)$$

The coefficients a_n $(n = 1, 2, \ldots, N)$ can be determined by the solution of a set of N simultaneous equations [96].

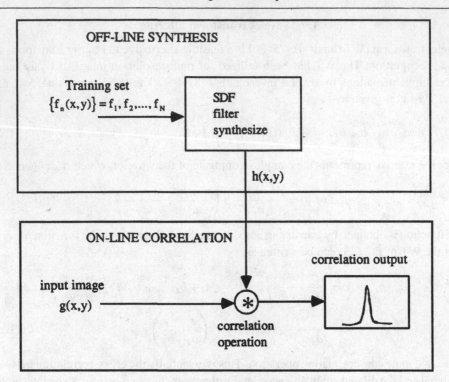

Fig. 1.19. Black-box representation SDF correlator.

The correlation peak value between $h(x, y)$ and an arbitrary training image $f_n(x, y)$ can be shown to be

$$h(x, y) \circledast f_n(x, y)|_{x=0, y=0} = \iint h^*(x, y) f_n(x, y)\, dx\, dy$$

$$= \iint \sum_{n=1}^{N} a_n^* f_n^*(x, y) f_m(x, y)\, dx\, dy$$

$$= \sum_{n=1}^{N} a_n^* R_{nm} = C, \tag{1.27}$$

where C represents a prespecified constant, which is independent of the perspective view of $f_n(x, y)$. In other words, the output correlation peak value from an SDF filter is independent of the training images. Figure 1.19 shows a flow chart for an SDF filtering system in which the SDF filter is off-line synthesized while the correlation is on-line processed.

There are techniques proposed to improve the performance of the classical SDF filter [97–99]. Nevertheless, it is not our intention here to describe all those techniques. Instead, two of the simplest approaches are mentioned: (1) Higher noise tolerance can be achieved if the variance of the SDF filter is minimized, as reported by Kumar [97], and (2) sharper correlation peaks can be obtained if the average correlation energy of the SDF filter is minimized, as advocated by Mahalanobis *et al.* [98]. Simulated annealing algorithms have also recently been applied to pattern recognition [100, 101]. These algorithms were previously used to design a bipolar SDF filter for detecting out-of-plane object rotation [102, 103]. The major advantages of the bipolar SDF filter are high discriminability, narrow dynamic range, high noise tolerance, and the fact that they can be easily displayed onto an electronically addressable SLM.

1.11 Phase-only filtering

The concept of the phase-only matched filter (POMF) probably originated with the optimal linearization of a Fourier hologram in which the phase distribution is preserved [104, 105]. This filter is also known as the *phase-preserving filter*. The phase distribution preserves the main features of the pattern. In fact, a bleached matched filter (a form of a phase-only filter) was frequently used in optical correlators in the past [17] because of the higher detection efficiency that can be obtained. To improve the light efficiency of an optical system, POMF's were recently readvocated by Horner and Giainino [106] and Horner and Leger [107].

Although the phase-only filter is not an optimal filter (in terms of maximum SNR), it does improve the light efficiency of an optical filtering system. With current advances in phase-modulating SLM's [108–123], phase-only spatial filters have become popular. The performance of a POMF can be described as follows:

A matched filter is defined as

$$H(p, q) = F^*(p, q) = |F(p, q)| \exp[-i\phi(p, q)]. \tag{1.28}$$

The phase-only filter is in fact an all-pass filter given by

$$H_\phi(p, q) = \exp[-i\phi(p, q)]. \tag{1.29}$$

The output correlation can be written as

$$g(x, y) = \mathcal{F}^{-1}[F(p, q)H_\phi(p, q)] = f(x, y) \circledast f'(x, y), \tag{1.30}$$

where $f(x, y) = \mathcal{F}^{-1}[H_\phi(p, q)] \sim f(x, y)$, which is similar to the input object. Because the phase-only filter eliminates the zero-order diffraction (most of the light energy is absorbed), the diffraction efficiency in principle can be as high as 100%.

To show that the POMF is *not* an optimum filter, we assume that the input object is embedded with additive Gaussian noise with zero mean. The output complex light field can be shown to be

$$g(x, y) = \mathcal{F}^{-1}\{[F(p, q) + N(p, q)]H_\phi(p, q)\}. \tag{1.31}$$

The output SNR is

$$\left(\frac{S}{N}\right)_\phi = \frac{1}{4\pi^2} \frac{\left| \iint H_\phi(p, q)F(p, q) \, dp \, dq \right|^2}{\iint |H_\phi(p, q)|^2 |N(p, q)|^2 \, dp \, dq}$$

$$= \frac{1}{4\pi^2} \frac{\left| \iint |F(p, q)| \, dp \, dq \right|^2}{\iint |N(p, q)|^2 \, dp \, dq}. \tag{1.32}$$

The output SNR for a complex matched filter is given by [124]

$$\left(\frac{S}{N}\right)_M = \frac{1}{4\pi^2} \iint \frac{|F(p, q)|^2}{|N(p, q)|^2}. \tag{1.33}$$

It is trivial to show that the POMF is *not* an optimum filter:

$$(S/N)_\phi \leq (S/N)_M. \tag{1.34}$$

The equality holds if and only if

$$\mathcal{F}[f(x, y)] = K \exp[i\phi(p, q)], \tag{1.35}$$

where K is an arbitrary constant. As an example, if

$$f(x, y) = \sum_{n=0}^{N} S(x + x_\eta, y + y_\eta), \tag{1.36}$$

then

$$F(p, q) = \sum_{n=0}^{N} \exp[i(px_\eta + qy_\eta)]. \tag{1.37}$$

To alleviate the stringent phase-modulation control required of the SLM, a binary POMF has become an *ad hoc* phase-only filter design [125–130] in which the phase is either 0 or $-\pi$. However, phase-quantization errors cause a reduction in correlation intensity and enhance the output noise level. Because the impulse response of a binary phase-only filter is a superposition of two edge-enhanced replicas of the original reference object (one appears in the original position and another is rotated 180° around the center of the reference scene), the binary POMF performs both correlation and convolution operations. In general, a position offset of the reference object from the center of the reference scene is necessary to reduce the effects of the superimposed convolution. If the input object is antisymmetric, a zero offset is required so that the two replicas will exactly overlap each other; otherwise the convolution term produces a peak intensity as high as the autocorrelation peak, which would cause detection ambiguity.

Even though the POMF improves light efficiency, the output SNR in practice is much lower than that of the conventional matched filter. As shown above, the POMF is an all-pass filter; it cannot suppress noise disturbances (or background clutter in the input scene), so the autocorrelation peak may be lost in output noise.

Because of the poor noise performance of the POMF, the ternary phase-amplitude matched filter is a compromise for better performance [131–133]. A ternary phase-amplitude matched filter is a three-value filter (e.g., -1, 0, and $+1$) in which the 0's are used to block the unwanted input noise. Even though the ternary matched filter behaves as a phase-only filter, the diffraction efficiency is greatly reduced. However, compared with the multilevel complex matched filter, the ternary matched filter has advantages in implementation with a ternary SLM [133].

1.12 Pattern recognition with neural networks

One of the important aspects of biological NN's is that they can recognize patterns more efficiently than a computer. One of the most frequently used neural net model is the Hopfield model [134, 135], which allows the desired output pattern to be retrieved from a distorted or partial input pattern. The model utilizes an associative memory retrieval process equivalent to an iterative thresholded matrix–vector outer-product expression given by

$$V_i \rightarrow 1 \quad \text{if } \sum_{j=1}^{N} T_{ij} V_j \geq 0,$$

$$V_i \rightarrow 0 \quad \text{if } \sum_{j=1}^{N} T_{ij} V_j < 0, \tag{1.38}$$

where V_i and V_j are binary output and binary input patterns, respectively, and the associative memory operation is written as

$$T_{ij} = \begin{cases} \sum_{m=1}^{N} (2V_i^m - 1)(2V_j^m - 1) & \text{for } i \neq j, \\ 0 & \text{for } i = j, \end{cases} \tag{1.39}$$

where V_i^m and V_j^m are the ith and the jth elements of the mth binary vector, respectively.

The Hopfield model depends on the outer-product operation for construction of the associated memory, which severely limits storage capacity [136, 137] and often causes failure in retrieving similar patterns. To overcome these shortcomings, NN models, such as back-propagation [138], orthogonal projection [139], multilevel recognition [139], interpattern association [140], moment invariants [141], and others, have been proposed [142–153]. One of the important aspects of neural computing must be the ability to retrieve distorted and partial inputs. To illustrate partial input retrieval, a set of English letters shown in Fig. 1.20(a) was stored in a Hopfield neural net. The positive and the negative parts of

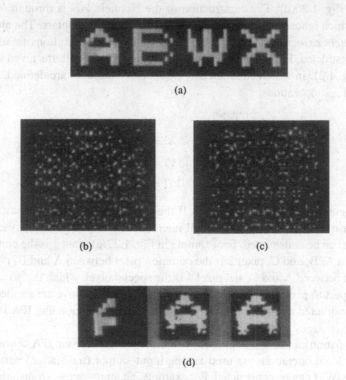

(a)

(b) (c)

(d)

Fig. 1.20. Results from a Hopfield model: (a) training set; (b), (c) positive and negative weight matrices, respectively; (d) reconstructed results.

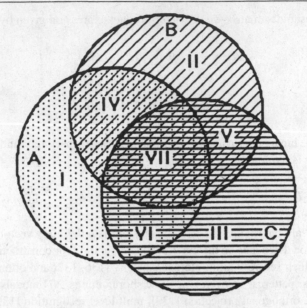

Fig. 1.21. Common and special subspaces.

the memory matrix are given in Figs. 1.20(b) and 1.20(c), respectively. If a partial image of A is presented to the Hopfield net, a reconstructed image of A converges by iteration; it is shown in Fig. 1.20(d). The construction of the Hopfield NN is through *intrapattern association*, which ignores the association among the stored exemplars. The alternative is called *interpattern association* (IPA) [140]. When simple logic operations are used, an IPA NN can be constructed. For example, use the three overlapping patterns given in the Venn diagram of Fig. 1.21, in which common and the special subspaces are defined. If one uses the following logic operations,

$$
\begin{aligned}
I &= A \wedge \overline{(B \vee C)}, \\
II &= B \wedge \overline{(A \vee C)}, & V &= (B \wedge C) \wedge \bar{A}, \\
III &= C \wedge \overline{(A \vee B)}, & VI &= (C \wedge A) \wedge \bar{B}, \\
IV &= (A \wedge B) \wedge \bar{C}, & VII &= (A \wedge B \wedge C) \wedge \bar{\phi}, & (1.40)
\end{aligned}
$$

then an IPA neural net can be constructed. If the interconnection weights are equal to 1, −1, and 0, for excitory, inhibitory, and null interconnections, respectively, then a tristate IPA neural net can be constructed. For example in Fig. 1.22(a) pixel 1 is the common pixel among patterns A, B, and C, pixel 2 is the common pixel between A and B, pixel 3 is the common pixel between A and C, and pixel 4 is the special pixel, which is also an exclusive pixel with respect to pixel 2. When the logic operations given above are applied, a tristate NN can be constructed as shown in Fig. 1.22(b), with the corresponding IPA IWM shown in Fig. 1.22(c).

Pattern translation can be accomplished with the heteroassociation IPA neural net [142]. When similar logic operations are used among input–output (translation) patterns, a heteroassociative IWM can be constructed. For example, an input–output (translation) training set is given in Fig. 1.23(a). With the logic operations, a heteroassociation neural net can be constructed as shown in Fig. 1.23(b); Fig. 1.23(c) is its IWM. To illustrate the optical

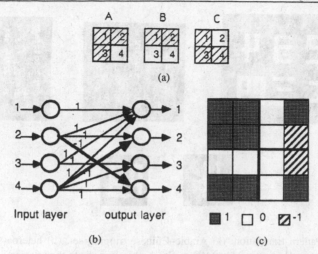

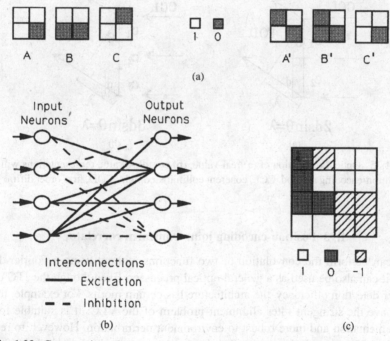

Fig. 1.22. Construction of an IPA NN: (a) three reference patterns, (b) one-layer neural net, (c) IWM.

Fig. 1.23. Construction of a heteroassociation IPA NN: (a) input–output training sets, (b) heteroassociation neural net, (c) heteroassociation IWM.

implementation, an input–output training set is shown in Fig. 1.24(a). The positive and the negative parts of the heteroassociation IWM's are depicted in Fig. 1.24(b). If a partial Arabic numeral 4 is presented to the optical neural net, a translated Chinese numeral is obtained, as shown in Fig. 1.24(c). Thus the heteroassociation neural net can indeed translate patterns.

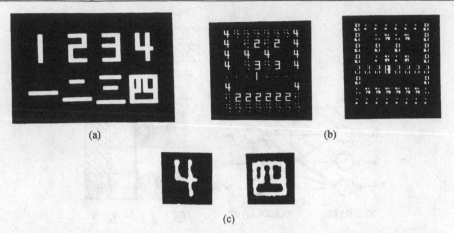

(a)   (b)

(c)

Fig. 1.24. Pattern translation: (a) Arabic–Chinese training set, (b) heteroassociation IWM (positive and negative parts), (c) partial Arabic numeric to the translated Chinese numeric.

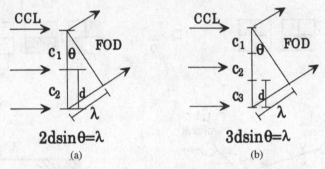

Fig. 1.25. Optical realization of (a) real-value, (b) complex-value representations with a position-encoding method. CCL, coherent colliminated light; FOD, first-order diffraction.

1.13 Position-encoding joint transform correlator

The JTC can perform the convolution of two functions without using a Fourier-domain filter, and it can also be used as a general optical processor. Even though the JTC usually has a lower detection efficiency, the architecture has certain merits. For example, the JTC does not have the stringent filter alignment problem of the VLC. It is suitable for real-time implementation and more robust to environment perturbation. However, to realize a spatial-domain filter in a JTC, complex function implementation is often needed [133, 154–157]. Gregory *et al.* [120] have proposed using two LCTV panels to obtain a full complex modulation so that complex representations can be realized. However, when a position-encoding technique is used [158, 159], it is possible to obtain complex-valued reference functions with an amplitude-modulated SLM.

It is well known that a real function can be decomposed into $c_1\phi_0 + c_2\phi_1$ and a complex function can be decomposed into $c_1\phi_0 + c_2\phi_{2/3} + c_3\phi_{4/3}$, where c_1, c_2, and c_3 are nonnegative coefficients and $\phi_k = \exp(i\pi k)$ are elementary phase vectors [158–162]. The decompositions can be optically realized with position encoding, as illustrated in Fig. 1.25 [163].

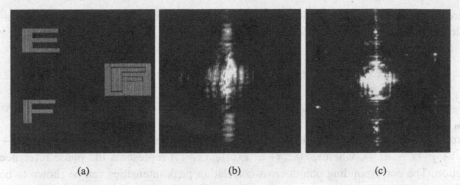

Fig. 1.26. Position-encoding JTC: (a) the position-encoded inputs, (b) first-order JTPS, (c) output correlation distribution.

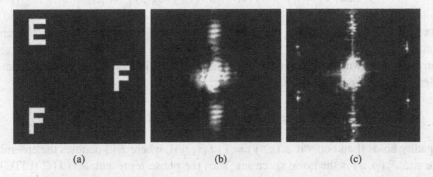

Fig. 1.27. Conventional JTC: (a) the inputs; (b) JTPS; (c) output correlation distribution.

A proof-of-concept experiment is shown in Fig. 1.26. Position-encoded letters E and F are shown in Fig. 1.26(a). Figure 1.26(b) shows the first-order JTPS, and the corresponding output correlation distribution is shown in Fig. 1.26(c). The autocorrelation peaks for detecting F can readily be seen, and they were measured to be approximately twice as high as the cross-correlation peaks. For comparison, Fig. 1.27(a) is presented to a conventional JTC (CJTC) without position encoding. Figure 1.27(b) shows the joint Fourier power spectrum, and the output correlation distribution is shown in Fig. 1.27(c). The CJTC fails to differentiate between the letters E and F. In short, when position encoding is used, complex-valued functions can be represented on an amplitude-modulating SLM, thereby improving the pattern discriminability.

1.14 Phase-representation joint transform correlator

Because of the availability of phase-modulating SLM's such as LCTV's [112], magneto-optic SLM's [133], deformable mirror arrays [164], and others [165, 166], phase-encoded inputs to a JTC are relatively convenient. The CJTC offers the avoidance of filter synthesis and critical system alignment but it suffers from poor detection efficiency for targets embedded in clutter, bright background noise, and multiple targets. A phase-encoded input can alleviate some of these shortcomings. Phase representation of an intensity $f(x, y)$ can be written as

$$pf(x, y) = \exp\{iT[f(x, y)]\}, \tag{1.41}$$

where $T[\ ]$ represents a monotonic real-to-phase transformation. Denote G_{min} and G_{max} as the lowest and the highest gray levels of an intensity object, respectively; then the phase transformation is given by

$$T[f(x,\ y)] = \frac{f(x,\ y) - G_{min}}{G_{max} - G_{min}}\pi. \tag{1.42}$$

If $G_{min} = 0$, Eq. (1.42) reduces to the formula defined by Kallman and Goldstein [167]. As G_{min} may not be zero in practice, it is more efficient to use a $[0, \pi]$ phase domain.

Applied to the JTC, the phase representation of the input can be written as $pf(x + a,\ y) + pr(x - a,\ y)$, where $pr(x,\ y) = \exp[i\varphi_r(x,\ y)]$ represents the phase reference function. The corresponding output cross-correlation peak intensities can be shown to be [168]

$$C(\pm 2a,\ 0) = \left| \iint pf(x,\ y)pr^*(x,\ y)\,\mathrm{d}x\,\mathrm{d}y \right|^2 = |A|^2, \tag{1.43}$$

for $\varphi_r(x,\ y) = T[f(x,\ y)]$, and A is a constant proportional to the size of the reference function.

It can also be shown that the phase-representation JTC is indeed an optimum correlator, regardless of non-zero-mean noise, that is [168],

$$\mathrm{SNR} \leq \frac{1}{4K\pi^2} \iint E[|PF_n(p,\ q)|^2]\,\mathrm{d}p\,\mathrm{d}q. \tag{1.44}$$

The equality holds if and only if $\varphi_r(x,\ y) = T[f(x,\ y)]$, where $E[\ ]$ denotes the ensemble average and $F_n(p, q)$ is the noise spectrum. Thus the phase-representation JTC (PJTC) is an optimal filtering system, independent of the mean value of the additive noise.

A CJTC often loses its pattern discriminability whenever a false target is similar to the reference function or the object is heavily embedded in a noisy background. For comparison, let the two images of an M60 (American) tank and a T72 (Russian) tank be embedded in a noisy background, as shown in Fig. 1.28. The M60 tank is used as the reference function. The corresponding output correlation distributions obtained with the CJTC and with the phase-representation JTC are shown in Fig. 1.29. The CJTC fails to identify the location of the M60 tank.

M60

Input scene

Fig. 1.28. Input objects to the JTC.

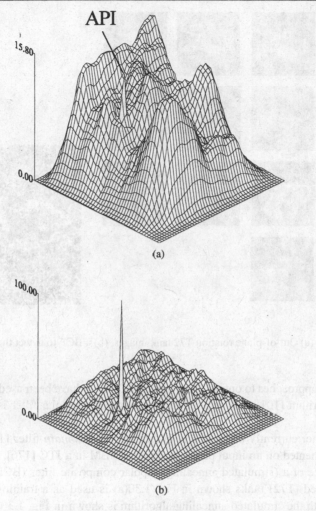

Fig. 1.29. Output correlation results (a) obtained from the CJTC, (b) obtained from a phase-representation JTC.

1.15 Composite filtering with the joint transform correlator

Classical spatial matched filters are sensitive to rotational and scale variances. A score of approaches to developing composite distortion-invariant filters [169] are available. Among them, the SDF filter [170] has played a central role for three-dimensional target detection. The original idea of the SDF filter can be viewed as a linear combination of classical matched filters, for which the coefficients of the linear combination are designed to yield equal correlation peaks for each of the distorted patterns [171, 172]. Because the dynamic range of an SDF filter is large, it is difficult to implement with currently available SLM's. On the other hand, the bipolar filter has the advantage of a limited dynamic range requirement and can be easily implemented with commercially available SLM's. Because the bipolar filter has uniform transmittance, it has the additional advantage of being light efficient.

Several attempts have been made in the construction of a bipolar SDF filter and binarization is not the best approach, as there is no guarantee that the SDF filter will be

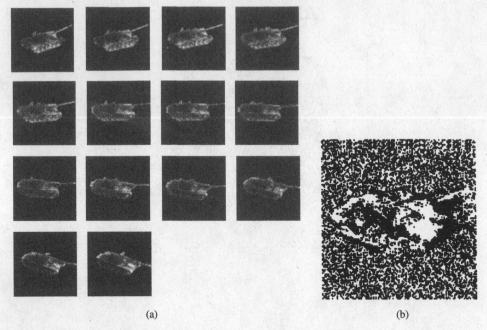

(a)   (b)

Fig. 1.30. (a) Out-of-plane rotation T72 tank images, (b) a BCF to detect the T72 tank.

valid. Iterative approaches to optimize the bipolar SDF filter have been used. The simulated annealing algorithm [101, 102] applied to the design of a bipolar filter is relatively new [173–176].

A bipolar filter currently advocated is in fact a *spatial-domain* filter [175] and can be directly implemented on an input phase-modulating SLM in a JTC [176]. To demonstrate the performance of a (simulated annealing) bipolar composite filter (BCF), a set of out-of-plane-oriented (T72) tanks shown in Fig. 1.30(a) is used as a training set. The BCF constructed with the simulated annealing algorithm is shown in Fig. 1.30(b). If the input scene to the JTC shown in Fig. 1.31(a) is used (a BCF for detecting the T72 tank is given on the right-hand side) then the output correlation distribution shown in Fig. 1.31(b) is obtained. The target T72 tanks are indeed extracted.

1.16 Non-zero-order joint transform correlator

One of the major advantages of using a hybrid optical architecture is the exploitation of the parallelism of optics and the flexibility of electronic computers. Although the JTC is robust to environmental perturbation [176], it suffers from poor detection efficiency, which is due to the zero-order diffraction [8].

Methods to remove the zero-order spectra have been reported by Cheng *et al.* [177] and Lu *et al.* [178]. We show a simple technique for zero-order removal by using a hybrid JTC [179]. Let us refer to the JTPS as given by

$$I(p, q) = |R(p, q)|^2 + |F(p, q)|^2 + R^*(p, q)F(p, q)\exp(-i2ap)$$
$$+ F^*(p, q)R(p, q)\exp(i2ap), \tag{1.45}$$

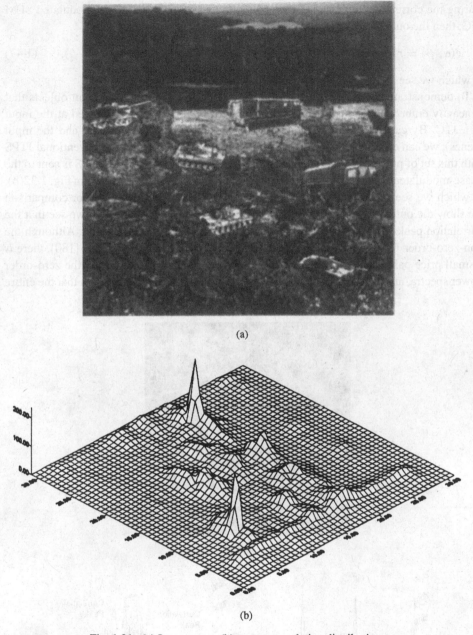

(a)

(b)

Fig. 1.31. (a) Input scene, (b) output correlation distribution.

in which two zero-order power spectra inherently exist. Because these zero-order terms are obtained from the power spectra of input objects, they can be removed from the recorded JTPS. For example, the spectral distributions of $|F(p, q)|^2$ and $|R(p, q)|^2$ can be prerecorded, by which they can be removed from the JTPS by means of the computer. Hence the nonzero JTPS would be

$$I(p, q) = R^*(p, q)F(p, q)\exp(-i2ap) + F^*(p, q)R(p, q)\exp(i2ap). \qquad (1.46)$$

During the correlation duty cycle, we send the nonzero JTPS to a phase-modulated SLM JTC; then the output result would be

$$g(\alpha, \beta) = r^*(\alpha - 2x_0, \beta) \otimes f(\alpha, \beta) + r(-\alpha - 2x_0, -\beta) \otimes f^*(-\alpha, -\beta), \qquad (1.47)$$

in which we see that the zero-order output diffractions have been avoided.

To demonstrate the performance of the non-zero-order JTC, a set of input objects that is heavily embedded in random noise, as shown in Fig. 1.32(a), is displayed at the input of a JTC. By capturing this set of input object spectra (i.e., the reference and the input scenes), we can obtain a non-zero-order JTPS by simply subtracting the conventional JTPS with this set of power spectra. Needless to say, when the non-zero-order JTPS is sent to the phase-modulated SLM, the correlation duty cycle can be obtained, as shown in Fig. 1.32(b), in which we see that two distinctive correlation peaks can be identified. For comparison we show the output result obtained from a CJTC in Fig. 1.32(c), in which we see that the correlation peaks have been heavily shadowed by the zero-order diffraction. Although the non-zero-order JTC outperformed the CJTC in terms of detection efficiency [180], there is a small price paid in the form of additional steps for capturing and storing the zero-order power spectra, and the subtraction of the JTPS. Nevertheless, it can be shown that the entire

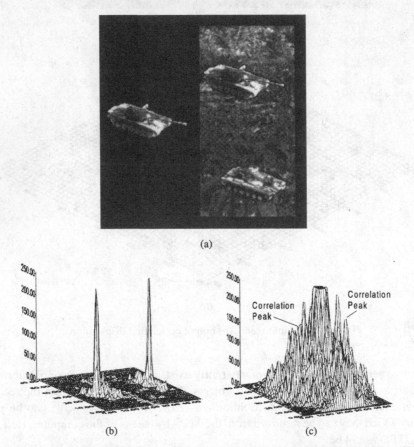

(a)

(b)   (c)

Fig. 1.32. Nonzero JTC: (a) input objects to the JTC, (b) output correlation profile obtained from the nonzero JTC, (c) output correlation profile obtained from the CJTC.

operation cycle can be designed within a fraction of a second. If one uses high-speed SLM's (e.g., ferroelectric SLM's), then the operating duty cycle can be as small as several tenths of a millisecond.

1.17 Summary and conclusions

The synthesis of Fourier-domain matched filters by use of holographic techniques was initiated by VanderLugt in 1964 [2], approximately three decades ago. Since then, VanderLugt correlators (VLC's) have been intensively studied, improved, and modified by advances in electronically addressable spatial light modulators (SLM's) and nonlinear photorefractive devices. A couple of years after the VanderLugt matched filter (1966), a different kind of correlator, called the joint transform correlator (JTC), was first demonstrated by Weaver and Goodman [3] and independently by Rau [4]. However, almost no application of the JTC was reported before 1984, despite its inclusion in several texts on optical processing. This may have been due the lack of adequate interface devices such as SLM's and CCD cameras. In 1984, Yu and Lu [5] were the first to take advantage of newly available SLM's and demonstrated a programmable JTC. This resulted in the rebirth of the JTC.

Pattern recognition can also be approached by use of simulated neural networks (NN's). The first optical implementation of NN's was proposed by Psaltis and Farhat in 1985 [6]. Since then, neuropattern recognition has seen a wide variety of optical implementations.

In this chapter, some of the basic optical architectures have been reviewed, namely VLC's, JTC's, and NN's, as they apply to optical pattern recognition. Pros and cons have been identified and demonstrated. Together with recent advances and improvements in interface devices, it is expected that optical pattern recognition technology will become a widespread practical reality.

In view of the great number of significant papers in the field of optical pattern recognition (after all, it has existed for over three decades), apologies are in order for omissions that surely occurred.

References

[1] E. Abbé, "Beiträge zur Theorie des Mikroskops und der mikroskopischen Wahrnehmung," Arch. Mikrosk. Anat. **9**, 413–463 (1873).

[2] A. VanderLugt, "Signal detection by complex spatial filtering," IEEE Trans. Inf. Theory **IT-10**, 139–145 (1964).

[3] C. S. Weaver and J. W. Goodman, "A technique for optically convolving two functions," Appl. Opt. **5**, 1248–1249 (1966).

[4] J. E. Rau, "Detection of differences in real distributions," J. Opt. Soc. Am. **56**, 1490–1494 (1966).

[5] F. T. S. Yu and X. J. Lu, "A real-time programmable joint transform correlator," Opt. Commun. **52**, 10–16 (1984).

[6] D. Psaltis and N. Farhat, "Optical information processing based on an associative-memory model of neural nets with thresholding and feedback," Opt. Lett. **10**, 98–100 (1985).

[7] F. T. S. Yu and S. Jutamulia, *Optical Signal Processing, Computing and Neural Networks* (Wiley-Interscience, New York, 1992), Chap. 2.

[8] F. T. S. Yu, Q. W. Song, Y. S. Cheng, and D. A. Gregory, "Comparison of detection efficiencies for VanderLugt and joint transform correlators," Appl. Opt. **29**, 225–232 (1990).

[9] T. C. Lee, J. Rebholz, and P. Tamura, "Dual-axis joint-Fourier transform correlator," Opt. Lett. **4**, 121–123 (1979).

[10] F. T. S. Yu, C. H. Zhang, Y. Jin, and D. A. Gregory, "Nonconventional joint transform correlator," Opt. Lett. **14**, 922–924 (1989).

[11] F. T. S. Yu, F. Cheng, T. Nagata, and D. A. Gregory, "Effects of fringe binarization of multiobject joint transform correlator," Appl. Opt. **28**, 2988–2990 (1989).

[12] F. T. S. Yu, E. Tam, and D. A. Gregory, "High efficient joint transform correlator," Opt. Lett. **15**, 1029–1031 (1990).

[13] S. Jutamulia, G. M. Storti, D. A. Gregory, and J. C. Kirch, "Illumination-independent high-efficiency joint transform correlator," Appl. Opt. **30**, 4173–4175 (1991).

[14] F. T. S. Yu, S. Jutamulia, T. W. Lin, and D. A. Gregory, "Adaptive real-time pattern recognition using a liquid-crystal TV based transform correlator," Appl. Opt. **26**, 1370–1372 (1987).

[15] E. C. Tam, F. T. S. Yu, D. A. Gregory, and R. D. Juday, "Autonomous real-time object tracking with an adaptive joint transform correlator," Opt. Eng. **29**, 314–320 (1990).

[16] E. C. Tam, F. T. S. Yu, A. Tanone, D. A. Gregory, and R. D. Juday, "Data association multiple target tracking using a phase-mostly liquid-crystal television," Opt. Eng. **29**, 1114–1121 (1990).

[17] A. D. Gara, "Real-time tracking of moving objects by optical correlation," Appl. Opt. **18**, 172–174 (1979).

[18] T. H. Chao and H. K. Lin, "Real-time optical holographic tracking of multiple objects," Appl. Opt. **28**, 226–231 (1989).

[19] D. Psaltis, M. A. Neifeld, and A. Yamamura, "Image correlators using optical memory disks," Opt. Lett. **14**, 429–431 (1989).

[20] T. Lu, K. Choi, S. Wu, X. Xu, and F. T. S. Yu, "Optical-disk-based neural network," Appl. Opt. **28**, 4722–4724 (1989).

[21] F. T. S. Yu, E. C. Tam, T. W. Lu, E. Nishihara, and T. Nishikawa, "Optical-disk-based joint transform correlator," Appl. Opt. **30**, 915–916 (1991).

[22] J. H. Rilum and A. R. Tanguay Jr., "Utilization of optical memory disks for optical information processing," in *OSA Annual Meeting*, Vol. 2 of 1988 OSA Technical Digest Series (Optical Society of America, Washington, DC, 1988), paper MI5.

[23] D. A. Jared, K. M. Johnson, and G. Moddel, "Joint transform correlation using optically addressed chiral smectic liquid-crystal spatial light modulators," in *OSA Annual Meeting*, Vol. 18 of 1989 OSA Technical Digest Series (Optical Society of America, Washington, DC, 1989), paper THI5.

[24] A. Yariv and P. Yeh, *Optical Waves in Crystal* (Wiley-Interscience, New York, 1984).

[25] R. A. Fisher, ed., *Optical Phase Conjugation* (Academic, New York, 1983).

[26] J. Feinberg, "Self-pumped, continuous-wave phase conjugator using internal reflection," Opt. Lett. **7**, 486–488 (1982).

[27] E. Ochoa, L. Hesselink, and J. W. Goodman, "Real-time intensity inversion using two-wave and four-wave mixing in photorefractive $Bi_{12}G_eO_{20}$," Appl. Opt. **24**, 1826–1832 (1985).

[28] D. W. Vahey, "A nonlinear coupled-wave theory of holographic storage in ferroelectric materials," J. Appl. Phys. **46**, 3510–3515 (1975).

[29] S. Wu, Q. Song, A. Mayers, D. A. Gregory, and F. T. S. Yu, "Reconfigurable interconnections using photorefractive holograms," Appl. Opt. **29**, 1118–1125 (1990).

[30] P. Yeh, A. E. T. Chiou, and J. Hong, "Optical interconnection using photorefractive dynamic holograms," Appl. Opt. **27**, 2093–2096 (1988).

[31] P. Yeh, *Introduction to Photorefractive Nonlinear Optics* (Wiley-Interscience, New York, 1993).

[32] D. M. Pepper, J. Auyeung, D. Fekete, and A. Yariv, "Spatial convolution and correlation of optical field via degenerate four wave mixing," Opt. Lett. **3**, 7–9 (1978).

[33] L. Pichon and J. P. Huignard, "Dynamic joint-Fourier-transform correlator by Bragg diffraction in photorefractive $Bi_{12}SiO_{20}$ crystals," Opt. Commun. **36**, 277–280 (1981).

[34] F. T. S. Yu, S. Wu, and A. W. Mayers, "Applications of phase conjugation to a joint transform correlator," Opt. Commun. **71**, 156–160 (1989).

[35] G. Gheen and L. J. Cheng, "Optical correlators with fast updating speed using photorefractive semiconductor materials," Appl. Opt. **27**, 2756–2758 (1988).

[36] D. Z. Anderson, D. M. Liniger, and J. Feinberg, "Optical tracking novelty filter," Opt. Lett. **12**, 123–125 (1987).

[37] M. Cronin-Golomb, A. M. Brenacki, C. Lin, and H. Kong, "Photorefractive time diffraction of coherent images," Opt. Lett. **12**, 1029–1031 (1987).

[38] J. E. Ford, Y. Fainman, and S. H. Lee, "Time integrating interferometry using photorefractive fanout," Opt. Lett. **13**, 1856–1858 (1988).

[39] F. T. S. Yu, S. Wu, S. Rajan, A. Mayers, and D. A. Gregory, "Optical novelty filtering using phase carrier frequency," Opt. Commun. **92**, 205–208 (1992).

[40] B. Loiseaux, G. Illiaquer, and J. P. Huignard, "Dynamic optical cross-correlator using a liquid-crystal light valve and a bismuth silicon oxide crystal in the Fourier plane," Opt. Eng. **24**, 144–149 (1985).

[41] M. G. Nicholson, I. R. Cooper, M. W. McCall, and C. R. Petts, "Simple computation model of image correlation by four-wave mixing in photorefractive media," Appl. Opt. **6**, 278–286 (1987).

[42] F. T. S. Yu, S. Wu, S. Rajan, and D. A. Gregory, "Compact joint transform correlator using a thick photorefractive crystal," Appl. Opt. **31**, 2416–2418 (1992).

[43] J. A. Davis, M. A. Waring, G. W. Bach, R. A. Lilly, and D. M. Cotrell, "Compact optical correlator design," Appl. Opt. **28**, 10–11 (1989).

[44] D. Liu and L. J. Cheng, "Real-time VanderLugt optical correlator that uses photorefractive GaAs,"Appl. Opt. **31**, 2675–5680 (1992).

[45] Q. He, P. Yeh, L. Hu, S. Lin, T. Yeh, T. Tu, S. Yang, and K. Ksu, "Shift-invariant photorefractive joint transform correlator using Fe:LiNbO$_3$ crystal plates," Appl. Opt. **32**, 3113–3115 (1993).

[46] R. Young and C. Chatwin, "Design and simulation of a synthetic discriminant function filter for implementation in an updateable photorefractive correlator," in *Optical Pattern Recognition III*, D. P. Casasent and T. Chao, eds., Proc. SPIE **1701**, 239–263 (1992).

[47] D. Liu, T. Chao, and L. Cheng, "Novelty filtered optical correlator using photorefractive crystal," in *Optical Pattern Recognition III*, D.P. Casasent and T. Chao, eds., Proc. SPIE **1701**, 239–263 (1992).

[48] M. Wen, S. Yin, P. Purwosumarto, and F. T. S. Yu, "Wavelet matched-filtering using a photorefractive crystal," Opt. Commun. **99**, 325–330 (1993).

[49] D. Liu, K. Luke, and L. Cheng, "GaAs-based photorefractive time-integrating correlator," in *Hybrid Image and Signal Processing III*, D. P. Casasent and A. G. Tescher, eds., Proc. SPIE **1702**, 205–209 (1992).

[50] C. Chang, H. Yau, Y. Tong, and N. Puh, "Rotational-invariant pattern recognition with the method of circular harmonics using a BaTiO$_3$ crystal," Opt. Commun. **87**, 219–222 (1992).

[51] S. Faria, A. Tagliaferri, S. Bos, and A. Paulo, "Photorefractive optical holographic correlation using a Bi$_{12}$TiO$_{20}$ crystal at $\lambda = 0.633 \mu$m," Opt. Commun. **86**, 29–33 (1991).

[52] S. Jutamulia, ed., *Optical Correlators*, Vol. 76 of SPIE Milestone Series (Society of Photo-Optical Instrumentation Engineers, Bellingham, WA, 1993).

[53] F. T. S. Yu, S. Wu, A. Mayers, and S. Rajan, "Wavelength-multiplexed reflection-type matched spatial filtering using LiNbO$_3$," Opt. Commun. **81**, 343–346 (1991).

[54] F. T. S. Yu, S. Yin, and Z. H. Yang, "Thick volume photorefractive crystal wavelength-multiplexed reflection-type matched-filter," Opt. Memory Neural Networks **3**, 207–214 (1994).

[55] S. Yin, H. Zhou, M. Wen, J. Zhang, and F. T. S. Yu, "Wavelength-multiplexed holographic construction using a Ce:Fe-doped LiNbO$_3$ crystal with a tunable visible-light diode laser," Opt. Commun. **101**, 317–321 (1993).

[56] K. Fukushima, "Visual feature extraction by a multilayered network of analog threshold elements," IEEE Trans. Syst. Sci. Cybern. **SSC-5**, 322–329 (1969).

[57] J. J. Hopfield, "Neural network and physical system with emergent collective computational abilities," Proc. Natl. Acad. Sci. USA **79**, 2554 (1982).

[58] T. Kohonen, *Self-Organization and Associative Memory* (Springer-Verlag, Berlin, New York, 1984).

[59] D. E. Rumelhart and D. Zipser, "Feature discovery by competitive learning," in *Parallel Distributed Processing*, D. E. Rumelhart and J. L. McClelland, eds. (MIT, Cambridge, MA, 1988), Vol. 1, Chap. 5.

[60] R. P. Lippmann, "An introduction to computing with neural nets," IEEE Trans. Acoust. Speech Signal Process. **4**, 4–22 (1987).

[61] G. A. Carpenter and S. Grossberg, "A massively parallel architecture for a self-organizing neural pattern recognition machine," Comput. Vision Graphics Image Process. **37**, 54–62 (1987).

[62] D. Psaltis and N. Farhat, "Optical information processing based on an associative memory model of neural nets with thresholding and feedback," Opt. Lett. **10**, 98–100 (1985).

[63] P. Lalanne, J. Taboury, J. C. Saget, and P. Chavel, "An extension of the Hopfield model suitable for optical implementation," in *Optics and the Information Age*, H. H. Arsenault, ed., Proc. SPIE **813**, 27 (1987).

[64] R. A. Athale, H. H. Szu, and C. B. Friedlander, "Optical implementation of associative memory with controlled nonlinearity in the correlation domain," Opt. Lett. **11**, 482 (1986).

[65] S. Wu, T. Lu, X. Xu, and F. T. S. Yu, "An adaptive optical neural network using a high resolution video monitor," Microwave Optical Technol. Lett. **2**, 252–257 (1989).

[66] F. T. S. Yu, T. Lu, X. Yang, and D. A. Gregory, "Optical neural network with pocket-sized liquid-crystal televisions," Opt. Lett. **14**, 863–865 (1990).

[67] K. M. Johnson, M. A. Handschy, and L. A. Pagano-Stauffer, "Optical computing and image processing with ferroelectric liquid crystals," Opt. Eng. **26**, 385–391 (1987).

[68] D. Z. Anderson and M. C. Erie, "Resonator memories and optical novelty filter," Appl. Opt. **26**, 434–438 (1987).

[69] X. M. Wang and G. G. Mu, "Optical neural network with bipolar neural states," Appl. Opt. **31**, 4712–4719 (1992).

[70] D. Casasent and D. Psaltis, "Scale-invariant optical correlation using Mullin transforms," Opt. Commun. **17**, 59–63 (1976).

[71] D. Casasent and D. Psaltis, "Scale-invariant optical transform," Opt. Eng. **15**, 258–264 (1976).

[72] Y. Sheng and H. H. Arsenault, "Experiments on pattern recognition using invariant Fourier–Mullin descriptors," J. Opt. Soc. Am. A **3**, 771–776 (1986).

[73] Y. Hsu and H. H. Arsenault, "Optical pattern recognition using circular harmonic expansion," Appl. Opt. **21**, 4016–4019 (1982).

[74] F. T. S. Yu, X. Li, E. Tam, S. Jutamulia, and D. A. Gregory, "Rotation invariant pattern recognition with a programmable joint transform correlator," Appl. Opt. **28**, 4725–4727 (1989).

[75] E. Freysz, B. Pouligny, F. Argoul, and A. Arneodo, "Optical wavelet transform of fractal aggregate," Phys. Rev. Lett. **64**, 7745–7748 (1990).

[76] I. Daubechies, "The wavelet transform, time-frequency localization and signal analysis," IEEE Trans. Inf. Theory **36**, 961–1005 (1990).

[77] S. G. Mallat, "A theory for multiresolution signal decomposition: the wavelet representation," IEEE Trans. Pattern Anal. Mach. Intell. **11**, 674–693 (1989).

[78] R. A. Haddad, A. N. Akansu, and A. Benyassine, "Time-frequency localization in transforms, subbands, and wavelets: a critical review," Opt. Eng. **32**, 1411–1429 (1993).

[79] O. Rioul and M. Vetterli, "Wavelets and signal processing," IEEE Signal Process. Mag. **8**(10), 14–38 (1991).

[80] M. B. Ruskai, G. Beylkin, R. Coifman, I. Daubchies, S. Mallat, Y. Meyer, and L. Raphael,

Wavelets and Their Applications (Jones and Bartelett, Boston, MA, 1992), Chap. IV, p. 241.

[81] Y. Zhang and Y. Li, "Optical determination of Gabor coefficients of transient signals," Opt. Lett. **16**, 1031–1033 (1991).

[82] Y. Zhang, Y. Li, E. G. Kanterakis, X. J. Lu, R. Tolimieri, and N. P. Caviris, "Optical realization of wavelet transform for a one-dimensional signal," Opt. Lett. **17**, 210–212 (1992).

[83] Y. Li and Y. Zhang, "Coherent optical processing of Gabor and wavelet expansions of one- and two-dimensional signals," Opt. Eng. **31**, 1865–1885 (1992).

[84] H. J. Caulfield and H. Szu, "Parallel discrete and continuous wavelet transform," Opt. Eng. **31**, 1835–1839 (1992).

[85] H. Szu, Y. Sheng, and J. Chen, "Wavelet transform as a bank of matched filters," Appl. Opt. **31**, 3267–3277 (1992).

[86] Special issue on wavelet transforms, Opt. Eng. **31**, 1823–1916 (1992).

[87] M. O. Freeman, A. Fedor, B. Bock, and K. Duell, "Optical wavelet processor for producing spatial localized ring-wedge-type information," in *Optical Information Processing Systems and Architectures IV*, B. Javidi, ed., Proc. SPIE **1772**, 241–250 (1992).

[88] B. Telfer and H. Szu, "New wavelet transform normalization to remove frequency bias," Opt. Eng. **31**, 1830–1834 (1992).

[89] H. Szu, B. Telfer, and A. Lohmann, "Causal analytical wavelet transform," Opt. Eng. **31**, 1825–1829 (1992).

[90] Y. Sheng, T. Lu, D. Roberge, and H. J. Caulfied, "Optical N^4 implementation of a two-dimensional wavelet transform," Opt. Eng. **31**, 1859–1864 (1992).

[91] X. Yang, H. Szu, Y. Sheng, and H. J. Caulfield, "Optical Haar wavelet transform," Opt. Eng. **31**, 1846–1851 (1992).

[92] F. T. S. Yu and G. Lu, "Short-time Fourier-transform and wavelet transform with Fourier-domain processing," Appl. Opt. **33**, 5262–5270 (1994).

[93] J. M. Combes, A. Grossmann, and Ph. Tchamitchian, eds., *Wavelets*, 2nd ed. (Springer-Verlag, Berlin, 1990).

[94] X. J. Lu, A. Katz, E. G. Kanterakis, and N. P. Caviris, "Joint transform correlator that uses wavelet transforms," Opt. Lett. **17**, 1700–1702 (1992).

[95] C. F. Hester and D. Casasent, "Multivariant technique for multiclass pattern recognition," Appl. Opt. **19**, 1758–1761 (1980).

[96] D. Casasent, "Unified synthetic discriminant function computational formulation," Appl. Opt. **23**, 1620–1627 (1984).

[97] B. V. K. Vijaya Kumar, "Minimum variance synthetic discriminant functions," J. Opt. Soc. Am. A **3**, 1579–1584 (1986).

[98] A. Mahalanobis, B. V. K. Vijaya Kumar, and D. Casasent, "Minimum average correlation energy filters," Appl. Opt. **26**, 3633–3640 (1987).

[99] R. D. Juday, "Optimal realizable filters and the minimum Euclidean distance principle," Appl. Opt. **32**, 5100–5111 (1993).

[100] S. Kirpatrick, C. D. Gelatt, and M. P. Vecchi, "Optimization by simulated annealing," Science **220**, 671–680 (1983).

[101] M. Kim, M. R. Feldman, and C. Guest, "Optimum encoding of binary phase-only filters with a simulated annealing algorithm," Opt. Lett. **14**, 545–547 (1989).

[102] S. Yin, M. Lu, C. Chen, F. T. S. Yu, T. Hudson, and D. McMillen, "Design of a bipolar composite filter using simulated annealing algorithm," Opt. Lett. **20**, 1409–1411 (1995).

[103] M. Lu, S. Yin, C. Chen, F. T. S. Yu, T. Hudson, and D. McMillen, "Optimum synthesis of a bipolar composite filter with simulated annealing," Appl. Opt. **35**, 2710–2720 (1996).

[104] F. T. S. Yu, "Linear optimization in synthesis of nonlinear spatial filters," IEEE Trans. Inf. Theory **IT-17**, 524–429 (1971).

[105] F. T. S. Yu, "Optimal linearization in holography," Appl. Opt. **8**, 2483–2487 (1969).

[106] J. L. Horner and P. D. Giainino, "Phase-only matched-filtering," Appl. Opt. **23**, 812–816 (1984).

[107] J. L. Horner and J. R. Leger, "Pattern recognition with binary phase-only filters," Appl. Opt. **24**, 609–611 (1985).

[108] N. Konforti, E. Marom, and S.-T. Wu, "Phase-only modulation with twisted nematic liquid-crystal spatial light modulators," Opt. Lett. **13**, 251–253 (1988).

[109] K. Lu and B. E. A. Saleh, "Theory and design of the liquid-crystal TV as an optical spatial phase modulator," Opt. Eng. **29**, 240–246 (1990).

[110] D. A. Gregory, J. A. Loudin, J. C. Kirsch, E. C. Tam, and F. T. S. Yu, "Using the hybrid modulating properties of liquid-crystal television," Appl. Opt. **30**, 1374–1378 (1991).

[111] E. C. Tam, F. T. S. Yu, S. Wu, A. Tanone, S.-D. Wu, J. X. Li, and D. A. Gregory, "Implementation of kinoforms using a continuous-phase SLM," *OSA Annual Meeting*, Vol. 15 of 1990 OSA Technical Digest Series (Optical Society of America, Washington, DC, 1990), p. 259.

[112] T. H. Barnes, T. Eiju, K. Matsuda, and N. Ohyama, "Phase-only modulation using a twisted nematic liquid-crystal television," Appl. Opt. **28**, 4845–4852 (1989).

[113] J. Amako and T. Sonehara, "Computer-generated hologram using TFT active matrix liquid-crystal spatial light modulator (TFT-LCSLM)," Jpn. J. Appl. Phys. **29**, L1533–L1535 (1990).

[114] J. Amako and T. Sonehara, "Kinoform using an electrically controlled birefringent liquid-crystal spatial light modulator," Appl. Opt. **32**, 4622–4628 (1991).

[115] A. Ogiwara, H. Sakai, and J. Ohtsubo, "Application of LCTV to non-linear speckle correlator," Opt. Commun. **86**, 513–522 (1991).

[116] L. B. Lesem, P. M. Hirsch, and J. A. Jordan Jr., "The kinoform: a new wavefront reconstruction device," IBM J. Res. Develop. **13**, 150–155 (1969).

[117] A. Tanone, Z. Zhang, F. T. S. Yu, and D. A. Gregory, "Phase-modulation depth for a real-time kinoform using a liquid-crystal television," Opt. Eng. **32**, 517–521 (1993).

[118] J. M. Florence and R. D. Juday, "Full complex spatial filtering with a phase mostly DMD," in *Wave Propagation and Scattering in Varied Media II*, V. K. Varadan, ed., Proc. SPIE **1558**, 487–498 (1991).

[119] D. A. Gregory, J. A. Loudin, J. C. Kirsch, E. C. Tam, and F. T. S. Yu, "Using the hybrid modulating properties of liquid-crystal television," Appl. Opt. **30**, 1374–1378 (1991).

[120] D. A. Gregory, J. C. Kirsch, and E. C. Tam, "Full complex modulation using liquid-crystal television," Appl. Opt. **31**, 163–165 (1992).

[121] J. C. Kirsch, D. A. Gregory, M. W. Thie, and B. K. Jones, "Modulation characteristics of the Epson liquid-crystal television," Opt. Eng. **31**, 963–970 (1992).

[122] Z. Zhang, G. Lu, and F. T. S. Yu, "Simple method for measuring phase-modulation in liquid-crystal televisions," Opt. Eng. **33**, 3018–3022 (1994).

[123] L. G. Nito, D. Roberge, and Y. Zheng, "Programmable optical phase-mostly holograms with coupled-midmodulation liquid-crystal television," Appl. Opt. **34**, 1944–1950 (1991).

[124] F. T. S. Yu, *Optical Information Processing* (Wiley-Interscience, New York, 1983), p. 15.

[125] D. Cottrell, R. Lilly, J. Davis, and T. Day, "Optical correlator performance of binary phase-only filters using Fourier and Hartley transforms," Appl. Opt. **26**, 3755–3761 (1987).

[126] D. L. Flannery, J. S. Loomis, and M. E. Mikovich, "Design elements of binary phase-only correlation filter," Appl. Opt. **27**, 4231–4235 (1988).

[127] M. S. Kim, M. R. Feldman, and C. C. Guest, "Optimum encoding of binary phase-only filters with a simulated annealing algorithm," Opt. Lett. **14**, 545–547 (1989).

[128] D. C. Wunsch II, R. J. Marks II, T. P. Caudell, and C. D. Capps, "Limitations of a class of binary phase-only filters," Appl. Opt. **31**, 5661–5667 (1992).

[129] R. R. Kallman, "The design of phase-only filters for optical correlators," AFATL-TR-90-63 (U.S. Air Force Armament Laboratory, Eglin Air Force Base, FL, July 1990).

[130] B. V. K. Kumar and R. D. Juday, "Design of phase-only, binary phase-only, and complex ternary matched filters with increased signal-to-noise ratios for colored noise," Opt. Lett. **16**, 1025–1027 (1991).

[131] D. L. Flannery, K. M. Lee, and D. H. Goldstein, "Performance comparison of ternary and minimum noise and correlation energy filters using realistic test images," Opt. Eng. **33**, 1785–1792 (1994).

[132] R. R. Kallman and D. H. Goldstein, "Phase-encoding input images for pattern recognition," Opt. Eng. **33**, 1806–1812 (1994).

[133] B. A. Kest, M. K. Gile, S. D. Lindell, and D. L. Flannery, "Implementation of a ternary phase-amplitude filter using a magneto-optic spatial light modulator," Appl. Opt. **28**, 1044–1046 (1989).

[134] J. J. Hopfield, "Neurons with graded responses have collective computational properties like those of two-state neurons," Proc. Natl. Acad. Sci. USA **81**, 3088–3092 (1984).

[135] D. W. Tank and J. J. Hopfield, "Simple neural optimization networks: an AID converter, signal decision circuit, and a linear programming circuit," IEEE Trans. Circuits Syst. **CS-33**, 533–541 (1986).

[136] Y. S. Abu-Mostafa and J. St. Jacques, "Information capacity of the Hopfield model," IEEE Trans. Inf. Theory **IT-31**, 461–464 (1985).

[137] R. J. McEliece, E. C. Posner, E. R. Rodemich, and S. S. Venkatesh, "The capacity of the Hopfield associative memory," IEEE Trans. Inf. Theory **IT-33,** 461–482 (1987).

[138] D. E. Rumelhart, G. E. Hinton, and R. J. Williams, "Learning representations by back-propagating errors," Nature (London) **323**, 533–536 (1986).

[139] T. Lu, S. Wu, X. Xu, and F. T. S. Yu, "Two-dimensional programmable optical neural networks," Appl. Opt. **28**, 4908–4913 (1989).

[140] T. Lu, X. Xu, S. Wu, and F. T. S. Yu, "A neural network model using inter-pattern association (IPA)," Appl. Opt. **29**, 284–288 (1990).

[141] F. T. S. Yu, Y. Li, X. Yang, and T. Lu, "Application of moment invariant pattern recognition to optical neural net," Optik **89**, 55–58 (1991).

[142] F. T. S. Yu, T. Lu, and X. Yang, "Optical implementation of hetero-association neural network with inter-pattern association model," Int. J. Opt. Comput. **1**, 129–140 (1990).

[143] F. T. S. Yu, X. Yang, and T. Lu, "Space-time sharing optical neural network," Opt. Lett. **16**, 247–249 (1991).

[144] X. Yang, T. Lu, and F. T. S. Yu, "Compact optical neural network using cascaded liquid-crystal televisions," Appl. Opt. **29**, 5223–5225 (1990).

[145] Y. Li, "Applications of moment invariants to neurocomputing for pattern recognition," Ph.D. dissertation (Pennsylvania State University, University Park, PA, 1990).

[146] R. O. Winder, "Threshold logic," Ph.D. dissertation (Princeton University, Princeton, NJ, 1962).

[147] M. K. Hu, "Visual pattern recognition by moment invariants," IRE Trans. Inf. Theory **IT-8**, 179–187 (1959).

[148] M. R. Teague, "Image analysis via the general theory of moments," J. Opt. Soc. Am. **70**, 920–930 (1980).

[149] D. Casasent and D. Psaltis, "Hybrid processor of computer invariant moments for pattern recognition," Opt. Lett. **5**, 395–397 (1980).

[150] Y. Li and F. T. S. Yu, "Variations of irradiance moments in neurocomputing for pattern recognition," Optik **86**, 141–143 (1991).

[151] T. Lu, F. T. S. Yu, and D. A. Gregory, "Self-organizing optical neural network for unsupervised learning," Opt. Eng. **29**, 1107–1113 (1990).

[152] R. P. Lippmann, B. Gold, and M. L. Malpass, "A comparison of Hamming and Hopfield neural nets for pattern classification," MIT Lincoln Lab. Tech. Rep. TR 769 (MIT, Cambridge, MA, 1987).

[153] X. Yang and F. T. S. Yu, "Optical implementation of hamming net," Appl. Opt. **31**, 3999–4003 (1992).

[154] F. Cheng, F. T. S. Yu, and D. A. Gregory, "Multitarget detection using spatial synthesis joint transform correlation," Appl. Opt. **32**, 6521–6526 (1993).

[155] G. Gheen, "Design considerations for low-clutter, distortion invariant correlation filters," Opt. Eng. **29**, 1029–1032 (1990).

[156] F. Cheng, X. Xu, S. Wu, F. T. S. Yu, and D. A. Gregory, "Restoration of blurring images due to linear motion using a joint transform processor," Microwave Optical Technol. Lett. **3**, 24–27 (1990).

[157] A. A. S. Awwal, M. A. Karim, and S. R. Jahan, "Improved correlation discrimination using an amplitude-modulated phase-only filter," Appl. Opt. **29**, 233–236 (1990).

[158] W. H. Lee, "Sampled Fourier-transform hologram generated by computer," Appl. Opt. **9**, 639–643 (1970).

[159] C. B. Burckhardt, "A simplification of Lee's method of generating holograms by computer," Appl. Opt. **9**, 1949–1951 (1970).

[160] W. Swindell, "A noncoherent optical analog image processor," Appl. Opt. **9**, 2459–2469 (1970).

[161] G. G. Yang and E. Leith, "An image deblurring method using diffraction gratings," Opt. Commun. **36**, 101–106 (1981).

[162] F. T. S. Yu, G. Lu, D. Zhao, and F. Cheng, "Implementation of complex gray leveled functions for a joint transform processing by using a decomposition method," in *OSA Annual Meeting*, Vol. 16 of 1993 OSA Technical Digest Series (Optical Society of America, Washington, DC, 1993), p. 46.

[163] F. T. S. Yu, G. Lu, M. Lu, and D. Zhao, "Application of position-encoding to a complex joint transform correlator," Appl. Opt. **34**, 1386–1388 (1995).

[164] L. J. Hornbeck, "Deformable-mirror spatial light modulators," in *Spatial Light Modulators and Applications III*, U. Efron, ed., Proc. SPIE **1150**, 86–102 (1989).

[165] R. Juday, S. E. Monroe Jr., and D. A. Gregory, "Optical correlation with phase-encoding and phase filtering," in *Spatial Light Modulators and Applications II*, U. Efron, ed., Proc. SPIE **825**, 149–156 (1987).

[166] C. Hester and M. Temmen, "Phase implementation of optical correlator," in *Hybrid Image and Signal Processing II*, D. P. Casasent and A. G. Tescher, eds., Proc. SPIE **1297**, 207–219 (1991).

[167] R. R. Kallman and D. H. Goldstein, "Phase-encoding input image for optical pattern recognition," Opt. Eng. **33**, 1806–1813 (1994).

[168] G. Lu, Z. Zhang, and F. T. S. Yu, "Phase-encoded input joint transform correlator with improved pattern discriminability," Opt. Lett. **20**, 1307–1309 (1995).

[169] H. J. Caulfield and R. Haimes, "Generalized matched-filtering," Appl. Opt. **19**, 181–183 (1980).

[170] D. Casasent, "Unified synthetic discriminant function computational formulation," Appl. Opt. **23**, 1620–1627 (1984).

[171] B. V. K. V. Kumar and Z. Bahri, "Efficient algorithm for designing a ternary valued filter yielding maximum signal to noise ratio," Appl. Opt. **28**, 1919–1925 (1989).

[172] G. Ravichandran and D. Casasent, "Minimum noise and correlation energy (MINACE) optical correlation filter," Appl. Opt. **31**, 1823–1833 (1992).

[173] N. Yoshikawa and T. Yatagai, "Phase optimization of a kinoform by simulated annealing," Appl. Opt. **33**, 863–868 (1994).

[174] M. S. Kim and C. C. Guest, "Simulated annealing algorithm for binary phase only filters in pattern classification," Appl. Opt. **29**, 1203–1208 (1990).

[175] F. T. S. Yu, M. Lu, G. Lu, S. Yin, T. D. Hudson, and D. McMillen, "Optimum target detection using a spatial-domain bipolar composite filter with a joint transform correlator," Opt. Eng. **34**, 3200–3207 (1995).

[176] P. Purwosumarto and F. T. S. Yu, "Robustness of joint transform correlator versus VanderLugt correlator," Opt. Eng. **36**, 2775–2780 (1997).

[177] F. Cheng, P. Andres, and F. T. S. Yu, "Removal of intra-class association in joint transform power spectrum," Opt. Commun. **99**, 7–12 (1993).

[178] G. Lu, Z. Zhang, S. Wu, and F. T. S. Yu, "Implementation of a non-zero-order joint-transform correlator by use of phase-shifting techniques," Appl. Opt. **36**, 470–483 (1997).

[179] C. T. Li, S. Yin, and F. T. S. Yu, "A non-zero-order joint transform correlator," Opt. Eng. **37**, 58–65 (1998).

[180] F. T. S. Yu, C. T. Li, and S. Yin, "Comparison of detection efficiencies for nonzero and conventional joint transform correlators," Opt. Eng. **37**, 52–57 (1998).

2

Hybrid neural networks for nonlinear pattern recognition

Taiwei Lu

2.1 Introduction

Numerous research and development efforts have been attempted to make a computer perform pattern recognition, as we humans do every day [1]. Many successful examples have been shown in the machine vision field for product inspection and automation in a controlled environment. Well-defined linear filtering algorithms have been developed to extract features from images and classify objects in real time with a microcomputer. Dedicated electronic processors have been developed to speed up the recognition process. However, many real-world problems, such as three-dimensional (3-D) vision of an arbitrarily oriented object in an uncontrolled environment, are still difficult for a computer to perform. Those problems require nonlinear feature extraction and classification in a complex nonlinear feature space and demand constant adaptation to a changing environment.

Figure 2.1 illustrates the processes of feature classification in a two-feature space; the feature distributions of three objects, A, B, and C, are shown. In Fig. 2.1(a), A, B, and C are well separated in the feature space. Two linear decision hyperplanes, D1 and D2, can be drawn to classify A, B, and C with 100% accuracy. However, in Fig. 2.1(b), the feature boundaries of A, B, and C are meshed, posing a problem for linear decision hyperspace. A decision error (shaded areas) could be made because of the nonlinear boundary condition of the feature space.

In the past decade, neural network (NN) technology has been developed to attack these problems. Artificial NN's are modeled after biological NN's [2, 3]. A true NN is a parallel processor that mimics some of the recognition and deductive functions of the human brain at a simple level. One of the most intriguing properties of a NN is its ability to learn dynamically the interconnection weights that correspond to a desired behavior of the network nonlinear processing units [4–10]. The innate parallel processing and the unique 3-D geometric capability of optics complement rather than compete with electronic systems by addressing problem domains that tax the capabilities of conventional electronic processors, for example, a parallel search of a large database and associative memory [11–13].

An important characteristic of artificial NN's is their capability for massive interconnection and parallel processing. Recently, specialized electronic NN processors and VLSI neural chips have been introduced into the commercial market. The number of parallel channels they can handle is limited because of the limited parallel interconnections that can be implemented with one-dimensional electronic wires. High-resolution pattern recognition problems can require a large number of neurons for parallel processing of an image.

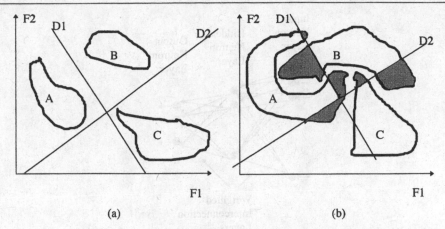

Fig. 2.1. Three classes, A, B, and C, in a feature space: (a) A, B, and C are linearly separable by two decision hyperplanes, D1 and D2; (b) A, B, and C are not separable by linear classification (the shaded areas are decision errors).

Because the parallel processing capability of optics permits these operations to take place at high speed, two-dimensional (2-D) optical processors are well suited for these problems. Optical processors are particularly useful in feature extraction and information reduction at the preprocessing layer, which requires massive interconnections. Optical engines are more efficient than electronic systems because the parallel searching operation does not require backtracking through the knowledge base. Associative memory has the ability to store associated information patterns (u, v) so that subsequent presentation of one pattern u recalls its paired pattern v. This processing technique is inherently parallel. Ideally, an associative memory compares a given input pattern simultaneously with all possible matching patterns [14–18].

This chapter describes the multilayer NN architectures for nonlinear pattern recognition. A holographic optical neural network (HONN) is introduced as an example. The HONN is based on high-resolution volume holographic materials and is capable of performing 3-D massive parallel interconnection of tens of thousand of neurons. A HONN with more than 16,000 neurons packaged in an attaché case has been developed. Rotation-, shift-, and scale-invariant pattern recognition operations have been demonstrated with this system. System parameters such as signal-to-noise ratio (SNR), dynamic range, and processing speed are discussed.

2.2 Neural network background

2.2.1 Neural networks for nonlinear transformation

A NN consists of layers of processors (neurons) that are interconnected, as shown in Fig. 2.2. Discrete data enter the system at the input neuron layer. The data element in each neuron signifies a particular transformation or feature of the input data. The set of neurons performs a nonlinear transformation, producing a result represented by the status of the output neuron layer. A simple function is executed by all neurons in each layer of the network. Each neuron receives input from the neurons of the preceding layer. A receiving neuron performs a weighted summation operation on all the inputs, compares this result with a predefined

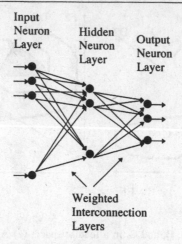

Fig. 2.2. Schematic representation of a multilayer NN.

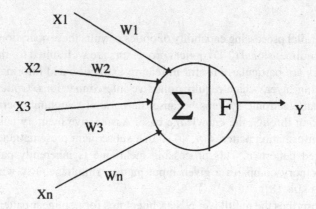

Fig. 2.3. Schematics of a single neuron.

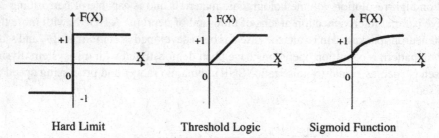

Fig. 2.4. Nonlinear threshold functions.

threshold, then generates an output by using a nonlinear function, according to the formula

$$y_j = F\left(\sum w_{ij}x_i - \theta_j\right), \tag{2.1}$$

where x_i ($i = 0, 1, 2, \ldots, N$) is the input to the jth neuron, w_{ij} is the weight associated with input neuron x_i, θ_j is a bias value, and F is a nonlinear function. This process in a neuron is shown in Fig. 2.3. F can be a hard limit, a threshold logic, or a sigmoid function, as shown in Fig. 2.4.

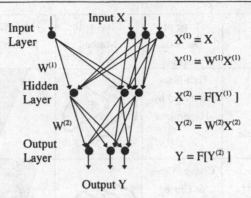

Fig. 2.5. Operational flow of a MLNN. It consists of a series of linear matrix–vector multiplication and nonlinear thresholding.

A multilayer feed-forward NN can be represented as follows:

$$y_j^{(k)} = F\left[\sum w_{ji} y_i^{(k-1)} - \theta_j\right],\qquad(2.2)$$

where y_j is the jth output neuron in the kth layer, $F[\,]$ is the nonlinear transfer function of each neuron, w_{ji} is the interconnection weight between $y_j^{(k)}$ and $y_i^{(k-1)}$, and θ_j is the bias constant in each neuron. In vector form, Eq. (2.2) can be expressed as

$$Y^{(k)} = WX^{(k-1)} - \Theta,\qquad(2.3a)$$

$$X^{(k)} = F\left[Y^{(k)}\right].\qquad(2.3b)$$

When $k = 1$, the network is a single-layer network. Assuming that $y_i^{(0)} = x_i$, where x_i is the ith input neuron, then the network can be represented as Eq. (2.1).

In vector form, Eq. (2.1) can be simply expressed as

$$Y = WX - \Theta,\qquad(2.4a)$$

$$Y = F(Y).\qquad(2.4b)$$

From Eqs. (2.4) we can see that a single-layer NN is a general matrix–vector product that can be trained to represent most linear transformations such as the Fourier transform, correlation, linear filtering, principal component analysis, and partial least-squares analysis [19].

However, a multilayer neural network (MLNN) performs linear transformations within a single-layer and nonlinear transfer functions between layers, as shown in Fig. 2.5. Thus a MLNN is a nonlinear processor that can be used to approximate any continuous nonlinear function with arbitrary desired accuracy. Hornik *et al.* proved that a network with only one hidden layer of sigmoid neurons is enough to have universal approximation properties [20]. Conventional signal processing methods are moving toward nonlinear processing, but they become cumbersome and limited in complex applications – unlike a MLNN, which can more easily fit complex nonlinear problems. The principal advantage of a MLNN stems from the universal NN architecture that enables training by example to search for a solution.

Figure 2.6 compares the classification capability of a linear classifier, a two-layer NN, and a three-layer nonlinear NN [3]. Neither a single-layer NN nor a linear classifier can separate two classes that have features mashed by a hyperplane. In Fig. 2.6, regions A and B are the distribution of features for two classes A and B, respectively. The thin lines are

Structure	Type of Decision Regions	Class with Mashed Regions	Most General Region Shapes
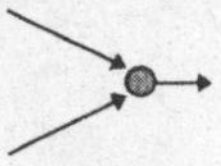 Linear, Single Layer	Half-Plane Bounded by Hyperplane	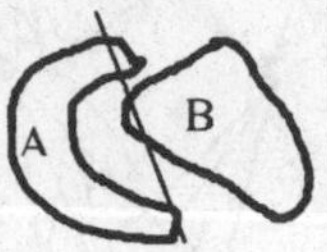	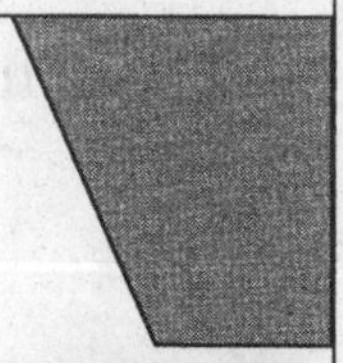
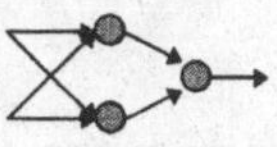 Two Layer	Convex Open or Closed Regions	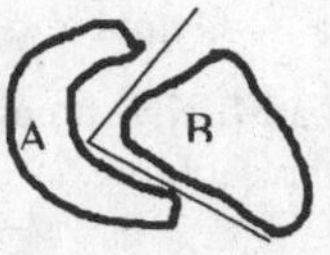	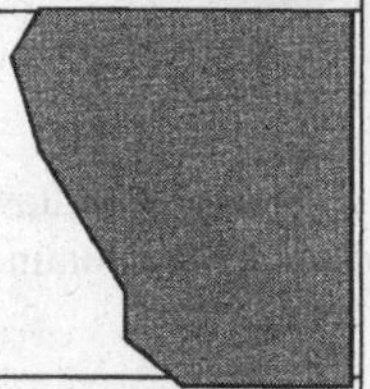
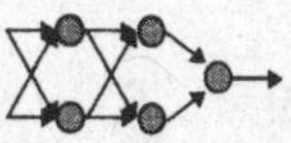 Three Layer	Arbitrary Limited by Number of Nodes	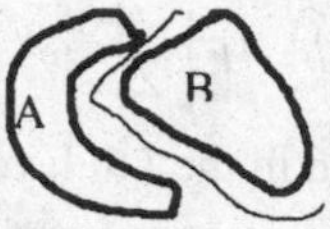	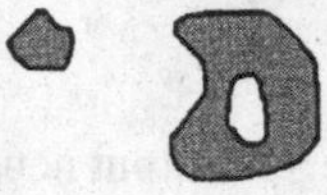

Fig. 2.6. Comparison of decision regions that can be formed by a linear classifier, a two-layer NN, and a three-layer NN.

the boundaries of the decision planes. We can see that, by proper training, the two- and the three-layer NN's can cut through the mashed areas of two classes and form a nonlinear boundary to separate the two classes. The error of misclassification can be reduced by the MLNN, compared with a linear classifier, which cannot reduce the error.

A NN can learn from the training examples to associate features of input images in order to reach an optimal output result. The adaptive learning, massive interconnection, and nonlinear classification capabilities of NN's make them generally more robust to noise and distortion and more sensitive to trained features for signal identification and classification.

2.2.2 Black box versus transparent box

Many users of commercial NN software packages view a NN as a black box. This is due mainly to the complexity of NN systems that demand that the user choose from among many options. For example, there are over 30 popular learning paradigms besides the most popular backpropagation method. The user may also be required to choose the number of layers, the number of neurons in each layer, the learning rate, momentum, transfer function, and other parameters. Each parameter will affect the result of the training process. It is difficult for a nonspecialist to set up the proper parameters in order to obtain a satisfactory result. For the most complex nonlinear problems it is better to customize a NN to ensure an optimal solution.

In fact, a NN is not intrinsically a black box, because it can be expressed by precise analytical formulations, such as Eqs. (2.1)–(2.4). Because NN's have many weights in each layer and each weight acts as a variable during training, it is difficult to describe the learning process and predict the convergence of the learning. Following training, the NN should be tested with blind data to ensure the effectiveness and reliability of the operation. Numerical simulation can be performed to evaluate the network, as the trained network is a set of linear equations with nonlinear transfer functions between stages (as shown in Fig. 2.5). Thus statistical methods can be applied to validate the NN performance once it is trained.

2.2.3 Hidden neurons

For many nonspecialists, the hidden neurons in the MLNN are mysterious. In fact, it was the emphasis on hidden neurons that revived NN research in the 1980's, as they are the key to nonlinear operations that can surpass conventional linear algorithms. No precise rule defines the optimal number of hidden neurons; for general applications, trial and error is the usual practice for selecting the optimal number. There are, however, some general guidelines that can be used in selecting the number of hidden neurons:

- If the network need not maintain scale, shift, or rotation invariance, one or two hidden layers will be sufficient for most nonlinear problems. Theoretically, one hidden layer can approximate any arbitrary nonlinear function, but sometimes two hidden layers are more efficient in learning a highly nonlinear data set.
- In most applications, training data are either difficult or expensive to acquire. If the number of hidden neurons is more than the number of training samples, then each hidden neuron remembers each single individual example. As a result, the network overfits the problem and gives poor generalization. Thus the number of hidden neurons must be smaller than the number of training samples. Indeed, because each weight is a variable during training, the number of weights in each layer of the network should also be smaller than the number of training examples.
- The optimal number of hidden neurons is related to the particular application and the complexity of the training samples. A good indicator of the best number of hidden neurons could be the number of key features that represent the signals or distinguish among classes.

2.2.4 Hybrid neural networks

NN's with a large number of hidden neurons are usually difficult to train because they create many local minima in searching for a global optimum point. The NN can become trapped in a local minimum for hours or forever.

Conventional algorithms, such as correlation, principal component analysis, partial least-squares analysis, fast Fourier transforms, and wavelet transforms are effective tools for identifying and enhancing special features in the information that may well be more compact and organized than the raw data itself [21, 22]. If this is the case, then a smaller NN will adequately perform a nonlinear transformation of the preprocessed data than would be required for the raw data. For example, data sets of 1024 elements can be reduced to 200 elements in the form of shapes, location, intensity, ratios, and slopes to be fed into 200 input neurons. This hybrid NN (HNN) will be highly efficient and robust. Complete data analysis and reporting can take as little as 30 ms on a PC and yet has been demonstrated to improve the threshold of detection of some instruments by much as 10 times [23–26].

2.3 Hybrid optical neural networks

To perform pattern recognition on a high-resolution image in parallel – for example, to recognize an image of 256×256 pixels – one must construct a NN with over 65,000 input neurons and up to 4.3×10^9 global interconnections in the first layer in order to associate the whole image. It is beyond state-of-the-art VLSI technology to fabricate a chip or a board of this interconnection capacity to perform parallel pattern recognition.

Optical technologies, by virtue of their inherent 3-D global interconnection capability, are good candidates for implementing this massive interconnection and parallel processing in the first layer of a MLNN. Photorefractive crystals and active devices have the potential of dynamic learning (similar to that of random access memory) [11], although power requirement, cross talk, and resolution require further development.

Passive holographic materials (similar to a read-only memory), such as dichromated gelatin, silver halide, and DuPont polymers, are well developed. They possess the properties of high resolution (>5000 line pairs/mm), high refractive-index modulation (>0.2), large recording area (>25 cm × 25 cm), and low cost. These holographic materials offer an ideal means of a massively parallel 3-D interconnection for large-scale NN implementations [12, 13]. The interconnections in the first layer may be fixed as a static associative memory by use of holographic implementations. The subsequent layers are much smaller than the first layer, and then can be implemented by an electronic NN for adaptive training and nonlinear recognition.

In this chapter, as an example, the development of a compact HNN system that consists of 1000 to 16,000 optical neurons and approximately 100 electronic neurons is shown. The HNN has been used for rotation-, shift-, and scale-invariant pattern recognition.

A HNN uses a combination of preprocessing algorithms (as the first layer) and NN algorithms for object identification (see Fig. 2.7). Preparation of the images begins with reading the images into the computer in a gray-scale format and digitizing them for computer processing. Next, the images go through a preprocessing stage in which smoothing, normalization, and edge enhancement are performed. Smoothing removes any noisy artifacts. Normalization improves the contrast of the images by eliminating background bias signals. Edge enhancement is the first operation that begins to extract some of the special features of each of the images.

The preprocessed signals are then sent to the next stage for NN training and identification. The input neurons are grouped to represent the feature windows from the preprocessing

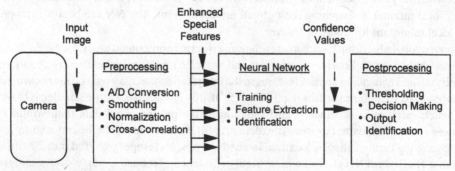

Fig. 2.7. Schematic diagram of a HNN-based pattern recognition system. A/D, analog-to-digital.

stage. A set of training examples is used to train the NN to produce correct responses from the output neurons such that only one output neuron responds high and the rest low for a certain image. The computation time for training the six pairs of patterns is less than 15 secs. on a 486-based microcomputer.

2.3.1 Basic architecture of the holographic optical neural network system

The operation of a one-layer 2-D NN can be expressed in the following matrix–vector product form:

$$Y_{ij}(t) = f\left[\sum_{l=0}^{N}\sum_{k=0}^{N}(T_{ij})_{lk}x_{lk}(t)\right], \qquad i, j = 1, 2, \ldots, M, \tag{2.5}$$

where $y_{ij}(t)$ and $x_{lk}(t)$ are the states of the ijth output neuron and the lkth input neuron, respectively, $(T_{ij})_{lk}$ is the set of interconnection weights between these pairs of neurons, f is the nonlinear threshold function, and $N \times N$ and $M \times M$ are the number of input and output neurons in the network, respectively.

From Eq. (2.5), if $M = N$, an $N \times N$ neural network has N^4 interconnections. For a 256×256 NN, $(256)^4$ or 4×10^9 interconnections must be defined.

In an optical implementation of a NN, training occurs though the modification of holographic interconnection gratings. The gratings represent weighted communication pathways between the optical neurons. Light emitted from an optical neuron is diffracted by a grating with a diffraction efficiency proportional to its interconnection weight. Diffracted light from the array of input sources is focused onto each output optical neuron, thereby implementing the weighted interconnection operation that is the foundation of almost all NN models. The weighted gratings are modified through a sequence of holographic exposures.

A high-density optical associative memory that uses a parallel N^4 interconnection weight matrix (IWM) has been developed by other researchers [12]. Difficulties with their architecture include the following: (a) most spatial light modulators (SLM's) are designed for near-normal incidence, (b) lenses used in the system must not produce aberrations, (c) the grid pattern design in SLM's causes high-order diffraction, and (d) the summation of coherent synaptic signals produces coherent interference noise. In this chapter a practical realization for an N^4 optical holographic IWM to eliminate all the above difficulties is proposed.

Figure 2.8 shows the setup for recording an N^4 IWM. A collimated beam illuminates an SLM at one particular incidence angle (near-normal, for example). A diffuser, placed in the image plane of a $4f$ imaging optical system, spreads the incident pattern (modulated by the SLM) over a wide angular range. Because most SLM's are protected by substrates or packing plates, a $4f$ image system faithfully images the SLM pixel array (which carries the pattern) onto the diffuser. A hologram plate with a mask is placed at a proper distance Z (which depends on the desired recording size and the angle of spread of the diffuser) behind the diffuser. With an additional reference beam, the pattern from the SLM (T_{ij}) will be holographically recorded in an element of the array. By changing the SLM pattern and moving the mask along both horizontal and vertical directions, one can easily fabricate an N^4 IWM [i.e., $(T_{ij})_{kl}$] without difficulties (a), (b), and (c) mentioned above. The weight information can be coded by varying the ON time of each SLM pixel. Thus, the longer the two-beam exposure, the stronger the grating strength and the higher the diffraction

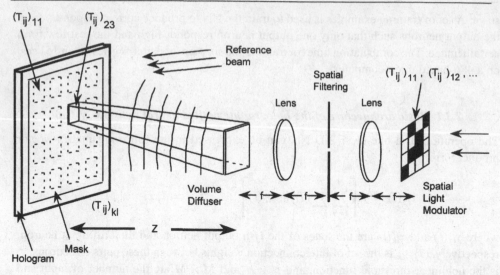

Fig. 2.8. Recording setup for an N^4 holographic IWM with a diffuser.

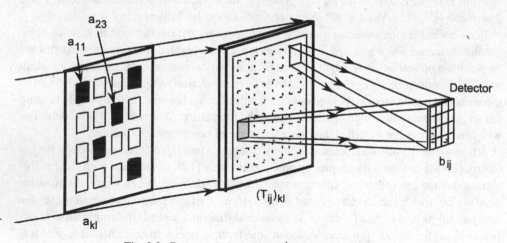

Fig. 2.9. Reconstruction of an N^4 holographic IWM.

efficiency. Note that this statement is true within a certain exposure range and that this range is determined by hologram characteristics such as coating thickness, concentration ratio, and processing procedures. The key element in the recording process is the design of the diffuser, because its characteristics, such as speckle size, directionality, and uniformity, are crucial to the performance of the N^4 holographic element array.

In the reconstruction process (see Fig. 2.9), an encoded reference beam (with pixel a_{kl}), which is conjugated to the reference beam in the recording process and represents the input information, illuminates the holographic array. Each pixel covers one holographic element in the array. This encoded reference beam can be realized by an SLM or can originate from an array of laser diodes with collimating lenses. A photodetector array that has the same packing density and pixel size as the SLM used in the recording process is placed at the diffuser. The beams diffracted from the holographic matrix elements are directed to the photodetector array.

Because the holographic array consists of many matrix elements, the outputs of these holographic elements add up pixel by pixel in the photodetector array. Thus the resulting signals detected by one of the detectors can be described as

$$b_{ij} = \sum_{kl} (T_{ij})_{kl} a_{kl}. \tag{2.6}$$

This equation describes the parallel interconnects of an $N \times N$ input array to an $N \times N$ output array through an N^4 holographic IWM.

2.3.2 Construction of an automatic recording system

An $N \times N$ 2-D NN requires an array of $N \times N$ holograms for interconnection. The NN is first trained on an electronic computer; the IWM is then recorded on a hologram plate by means of an automatic recording system. An automatic holographic recording system is designed that synchronizes the sequential display of each submatrix on the liquid-crystal television (LCTV), the shutter exposure time, and the movement of the mask, recording an array of holographic interconnection weights. Figure 2.10 illustrates an automatic holographic interconnection array recording system. The light from an argon laser is divided into two beams, an object beam and a reference beam, both of which are expanded and collimated by a variable beam splitter. The object beam illuminates an LCTV that displays a submatrix of the IWM. The submatrix is imaged onto a diffuser and is then diffracted uniformly on the holographic recording plate. An $X–Y$ translation stage holds a mask in which a small window is open for recording in a certain position. The experimental setup of the system is shown in Fig. 2.11. There are three key components in the system: an LCTV display, an $X–Y$ translation stage, and a shutter controller. The shutter controller can be interfaced to a PC through an RS232 serial card. The $X–Y$ translation stage can be programmed by the PC by means of an IEEE-488 interface card.

Once the recording of the IWM is completed, the HONN can readily be used to perform high-speed parallel operations for pattern recognition. The HONN for image reconstruction and recognition is shown schematically in Fig. 2.12. The input mask behind the holographic interconnection array ensures that each pixel of the input pattern is superimposed

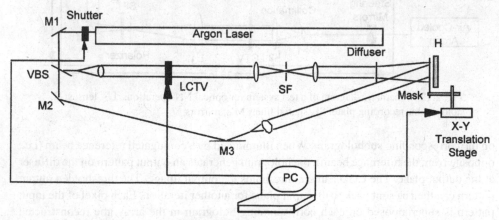

Fig. 2.10. Automatic holographic recording system. M's, mirrors; VBS, variable beam splitter; SF, spatial filter; H, holographic recording plate.

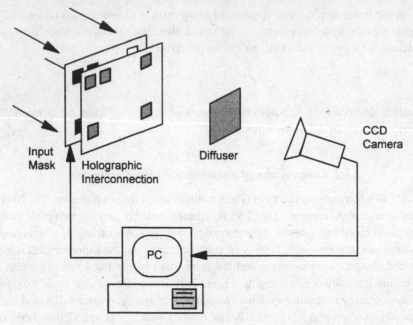

Fig. 2.11. Reconstruction and recognition system.

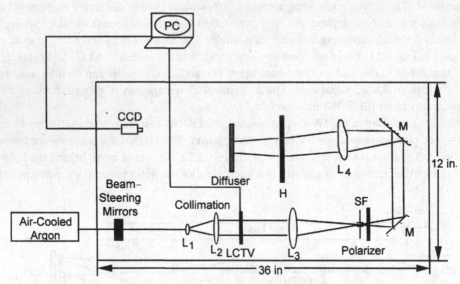

Fig. 2.12. Schematic diagram of a test system for optical NN operations. L's, lenses; H, holographic recording plate; SF, spatial filter; M's, mirrors.

on the corresponding subhologram. When illuminated by a conjugated reference beam (i.e., opposite from the reference beam), the hologram generates an output pattern on the diffuser at the output plane. The CCD camera thresholds the output image. The thresholded output pattern can then be sent back to the input plane for another iteration. Each pixel of the input pattern is superimposed on each corresponding hologram in the array, the reconstructed output pattern is imaged on a diffuser, and a CCD camera is used to pick up the output image for thresholding in the microcomputer.

2.4 Construction of holographic optical neural network systems

A series of prototypical HONN systems were built for evaluation: a benchtop system, an attaché-case-sized portable system, and a lunchbox-sized compact system. These HONN systems contain from 1000 to 16,000 optical neurons for pattern recognition.

2.4.1 Benchtop demonstration system

A computer-controlled system was constructed for HONN operations (see Fig. 2.12). The system size was 36 in. × 12 in. × 9 in. All optical components were positioned on a bread-board. The light source was an air-cooled argon laser operating at 514.5 nm with 20-mW output power. The collimated beam was spatially modulated by an LCTV, modified by removal of the polarizer and the analyzer (low-quality sheet polarizers). The polarizer was replaced with a high-quality Glan–Thompson polarizer, because the argon laser light source had a polarization ratio of 500:1. As a result of the addressing scheme of the LCTV and the high-frequency grid incorporated into the display, multiple diffraction orders of light were produced when the LCTV was illuminated with collimated laser light. Severe cross talk resulted. A $4f$ optical imaging system was used to eliminated the high diffraction orders. The $4f$ system consisted of a 75-mm-diameter, 200-mm focal-length lens (L3), a 0.8-mm-diameter pinhole mounted on a three-axis translation stage from Newport Corporation, a 20× objective lens (NA 0.2), and a 75-mm-diameter, 400-mm focal-length lens (L4).

The polarizer, pinhole, and $4f$ system were aligned so as to allow only the zero order to pass through the system. The illumination patterns were generated by the LCTV pattern that illuminated the hologram array. A diffracted pattern was displayed on the diffuser. The image from the diffuser was captured by the CCD camera connected to the PC.

2.4.2 Portable demonstration system

The first-generation test setup was modified, which led to a second-generation portable HONN system packaged in an attaché case (see Fig. 2.13). A laser diode (10-mW, 677-nm operating wavelength) is used as a power source in place of the bulky air-cooled argon laser source. An LCTV is used as the input SLM. The LCTV has a resolution of 320 × 220 pixels and a contrast ratio of 70:1. The laser light is expanded and collimated by the collimation optics and then passed through an input SLM, a polarizer, a $4f$ system consisting of two Fourier lenses, and a spatial filter. The $4f$ system has two purposes: to enlarge the input image to twice the original size and image it onto the hologram array, and to filter out the high-frequency grid pattern imposed by the input LCTV.

The optical path is folded several times by three mirrors in order to fit into the 18 cm × 28 cm × 6 cm case. The hologram array is recorded with a four-dimensional (4-D) IWM. An interpattern association (IPA) NN model [22] is used to train the IWM with a set of training patterns. The input pattern is imaged onto the hologram array, multiplying the 4-D IWM by the 2-D input vector. An output diffraction pattern is generated, displayed on the output diffuser, and collected by the output CCD camera array. The IWM contains $(32 \times 32)^2$ weights.

Software was developed to test the hologram's performance, primarily in terms of its robustness to noise. The software, based on the IPA NN model, was used to train a NN of 64×64 input neurons. Once the hologram was placed in the portable NN unit, its performance was tested by an increase in the noise level superimposed on the original object.

Fig. 2.13. Second-generation portable HONN.

The advantage of using the IPA model for HONN learning is that it has simple interconnection weights, in that only three gray levels are required for the hologram recording; i.e., -1 for inhibition, 0 for no connection, and $+1$ for excitation. The negative value can be encoded in the hologram by phase encoding or polarization encoding or by adding a bias. In the hologram array generation here, a spatial multiplexing technique is used to separate the positive and the negative output locations and to perform subtraction in the postprocessing stage. Another advantage of the IPA model is its fast training capability. The NN of 10,000 neurons can be trained typically in less than 5 min on a PC.

2.4.3 Compact lunchbox demonstration system

The attaché case HONN system was further reduced to a lunchbox-sized system (see Fig. 2.14). A laser diode (10-mW, 670-nm operating wavelength) is used as the light source. It is compact, readily available, stable, and has enough power for diffraction so the CCD camera array operates above threshold. An anamorphic prism pair converts the laser beam from a solidly elliptical shape to a somewhat circular shape. The beam is then expanded by a $20\times$ magnification objective lens (NA 0.2) and collimated with a planoconvex lens that is coated with an antireflection coating, a diameter of 50.8 mm and a focal length of 62.9 mm. An LCTV (resolution 480×440 pixels, contrast ratio $>100:1$, frame rate 30 frames/s) is used as the input SLM. A 50×50 hologram array is in close contact with the SLM. The hologram array constitutes the interconnection between the input and the output neurons. Because the SLM and the hologram array are in close contact, the grid structure in the LCTV does not affect the input pattern on the hologram array. Thus the system was made compact by the elimination of the $4f$ system of the previous design. The diffracted beam from the hologram array is imaged onto a holographic mirror, and the deflected output

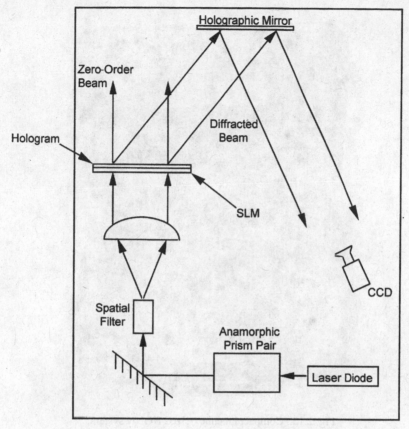

Fig. 2.14. Design of a compact HONN system.

image from the holographic mirror is captured by the CCD camera array. The operation of
the system can be represented by an input image $x(i, j)$ on the LCTV multiplied by the
IWM $T(l, k, i, j)$ on the hologram array. The summation in the output holographic mirror
is imaged onto the CCD array in accordance with Eq. (2.5).

The third-generation HONN system is packaged in a 9 in. $\times$ 12 in. $\times$ 5 in. lunchbox,
shown in Fig. 2.15.

2.5 Holographic optical neural network for pattern recognition

The HONN systems have been used to perform pattern recognition in various simulated
platforms. This section presents static pattern recognition with the attaché case HONN
system (32×32 neurons); real-time target detection, tracking, and identification with a
lunchbox HONN system (64×64 neurons); and a lunchbox HONN system (128×128
neurons) for image identification.

2.5.1 Hybrid holographic optical neural network hybrid for
distortion-invariant pattern recognition

The HONN system is designed to have an array of input neurons and a certain number
of intermediate and output neurons. The number of intermediate and output neurons is
determined by the number of training sets and the specific applications. For example, a

Fig. 2.15. Compact lunchbox-sized HONN system.

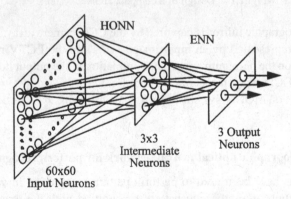

Fig. 2.16. Three-layer $(60 \times 60 - 3 \times 3 - 3)$ HONN architecture for rotation-invariant pattern recognition; ENN, electronic NN.

multilayer (one hidden layer) hybrid NN $(60 \times 60 - 3 \times 3 - 3)$ has been designed to perform rotation-invariant pattern recognition, as shown in Fig. 2.16. The first layer is implemented on a HONN system and the second layer in an electronic network.

An IPA is used to train the first NN layer. A tank, a plane, and a helicopter are used as three classes of objects. Input images of each object rotated in $10°$ increments around its center of mass are taken by a CCD camera. A total of $36 \times 3 = 108$ edge-enhanced input images are used as training patterns for the NN. Figure 2.17 shows some of the three training objects in

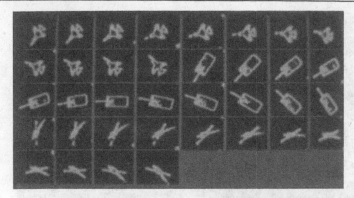

Fig. 2.17. Some of the input training patterns of a 60×60 HONN: a plane, a tank, and a helicopter, rotated at $10°$ increments in the $0°–120°$ range.

Fig. 2.18. IWM trained by an IPA NN model.

the $0°–120°$ angular range. The IPA model extracts the special features of each object and builds positive interconnection weights to enhance its own special features and negative weights to inhibit those of the other objects. Logic operations AND, OR, and NOT are applied to the binary training patterns to extract the special features. The IPA model constructs the IWM in a single iteration, making large-scale NN learning on a PC very fast. The average learning time for a network with 3600 input neurons and 9 output neurons is less than 2 min on an Intel 80486-based computer. After training by the IPA model, a 4-D IWM is formed, as shown in Fig. 2.18. The IWM consists of 3×3 submatrices, corresponding to 3×3 output neurons, and each submatrix has 60×60 elements. The nine output neurons are trained to represent the three objects – tank, plane, and helicopter – in three angular ranges, $0°–120°$, $120°–240°$, and $240°–360°$.

Figure 2.18 shows three states assigned to the IWM in the course of the NN training: excitation (1, white), inhibition (-1, black), or no connection (0, light gray). The first column of the IWM corresponds to the three output nodes for the planes ($0°–120°$, $120°–240°$, and $240°–360°$, respectively), and the second and the third columns to the tanks and the helicopters, respectively. The nine submatrices can be regarded as nine rotation-invariant filters for the three objects, each in a $120°$ range. The NN is designed to perform rotation-invariant pattern recognition for an arbitrary degree of rotation, although the network was

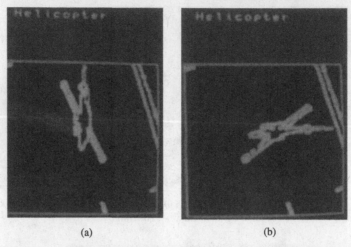

(a)   (b)

Fig. 2.19. Results of using a 64×64 NN for rotation-invariant pattern recognition of a helicopter (in two positions).

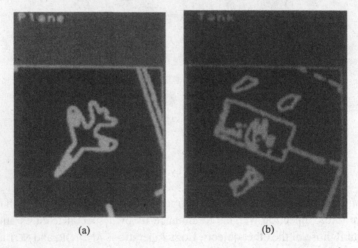

(a)   (b)

Fig. 2.20. Results of using a 64×64 NN for rotation-invariant pattern recognition of (a) a plane, (b) a tank.

trained with $10°$ intervals. The angular discontinuity (gap) in the nine submatrices can be seen in Fig. 2.18. However, the initial output recognition results were quite satisfactory, with a success rate of batter than 80%. In order to eliminate the discontinuity problem, swelling was performed after the edge enhancement of the training objects, making the gaps in the submatrices much narrower than those in Figs. 2.19 and 2.20 and increasing the success rate to over 90%.

The rotation-invariant operations were tested with input objects in various noisy environments and arbitrary angular positions. The test results showed that the NN gave correct recognition results in over 80% of the trials. As seen in Figs. 2.19 and 2.20, input objects such as a helicopter, a tank, and a plane were used as unknown patterns. A CCD camera captured the input object and sent the image to a PC through a framegrabber. The computer edge enhanced and thresholded the input image, as shown in Figs. 2.19 and 2.20. The preprocessed image was then sent to the input of the HONN for rotation-invariant recognition. The hybrid NN has two layers: the first connects a network of 64×64 input neurons to 12 hidden neurons, and the second connects those 12 to 3 output neurons. The first layer was trained by an IPA model. The second layer was first trained by a MAXNET NN model (a competitive learning model [3]), and then further trained with a backpropagation learning model. Although backpropagation training generally requires a very long time, the training time was substantially reduced when the MAXNET model was used first.

The system took 1 s to recognize an object. It then output the name of the input object above the input image, as shown in Figs. 2.19 and 2.20. The input objects were rotated at arbitrary angles, with a successful recognition rate of 80% to 90%. The network's ability to recognize the object in a noisy environment was object dependent. For example, because the features of the tank are more distinctive than those of the helicopter and the plane, the tank was easily recognized in high noise. Optimization during training should ensure uniform performance.

In another experiment, the HONN system was used to perform real-time object detection, tracking, and identification. The system setup is illustrated in Fig. 2.21. The input objects are on a rotational platform; a remote-controlled tank and several planes are the objects to be recognized. The input scene is shown in Fig. 2.22. A CCD camera took a live input image and sent the image to an electronic preprocessor. The input image was first adaptively thresholded in terms of the background light intensity. The thresholded image was then segmented, and isolated objects were located.

The electronic preprocessor also calculated the first-order moment of each object to find its center and performed edge enhancement. Then the edge-enhanced objects were sent sequentially to the HONN rotation-invariant pattern recognition system for recognition processing. The input objects were classified as tanks, planes, and unknown objects.

A C language program was written to control the system. Figure 2.23 shows the output results. The tank and the planes were correctly recognized at a frame rate of approximately 1 frame/s. The processing speed generally depends on the number of objects detected in

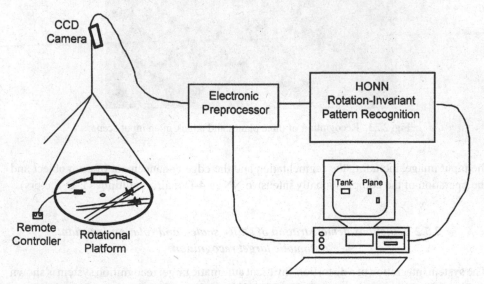

Fig. 2.21. Layout of HONN rotation-, shift-, and scale-invariant pattern recognition system.

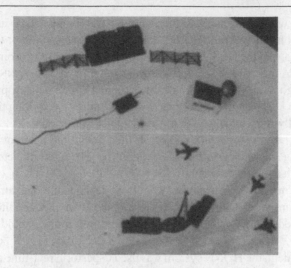

Fig. 2.22. Input scene.

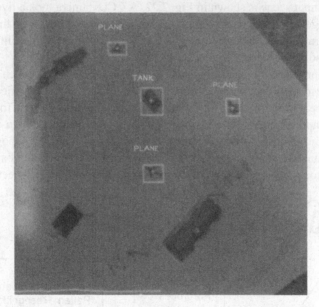

Fig. 2.23. Recognition of three planes and a tank in an input scene.

the input image, including the segmentation and the edge enhancement of each object and the operation of the computationally intensive NN (a 4-D matrix multiplies 2-D vectors).

2.5.2 Lunchbox demonstration of shift-, scale-, and rotation-invariant automatic target recognition

The system integration of a distortion-invariant automatic target recognition system is shown in Fig. 2.24. The system consists of a HONN, two framegrabbers, a VCR, and a video monitor.

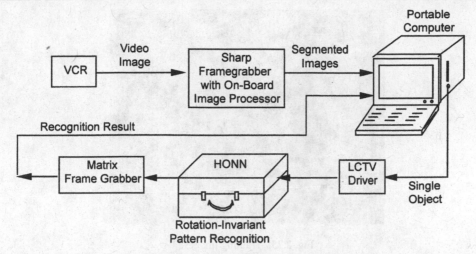

Fig. 2.24. HONN system for distortion-invariant target recognition.

The VCR played a tape of video images of image sensors. A Sharp framegrabber digitized the video image and performed on-board image processing on the image. The image was properly thresholded and segmented into a set of objects of interest. The segmented images were sent to an Intel 486-based PC.

The computer displayed one object of interest at a time, which was transferred though an LCTV driver and displayed on an LCTV, the input device in the HONN box. The HONN box contained a 128 × 128 holographic array that stores a NN IWM trained to identify tanks and planes, regardless of their orientations. A CCD camera was used as the output detector. The result of the HONN pattern recognition was then transferred back to the computer, and the appropriate label, tank or plane, was mapped onto each object. A question mark was used to label unknown objects.

Because commercially available devices (an LCTV and a CCD camera) were used as the input–output devices, the cost of the system is low. But the speed of this HONN is limited to 30 frames/s. The overall system throughput, including preprocessing and postprocessing, is approximately 2–5 frames/s. When multiple digital signal processing chips are used for preprocessing and postprocessing and high-speed input–output devices are used for the HONN system, the overall throughput could be increased to over 100 frames/s.

The images of four planes were used as the training patterns for the HONN system; their edge-enhanced versions are shown in Fig. 2.25. Each image was rotated a total of 20° in increments of 2.2°, giving a total of 45 images. The images were of four aircraft.

An IPA learning model was used to train the NN on a PC. The system combines electronic NN's and optical NN's into a hybrid system. It consists of an array of 128 × 128 input neurons, six hidden layers of neurons, and three output neurons. The connection between the 128 × 128 input neurons and the six hidden layers is necessarily optical, because of the large and complex relationships to be established. The connection between the six hidden neurons and the three output neurons is electronic. The IWM of the first layer is shown in Fig. 2.26.

After training, each image of the aircraft produced two output images, one positive and the other a negative image. A total of six trained images was recorded on the hologram, as illustrated in Fig. 2.26. Each image was recorded in a different part of the reconstruction image focal plane so that reconstruction of the hologram with a plane wave produced a

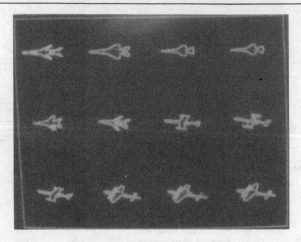

Fig. 2.25. Training images.

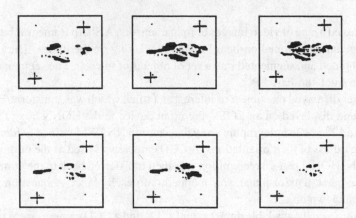

Fig. 2.26. Six submatrices of the first layer defined by the IPA.

diffraction pattern with six convergent points, evenly spaced. A smaller NN was developed for postprocessing the diffraction pattern intensity distribution for final aircraft identification. At this stage, the user can define the desired threshold value for recognition. Output values above the threshold are recognized as positive identification, whereas output values below the threshold are considered negative. When an unknown aircraft image is input, the NN generates the image on an LCTV, which modulates the laser beam illuminating the hologram. The hologram, which contains the IWM, produces an output diffraction pattern corresponding to the input image. A distinct diffraction pattern is produced for each input aircraft.

The optical NN was tested for high-speed parallel target recognition. In the experimental demonstration, images of three aircraft were used as the input patterns. These input patterns were superimposed on the holographic IWM, and the optical network recognized these patterns at the output plane.

A smaller NN, such as a backpropagation model, could have been used here for postprocessing. Because there were a small number of possible outputs for postprocessing, a set of three rules was simply applied to the diffraction intensity, depending on the image

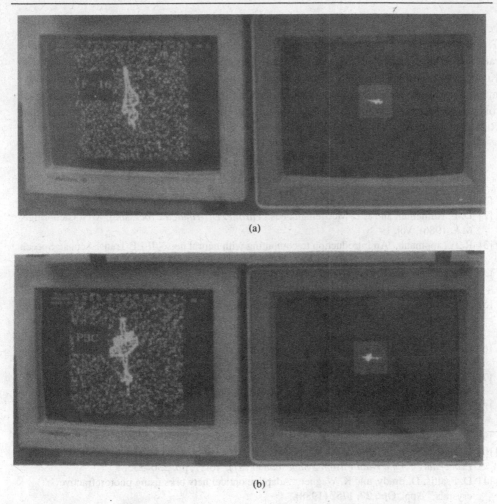

(a)

(b)

Fig. 2.27. Identification in a noisy environment (SNR 1:1): (a) F-16, (b) P3C.

input into the LCTV. The three rules were chosen to identify one aircraft uniquely from the others based on unique intensity distributions.

Figure 2.27 shows the HONN's capability to identify the correct aircraft automatically in a noisy environment. Here identification with a SNR of 2:1 is demonstrated. The HONN is robust enough to accommodate SNR's as low as 1:1.

2.6 Conclusions

A neural network can be used to perform nonlinear transformations. It is a powerful tool that is able to extract nonlinear features, perform complex decision making, and adaptively learn a changing environment. A compact, rugged, lightweight, and low-powered hybrid holographic neural network system has been used to perform robust rotation-, scale-, and shift-invariant pattern recognition. Optical experiments have been demonstrated with HONN systems of up to 128×128 neurons. The hybrid neural networks have the potential for high-speed nonlinear pattern recognition on high-resolution images.

Acknowledgments

The HONN research was performed when the author was with the Physical Optics Corporation, Torrance, California. The author acknowledges David Mintzer, Freddie Lin, Andrew Kostrzewski, Hung Chou, and Jenkins Chen for their contribution to the HONN development. The author also acknowledges the sponsorship of the U.S. Air Force Rome Laboratory through contract F30602-93-C-0136. Special thanks are due to Robert Kaminski and Richard Fedors for their suggestions and comments.

References

[1] D. Marr, *Vision* (Freeman, San Francisco, 1982).

[2] D. E. Rumelhart and J. L. McClelland, eds., *Parallel Distributed Processing* (MIT, Cambridge, MA, 1986), Vol. 1.

[3] R. P. Lippmann, "An introduction to computing with neural nets," IEEE Trans. Acoust. Speech Signal Process. **4**, 4–22 (1987).

[4] H. A. Castro, S. M. Tam, and M. Holler, "Implementation and performance of an analog nonvolatile neural network," submitted to IEEE AICSP.

[5] A. V. Forman et al., "Multisensor target recognition system (MUSTRS)," in *Proceedings of the Third Automatic Target Recognition System and Technology Conference*, GACIAC PR 93-01 (1993).

[6] C. Bowman, "Artificial neural network approaches to target recgnition," in *Proceedings of the IEEE Digital Avionic Systems Conference* (IEEE, New York, 1988), pp. 847–857.

[7] N. Farhat, "Smart sensing and recognition based on models of neural networks," Final Rep. to the U.S. Army Research Office (U.S. Army Research Office, 1990).

[8] C. C. Piazza, "Modified backward error propagation for tactical target recognition," Masters thesis (U.S. Air Force Institute of Technology, 1988).

[9] J. Llinas and E. Waltz, *Multisensor Data Fusion* (Artech House, Norwood, MA, 1990).

[10] D. Marchette and C. Priebe, "An application of neural networks to a data fusion problem," in *Proceedings of the Data Fusion Symposium* (1987), Vol. 1, pp. 230–235.

[11] D. Psaltis, D. Brady, and K. Wagner, "Adaptive optical networks using photorefractive crystals," Appl. Opt. **27**, 1752 (1988).

[12] H. J. Caulfield, "Parallel N^4 weighted optical interconnects," Appl. Opt. **26**, 4039 (1987).

[13] F. Lin, "Practical realization of N^4 optical interconnections." Appl. Opt. **29**, 5226–5227 (1990).

[14] T. Lu and F. Lin, "Experimental demonstration of large-scale holographic optical neural network," in *Proceedings of the International Joint Conference on Neural Networks* (1991), Vol. 1, pp. 535–540.

[15] A. A. Kostrzewski, T. Lu, H. Chou, and F. Lin, "Portable holographic neural system for distortion-invariant pattern recognition and tracking in controlled environment" in *Optical Pattern Recognition III*, D. Casasent and T. H. Chao, eds., Proc. SPIE **1701**, 298–307 (1992).

[16] D. Lu, Mintzer, A. Kostrzewski, and F. Lin, "Compact holographic optical neural network system for real-time pattern recognition," Opt. Eng. **35**, 2122–2131 (1996).

[17] A. Lu, Kostrzewski, H. Chou, S. Wu, and F. Lin, "Performance evaluation of a holographic optical neural network (HONN) system," in *Photonics for Computers, Neural Networks, and Memories*, S. T. Kowel, W. J. Miceli, and J. A. Neff, eds., Proc. SPIE **1773**, 46–52 (1992).

[18] T. Lu, F. Lin, H. Chou, A. Kostrzewski, and J. Chen, "Large-scale neural network model for multi-class pattern recognition," in *Photonics for Processors, Neural Networks, and Memories*, J. L. Horner, B. Javidi, S. T. Kowel, and W. J. Miceli, eds., Proc. SPIE **2026**, 403–414 (1993).

[19] T. Lu and J. M. Lerner, "Spectroscopy and hybrid neural network analysis," Proc. IEEE **84**, 895–905 (1996).

[20] K. Hornik, M. Stinchcombe, and H. White, "Multilayer feed-forward neural networks are universal approximators," Neural Networks **2**, 359–366 (1989).

[21] Y. Sheng, T. Lu, and H. J. Caulfield, "Optical N^4 implementation of 2-D wavelet transform," Opt. Eng. **31**, 1859–1864 (1992).

[22] T. Lu, X. Xu, S. Wu, and F. T. S. Yu, "Neural network model using interpattern association," Appl. Opt. **29**, 284–288 (1990).

[23] J. M. Larner and T. Lu, "Practical neural nets for spectroscopic analysis," Photon. Spectra Mag. (8), 93–95 (1993).

[24] D. Mintzer, S. Zhao, and T. Lu, "Custom neural networks aid spectroscopic analysis," Environ. Test. Mag. **5**(1), 32–35 (1996).

[25] T. Lu, F. T. S. Yu, and D. A. Gregory, "Self-organizing optical neural network for unsupervised learning," Opt. Eng. **29**, 1107–1113 (1990).

[26] T. Lu, F. Lin, A. Kostrzewski, and J. M. Lerner, "Large-scale holographic neuron system for multi-spectral sensor fusion and high-speed signal processing," in *Proceedings of the International Symposium on Substance Identification* (1993).

3

Wavelets, optics, and pattern recognition

Yao Li and Yunlong Sheng

3.1 Introduction[‡]

What is a wavelet? Why is it interesting? And how can it be used to solve problems? These are the three questions this chapter is trying to deal with. To answer the first question, a simple quotation of the following sentence will suffice: the wavelet is "a versatile tool with very rich mathematical contents and great potential for application" [1]. The answer to the second question may take a little longer, as people who have different interests in the wavelet may give you a different reason. For example, wavelets can be viewed as a new basis for representing signals and functions. Wavelets can also be used as a new technique for signal analysis and synthesis. Additionally, wavelets are becoming a novel mathematical subject. As can be expected, to find satisfactory answers to the third question, additional efforts are needed because the scope of wavelet techniques is expanding rapidly. Many books and articles are now available to teach how wavelets can be used for various applications. Thus what we hope to achieve in this chapter is to introduce basic wavelet concepts to readers with a general optical signal processing background. Incidentally, to answer the sequence of questions of what? why? and how? requires a coarse-to-fine filtering process, which is itself a waveletlike process.

3.2 Historical background

Historically speaking, the Fourier transform might have played the most dominant role in the theory and practice of information processing [2]. A periodic function $f(t)$ can be expressed in terms of the Fourier series defined as

$$f(t) = \sum_{n=-\infty}^{\infty} C_n \exp(jnt), \tag{3.1}$$

where

$$C_n = (1/2\pi) \int_0^{2\pi} f(t) \exp(-jnt)\, dt \qquad \text{with} \int_0^{2\pi} |f(t)|^2\, dt < \infty,$$

and $n = \cdots -1, 0, 1, \ldots$. The key concept behind the Fourier series is the superposition of integral dilations of a kernel function $w_n(t) = \exp(jnt)$, a set of sinusoidal waves. The extension from this series to the integral Fourier transform can be made by assuming that the period of $f(t)$ approaches infinity so that discrete summation becomes continuous

[‡] Portions reprinted, with permission, from Proc. IEEE **84**, 720–732 (1996). © 1996 IEEE.

integration. To perform the Fourier analysis, the kernel on which a function is expanded contains a set of everlasting monochromatic waves, which in reality is impossible to handle. A truncation of the integration to a finite interval inevitably causes a time-domain resolution problem. Another inconvenient feature for many is that this linear transform washes out the time coordinate when its frequency spectrum is obtained. To complement this shortcoming, in 1932 Wigner introduced a two-dimensional (2-D) time–frequency joint representation commonly referred to as the Wigner distribution, $WD(t, \omega)$ [3]:

$$WD(t, \omega) = \int_{-\infty}^{\infty} f(t - \tau/2)f^*(t + \tau/2)\exp(-j\omega\tau)\,d\tau, \tag{3.2}$$

where * denotes a complex conjugate. The use of the second order of the signal earned the Wigner distribution a name: a bilinear transform. Several difficulties arose. The single-step inverse transform is not defined, oversampling of the signal has to be performed, and most seriously, the self-multiplication causes cross terms known as Wigner interference. Remedies to these problems are now available but at a cost of convenience.

A slight modification of Eq. (3.2) in the form of

$$WD_c(t, \omega) = \int_{-\infty}^{\infty} f(t - \tau/2)g^*(t + \tau/2)\exp(-j\omega\tau)\,d\tau \tag{3.3}$$

is called a cross Wigner distribution, where $g(t)$ is a reference function. This minor change, however, has at least one significant impact: the transform becomes linear to the signal. A second impact that might be even more significant was created by the transform defined as

$$G(t, \omega) = \int_{-\infty}^{\infty} f(\tau)g(\tau - t)\exp(-j\omega\tau)\,d\tau, \tag{3.4}$$

where $g(t)$ is a called a window function. The transform is commonly referred to as the Gabor transform. Besides the integration arrangement difference of the convolutionlike and the correlationlike processes between Eqs. (3.3) and (3.4), the real novelty of the Gabor transform is that it uses a well-localized Gaussian function (in both time and frequency domains) as its window $g(t)$, so that a long-duration signal can be analyzed piecewise with fairly good time and frequency resolutions. In general, the Gabor transform that can use many other non-Gaussian localized windows has also been called the cross-ambiguity function, the windowed Fourier transform, or the short-time Fourier transform. Reconstruction of the original signal from its Gabor transform is possible with the so-called Gabor representation, which is defined as [4]

$$f(t) = \int_{-\infty}^{\infty} \int_{-\infty}^{\infty} G(\tau, \omega)g(\tau - t)\exp(j\omega\tau)\,d\tau\,d\omega. \tag{3.5}$$

In fact, the Gabor representation was proposed in 1946 by Gabor himself; he intuitively outlined a procedure to represent a signal linearly with a set of mostly localized Gabor basis functions. On the other hand, Eq. (3.4), or what is now known as the Gabor transform, was the work of Bastiaans [5]. Bastiaans, in the late 1970's, mathematically detailed a Zak-transform-based biorthonormalization procedure to make the Gabor representation practical for implementation.

The Gabor transform has been widely used for various signal and image analysis problems. Like everything else, once something became general purpose, problems arose. One serious problem with the Gabor transform was its prespecified resolutions in the time and the frequency domains linked fundamentally by the Heisenberg's uncertainty principle. For those applications for which spectra are wide but it is difficult to prespecify the analyzing resolution, the Gabor transform is difficult to apply. The advent of digital computing and VLSI technology made this problem look even more serious, as the prespecified digital sampling period could be too large to catch sudden changes of a signal or too small to avoid useless details of a signal. It was this user inconvenience that prompted research that led to the invention of the wavelet transform concept.

The concept of the wavelet transforms was formalized when Grossman and Morlet were investigating methods for processing wideband seismic data [6]. Although the phrase wavelet transform was officially coined to the locality-oriented multiresolution-based transform in 1984, some basic concepts can be traced back to almost a century ago [7]. Similar concepts were also found in the area of a pyramidal processing method and in the quadrature-mirror filtering approach in the speech processing and coding community [8, 9]. The work by Daubechies in late 1980's, which first identified some key mathematical features of wavelets, opened a new avenue to the investigations of wavelet transforms [10].

The sections below serve partially as a brief description of the basics of wavelet transforms and, more importantly, as an overview of the recent progress in the use of wavelet processing concepts in various optics and pattern recognition research. More specifically, in Section 3.3, the wavelet transforms and related concepts and parameters are defined and described. In Section 3.4, a discussion of wavelets in general optics is carried out to illustrate the fact that various optical phenomena can be explained with the wavelet or waveletlike concepts. In Section 3.5, the discussion focus is shifted to wavelets for optical signal processing and pattern recognition. Various demonstrated optical wavelet transform methods, their advantages and drawbacks, and proposed wavelet-based feature extraction and pattern classification methods are reviewed.

3.3 Wavelet transforms: definitions and properties

3.3.1 Continuous wavelet transform

Let us first examine the Gabor-transform expression of Eq. (3.4). The multiplicative term $g(\tau - t)\exp(-j\omega\tau)$ appears functionally redundant if there is a way to use a single finite-duration kernel to track both time and frequency information. Because the scale of a function is related to its frequency content, the change of the window function's shift and scale might be sufficient to track both the time- and the frequency-domain information of a signal. This was indeed the idea of Grossmann and Morlet when they were formulating the framework of the wavelet transform in early 1980. Accordingly, Grossmann and Morlet selected a finite-duration window function $\psi(x)$ to be a mother wavelet that can generate a family of daughter wavelets by shifts and dilations, i.e.,

$$\psi_{a,b}(x) = \frac{1}{\sqrt{a}}\psi\left(\frac{x-b}{a}\right), \tag{3.6}$$

where the coefficient $1/\sqrt{a}$ is used to equalize the energy contained in each daughter wavelet. Here the variable x instead of t is used, primarily because our remaining discussions are centered around spatial-domain signals and images.

With a family of wavelets $\psi_{a,b}(x)$, the continuous wavelet transform is defined as the inner product of the signal and these wavelets, i.e.,

$$\mathrm{WT}[f(x)] = W(a, b) = \langle f(x), \psi_{a,b}^{*}(x)\rangle$$

$$= \int_{-\infty}^{\infty} \frac{1}{\sqrt{a}}\psi^{*}\left(\frac{x-b}{a}\right)f(x)\,\mathrm{d}x. \tag{3.7}$$

In this way, the everlasting sinusoidal waves are no longer needed. To make sure there is no information loss in this transform, the signal reconstruction or the inverse wavelet transform has to be defined. It was shown that the inverse wavelet transform is in the form of

$$
\begin{aligned}
\mathrm{WT}^{-1}[W(a, b)] &= f(x) \\
&= \frac{1}{C_\Psi} \int_{-\infty}^{\infty} \int_0^{\infty} W(a, b) \psi_{a,b}(x) \frac{\mathrm{d}a\,\mathrm{d}b}{a^2},
\end{aligned}
\tag{3.8}
$$

with an admissible condition for $\psi(x)$:

$$
C_\Psi = \int_0^{\infty} \frac{|\Psi(u)|^2}{u}\,\mathrm{d}u < \infty,
\tag{3.9}
$$

where $\Psi(u)$ is the Fourier transform of $\psi(x)$. Any $\psi(x)$ that satisfies this admissibility condition can serve as a mother wavelet. The admissible condition implies that $|\Psi(0)|^2 = 0$ or, in the space domain, $\int_{-\infty}^{\infty} \psi(x)\,\mathrm{d}x = 0$. Thus a qualified wavelet should have a zero mean or should behave like a bandpass filter in the Fourier frequency domain. The constant C_Ψ also provides a signature about the orthogonality of a selected wavelet, which is discussed below.

Unlike the Gabor transform and Wigner distribution, a wavelet transform operates at a constant Q or quality factor, which refers to the central frequency divided by the bandwidth of a filter, that is,

$$
Q = \frac{\langle |u| \rangle}{\langle |\Delta u| \rangle}
\tag{3.10}
$$

is a constant in the wavelet transform. At low frequency, the transform offers a low spatial resolution. At high frequency, it has a high spatial resolution. Finally, wavelet transforms are invertible. For most purposes this combination of properties makes wavelet transforms superior to either Gabor transforms or Wigner distributions. Just as for Fourier transforms

$$
\int_{-\infty}^{\infty} |f(x)|^2\,\mathrm{d}x = \int_{-\infty}^{\infty} |F(\omega)|^2\,\mathrm{d}\omega
\tag{3.11}
$$

represents the total power P_t in the system, for wavelets we have

$$
P_t = \frac{1}{C_\Psi} \int_{-\infty}^{\infty} |W(a, b)|^2 \frac{\mathrm{d}a\,\mathrm{d}b}{a^2}.
\tag{3.12}
$$

3.3.2 Discrete wavelet transform and the frame

Similar to the continuous wavelet transform definition of Eq. (3.7), the discrete wavelet transform can be defined as [11]

$$
\begin{aligned}
W_{m,n} &= \int_{-\infty}^{\infty} f(x) \psi_{m,n}(x)\,\mathrm{d}x \\
&= \int_{-\infty}^{\infty} f(x) a_0^{-m/2} \psi\left(\frac{x - n b_0 a_0^m}{a_0^m}\right)\mathrm{d}x,
\end{aligned}
\tag{3.13}
$$

where $a_0 > 1$ and $b_0 > 0$. It is this discretization process that generates substantial interest within the mathematical community. As long as the admissible condition is satisfied, the continuous wavelet transform guarantees that there is no information loss, or 100% signal recovery is ensured. However, when the sampling is sparse, a numerically stable signal recovery from the discrete wavelet transform can be achieved for only some special choices of the wavelet $\psi(x)$. A tighter condition ought to be specified. Research has shown that the stability condition for a discrete wavelet transform can be described with a useful mathematical concept, a frame. A family of functions (wavelets in this case) $\psi_m(x)$ for all integers m and in a Hilbert (a measurable and square integrable) space can form a frame if there are two constants A and B with $0 < A < B < \infty$ so that, for all signals $f(x)$ in the Hilbert space, the relation

$$A\|f(x)\|^2 \le \sum_m |\langle f(x), h_m(x)\rangle|^2 \le B\|f(x)\|^2, \tag{3.14}$$

where

$$\|f(x)\| = \int_{-\infty}^{\infty} |f(x)|^2 \, dx$$

is satisfied. The two constants A and B are called the frame bounds. Once relation (3.14) is satisfied, signal reconstruction with the following inverse transform is possible:

$$f(x) = \sum_{m=-\infty}^{\infty} \sum_{n=-\infty}^{\infty} W_{m,n} a_0^{-m} \psi \left(\frac{x - nb_0 a_0^m}{a_0^m} \right). \tag{3.15}$$

3.3.3 Other important wavelet-related concepts

A family of wavelets is said to be orthogonal if the inner product satisfies the condition [12]

$$\langle \psi_{m,n}, \psi_{i,j} \rangle = \delta_{m,i}\delta_{n,j} \quad \text{for all } m, n, i, j. \tag{3.16}$$

They are semiorthogonal if the inner products satisfy

$$\langle \psi_{m,n}, \psi_{i,j} \rangle = \delta_{m,i} \quad \text{for all } m, n, i, j. \tag{3.17}$$

Otherwise this family of wavelets is nonorthogonal. A necessary condition of an orthogonal wavelet is that the corresponding frame bounds converge to a tight frame $A = B = 1$. Similarly, $C_\Psi = 1$ in Eq. (3.9) is another way of ensuring the wavelet orthogonality.

A wavelet function is said to be dyadic if $a_0 = 2$ and its frequency spectrum satisfies [12]

$$\sum_{m=-\infty}^{\infty} |\psi(2^j u)|^2 = 1. \tag{3.18}$$

The dyadic wavelets are often used for their convenience in terms of binary digital implementation. The corresponding mathematical properties of the binary dilations are also relatively less complicated.

After the binary-based dilation and translation are specified, a dual $\tilde{\psi}(t)$ of a wavelet $\psi(t)$ can be defined as [1]

$$\tilde{\psi}^*(u) = \frac{\psi(u)^*}{\sum_{j=-\infty}^{\infty} |\psi(2^{-j} u)|^2}. \tag{3.19}$$

$\psi(x)$ is said to be self-dual, and $\psi(x) = \tilde{\psi}(x)$. An orthogonal wavelet is self-dual.

One other important and often-mentioned term in the wavelet community is the scaling function $s(x)$, which has a close tie to orthogonal wavelet generation. A scaling function $s(x)$ that is used to characterize a multiresolution approximation completely is a function in the Hilbert space such that by denoting $s_{2^{-m}}(x) = 2^m s(2^m x)$, the set $\sqrt{2^{-m}} s_{2^{-m}}(x - 2^{-m} n)$ for all m, n forms an orthonormal basis. The scaling function, which does not have to be zero mean, can be used to form a family of orthonormal wavelets through a conjugate discrete filter whose impulse response $p(n)$ and the corresponding Fourier series $P(u)$ are defined as [13]

$$p(n) = \langle s_{2^{-1}}(x), s(x - n) \rangle, \qquad P(u) = \sum_{n=-\infty}^{\infty} p(n) \exp(-jnu). \qquad (3.20)$$

The function $P(u)$ satisfies the conditions

$$|P(0)| = 1, \qquad |P(u)|^2 + |P(u + \pi)|^2 = 1. \qquad (3.21)$$

Thus a wavelet constructed by the scaling function and the associated conjugate filter in the form of

$$\psi(u) = \exp(-ju/2) P^* \left(\frac{u + \pi}{2} \right) S\left(\frac{u}{2} \right) \qquad (3.22)$$

can form a family of orthonormal wavelets.

Orthogonal wavelets have other interesting mathematical properties, such as the fact that their symmetry is related to whether they have a compact support. Those interested readers can refer to the texts by Daubechies [11] and Chui [1].

3.4 Wavelets in general optics

Starting from this section and on, the discussion focus is shifted to the interrelations among the wavelet, optics, and optical pattern recognition. It is well known that optics is an old but venerable subject involving generation, propagation, processing, and detection of signals in a certain electromagnetic spectrum range. It appears that the wavelets might have some intimate relations with all four mentioned aspects of optics. To our knowledge, the relations of the wavelets to at least three of the four aspects of optics (except the light-generation process) have been identified. In this section some key findings of such relations are briefly summarized.

3.4.1 Wavelets in diffraction

Fourier optics became an independent area of study in optical science primarily because of advances in diffraction theory [14]. Speaking of diffraction, one could not discuss it without mentioning Huygens. As early as 1678, Huygens had intuitively interpreted light propagation as a wavelike motion of newly constructed envelopes generated by treating the previous wave front as the source of some secondary waves – wavelets. However, this diffraction wavelet appeared illusively independent of the mathematical wavelets developed some 300 years later. This independence has recently been questioned by Onural, who reexamined the well-known Huygens–Fresnel diffraction formula by looking for the missing link [15]. According to the diffraction theory, the diffraction pattern of a one-dimensional

(1-D) object $f(x)$ at a distance z away can be predicted by

$$f_z(u) = \frac{1}{j\lambda z} \exp\left(\frac{j2\pi z}{\lambda}\right) \int_{-\infty}^{\infty} f(x) \exp\left[\frac{j\pi}{\lambda z}(u-x)^2\right] dx. \qquad (3.23)$$

Onural rewrote Eq. (3.23) as

$$f_z(u) = f(x) * K_a \psi\left(\frac{x-b}{a}\right), \qquad (3.24)$$

where $*$ denotes convolution, $a = \sqrt{z\lambda/\pi}, b = u$, and $K_a = \exp\{j[2(a\pi/\lambda)^2 - \pi/2]\}/ (\pi a^2)$. Equation (3.24) indicates that the Fresnel diffraction, which is based on the Huygens wavelet theory, does exhibit similar features characterized by the modern mathematical wavelet. Originating from the same 2-D object, a set of different diffraction patterns produced over a distance range forms a set of wavelet transformed images. These diffracted images can also be used to reconstruct the original image with the inverse formula,

$$f(x) = 1/C \int_z \frac{1}{j\lambda z} \exp\left(\frac{-j2\pi z}{\lambda}\right) \int_{-\infty}^{\infty} f_z(u) \exp\left[\frac{-j\pi}{\lambda z}(x-u)^2\right] du \, dz, \qquad (3.25)$$

which gives rise to the underlining principle of holography or phase conjugation. The significance of this research is that it links, for the first time, the physical as well as the abstracted propagation problems to the wavelet transform and points to a new direction to interpret various physical phenomena and to use this interpretation as guidance for solving various inverse problems. However, some key questions, such as why this wavelet that appears to violate the admissible condition for a mathematical wavelet can still function, have yet to be answered in a satisfactory way.

3.4.2 Wavelets in early vision interpretation

One important reason for optics to be regarded as an independent area of study apart from general electromagnetics is that it deals with a spectrum that is visible to human eyes. A fundamental understanding of vision can provide invaluable insights into future designs of optical smart sensors and detectors. Generated, propagated, and processed optical signals will eventually be viewed directly or indirectly by human eyes. Over the past 25 years, significant research has been devoted to understanding the response of the human visual system. Various experiments have indicated that a retinal image is decomposed and processed in separate linear frequency filter channels. To study these filters, a quantity known as the contrast sensitivity function, or $\text{CSF}(u)$, is defined as [16]

$$\text{CSF}(u) = \frac{1}{C_t(u)}, \qquad (3.26)$$

where $C_t(u)$ is the threshold contrast at a given frequency u or

$$C_t(u) = \frac{L_{\max}(u) - L_{\min}(u)}{L_{\max}(u) + L_{\min}(u)}. \qquad (3.27)$$

$C_t(u)$ is the minimum contrast necessary to distinguish a sinusoidal grating from a uniform background. An adaptation technique is often used in the experiments with which a grating of a frequency u_0 at the threshold contrast is displayed to an observer for a long period of time so that his or her visual sensitivity for the same kind of stimuli decreases. It is interesting that when similar contrast stimuli at a substantially different frequency are shown to the observer, his or her sensitivity is not affected. The explanation is that when a visual system is adapted to a frequency u_0, its sensitivity for any stimuli whose frequency falls outside a band around u_0 is not affected. This suggests that at certain stage in the visual processing system, signals with different frequencies are processed independently. Campbell and Robson, who first performed a set of experiments in this area, suggested that the retinal image is decomposed through a set of independent bandpass linear filters, each of a narrow bandwidth [17]. Later, an experiment by Nachmais and Weber determined that the bandwidth relation between two consecutive filters is ~ 1 octave, i.e., they have the same width on a logarithmic scale [18]. Other experiments reported later showed that the bandwidths of different frequency channels range from 0.6 to 2.0 octaves and with a Gaussian envelope transfer function [19, 20]. Thus, to some extent, these research results indicated that the visual system behaves similarly to Gabor wavelet filtering when a set of self-similar bandpass filters of increasing scales is used.

3.4.3 Wavelets in binocular vision

In addition to visual processing at the cortex cells after retinal imaging, whose functionality exhibits some similarities to the features of wavelets, other parts of overall visual processing may also have something to do with wavelets. One such possibility may be identified as we question the role of the binocular imaging system used by a human being as well as by many types of animals. A binocular imaging system has been well understood for its role in coping with a stereo scene [21]. What has been lacking in the discussion is the fact that it could also function as sequential bandpass filtering, which is useful for performing wavelet processing. When looking at something, we feel comfortable in directly facing the object being viewed rather than looking at it at an angle. Also, when an interesting object cannot be immediately identified, we sometimes subconsciously move to closer or farther away to have a set of different views. This is especially true when we try to solve some graphical puzzles or deal with some optical illusion problems. An object is often moved back and forth a few times from our eyes before we can gather enough information to figure out what it is. This combination of the coarse-to-fine resolution procedure and the binocular imaging system can somehow be interpreted by use of wavelet concepts. This point of view can be best explained with the geometry of Fig. 3.1. An object is located on the symmetry axis (an

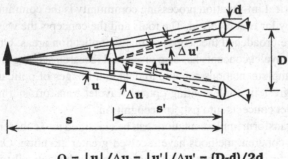

$$Q = |u|/\Delta u = |u'|/\Delta u' = (D-d)/2d$$

Fig. 3.1. Binocular imaging system acting as a constant-Q filter. d, pupil diameter; D, spacing between the two pupils; s and s', object distances; u and u', the corresponding center spatial frequencies of the object allowed to pass the pupils; Δu and $\Delta u'$, the corresponding passing bands of the object's spatial spectrum.

axis that is perpendicular to and bisects the line connecting the centers of the two lenses)
of a binocular imaging system. At any given position, only a cone of its diffracted beams
enters each pupil. Let us denote the aperture of each pupil by d, the distance between the
two pupils by D, the distance between the object point and the plane where both pupils are
situated by s, and the illumination wavelength by λ. Then the center spatial frequency u
and the bandwidth Δu that are allowed to pass the aperture of each pupil (a bandpass filter)
are [22]

$$|u| = \frac{\pi(D-d)}{s\lambda}, \qquad \Delta u = \frac{2\pi d}{s\lambda}. \tag{3.28}$$

Clearly, both the positive and the negative components of the spatial frequencies are filtered
symmetrically. When the viewing distance is shortened by a half, both the center frequency
and the bandwidth are increased by a factor of 2 so that the Q factor remains unchanged,
i.e., $Q = |u|/\Delta u = (D-d)/2d$. This constant-$Q$ filtering is clearly a signature of the
wavelet processing. Thus a binocular imaging system can perform a sequential wavelet
transform by scanning the object along the direction approaching the object. The flattop
bandpass filters under coherent illumination will change to the triangularly shaped filters (by
autocorrelation) under incoherent illumination [23]. It is thus possible to perform an incoherent wavelet transform by use of the frequency-domain synthesis. To incorporate wavelets
of other than triangular function, each lens can be masked by a density-modulated transparency whose autocorrelation function is the wavelet function in the frequency-domain
representation. The various images formed when the binocular imaging system is moved
toward the object represent the wavelet transform results of the object. Because the autocorrelation function is always symmetrical, this method is difficult to use to incorporate the
orthogonal wavelets with compact support, which inevitably results in asymmetric shapes.
Nevertheless, the described model may help us understand the complicated human vision
system. It may also be useful in helping to design efficient artificial-intelligence robot
eyes.

3.5 Optical wavelet transforms

Now let us further zoom our discussion focus to the subject at hand of this book: wavelets in
optical signal processing and pattern recognition. Compared with other research communities in optics, the optical information processing community is the community in which the
wavelet concept is by far best received. The tools and the concepts the wavelets brought to
this community have broadened the scopes of many application areas. On the other hand,
the introduction of wavelet concepts has also prompted research seeking for various optical
wavelet transform implementations suitable to data and images of optical formats. Let us
first review the progress in implementing optical wavelet transforms, a necessary step to
incorporating wavelet concepts into pattern recognition.

Optical wavelet transform implementations can be classified into coherent and incoherent
approaches, but the coherent methods have received greater attention. One key operation
of the wavelet transform is the correlation that can easily be optically implemented for
both 1-D data and 2-D images. The relatively difficult part is the coordinate dilation that
transfers the 1-D signal into 2-D, and the 2-D image into its four-dimensional (4-D) wavelet
transform formats. Various direct and inverse optical wavelet transform methods that have
been suggested are briefly reviewed.

3.5.1 Coherent optical wavelet transforms

In principle, coherent methods can be applied to optical correlations and convolutions involving any signals, real or complex. For the convenience of discussing an optical inverse wavelet transform below, let us assume a coordinate transformation $a' = 1/a$. Correspondingly, the daughter wavelets and the wavelet transform can then be rewritten as

$$\psi_{a',b}(x) = \sqrt{a'}\psi[a(x - b)], \tag{3.29}$$

$$W(a', b) = \sqrt{a'} \int_{-\infty}^{\infty} f(x)\psi^*[a'(x - b)]\,\mathrm{d}x. \tag{3.30}$$

The frequency-domain representations of the daughter wavelets and the wavelet transform are

$$\Psi_{a',b}(u) = \int_{-\infty}^{\infty} \psi_{a',b}(x)\exp(-j2\pi ux)\,\mathrm{d}x$$

$$= \frac{1}{\sqrt{a'}}\Psi\left(\frac{u}{a'}\right)\exp(-j2\pi bu), \tag{3.31}$$

$$W(a', b) = \int_{-\infty}^{\infty} F(u)\Psi_{a',b}^*(u)\exp(j2\pi ub)\,\mathrm{d}u. \tag{3.32}$$

Thus the wavelet transform may be implemented by an optical correlator with a bank of frequency-domain filters $\Psi_{a',b}^*(u)$.

Figure 3.2 shows a 2-D optical correlator capable of performing the wavelet transform of a 1-D signal [24–26]. The input signal $f(x)$ is displayed on an acousto-optic modulator, illuminated by a parallel laser beam. A cylindrical lens performs the 1-D Fourier transform of $f(x)$ along the x axis. The Fourier spectrum $F(u)$ is along the u axis in the Fourier plane. A bank of 1-D continuous wavelet transform filters $\Psi(u/a')$ is placed as a set of horizontal strips in the Fourier plane. The scale factor a' varies along the vertical axis v. After the Fourier plane a spherocylindrical lens ensemble performs the inverse Fourier transform along the horizontal direction and images the Fourier-plane strips onto the output plane along the vertical direction. In the output plane each strip $v = a'_i$ is the wavelet transform $\mathrm{WT}(a'_i, b)$

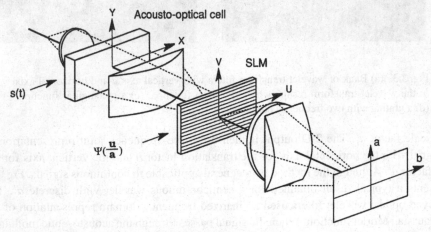

Fig. 3.2. 2-D optical correlator that uses a bank of strip filters for a wavelet transform of a 1-D signal.

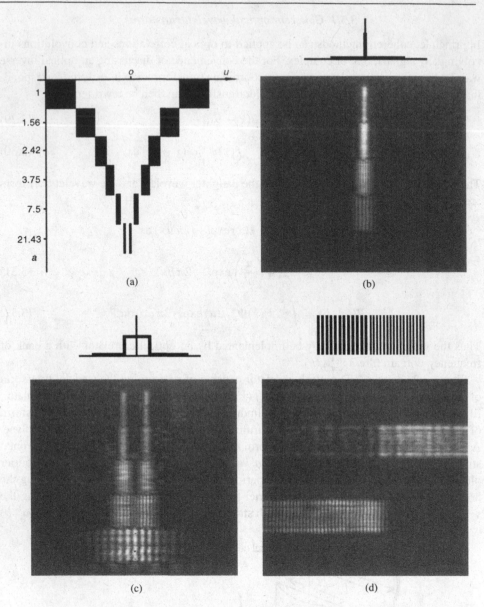

Fig. 3.3. (a) Bank of wavelet transform filters with vertical axis a and horizontal axis
b; the wavelet transform results with (b) a delta function, (c) a rectangular function,
(d) a grating with two frequency components.

for a scale factor a_i'. The 2-D output is then the space–frequency joint representation of
the signal with the horizontal axis for the translation factor b and the vertical axis for the
scale factor a'. Although the method is in general applicable to continuous signals, Fig. 3.3
represents a typical result obtained for a semicontinuous wavelet with discrete a'. The
employed optical wavelet filters used a binarized frequency-domain representation of the
cos-Gaussian Morlet function. When the signal passes through the acousto-optic modulator
continuously, the use of a pulse-laser source synchronized to the acousto-optic modulator
can avoid the loss of information at the boundaries of the signal frames [24].

The 2-D optical wavelet transform requires the application of a bank of 2-D wavelet transform filters of different scale factors. A variety of optical 2-D wavelet correlators has been proposed. The wavelet transform filter bank can be introduced in a time sequence into the real-time optical correlator [27–35]. It can also be inserted into a multichannel optical correlator, in which a holographic grating can be used to split the beam and generate an array of Fourier spectra of the input image, so that the array of multiscale wavelet transform filters can perform the wavelet transforms [27, 32]. Here the 2-D plane is partitioned into various regions to incorporate various daughter wavelets. The partition can be perform in either rectangular or polar coordinates. Another approach is the use of a spatially multiplexed holographic filter in a conventional optical correlator. The impulse response of the multiplexed filter is the wavelet functions of different scales distributed in the space [33].

The wavelet transform filters can be optically recorded or computer generated. For optical recording, Haar wavelets have been generated with a binary phase-only spatial light modulator [34]. Wavelet transform filters are then recorded with reference beams as holographic filters. Wavelet transform filters can be generated digitally and then encoded as the computer-generated holograms. Other real-valued wavelets can be generated by spatial light modulators that use amplitude-phase-coupled modulation [35]. Also, a complex amplitude-modulation spatial light modulator is now available [36]. The most frequently mentioned continuous wavelets, such as the Morlet and the Mexican-hat wavelets, are real valued and symmetrical. Their Fourier transforms are positively valued, and their wavelet transform filters can be fabricated simply as photographic masks. Because the correlation is shift invariant, the optical wavelet transform is independent of the origin of the coordinate system in which the wavelets are defined. We can shift the Haar wavelets in the space domain such that the Haar wavelets become odd and the Haar wavelet transform filters are purely imaginary and positive.

A square-law detector in the output of the optical correlator can record only the intensity of the wavelet transform. To recover the phase information, an approach that uses the Smartt interferometer was proposed [37]. The interference pattern between the output beam and a plane wave from a small pinhole at the center of the wavelet transform filter records the phase information of the wavelet transform. The processor has three correlator channels that accommodate the wavelet transform filters with the pinhole, without the pinhole, and a spatial filter with only the pinhole in each channel. The amplitude and the phase of the wavelet transform can be recovered from the three output intensities followed by a simple calculation.

An alternative coherent optical approach is the joint transform optical correlator [38, 39], which possesses some advantages, e.g., it can avoid complex wavelet encoding in the frequency domain. Figure 3.4 shows a typical joint wavelet transform correlator. An input image $f(x, y)$ and a wavelet function $\psi_{a'_x a'_y b_x b_y}(x, y)$ are displayed side by side on spatial light modulator SLM$_1$. The CCD camera, which is a square-law detector device located at the back focal plane of Fourier-transform lens L$_1$, records the interference pattern $|G(u, v)|^2$ of the Fourier transforms of both inputs:

$$
\begin{aligned}
|G(u, v)|^2 = {} & |F(u, v)|^2 + \left| \Psi_{a'_x a'_y b_x b_y}(u, v) \right|^2 \\
& + F(u, v) \Psi^*_{a'_x a'_y b_x b_y}(u, v) \exp[j2p(b_x u + b_y v)] \\
& + F^*(u, v) \Psi_{a'_x a'_y b_x b_y}(u, v) \exp[-j2p(b_x u + b_y v)].
\end{aligned}
\tag{3.33}
$$

A second SLM (SLM$_2$) is used to display Eq. (3.33) so that an inverse Fourier transform can be performed through the second Fourier-transform lens, L$_2$. The obtained

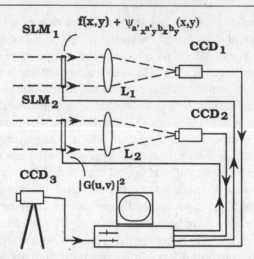

Fig. 3.4. Joint-transform-correlator-based coherent optical wavelet transform setup. SLM, spatial light modulator; L, Fourier-transform lens; CCD, charge-coupled device camera. A joint spectrum of the input image and the wavelets is formed at CCD_1. The intensity of the joint spectrum is displayed at SLM_2. CCD_2 records the wavelet transformed signals.

Fourier-transform result can be written as

$$g(x, y) = |f(x, y)|^2 + \left|\psi_{a'_x a'_y b_x b_y}(x, y)\right|^2$$
$$= f(x, y) * \psi^*\left[a'_x(x - b_x), a'_y(y - b_y)\right]$$
$$+ f^*(x, y) * \psi\left[a'_x(x + b_x), a'_y(y + b_y)\right]. \tag{3.34}$$

The first and the second terms are the dc component and the intensity of the wavelet function, respectively. However, the last two terms are the cross correlations that correspond to the wavelet transforms in the y direction shifted from the origin of the coordinate. The joint wavelet transform correlator is easy to implement and requires no precomputing and encoding. However, this method requires a greater space–bandwidth product than that of the on-axis optical correlator.

3.5.2 Coherent optical inverse wavelet transforms

The main reason to introduce the coordinate transformation of $a' = 1/a$ is for the convenience of performing an optical inverse wavelet transform. The original definition of the inverse wavelet transform of Eq. (3.10) has a coefficient of $1/a^2$ in the double integration. After this coordinate transformation, the inverse wavelet transform is written as [13]

$$f(t) = \frac{1}{C'_\psi} \int_0^\infty \int_{-\infty}^\infty W(a', b)\sqrt{a'}\,\psi[a'(x - b)]\,\mathrm{d}b\,\mathrm{d}a'. \tag{3.35}$$

Note that this coordinate change implies fine-to-coarse signal filtering rather than conventional coarse-to-fine signal filtering. The difference between the order of such filtering that may influence the design of digital filters in electronic implementations has no impact on optical implementation because the scale parameter, a or a', will be displayed in parallel in space. Equation (3.35) uses linear integration with respect to a' and thus can be optically

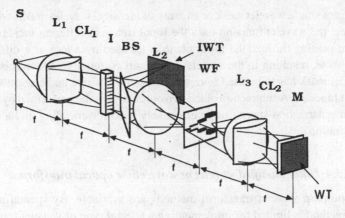

Fig. 3.5. Schematic of a coherent optical wavelet transform processor. BS, beam splitter; CL's, cylindrical lenses; f's, lenses' focal lengths; I, acousto-optic modulator for 1-D input signal; L's, spherical lenses; M, semitransparent mirror; S, coherent optical point source; WF, wavelet bandpass filter; WT, direct wavelet transform result plane; IWT, inverse wavelet transform result plane.

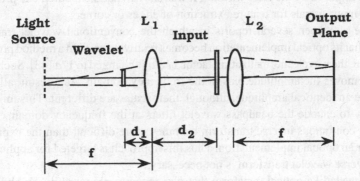

Fig. 3.6. Shadow-casting system for implementing an incoherent optical wavelet transform for 1-D or 2-D signals. d_1, the distance between the wavelet mask and collimating lens L_1; d_2, the distance between L_1 and the shadow-casting lens L_2; f, the corresponding focal length.

implemented by convolving the wavelets with the wavelet transformed signal. In terms of the physical systems for performing such transforms, either an extra stage of optical correlator can be added to the system or the original optical setup can be modified to suit this need [24]. In the latter case, a flat-end reflector or a photorefractive phase-matching mirror with complex signals can be used at the output that feeds the obtained wavelet transformed signal back to the setup (see Fig. 3.5). The mirror-reflected signal is then filtered by the wavelet filter bank in a reversed direction to implement a convolution instead of a correlation operation before the final Fourier transform and integration operation. This reconstructed signal can be separated by a beam splitter.

3.5.3 Incoherent optical wavelet transforms

Incoherent optical wavelet processors are shadow-casting systems, as shown in Fig. 3.6 [40, 41]. The wavelet transform correlation is performed in the space domain. A point light source array with a single wavelet mask or an array of identical wavelet masks with a lenslet array can project the wavelet function onto the input image at different incidence angles. The light beam passing through the input plane is focused by a lens at a different point on the output plane, resulting in the wavelet transform correlation. When this method is used, one optical mask has to be used to represent both a positive and a negative part of a bipolar wavelet function. A subtraction of the two outputs has to be performed by electronics unless the light-polarization states can be effectively used for encoding and a subsequent subtraction operation.

3.5.4 Other modified wavelet or waveletlike optical transforms

Other special-purpose wavelet transform methods are available. By special purpose we mean that the method is limited for implementing a special type of wavelet transform. An example in this class is the implementation of optical Haar wavelets by Fresnel zone plates [42]. This implementation takes advantage of the bipolar and square-aperture property of the Haar wavelet, which matches well with that of a zone plate. A conventional zone plate is slightly modified to introduce a sudden π-phase shift at its center to accommodate the requirements imposed by the Haar wavelet. It was shown that the method can process both 1-D and 2-D input signals for feature extraction of edges or corners.

There have also been several reports in which the conventional wavelet transform is modified so that its optical implementation becomes relatively easy. In a method proposed by Telfer and Szu, the wavelet normalization factor $1/\sqrt{a}$ is changed to $1/a$ [43]. Such a change effectively removes the amplitude bias in the frequency domain. As a result, all daughter wavelets have an identical amplitude, although their scales are different. This modification makes it easy to encode the bandpass wavelet filters at the frequency domain. However, the derivable continuous inverse transform becomes more difficult than the expression of Eq. (3.35) for an optical implementation. Thus this approach is targeted for applications for which the inverse wavelet transform is not necessary.

Another waveletlike optical transform that was recently proposed by Yu and Lu is intended to ease the optical implementation of an inverse transform while still preserving the frequency-domain amplitude bias-free property of the direct transform implementation [44]. It has been shown that with a semicontinuous wavelet definition in which the shift variable of the wavelet remains continuous but the scale variable becomes discrete, it is possible to define a waveletlike transform through the use of a scaling function and a scaling transform [45]. Perfect reconstruction is possible with the satisfaction of a biorthogonal condition. The resulting inverse transform is a summation of the results of the waveletlike transform with an additional low-pass-filtered input signal. This method, in many aspects, is identical to procedures used in subband codings and pyramidal image processing [46, 47]. The unique advantage of this approach is that, for applications in which semicontinuous wavelet coefficients are used, once a biorthogonal wavelet function is obtained, an optical implementation for both the direct and the inverse transform becomes straightforward.

3.5.5 Advantages and limitations of optical wavelet transforms

To be fair, just because something can be done optically does not imply that it should be done optically. Optical schemes have some inherent advantages when one is dealing with continuous signals. But it also suffers from many fundamental limitations. Some of these advantages and limitations are briefly outlined.

The orthonormal discrete wavelet series decomposition and reconstruction are currently computed digitally within the multiresolution analysis framework by recurring two discrete conjugate quadrature mirror filters. As mentioned above, the tree algorithm operating on a discrete wavelet transform requires only $O(N \log_2 L)$ operations. Fast wavelet-transform integrated circuits are also being developed [48].

In terms of computation applications, optics has had a difficult time competing with electronics in the digital domain. The strength of optics is its capability of dealing with continuous signals. An optical continuous wavelet transform could deliver something that cannot otherwise be accomplished by digital electronics. It has been shown that the orthonormal wavelet decomposition is based on only discrete translations and dilations. The discrete wavelet transform is not shift invariant. A slight shift can result in drastic changes in obtained wavelet coefficients. This could be a drawback in some applications that prefer or require shift invariance. More specifically, an orthogonal wavelet transform has no redundancy in its signal representation. The redundancy can help reduce sensitivity to noise in many applications. In addition, most digital wavelet-transform algorithms are limited to dyadic frequency sampling. Optical continuous wavelet transforms, on the other hand, can be shift invariant. The architecture of Fig. 3.5, for example, can be used to implement wavelet and inverse wavelet transforms with continuous a' and b. In principle, any wavelet function can be encoded by either a computer-generated hologram or a complex amplitude-modulation spatial light modulator. On the other hand, in general such continuous wavelet transforms cannot be implemented by digital electronics unless their mathematical close forms can be analytically derived for arbitrary input signals. The failure to guarantee such a continuous transform implementation is mainly due to the fact that there is no adequate sampling theorem to prevent missampling in such cases.

A second advantage associated with an optical wavelet transform is that it can vary the scale parameter rather arbitrarily, not just stick with factors of 2. Also, the digital tree algorithms can be extended to two or more dimensions only through the use of separable wavelets. This limitation inhibits many wavelet transform implementations from being able to extract directional information of an image. Quite opposite to this limitation, the optical 2-D wavelet transform allows the processing of nonseparable wavelets with or without directional selectivity [49].

As a direct result of being able to generate continuous wavelet transforms, optical methods may hold a competitive edge in terms of implementing the adaptive wavelet transform. The adaptive wavelet transform and the matching pursuits [50–52] tend to use the best basis functions to signal decomposition. The basis is selected from a library of dictionary waveforms to minimize energy or entropy. Most adaptive wavelets have fixed shapes with varying shift and dilation parameters. For example, the so-called supermother wavelet used in an adaptive wavelet transform is a linear combination of various wavelets with different scales and shifts. It has recently been proved through the use of the Schwartz inequality that if individual mother wavelets are admissible, the composed supermother wavelet will also be admissible [53]. These adaptive wavelets are continuous and redundant, with the shape adaptively chosen for particular applications. It is believed that, through allowing the wavelet transform to select its own linear transform kernels, the data-driven adaptivity can help enhance the signal-to-noise ratio and increase the robustness of the transform. An adaptive wavelet transform can be easily implemented by an optical wavelet-transform processor with a feedback loop [52, 54]. A recent study shows that when two Haar wavelets are combined at different resolutions that can be implemented optically, clutter in the image background can be extracted and removed [55].

There are various technical and fundamental limitations associated with optical wavelet transforms. Similar to many other optical transforms, input–output problems also exist with any optical wavelet implementation. Thus the advantage of on-the-fly transformation cannot be claimed with an optical implementation unless this input–output problem is solved.

It is not possible to implement a continuous optical wavelet transform for 2-D inputs. It is trivial to show that this is so because such a transform requires 4-D space for outputting its results. What optics can do in this case is to display 2-D continuous shift parameters with a set of 2-D discrete scaling parameters or any other mapping of 4-D to 2-D [29]. Unless other parallel operations, such as feature extraction or matched filtering, are performed in conjunction with the wavelet transform, fundamentally optics is not better than electronics if only the wavelet and the inverse wavelet transforms are implemented for 2-D signals.

Even for optical wavelet transforms of 1-D signals, various technological limits seriously threaten the practical applications of the proposed concepts. First, the present device and material technology cannot effectively support the accurate representations of wavelets. For example, the frequency-domain implementation of a wavelet transform requires a bank of wavelets whose scales vary linearly in continuous wavelet transform and exponentially in discrete wavelet transform. In order to generate multiresolution analysis with identical energy in each wavelet band, the magnitudes of the daughter wavelets, must vary accordingly. Especially for discrete wavelets, which are efficient for signal representation, such variations in magnitude and shape demand a high dynamic range and a high space–bandwidth product. With the current spatial light modulator technology, perhaps only 3 or 4 dyadic wavelet bands can be reasonably generated for wavelets whose frequency-domain representations are not ideal bandpass filters. The situation is similar to that which occurs when the wavelet transform is implemented by other approaches, such as a joint transform correlator or a holographic method, the only difference being that such demands on a high dynamic range and a high space–bandwidth product may appear in places other than the optical frequency domain. To overcome or bypass this problem, optical transforms with modified frequency-domain amplitude bias-free wavelets must be used. However, the relief comes at the expense of sacrificing the convenience of performing its inverse transform [43].

The last limitation discussed here is the sensitivity of alignment of such an optical wavelet processor. It is agreed that a digital wavelet implementation is not shift invariant. In some severe cases, a minor shift of input can lead to a set of drastically different wavelet parameters at the output. However, if an inverse wavelet transform is to be followed, the original signal can still be reconstructed. With this property in mind, optics is said to be advantageous for performing pattern recognition rather than image representation tasks, as such tasks require no inverse transforms. However, a space shift-invariant optical system may be frequency shift variant. This implies that errors must occur unless a bank of bandpass filters representing daughter wavelets is perfectly inserted into an optical system. For example, even for a continuous wavelet transform, shifts such as δa and δb imply only that

$$W(a' + \delta a', b + \delta b) = \int_{-\infty}^{\infty} f(x)\sqrt{a' + \delta a'}\,\psi^*[(a' + \delta a')(x - b - \delta b)]\,dx. \quad (3.36)$$

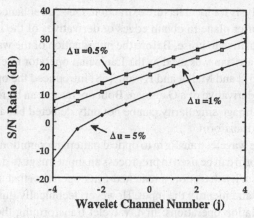

Fig. 3.7. Individual wavelet channel's signal-to-noise (S/N) performance versus frequency-domain wavelet misalignment Δu.

However, this shift invariance is not preserved in frequency-domain misalignment along the u direction [29], that is,

$$W(a', b) = \int_{-\infty}^{\infty} F(u)\Psi^*\left(\frac{u + \delta u}{a'}\right)\exp(jub)\,du$$

$$\neq W(a' + \delta u, b) \tag{3.37}$$

The effect of this misalignment is signal, wavelet shape, and wavelet band dependent. In general, the effect is much more severe for daughter wavelets with narrow passbands than with wide passbands. A plot of the signal-to-noise measure for such a commonly occurring misalignment for dyadic wavelets is shown in Fig. 3.7. Thus extra care must be exercised when the bandpass filters are placed in an optical wavelet processor.

3.6 Optical wavelet transforms for pattern recognition

As mentioned above, an optical continuous wavelet transform is inherently shift invariant and therefore is particularly useful for some image processing and pattern recognition applications. In general, signals are localized but noise is global. The wavelet transform offers a way of matching on a local basis, which requires a small number of expansion coefficients, each suffering less noise contamination. The strength of an optical wavelet transform seems to be feature detection, extraction, and classification rather than signal representation aspects. A perfect signal reconstruction is not, in general, required in pattern recognition applications. Some particular feature extraction or classification parts of pattern recognition applications for which optical wavelet transforms can fit suitably are briefly summarized.

3.6.1 Wavelet matched filters

Edges are often the most informative features of an image. A human being can recognize a complex object from a drawing that outlines only its edges. Most pattern recognition, image registration, motion estimation, and image fusion algorithms are based on edges, contours, and local features of the images. The wavelet transform is a multiscale local operation. It is superior to Fourier analysis for local feature extraction. One can enhance the high-frequency components in the Fourier plane to obtain edges or derivatives of the image. However, the high-pass filters are sensitive to noise. Before the introduction of the wavelet transform, the second derivative of Gaussian was used as the Laplacian operator to extract edges by zero crossing [56]. Canny [57] and Mallat and Hwang [58] introduced the optimal edge detector that is close to the first derivative of a Gaussian. Both operators can be formalized and refine by the wavelet theory. Image singularity can be not only detected but also characterized by the multiscale wavelet transform.

An application of the wavelet transform to optical pattern recognition was then suggested, in that a wavelet transform can be used to preprocess an input image to detect edges or shapes of interest. Following such a detection, a matched-filter operation that uses prestored data is performed to compare and identify the input. However, technically this procedure involves two consecutive correlation operations: first, wavelet transforming the input, and second, performing the matched filtering. The so-called wavelet matched filtering was demonstrated to simplify the procedure by combining the wavelet preprocessing and pattern recognition operations into a single step [59–61]. The filter is defined in the Fourier plane as the product of the matched spatial filter $T^*(u, v)$ and the square modulus of the wavelet transform filter $|\Psi(a_x u, a_y v)|^2$, i.e., $T^*(u, v)|\Psi(a_x u, a_y v)|^2$. For an input $f(x, y)$ with the Fourier transform $F(u, v)$ the correlation of the wavelet matched filter is

$$\int_{-\infty}^{\infty} \int_{-\infty}^{\infty} F(u, v)\Psi^*(a_x u, a_y v)T^*(u, v)\Psi(a_x u, a_y v) \exp[j(xu + yv)] \, du \, dv$$

$$= \int_{-\infty}^{\infty} \int_{-\infty}^{\infty} W_f(a_x, a_y, x'y')W_t(a_x, a_y, x' - x, y' - y) \, dx' \, dy', \tag{3.38}$$

where W_f denotes WT[f]. Thus Eq. (3.38) implies the correlation $W_f * W_t$ between two edge-enhanced images. The wavelet matched filter has a better discrimination capability than that of the classical matched spatial filter against clutter. When the bandpass wavelet transform filter is incorporated, the wavelet matched filter is more robust to noise compared with the phase-only matched filter, which is a high-pass filter. The performance of the wavelet matched filters depends on the wavelet scale. At the design stage, multiple wavelet scales are applied and the optimal scales (a_x, a_y) are determined in terms of the noise sensitivity and the discrimination capability of the filters. Only the optimal wavelet scales will be used in a 2-D optical implementation. The wavelet $|\Psi(a_x u, a_y v)|^2$ is simply an optical mask used as a bandpass filter in the Fourier plane. Figure 3.8 shows a programmable optical correlator with two cascaded liquid-crystal spatial light modulators that provide a full-range continuous complex amplitude modulation for the implementation of the wavelet matched filter. Figure 3.9 shows the optical correlation result with such a filter used to detect the letter E located in the upper right-hand corner of the input scene against other input images.

3.6.2 Adaptive composite wavelet matched filters

Both the wavelet transform and the matched filtering are linear operations and therefore can be replaced by a linear combination of the wavelets and a linear combination of the matched filters, respectively. When the images to be recognized and rejected are known a priori, a composite wavelet matched filter can be designed [62]. Let $\{t_n(x, y)\}$, where $n = 1, 2, \ldots, N$, be a set of training images with the desired output values c_n. The composite

$$F(u,v)\, |\Psi(a_x u,\, a_y v)|^2$$

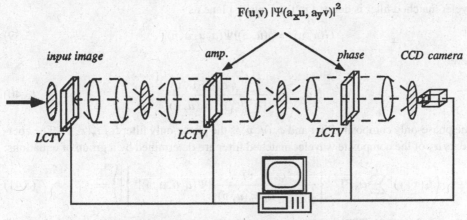

Fig. 3.8. Schematic of a programmable complex modulation optical correlator for implementing an optical wavelet matched filter. LCTV's, liquid-crystal televisions that serve as input devices for an input image, the amplitude, and phase portion of the filter function.

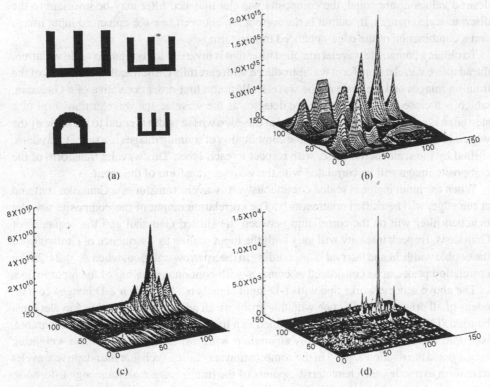

Fig. 3.9. (a) Input scene in which the letter E in the top right-hand corner is to be recognized. The correlation output with (b) a conventional matched filter, (c) a 1-D wavelet matched filter, (d) a 2-D wavelet matched filter. The 1-D and separable 2-D Mexican-hat wavelets are used.

wavelet matched filter is defined in the Fourier plane as

$$G(u, v) = \phi(u, v)|\Psi(a_x u, a_y v)|^2,\tag{3.39}$$

where

$$\phi(u, v) = \frac{\sum_{n=1}^{N} \alpha_n \phi_n(u, v)}{\left|\sum_{n=1}^{N} \alpha_n \phi_n(u, v)\right|}\tag{3.40}$$

is the phase-only composite filter and $\phi_n(u, v)$ is the phase-only filter for $t_n(x, y)$. The coefficients α_n of the composite wavelet matched filter are determined by a group of equations:

$$\left\langle t_k(x, y) \left| \sum_{n=1}^{N} \alpha_n \mathrm{FT}^{-1} \left\{ \frac{\phi_n(u, v)}{\left|\sum_{n=1}^{N} \alpha_n \phi_n(u, v)\right|} |\Psi(a_x u, a_y v)|^2 \right\} \right| \right\rangle = c_k.\tag{3.41}$$

For $k = 1, 2, \ldots, N$, $\mathrm{FT}^{-1}\{\}$ denotes the inverse Fourier transform and the angle brackets denote the inner product. Equation (3.41) can be solved for α_n by iteration with the delta learning rule [63].

The composite matched filter is useful for distortion-invariant pattern recognition. If, for example, the training images $t_n(x, y)$ are the scaled versions of the same image and the desired values c_n are equal, the composite wavelet matched filter may be invariant to the discrete scale changes. Its output is the correlation between an edge-enhanced input image and a combination of the edge-enhanced training images.

To obtain a composite wavelet matched filter that is invariant to continuous scale variation, the adaptive wavelet transform was introduced with carefully chosen scale increments of the training images and the scale of the wavelet. With the first-order derivative of a Gaussian, which is a close-to-optimal step-edge detector as the wavelet, the wavelet transform of a step edge is a Gaussian function centered at the edge whose width is equal to the scale of the wavelet a. The wavelet transform of the combination of training images is a set of Gaussians, shifted by the scale increment Δ with respect to each other. This wavelet transform of the composite image will be correlated with the wavelet transform of the input.

When the input image is scaled continuously, its wavelet transform, a Gaussian centered at the edge, will be shifted continuously. The correlation output of the composite wavelet matched filter will be the correlation between the shifted Gaussian and the sequence of Gaussians. Its peak intensity will vary with the input scaling as a sequence of Gaussians of the double width $2a$ and interval Δ. According to the Sparrow criterion, when $\Delta \leq 2\sqrt{2}a$ the correlation peak can be considered as constant with continuous scaling of the input image.

The above scheme works fine with 1-D input signals but fails when 2-D images contain edges of all orientations. To cope with this problem, an adaptive supermother wavelet must be used. The supermother wavelet is defined as a linear combination of two 2-D separable, isotropic wavelets. One is rotated by an angle θ with respect to the other. The weighting factor β and the rotation angle θ in the combination are chosen such that the adaptive wavelet transform extracting the characteristic points of the image has a maximum signal-to-noise ratio or a minimum variance. We thus define a criterion $R(\beta, \theta)$,

$$R(\beta, \theta) = \frac{1}{\left\{\iint [W_f(x, y)]^2 \, \mathrm{d}x \, \mathrm{d}y\right\}^{1/2}},\tag{3.42}$$

to be maximized by adjusting its parameters β and θ. In this case, the energy of the wavelet transform is concentrated on the points where the wavelet transform has maximum

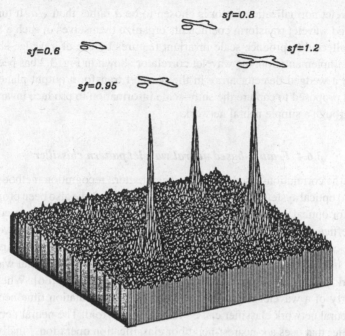

Fig. 3.10. Three-dimensional plot of the optical correlation output intensity of an adaptive composite wavelet matched filter, which is designed to recognize the images of aircraft of a continuous scale 0.8–1.2. The input scene is shown at the top with the image scale factors sf = 0.6, 0.8, 0.95, and 1.2.

intensities. The filter is invariant to continuous scale variation [54]. Figure 3.10 shows the optical correlation output of the adaptive composite wavelet matched filters. The filters were designed for detecting an image of aircraft with a scale factor sf range of 0.8–1.2. As can be observed, the image of a scale factor sf = 0.95 was not in the training set, but the filter still yielded a high correlation peak with this input. On the other hand, the image of a scale factor sf = 0.6 was not in the scale range and was not extracted by the filter.

3.6.3 Scale-invariant data classifications

It was recently shown that the invariant property of the analog wavelet transform is based on the linear superposition principle of the intrinsic scaling law of the space–scale joint representation. The wavelet transform of various scales of an identical signal is computed with this concept. For example, let

$$f_i''(x) = f_i(\gamma_i x); \qquad i = 1, 2, \ldots, \tag{3.43}$$

where the unknown scales γ's (suppressing class index i) are equivalent to the unknown frequency compaction or hopping of similar waveforms $f(x)''$. The wavelet coefficients can be computed as [64, 65]

$$W''(a, b) = \int_{-\infty}^{\infty} f''(x) \frac{1}{a} \psi^* \left(\frac{x - b}{a} \right) dx$$

$$= \int_{-\infty}^{\infty} f''(x'') \frac{1}{a''} \psi^* \left(\frac{x'' - b''}{a''} \right) dx''$$

$$= \mathrm{WT}(\gamma a, \gamma b), \tag{3.44}$$

where the wavelet normalization factor is chosen to be a rather than $\sqrt{a}$. It turns out that the scale-related wavelet transform coefficients organize themselves in such a fashion that they can be collected to produce scale-invariant features by use of a wedge-shaped filter. For an optical implementation, the wavelet correlator shown in Fig. 3.2 has been modified to incorporate a wedged detector array in the wavelet transform output plane [66]. This filter has been proposed to capture the shift–scale information to produce invariant pattern recognition through a simple neural network.

3.6.4 Feature-based neural wavelet pattern classifier

In addition to the correlation-based optical wavelet pattern recognition methods described, more advanced optical systems consisting of neural networks have also been proposed. Neural networks for optical pattern recognition have long been studied. It was recognized that unless some efficient deformation-invariant schemes are incorporated, the overall performance will be computation intensive. Wavelet transforms can be used as feature-extraction tools that can help reduce a vast amount of redundant information. Optical wavelet-based neural network pattern classification systems have been proposed [66]. When the localization property of a wavelet preprocessor is used, the computation time needed for the subsequent neural network classifier can be reduced manyfold. The neural network serves as a preclassifier that uses a k-nearest-neighbor classification operation. Finally, an optical amplitude-modulated phase-only filter can be used for final classification. This combination of optical wavelet preprocessing, neural network preclassification, and optical correlator postprocessing was demonstrated to possess advantages, such as fast convergence, parallel operation capability, and robust performance against blurred inputs.

3.7 Concluding remarks

Wavelets certainly presented their attractiveness to the signal processing community in the latter part of the 20th century. It is therefore well worth our time to ask what meaningful roles wavelets can play in optics and pattern recognition. This is indeed the purpose of this chapter. The intention here is definitely not to lead readers to believe that digital electronic wavelet processing will soon be outdated by its analog and continuous optical counterparts. Instead, what is needed is a fair assessment on the status of using wavelet concepts in various areas of optics research. Nevertheless we remain optimistic about the future of wavelet applications to optics and pattern recognition and believe that as this new concept gradually penetrates through the community, more and more effects of this locality-oriented transform will be created.

References

[1] C. K. Chui, *An Introduction to Wavelets* (Academic, San Diego, 1992).
[2] J. D. Gaskill, *Linear Systems, Fourier Transforms, and Optics* (Wiley, New York, 1978).
[3] E. Wigner, Phys. Rev. **40**, 749–753 (1932).
[4] D. Gabor, Proc. Inst. Electr. Eng. **93**, 429–457 (1946).
[5] M. J. Bastiaans, Proc. IEEE **68**, 538–545 (1980)
[6] A. Grossman and J. Morlet, SIAM Soc. Ind. Appl. Math. J. Math. **15**, 723–736 (1984).
[7] A. Haar, Math. Ann. **69**, 331–371 (1891).
[8] P. J. Burt and E. H. Adelson, IEEE Trans. Commun. **COM-31**, 532–540 (1983).

[9] S. W. Foo and L. F. Turner, Proc. Inst. Electr. Eng. Part G **129**, 61–67 (1982).

[10] I. Daubechies, Comm. Pure Appl. Math. **41**, 909–996 (1988).

[11] I. Daubechies, *Ten Lectures on Wavelets* (Society for Industrial and Applied Mathematics, Philadelphia, 1992).

[12] S. Mallat, IEEE Trans. Inf. Theory **37**, 1019–1033 (1991).

[13] S. Mallat, IEEE Trans. Acoust. Speech Signal Process. **37**, 2091–2110 (1989).

[14] M. Born and E. Wolf, *Principles of Optics,* 5th ed. (Pergamon, London, 1975), Chap. 8, pp. 370–387.

[15] L. Onural, Opt. Lett. **18**, 846–848 (1993).

[16] M. D. Levine, *Vision in Man and Machine* (McGraw-Hill, New York, 1985).

[17] F. Campbell and J. Robson, J. Physiol. **197**, 437–441 (1966).

[18] J. Nachmais and A. Weber, Vision Res. **15**, 217–223 (1975).

[19] B. Andrew and P. Pollen, J. Physiol. **287**, 167–176 (1979).

[20] M. Webster and R. De Valois, J. Opt. Soc. Am. A **2**, 1124–1132 (1985).

[21] D. Marr and T. Poggio, Proc. R. Soc. London Ser. B **204**, 301–328 (1979).

[22] Y. Li, in *International Conference on Optical Information Processing*, Y. V. Gulyaev and D. R. Pape, eds., Proc. SPIE **2051**, 130–141 (1993).

[23] J. W. Goodman, *Introduction to Fourier Optics* (McGraw-Hill, New York, 1968).

[24] Y. Zhang, Y. Li, E. G. Kanterakis, A. Katz, X. J. Lu, R. Tolimieri, and N. P. Caviris, Opt. Lett. **17**, 210–212 (1992).

[25] H. Szu, Y. Sheng, and J. Chen, Appl. Opt. **31**, 3267–3277 (1992).

[26] H. H. Szu, B. Telfer, and A. W. Lohmann, Opt. Eng. **31**, 1825–1829 (1992).

[27] E. Freysz, B. Pouligny, F. Argoul, and A. Arneodo, Phys. Rev. Lett. **64**, 745–748 (1990).

[28] M. O. Freeman, Opt. Photonics News **4**(8), 8–14 (1993).

[29] Y. Li and Y. Zhang, Opt. Eng. **31**, 1865–1892 (1992).

[30] Y. Sheng, D. Roberge, and H. Szu, Opt. Eng. **31**, 1840–1845 (1992).

[31] H. J. Caulfield and H. H. Szu, Opt. Eng. **31**, 1835–1839 (1992).

[32] Y. Sheng, T. Lu, D. Roberge, and H. J. Caulfield, Opt. Eng. **31**, 1859–1964 (1992).

[33] D. Mendlovic and N. Konforti, Appl. Opt. **32**, 6542–6546 (1993).

[34] T. Burns, K. H. Fielding, S. K. Rogers, S. D. Pinski, and D. W. Ruck, Opt. Eng. **31**, 1852–1858 (1992).

[35] T. H. Chao, A. Yacoubian, B. Lau, and W. J. Miceli, in *Optical Computing*, Vol. 10 of 1995 OSA Technical Digest Series (Optical Society of America, Washington, DC, 1995), paper OWB5-3.

[36] Y. Sheng, L. Gonsalves, and D. Roberge, in *Spatial Light Modulators and Applications*, Vol. 9 of 1995 OSA Technical Digest Series (Optical Society of America, Washington, DC, 1995), paper LTuC2-1.

[37] Y. Zhang, E. Kanterakis, A. Katz, and J. M. Wang, Appl. Opt. **33**, 5279–5286 (1994).

[38] X. J. Lu, A. Katz, E. G. Kanterakis, and N. P. Caviris, Opt. Lett. **17**, 1700–1702 (1992).

[39] W. Wang, G. Jin, Y. Yan, and M. Wu, Appl. Opt. **34**, 370–376 (1995).

[40] X. Yang, H. Szu, Y. Sheng, and H. J. Caulfield, Opt. Eng. **31**, 1846–1851 (1992).

[41] D. X. Wang, J. W. Tai, and Y. X. Zhang, Appl. Opt. **33**, 5271–5274 (1994).

[42] Z. He and S. Sato, Opt. Lett. **19**, 686–688 (1994).

[43] B. Telfer and H. H. Szu, Opt. Eng. **31**, 1830–1834 (1992).

[44] F. T. S. Yu and G. Lu, Appl. Opt. **33**, 5262–5270 (1994).

[45] S. G. Mallat, IEEE Trans. Pattern Anal. Mach. Intell. **11**, 674–693 (1989).

[46] M. J. Smith and T. P. Barnwell, IEEE Trans. Acoust. Speech Signal Process. **34**, 434–441 (1986).

[47] G. Eichmann, A. Kostrzewski, B. Ha, and Y. Li, Opt. Lett. **13**, 431–433 (1988).

[48] H. H. Szu, C. C. Hsu, P. A. Thaker, and M. E. Zaghloul, Opt. Eng. **33**, 2310–2325 (1994).

[49] T. H. Chao, E. Hegblom, and B. Lau, in *Optical Characterization Techniques for High-Performance Microelectronic Device Manufacturing*, J. P. Mathur, J. Lowell, and R. T. Chen, eds., Proc. SPIE **2237**, 268–274 (1994).

[50] R. R. Coifman and M. V. Wickerhauser, IEEE Trans. Inf. Theory **38**, 713–718 (1992).

[51] S. G. Mallat and Z. Zhang, IEEE Trans. Signal Process. **41**, 3397–3415 (1993).

[52] T. H. Chao, B. Lau, and W. J. Meceli, in *Wavelet Applications*, H. H. Szu, ed., Proc. SPIE **2242**, 382–388 (1994).

[53] H. H. Szu and B. A. Telfer, *Opt. Eng.* **33**, 2111–2124 (1994).

[54] D. Roberge and Y. Sheng, in *Wavelet Applications for Dual Use*, H. H. Szu, ed., Proc. SPIE **2491**, paper 42 (1995).

[55] D. P. Casasent, J. S. Smokelin, and A. Ye, Opt. Eng. **31**, 1893–1898 (1992).

[56] D. Marr and E. Hildreth, Proc. R. Soc. London Ser. B **7**, 187–217 (1980).

[57] J. Canny, "A computational approach to edge detection," IEEE Trans. Pattern Anal. Machine Intell. **PAMI-8**, 679–698 (1986).

[58] S. Mallat and W. L. Hwang, "Singularity detection and processing with wavelets," IEEE Trans. Inf. Theory **38**, 617–643 (1992).

[59] Y. Sheng, D. Roberge, H. Szu, and T. Lu, Opt. Lett. **18**, 299–301 (1993).

[60] D. Roberge and Y. Sheng, Appl. Opt. **33**, 5287–5293 (1994).

[61] M. Wen, S. Yin, P. Purwosumarto, and F. T. S. Yu, Opt. Commun. **99**, 325–330 (1993).

[62] D. Roberge and Y. Sheng, Opt. Eng. **33**, 2290–2295 (1994).

[63] D. A. Jared and D. J. Ennis, Appl. Opt. **28**, 233–239 (1989).

[64] H. H. Szu, X.-Y. Yang, B. A. Telfer, and Y. Sheng, Phys. Rev. E **48**, 1497–1501 (1993).

[65] H. H. Szu and H. J. Caulfield, Appl. Opt. **33**, 308–310 (1994).

[66] K. M. Iftekharuddin, T. D. Schechinger, K. Jemili, and M. A. Karim, Opt. Eng. **34**, 3193–3199 (1995).

4

Applications of the fractional Fourier transform to optical pattern recognition

David Mendlovic, Zeev Zalevsky, and Haldun M. Ozaktas

4.1 Preface

Pattern recognition plays a major role in machine vision, robotics, automation, and image understanding. In particular, because of its high parallel processing capabilities, optics might be an excellent tool for achieving advantageous pattern recognition systems. During the recent four decades, intensive activities have been performed in order to demonstrate attractive optical pattern recognition systems. Nevertheless, final results, although they were quite impressive, did not succeed in massively introducing optics to the practical optical pattern recognition field.

One explanation for that phenomenon is the fact that optical pattern recognition is not flexible enough, especially compared with digital signal processing systems. Moreover, optical pattern recognition systems are commonly restricted to perform only correlation-type operations.

A recent approach to introducing more flexibility into optical pattern recognition systems is the use of the fractional-Fourier-transform operation that leads to so-called fractional correlation systems.

As a generalization of the conventional correlation, fractional correlation is a much more flexible tool that handles various types of missions with high efficiency. In this chapter, after the introduction of the fractional correlation and its optical implementations, a detailed performance analysis of it is given with respect to standard performance criteria. Then the space-variance–invariance property of the fractional correlation operation is discussed, and a real-time control of the space-invariance property is proposed. This resulted in a more flexible tool that is a fractional correlator with multiple fractional orders (the localized fractional processor). A particular example of the multiple fractional order processor is the anamorphic fractional processor that has also been demonstrated experimentally. The final stage toward higher flexibility is the fractional joint transform correlator that provides a real-time ability in a wider sense.

4.2 Introduction

One of the first practical optical approaches for performing correlation is the well-known VanderLugt 4-f coherent configuration [1], its analogous incoherent system [2] or the joint transform correlator (JTC) [3, 4]. Because conventional correlation is a shift-invariant operation, shifts of the input pattern provide a shifted correlation output plane with no effect on the field distribution, and pixels located close to the center have exactly the same effect as pixels located in the outer area.

In several pattern recognition applications, the shift-invariance property within the entire input plane is not necessary and can even be disturbing. An example is the case in which the object is to be recognized only when its location is inside a certain area and rejected otherwise, e.g., a label that has been affixed on the incorrect place during manufacturing. Several approaches for obtaining such space-variance detection have been suggested. The first approach is the tandem component processor that trades the shift invariance with high efficiency and a high peak-to-correlation-energy (PCE) ratio [5]. A different approach is based on a coded phase processor that multiplexes many filters and yet keeps the space–bandwidth product of the ordinary single-filter correlator [6]. Recently a space-variant Fresnel-transform correlator [7], which is closely related to a lensless intensity correlator [8], was suggested.

A similar approach is the fractional correlator (FC), whose optical implementation is made with a setup similar to that of the VanderLugt correlator [9, 10]. In contrast to the solution of using an appropriate input pupil that is open in the desired location, the FC does not require any additional equipment for its optical implementation. The FC itself selects the area of interest within the input scene. An additional example of the necessity of the FC is the case in which the recognition should be based mainly on the central pixels and less on the outer pixels (for example, in systems whose spatial resolution is improved in the central pixels and thus the central region of pixels is more reliable for the recognition process). An important application for the FC might be the detection of localized objects by use of a single cell detector, eliminating the need for a CCD array detector.

Fractional correlation is a generalization of the conventional correlation operation, and it is based on the fractional Fourier transform (FRT) [11]. The FRT operation is useful for various spatial filtering and signal processing applications [12, 13], that is, it is defined through a transformation kernel, as illustrated in Ref. 13:

$$[\mathcal{F}^p u(x)](x) = \int_{\infty}^{-\infty} B_p(x, x') u(x') \, dx', \tag{4.1}$$

where $B_p(x, x')$ is the kernel of the transformation and p is the fractional order:

$$B_p(x, x') = \frac{\exp\left\{-i\left[\frac{\pi \operatorname{sgn}(\sin \phi)}{4} - \frac{\phi}{2}\right]\right\}}{|\sin \phi|^{\frac{1}{2}}} \exp\left(i\pi \frac{x^2 + x'^2}{\tan \phi} - 2i\pi \frac{xx'}{\sin \phi}\right), \qquad \phi = \frac{p\pi}{2}. \tag{4.2}$$

This kernel has two optical interpretations: one is propagation through a gradient-index medium [11] and the second is a rotation operation applied over the Wigner plane [14]. Both definitions were shown to be fully equivalent in Ref. 15.

Based on the conventional correlation definition (performing a Fourier transform of both objects, taking the complex conjugate of one of the objects, multiplying the results, and finally performing an inverse Fourier transform), a definition of the FC was formulated: Performing a fractional Fourier transform of both objects, taking the complex conjugate of one of the objects, multiplying the results, and finally performing an inverse conventional Fourier transform. This definition for the FC is not the only definition, but it is one that was found to have many useful properties and successful applications [9]. The FC operation allows us to control the amount of the shift-variance property of the correlation. This property is based on the shift variance of the FRT, and it is more significant for the fractional orders of

$p \approx 0 + 2N$ and less for $p \approx 1 + 2N$ (N is any integer). Note that because the conventional Fourier transform is a special case of the FRT (a FRT with a fractional order of 1 becomes the conventional Fourier transform), the FC is automatically at least as good as the ordinary correlator because it includes the other as a special case and adds another additional degree of freedom. In Section 4.2 the performances of the FC are evaluated according to three main criteria: the signal-to-noise ratio (SNR) [16], the PCE ratio [17], and the Horner efficiency [18].

In the shift-variant correlation approaches, in order to obtain a correlator with a certain space variance, the length of the optical setup or the focal lengths of the lenses should be designed properly (in order to obtain the appropriate fractional orders that correspond to a certain amount of space variance). The filter placed in the fractional Fourier domain contains the FRT of a certain fractional order of the reference object. In order to change the amount of space variance, the fractional orders of the correlator should be changed too. This is done by changing the distances and the focal lengths of the lenses. Because the filter contains a different FRT order of the reference object, it must be recalculated. In Section 4.3, this necessity is overcome by the introduction of a different setup for performing a partial space-variant correlation [19]. In this approach, in order to change the amount of the space variance, one should change only the longitudinal location of the filter. The distances between the optical elements and the focal lengths of the lenses should not be changed. Also, the filter should not be recalculated or reencoded. It remains the same filter, and only its longitudinal location is changed.

Another area for which the FC has a promising potential is noise removal. One can assume an object that appears with several islands of chirp noise, each one with different parameters. Thus, a FRT with a varied order (as a function of the localization at the input plane), is required. This procedure is coined here as the localized FRT (LFRT) [20]. The LFRT may be also used in problems of pattern recognition in which a different amount of shift invariance is needed in different areas of the input plane. A common case in which different space-variant processing is required is related to fingerprint recognition [21]. The fingerprint is a pattern whose space variance is changed with the spatial location. The central region of the pattern is more or less constant, whereas the outer region of the fingerprint is changed from instant to instant because one never presses a finger with equal force. Thus, in order to recognize or reconstruct those prints, one requires a processor whose spatial shift variance is changeable. The amount of invariance needed for spatial shifting must be small in the center, but in the outer regions of the print, increasing shift invariance is required from the processor. In this practical case, a LFRT processor should be helpful. The LFRT processor as well as its possible applications are discussed in Section 4.4.

Recently the FRT operation has been also extended to the anamorphic case [22, 23]. This provides the possibility of independently varying the space variance of the system in two perpendicular directions. In Section 4.5, a flexible system for obtaining an anamorphic-based fractional correlation is proposed. It is based on an adjustable-scale anamorphic FRT transformer, followed in cascade by a second transformer that, depending on the codification of the filter, can be amorphic or anamorphic [24]. The system is employed for space-variant processing, which implements multiple targets to be detected in different zones of the image.

As mentioned above, another approach for performing an optical spatial image process is the JTC. This scheme does not contain any filter and thus provides some advantages compared with the conventional VanderLugt 4-f configuration, in which the filter should be generated in a complex process, and must be aligned with high accuracy. When the shift invariance of the input plane is not important, a fractional JTC configuration may be used [25]. This configuration is analyzed in Section 4.6. The advantages of such a configuration in the optical pattern recognition field are similar to the advantages of the conventional JTC approach, but in addition the amount of the shift invariance may be controlled.

4.3 Fractional correlator performance analysis

In order to indicate the advantages of the FC configuration, its performance has been examined. This examination was done according to common performance criteria, which are described below.

4.3.1 Performance criteria

In this subsection three main performance criteria are presented: the SNR [16], the PCE ratio [17], and the Horner efficiency [18]. The advantage of using the above criteria over other common criteria such as the peak-to-maximum-sidelobe ratio [26] is that they may be easily analyzed mathematically.

The SNR measures the sensitivity of the autocorrelation peak to additive noise at the input plane. Mathematically,

$$\text{SNR} = \frac{|E[C_{u,u+n}(0)]|^2}{\text{VAR}[C_{u,u+n}(0)]}, \tag{4.3}$$

where u is the input signal, n is an additive noise, and $C_{u,v}$ is the correlation between u and v where in Eq. (4.3) $v = u + n$. E denotes the expected value operator, and VAR is the variance over the ensemble $n(x)$ [27].

The SNR measure considers only the average and the variance of the correlation peak but not the shape of the correlation output. The shape is estimated by the PCE criterion that measures the sharpness of the correlation peak. Its mathematical definition is

$$\text{PCE} = \frac{|C_{u,v}(0)|^2}{E_c}, \tag{4.4}$$

where E_c is the energy of the correlation signal defined by

$$E_c = \int_{-\infty}^{\infty} |C_{u,v}(x)|^2 \, dx. \tag{4.5}$$

From Eqs. (4.5) and (4.4), it is seen that for sharp correlation peaks, the value of the PCE is large.

The criterion that measures the light efficiency is the Horner efficiency criteria. It describes the ratio between the total light energy in the output plane and the total light energy in the input plane:

$$\eta = \frac{\int_{-\infty}^{\infty} |C_{u,v}(x)|^2 \, dx}{\int_{-\infty}^{\infty} |u(x)|^2 \, dx}. \tag{4.6}$$

However, Horner later recommended another definition for η [28]:

$$\eta = \frac{|C_{u,v}(0)|^2}{\int_{-\infty}^{\infty} |u(x)|^2 \, dx}. \tag{4.7}$$

This definition is much more relevant for correlators because the significant output of a correlator is the energy of the correlation peak itself $[|C_{u,v}(0)|^2]$ and not the energy of the entire correlation plane.

4.3.2 Performance optimization in conventional correlators

Various filtering configurations exist in the literature [1, 17, 29], and each optimizes another performance criterion.

The matched filter (MF) H_{MF} optimizes the SNR measure:

$$H_{MF}(v) = \alpha \frac{U^*(v)}{P_n(v)}, \tag{4.8}$$

where α is a constant that does not affect the SNR, P_n denotes the power spectral density of the noise $n(x)$, and U is the Fourier transform of the reference object. For the special case of white noise, $P_n = N_0$ (a constant), one obtains

$$H_{MF}(v) = \beta U^*(v), \tag{4.9}$$

where $\beta = (\alpha/N_0)$ (constant).

The filter type that is optimal according to the PCE measure is the inverse filter (IF), which is defined by

$$H_{IF}(v) = \frac{U^*(v)}{|U(v)|^2}. \tag{4.10}$$

Intuitively, when the IF is used and the input is the reference object, the correlation signal is a delta function (by design); thus the PCE ratio is the highest. However, the main disadvantage of the IF is that it may contain infinite values.

The phase-only filter (POF) is defined by

$$H_{POF}(v) = \frac{U^*(v)}{|U(v)|} \tag{4.11}$$

and is known to optimize the Horner efficiency measure [17]. The intuitive explanation is that the POF does not attenuate the energy passing through the filter, as $|H| = 1$.

4.3.3 Performance optimization in fractional correlators

In order to investigate which filtering configuration is optimal in the FC case, the fractional power filter (FPF) term [17] has to be introduced. Note that in this term there is no connection with the FRT although the name FPF could be confusing. The FPF definition is a generalization of the various filters (MF, IF, and POF) presented above:

$$H_{FPF}(v) = |U(v)|^s \exp[-i\theta(v)] = |U|^{s-1}U^*, \tag{4.12}$$

where $U(v) = |U(v)| \exp[i\theta(v)]$. The value of the s parameter can be any real number. The MF, POF, and IF are obtained with $s = +1, 0$, and -1, respectively.

For the FC case we used the schematic sketch of Fig. 1 for the configuration of the FC. We chose the parameters of $p_1 = p$, $p_2 = -p$, and $p_3 = -1$.

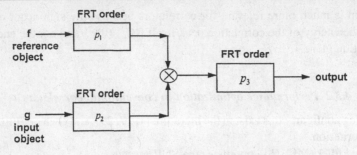

Fig. 4.1. Schematic sketch of the FRT filtering system.

Note that, for real input objects, choosing $p_2 = -p$ is equivalent to performing a FRT with the fractional order of p and applying the complex conjugate over the transformed function.

The FRT of the reference function $u(x)$ is denoted by $\mathcal{F}^p u(x) = u_p(x_p)$. The resultant FPF filter is given by

$$H(x_p) = |u_p(x_p)|^{s-1} u_p^*(x_p). \tag{4.13}$$

The SNR measure can be obtained with Eq. (4.3) as

$$\text{SNR} = \frac{\left| \int_{-\infty}^{\infty} H(x_p) u_p(x_p) \, dx_p \right|^2}{\int_{-\infty}^{\infty} P_n(x_p) |H(x_p)|^2 \, dx_p}. \tag{4.14}$$

While dealing with input white noise $n(x)$, we find that the obtained noise energy density $P_n(x_p)$ is also white (see Subsection 4.3.6). Thus assuming $P_n(x_p) = N_0$ and using Eq. (4.13) result in

$$\text{SNR} = \frac{\left[\int_{-\infty}^{\infty} |u_p(x_p)|^{s+1} \, dx_p \right]^2}{N_p \int_{-\infty}^{\infty} |u_p(x_p)|^{2s} \, dx_p}. \tag{4.15}$$

In order to maximize the SNR expression with respect to s, $(\text{dSNR})/\text{d}s$ is set to zero:

$$\int_{-\infty}^{\infty} |u_p(x_p)|^{2s} \, dx_p \frac{\text{d}}{\text{d}s} \left[\int_{-\infty}^{\infty} |u_p(x_p)|^{s+1} \, dx_p \right]^2$$

$$= \left[\int_{-\infty}^{\infty} |u_p(x_p)|^{s+1} \, dx_p \right]^2 \frac{\text{d}}{\text{d}s} \int_{-\infty}^{\infty} |u_p(x_p)|^{2s} \, dx_p. \tag{4.16}$$

After differentiation with respect to s, the following result is obtained:

$$\frac{\int_{-\infty}^{\infty} |u_p|^{2s} \, dx_p}{\int_{-\infty}^{\infty} |u_p|^{s+1} \, dx_p} = \frac{\int_{-\infty}^{\infty} |u_p|^{2s} \ln(|u_p|) \, dx_p}{\int_{-\infty}^{\infty} |u_p|^{s+1} \ln(|u_p|) \, dx_p}. \tag{4.17}$$

The condition that maximizes the SNR [Eq. (4.17)] is satisfied for $s = 1$, which is the MF $H = u_p^*$.

The light efficiency of the FPF can be obtained from Eqs. (4.6) and (4.13) as

$$\eta = \delta \frac{\int_{-\infty}^{\infty} |u_p(x_p)|^{2(1+s)} \, dx_p}{\int_{-\infty}^{\infty} |u_p(x_p)|^2 \, dx_p}, \tag{4.18}$$

where δ is a constant that is chosen so that the maximal magnitude of the FPF will be 1 (the filter is a passive device). After Eq. (4.18) is rewritten,

$$\eta = \delta \frac{\int_{-\infty}^{\infty} |H(x_p)u_p(x_p)|^2 \, dx_p}{\int_{-\infty}^{\infty} |u_p(x_p)|^2 \, dx_p} = \delta \frac{\int_{-\infty}^{\infty} |H(x_p)|^2 |u_p(x_p)|^2 \, dx_p}{\int_{-\infty}^{\infty} |u_p(x_p)|^2 \, dx_p}, \tag{4.19}$$

where H is the filter distribution. It is easy to see that if H is a passive filter, any filter with $|H| = 1$ (not just a POF) maximizes the light efficiency of Eq. (4.6). However, when the second definition of η [Eq. (4.7)] is used,

$$\eta = \delta \frac{\left| \int_{-\infty}^{\infty} |u_p(x_p)|^{(s+1)} \, dx_p \right|^2}{\int_{-\infty}^{\infty} |u_p(x_p)|^2 \, dx_p}, \tag{4.20}$$

rewriting Eq. (4.20) leads to

$$\eta = \delta \frac{\left| \int_{-\infty}^{\infty} H(x_p)u_p(x_p) \, dx_p \right|^2}{\int_{-\infty}^{\infty} |u_p(x_p)|^2 \, dx_p}$$

$$= \delta \frac{\left| \int_{-\infty}^{\infty} |H(x_p)| \exp[i\phi_H(x_p)] |u_p(x_p)| \exp[i\phi_u(x_p)] \, dx_p \right|^2}{\int_{-\infty}^{\infty} |u_p(x_p)|^2 \, dx_p}, \tag{4.21}$$

where ϕ_H is the phase of H and ϕ_u is the phase of u_p. Under the condition of the passivity of H, the numerator is maximized if and only if $\phi_H = -\phi_u$ and $|H| = 1$, i.e., H is a POF.

The PCE of the FPF is obtained when Eqs. (4.4) and (4.13) are combined:

$$\text{PCE} = \frac{\left[\int_{-\infty}^{\infty} |u_p(x_p)|^{s+1} \, dx_p \right]^2}{\int_{-\infty}^{\infty} |u_p(x_p)|^{2(s+1)} \, dx_p}. \tag{4.22}$$

After setting the derivative of Eq. (4.22) with respect to s to zero, one obtains

$$\frac{\int_{-\infty}^{\infty} |u_p|^{2(s+1)} \, dx_p}{\int_{-\infty}^{\infty} |u_p|^{s+1} \, dx_p} = \frac{\int_{-\infty}^{\infty} |u_p|^{2(s+1)} \ln(|u_p|) \, dx_p}{\int_{-\infty}^{\infty} |u_p|^{s+1} \ln(|u_p|) \, dx_p}. \tag{4.23}$$

The above condition is satisfied for $s = -1$ (the IF) for any $|u_p(x_p)| > 0$. This result agrees with the conventional correlator, for which the IF is designed to generate a delta function at the correlation plane.

However, one should note that in the fractional case the PCE measure is not significant. The PCE is a measure for the peak sharpness; however, the FC is shift variant and thus the shape of the peak is irrelevant. In most cases the FC configuration may work with a single detector. The FC cannot be used for localization of the input object; it can tell only if the input object exists in some specific region. The only exception is for p values that are close to 1. Then the shift variance is very small and the peak sharpness is relevant.

4.3.4 Signal-to-noise ratio comparison between a fractional correlator and a conventional correlator

Here a comparison of the SNR performances for the FC and the conventional correlator is performed. In the derivations below, white input noise is assumed. According to the

Cauchy–Schwartz inequality, one may conclude that the correlation's peak magnitude for a matched filter $H = u_p^*$ is

$$C_{u,v}^p(0) = C_{u,v}^1(0), \tag{4.24}$$

where u and v are any arbitrary functions. Note that the SNR is a function of the correlation peak only and not of the whole correlation signal. When Eqs. (4.3) and (4.24) are combined it is easy to show that the SNR remains unchanged when the fractional order of the FC is varied:

$$\mathrm{SNR}_p = \mathrm{SNR}_1. \tag{4.25}$$

Equation (4.25) states an important feature: The SNR performance of the FC is exactly the same as that of the conventional correlator. Thus the shift-variance property of the correlator is achieved without decreasing the SNR performances.

4.3.5 Fractional correlator performance with additive colored noise

Any colored noise $n_c(x)$ can be constructed from white noise $n_0(x)$ with a power spectrum of $Sn_0(v) = N_0$ convolving it with a linear time-invariant filter $h(x)$. The obtained power spectrum is [27]

$$Sn_c(v) = E[|N(v)H(v)|^2] = E[|N(v)|^2]|H(v)|^2 = N_0|H(v)|^2, \tag{4.26}$$

where $N(v) = \mathcal{F}^1 n_0(x)$ and $H(v) = \mathcal{F}^1 h(x)$. $E[\cdot]$ denotes averaging over the ensemble of the noise samples [27]. The relation between the statistical autocorrelation function and the power spectrum of the colored noise is an inverse Fourier transform:

$$Rn_c(x_1, x_2) = Rn_c(\tau = x_1 - x_2) = \mathcal{F}^{-1} Sn_c(v). \tag{4.27}$$

To find the noise spectrum at the filter plane, let $n_p(x_p)$ denote the FRT of one sample of $n_c(x)$. According to Eq. (4.2) the relation between n_p and n_c is

$$n_p(x) = \int_{-\infty}^{\infty} n_c(x_0) \exp\left(i\pi \frac{x^2 + x_0^2}{T} - 2\pi i \frac{x x_0}{S} \right) dx_0, \tag{4.28}$$

where

$$S = \sin \phi \qquad T = \tan \phi. \tag{4.29}$$

To find the power spectrum $Sn_p(v)$, one needs to write the autocorrelation function Rn_p:

$$Rn_p(x_1, x_2) = E\big[n_p(x_1) n_p^*(x_2)\big]. \tag{4.30}$$

By using Eqs. (4.28) and (4.29), one obtains

$$Rn_p(x_1, x_2) = \int_{-\infty}^{\infty} \int_{-\infty}^{\infty} E\big[n_c(x_1) n_c^*(x_2)\big] \exp\left(i\pi \frac{x_1^2 + x_0^2 - x_2^2 - x_0'^2}{T} \right.$$
$$\left. - 2\pi i \frac{x_1 x_0 - x_2 x_0'}{S} \right) dx_0 \, dx_0'. \tag{4.31}$$

Note that $E[n_c(x_1)n_c^*(x_2)] = Rn_c(x_1 - x_2)$, and thus

$$Rn_p(x_1, x_2) = \int_{-\infty}^{\infty} \int_{-\infty}^{\infty} Rn_c(x_1 - x_2)$$

$$\times \exp\left(i\pi \frac{x_1^2 + x_0^2 - x_2^2 - x_0'^2}{T} - 2\pi i \frac{x_1 x_0 - x_2 x_0'}{S}\right) dx_0 \, dx_0'. \quad (4.32)$$

From Eq. (4.32) it is clear that Rn_p cannot be written as $Rn_p(x_1 - x_2)$, and thus Rn_p is a nonstationary random process. Because the usual Fourier-transform relation between the autocorrelation and the power spectrum does not hold for a nonstationary process, the following relation should be used instead [27]:

$$Sn_p(\nu) = \int_{-\infty}^{\infty} \langle Rn_p(x_1, x_1 + \Delta x)\rangle \exp(-i2\pi\nu\Delta x) \, d\Delta x, \quad (4.33)$$

where $\langle \cdot \rangle$ denotes averaging over x_1. This relation was used for calculating the noise power distribution for various fractional orders by means of a computer simulation. The input colored-noise shape is a low-pass noise with a Gaussian spectral shape. Figure 4.2 illustrates the power distribution after the FRT (at the filter plane) for various fractional orders.

The result was obtained by the averaging of 100 random colored-noise vectors (because the vector is finite for $p = 1$, no fine Gaussian was obtained). The purpose of Fig. 4.2 is to illustrate that the increment of the fractional order p narrows the spectral width of the noise.

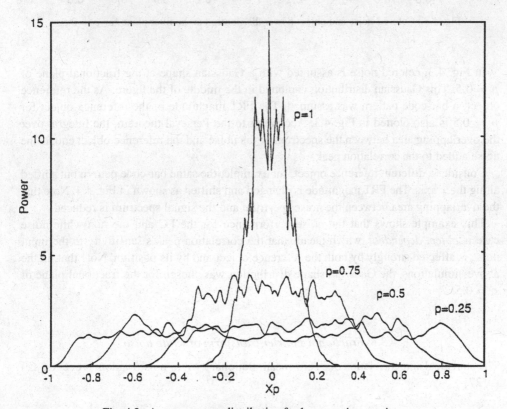

Fig. 4.2. Average power distribution for low-pass input noise.

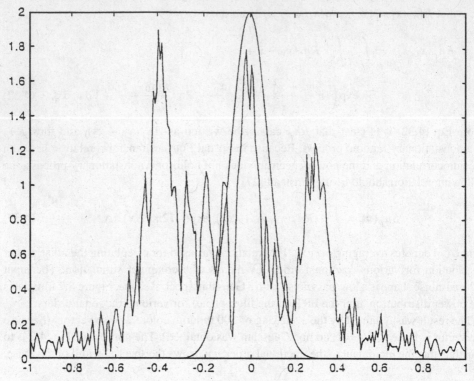

Fig. 4.3. Colored-noise and object power distribution at the filter plane for $p = 0.5$.

In Fig. 4.3, colored noise is assumed with a Gaussian shape at the fractional plane of $p = 0.5$. This Gaussian distribution is plotted in the middle of the figure. As the reference object, a bar-code pattern was assumed. The FRT magnitude of the reference object for $p = 0.5$ is also plotted in Fig. 4.3. According to the Perseval theorem, the integral over the overlapping area between the spectrum of the noise and the reference object equals the noise added to the correlation peak.

Consider a different reference object, for example, the same bar-code pattern but shifted along the x axis. The FRT magnitude is changed and shifted as shown in Fig. 4.4. Note that the overlapping area between the noise spectrum and the signal spectrum is reduced.

This example shows that the noise performance for the FC and the non-white-noise case is *object dependent*, which means that the correlation peak's sensitivity to the input noise is affected strongly by both the reference object and by its position. Note that, in the above simulations, the Gaussian shape distribution was chosen for the fractional plane of $p = 0.5$.

4.3.6 Fractional Fourier transform of white noise

The statistical autocorrelation of a zero-mean, stationary white-noise random process $n_0(x)$ is [27]

$$Rn_0(x_1, x_2) = N_0\delta(x_1 - x_2). \tag{4.34}$$

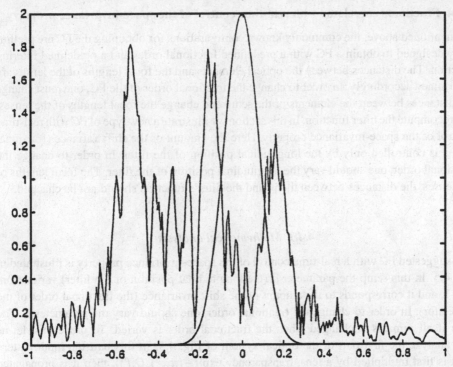

Fig. 4.4. Colored-noise and shifted object power distribution at the filter plane for $p = 0.5$.

The FRT of the white-noise input is denoted by $n_p(x)$. The statistical autocorrelation of $n_p(x)$ is

$$Rn_p(x_1, x_2) = E[n_p(x_1)n_p^*(x_2)]. \tag{4.35}$$

By using the FRT definition [Eqs. (4.2) and (4.29)], and changing the order of integration and expectation, one obtains

$$E[n_p(x_1)n_p^*(x_2)]$$

$$= \int_{-\infty}^{\infty} \int_{-\infty}^{\infty} Rn_0(x, x') \exp\left(i\pi \frac{x^2 + x_1^2 - x'^2 - x_2^2}{T} - 2i\pi \frac{xx_1 - x'x_2}{S}\right) dx\, dx'. \tag{4.36}$$

When Eq. (4.34) is used, Eq. (4.36) becomes

$$E[n_p(x_1)n_p^*(x_2)] = N_0 \int_{-\infty}^{\infty} \exp\left(i\pi \frac{x_1^2 - x_2^2}{T} - 2i\pi x \frac{x_1 - x_2}{S}\right) dx, \tag{4.37}$$

which yields

$$Rn_p(x_1, x_2) = N_0\delta(x_1 - x_2). \tag{4.38}$$

With a similar derivation it is easy to show that for a zero-mean input noise, the output is zero mean as well:

$$E[n_p(x)] = 0. \tag{4.39}$$

Thus from Eqs. (4.38) and (4.39) it is clear that $n_p(x)$ is zero mean, stationary, and white.

4.4 Fractional correlator with real-time control of the space-invariance property

As mentioned above, the commonly known configurations for obtaining the FC are preliminary designed to obtain a FC with a predefined fractional order and a predefined filtering function. The distances between the optical elements and the focal lengths of the lenses are determined accordingly. In order to change the fractional order of the FC, one must change the distances between the elements of the setup and change the focal lengths of the lenses, and recompute the filter function. In this section we illustrate a new type of FC with real-time control of the space-invariance property. Here the amount of the shift variance (fractional order) is controlled only by the longitudinal position of the filter. In order to change the fractional order, one should vary the longitudinal position of the filter. The focal lengths of the lenses, the distances between them, and the filter's function should not be changed.

4.4.1 Mathematical analysis

The suggested FC with a real-time control of the space-invariance property is illustrated in Fig. 4.5. In this setup the parameter a (the longitudinal position of the filter) varies from 0 to 1, and it corresponds to the amount of the shift invariance (the fractional order of the correlator). In order to change the fractional order, one should vary the parameter a. The filter itself remains unchanged when the fractional order is varied. To prove this, let us first analyze the first part of the optical system of Fig. 4.5. In this part the input pattern $g(x)$ is first multiplied by a lens, transparency $\exp[(-i\pi x^2)/(\lambda f)]$, then it is propagated (with the Fresnel diffraction formula) by a distance af in the free space (f is the focal length of the lens). Now the field distribution is multiplied by the filter $F(u)$. $F(u)$ is the conventional Fourier transform of the reference impulse response. The result is again propagated a distance of $(1-a)f$ in the free space with the Fresnel diffraction formula. Another multiplication with a lens, $\exp[(-i\pi v^2)/(\lambda f)]$, is done. The second part of the system is simply a Fourier transform over the output of the first part.

After the above-mentioned mathematical analysis, the field distribution at the output of the first part of the system (at the Fourier plane) is

$$D(v) = C \exp\left[\frac{i\pi a v^2}{\lambda(1-a)f}\right] \int_{-\infty}^{\infty} F(u)Z(u)$$

$$\times \exp\left[\frac{i\pi u^2}{\lambda a(1-a)f}\right] \exp\left[-2\pi i \frac{uv}{\lambda(1-a)f}\right] du, \qquad (4.40)$$

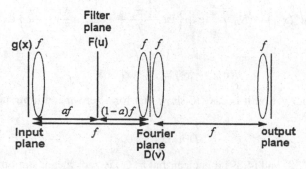

Fig. 4.5. Suggested optical filtering system.

where C is a constant and

$$Z(u) = \int_{-\infty}^{\infty} g(x) \exp\left[\frac{i\pi x^2\left(\frac{1}{a} - 1\right)}{\lambda f}\right] \exp\left[-2\pi i \frac{ux}{\lambda af}\right] dx. \tag{4.41}$$

Note that in Eq. (4.2) a normalized coordinate set x_r (that has no physical dimensions) was used:

$$x_r = \frac{x_{\text{ph}}}{\sqrt{\lambda f}}, \tag{4.42}$$

where x_r is the normalized coordinate and x_{ph} is the physical coordinate. Note that the normalized set is good for mathematical analysis. However, because here we are dealing with physical implementation, the physical set must be used. A comparison between Eqs. (4.2) (in physical coordinates) and (4.41) shows that $\exp\{[i\pi(1-a)u^2]/[\lambda a(1-2a-2a^2)f]\}Z(u)$ is a scaled FRT of $g(x)$:

$$\exp\left[\frac{i\pi(1-a)u^2}{\lambda a(1-2a-2a^2)f}\right] Z(u) = \mathcal{F}^{p_1} g(x) = G_{p_1}(su); \tag{4.43}$$

the fractional order is notated by p_1, and s is the scaling factor. The relation between the a parameter and the fractional order p_1 can be easily derived again from the comparison between Eqs. (4.2) and (4.41):

$$\tan\frac{p_1\pi}{2} = \frac{1}{\frac{1}{a} - 1} = \frac{a}{1-a}, \tag{4.44}$$

and the scaling factor s is found to be

$$s = \frac{\sin\frac{p_1\pi}{2}}{a} = \frac{1}{\sqrt{1 - 2a + 2a^2}}. \tag{4.45}$$

Note that the quadratic exponent $\exp\{[i\pi(1-a)u^2]/[\lambda a(1-2a-2a^2)f]\}$ that multiplies $Z(u)$ in Eq. (4.43) was determined to fit the quadratic exponent $\exp\{(i\pi x^2)/[\lambda f \tan(p_1(\pi/2))]\}$ appearing in Eq. (4.2) (after the scaling factor $x = su$ was substituted).

Thus Eq. (4.41) may be rewritten as

$$D(v) = C' \exp\left[\frac{i\pi a v^2}{\lambda(1-a)f}\right] \int_{-\infty}^{\infty} F(u)G_{p_1}(su) \exp\left(\frac{i\pi L u^2}{\lambda f}\right) \exp\left[-2\pi i \frac{uv}{\lambda(1-a)f}\right] du, \tag{4.46}$$

where C' is a constant and

$$L = \frac{1}{a(1-a)} - \frac{1-a}{(1-2a+2a^2)a} = \frac{a}{(1-a)(1-2a+2a^2)}. \tag{4.47}$$

Rewriting Eq. (4.47) leads to

$$D(v) = C'' \exp\left[\frac{i\pi a v^2}{\lambda(1-a)f}\right] \int_{-\infty}^{\infty} F(u)G_{p_1}(su)$$

$$\times \exp\left[\frac{i\pi(su)^2\frac{L}{s^2}}{\lambda f}\right] \exp\left[-2\pi i \frac{su\frac{v}{s}}{\lambda(1-a)f}\right] d(su), \tag{4.48}$$

where C'' is a constant. After changing the integration variables to $u' = su$, one obtains

$$D(v) = C'' \exp\left[\frac{i\pi a v^2}{\lambda(1 - a)f}\right] \int_{-\infty}^{\infty} F\left(\frac{u'}{s}\right) G_{p_1}(u')$$

$$\times \exp\left[\frac{i\pi u'^2 \frac{L}{s^2}}{\lambda f}\right] \exp\left[-2\pi i \frac{u'v}{\lambda v(1 - a)f}\right] du'. \qquad (4.49)$$

Another comparison between Eqs. (4.2) (in physical coordinates) and (4.48) shows that Eq. (4.48) is an exact FRT of the order of p_2 over the multiplication between the scaled filter $F(u/s)$ and $G_{p_1}(u)$ [the FRT of the order of p_1 over $g(x)$]:

$$\tan\frac{p_2\pi}{2} = \frac{s^2}{L} = \frac{1 - a}{a}; \qquad (4.50)$$

thus

$$\tan\frac{p_2\pi}{2} = \frac{1}{\tan\frac{p_1\pi}{2}}, \qquad (4.51)$$

which means that

$$p_2 = 1 - p_1. \qquad (4.52)$$

Thus a FRT of the order of p_2 is applied over the multiplication between $F(u/s)$ and $G_{p_1}(u)$ [the FRT of the order of p_1 of $g(x)$]. Because $F(u/s) = \mathcal{F}^1 f(sx)$, the setup performs a fractional correlation between the scaled reference object $f(sx)$ and the input object $g(x)$. Because the second part of the optical setup performs a conventional Fourier transform, the flow chart of the system can be illustrated, as was done in Fig. 4.1, choosing $p_1 = 1$, $p_2 = p$, and $p_3 = 2 - p$.

4.4.2 Interpretations

The most general schematic sketch for the fractional correlator is shown in Fig. 4.1. For this configuration a condition was derived for the fractional correlator that is optimal according to the PCE criteria [30]:

$$\frac{1}{T_1} + \frac{1}{T_2} + \frac{1}{T_3} = 0, \qquad (4.53)$$

where

$$T_k = \tan\phi_k, \qquad \phi_k = p_k(\pi/2), \qquad k = 1, 2, 3. \qquad (4.54a)$$

According to Subsection 4.4.1, which illustrated the equivalence between the suggested optical setup and the flow chart of Fig. 4.1 (for $p_1 = 1$, $p_2 = p$, and $p_3 = 2 - p$), one can easily see that the condition of Eq. (4.53) is fulfilled. To examine the applicability of a FC with parameters $p_1 = p$, $p_2 = 1$, and $p_3 = 2 - p$, let us first investigate the mathematical expression of the correlation plane. If the input object is indicated by $g(x)$ and the reference object by $f(x)$, the mathematical expression for the output of general FC (see Fig. 4.1) is, according to [30]:

$$|V(x)| = \left|\int g(x_1) f\left(\frac{-x_1}{\sin\phi} - \frac{x}{\sin\phi}\right) \exp\left[i\pi\left(\frac{x_1^2}{\tan\phi}\right)\right] dx_1\right|. \qquad (4.54b)$$

Assuming that the input is shifted and illuminated by a converging spherical wave, one obtains

$$g(x) = u(x - x_0) \exp\left[-i\pi \frac{(x - x_0)^2}{\tan\phi}\right],$$ (4.54c)

where u is the object we wish to recognize and x_0 is the relative lateral shift between the reference object encoded in the filter and the input pattern. Assuming also that the filter is $f(x) = u^*(-x \sin\phi)$ one easily finds that

$$|V(x)| = \left|\int u(x_1 - x_0)u^*(x_1 + x) \exp\left(\frac{2\pi i x_1 x_0}{\tan\phi}\right) dx_1\right|.$$ (4.54d)

Observing the last expression reveals that the ideal peak is obtained if $x_0 = 0$. The location of this peak is on $x = 0$ and it will not depend on the fractional order. However, if x_0 is not zero a disturbing phase factor appears in the exponential term that will attenuate the peak if the shift x_0 is too big.

Another ability of this configuration is as a noise-removing filter for an additive chirp-type noise. This time the input should be illuminated by a plane wave (and not a converging spherical wave) and $F(u)$ should be a notch filter (all pass filter that blocks only the axis center coordinate in order to remove the delta obtained in the filter plane due to the transformed chirp-type noise).

4.5 Localized fractional processor

The FC discussed so far was based on the FRT with a uniform fractional order applied over the reference and the input functions. In this section we introduce a localized FRT (LFRT), i.e., a FRT whose fractional order is space dependent, and thus the amount of shift variance–invariance is also spatially controlled. Such a transformation may be implemented optically in FC configurations, that achieving both shift-variant noise removal (for nonstationary noises whose statistical properties are varied with the spatial position) and image detection (for many detection applications that could be implemented with better efficiency by shift-variant or partially shift-variant processors).

4.5.1 Mathematical definitions

The basic assumption of the LFRT is that there is a full spatial separation between the areas that require different FRT orders. Thus we suggest, as a preliminary action, splitting up the different areas into different spatial locations. Assuming an input signal $f(x, y)$ and writing it as a separation of independent areas (as shown in Fig. 4.6) gives

$$f(x, y) = \sum_{i=1}^{i=N} f(x, y)A_i(x, y).$$ (4.55)

Each spatial area that is supposed to be transformed with a different fractional order is called a channel. The separation between the channels is achieved by the encoding of every area $A_i(x, y)$ with a different spatial frequency. Thus, after applying the FRT operation, one obtains

$$\text{OUT}(u, v) = \sum_{i=1}^{i=N}\left[G_i(u, v)\int_{-\infty}^{\infty} f(x, y)A_i(x, y)B_{p_i}(u, v; x, y)\, dx\, dy\right],$$ (4.56)

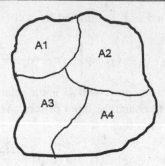

Fig. 4.6. Writing the signal as a separation of independent areas.

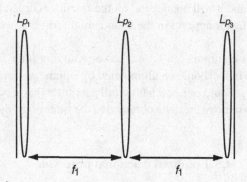

Fig. 4.7. Optical setup for obtaining a FRT with fixed distances and varying orders.

where $B_{p_i}(u, v; x, y)$ is the two-dimensional FRT kernel, p_i is the space-variant fractional order, and $G_i(u, v)$ is the independent output area in which the LFRT is obtained.

In the proposed processor we use the FRT optical configuration that is presented in Fig. 4.7. Here, the distances are fixed for all fractional orders and only the focal lengths of the lenses are varied for changing the FRT order.

The validity of this setup was proved in Ref. 31, in which Wigner terminology was used. In this reference it was also shown that the focal lengths of the three lenses are related to the desired fractional order as

$$L_{p_1} = \frac{f_1}{\tan\frac{\phi}{2} + 1}, \tag{4.57}$$

$$L_{p_2} = \frac{f_1}{\sin\phi + 2}, \tag{4.58}$$

$$L_{p_3} = L_{p_1}, \tag{4.59}$$

where f_1 is a scaling constant of the FRT that exists in the physical coordinates $\{x_r = [x_{ph}/(\lambda f_1)^{1/2}]\}$ and ϕ is related to the fractional order p as $\phi = [(p\pi)/2]$.

Figure 4.8 illustrates the optical setup used to obtain the LFRT. In this setup three filter plates are drawn. The input plate is divided into areas; in each area a different fractional order is to be applied. Every area is also multiplied by a quadratic-phase filter (a lens) of $\exp[(-i\pi x^2)/(\lambda L_{p_1})]$. L_{p_1} is calculated according to Eq. (4.57), and the ϕ value in the equation is determined according to the fractional order applied in that specific spatial area

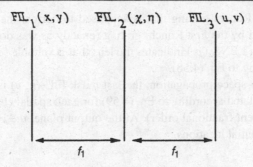

Fig. 4.8. Optical setup for obtaining the LFRT.

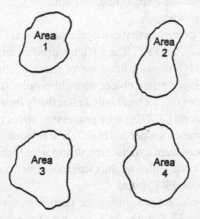

Fig. 4.9. Separation of the different areas $A_i(x, y)$ to the different channels.

$A_i(x, y)$. Hence the input mask plate is

$$\text{FIL}_1(x, y) = \sum_i A_i(x, y) R_{i_1}(x, y) \text{LENS}_1(p_i), \tag{4.60}$$

where $R_{i_1}(x, y)$ is the Ronchi grating and $\text{LENS}_1(p_i)$ is the lens suitable for the specific spatial area according to Eq. (4.57). Note that the mask is plotted as a single diffractive optical element. The Ronchi grating $R_{i_1}(x, y)$ is added to the mask in order to obtain spatial separation between the different areas that are about to be processed by the different FRT orders. A full spatial separation is needed so that information on the different spatial areas will not be mixed. The Ronchi grating separates the different areas into different channels by aiming the information of each area to a different spatial region. One controls the amount of separation between the different channels by changing the direction and the rate of the lines in the grating $R_{i_1}(x, y)$. Note that the spatial separation will be obtained after the free-space propagation.

When the light hits the second mask $\text{FIL}_2(\xi, \eta)$, the information of the different areas $A_i(x, y)$ is already spatially separated into different channels, as illustrated in Fig. 4.9.

Thus the function of the second mask is

$$\text{FIL}_2(\xi, \eta) = R_{i_2}(\xi, \eta) \text{LENS}_2(p_i) \tag{4.61}$$

per each spatial channel.

$R_{i_2}(\xi, \eta)$ is another Ronchi grating that is supposed to correct the undesirable linear-phase factor introduced by the first Ronchi grating (exactly as was done in the multifacet method [32]). The term LENS$_2(p_i)$ indicates the lens that is suitable for a specific channel and calculated according to Eq. (4.58).

After additional free-space propagation, the last mask FIL$_3(u, v)$ is placed. This mask consists of a lens calculated according to Eq. (4.59) for each spatial channel (each channel corresponds to a different fractional order). At the output plane, the FRT of each order is obtained in different spatial locations.

4.5.2 General applications

The applications of the suggested transformation could be significant in deterministic as well as in statistic signal processing.

A reconstruction of a FRT processor that can deal with several FRT orders simultaneously expands the application list of the FRT. The FRT by itself is an optimal tool for filtering chirp noise because in the FRT domain this type of noise becomes a delta function that can be easily filtered [12]. However, an object with chirp noise that, in different locations, appears with different densities cannot be handled effectively by a fixed FRT order. Another important application of a varying FRT's order processor relates to an application in which different amounts of shift variance are needed in different regions of the object. For example, an envelope for which one may want to detect the stamp with high shift variance (fractional order close to 0) and the zip code with low shift variance (fractional order close to 1) needs a processor with more than one FRT order.

Note that the alternative solution of splitting the input into several channels, while each channel handles one specific FRT order, is expensive if a space–bandwidth product or intensity limits exist for the system. Also, if after the processing one wants to compose back the image, it becomes much more complicated when one is dealing with multichannel systems. Furthermore, the option of using a cascade filter (which for some applications might be effective) usually does not have good results. Figure 4.10 illustrates such a case; it presents a Wigner chart. The signal is plotted as the filled shapes and the noise is represented by the unfilled shapes. A cascade application of a FRT will cause a rotation of the total chart

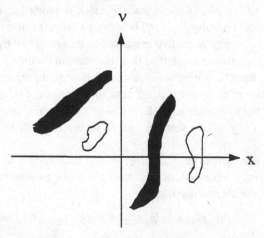

Fig. 4.10. Wigner chart of a case in which the LFRT is useful.

[14]. For any possible rotation, no separation between all parts of the signal and the noise is possible. However, if a LFRT is applied, the separation may be achieved. The optimal method for this case is to apply a different-order FRT for the regions $x < 0$ and $x > 0$. For the region $x < 0$ one should rotate the Wigner chart at an angle that is required for separation between the signal and the noise in this region. For $x > 0$ a different rotating angle should be applied in order to perform the separation. After the filtering of the noise, a perfect reconstruction of the signal will be obtained.

An additional possible application of the LFRT is to repeat the derivation done in Section 2 for the FC performance optimization. In those derivations it was shown that the optimal filter for the FC case is object dependent and that the fractional order p can optimize performance criteria as the SNR or the PCE ratio. Here we have another degree of freedom, which is the spatial dependence of $p: p(x, y)$. Thus, by the design of an optimal spatially dependent filter, the performances might be improved, as p is no longer a constant, but a function.

4.5.3 Application for pattern recognition

The most common case in which different spatial shift-variant processing is required relates to fingerprint recognition [21]. The fingerprint is a pattern whose spatial variance is changed with the spatial location. Its central region is more or less constant, whereas the outer one is changed from instant to instant as one never presses his or her finger with equal force. Thus, in order to recognize or reconstruct those prints, a processor whose spatial shift variance is changed is required. Because of the physical construction of the filter, a small shift invariance is needed in the center, but in the outer regions of the print, an increasing shift invariance is required. In this practical case a LFRT processor should be helpful, because an efficient recognition of the fingerprints can be applicable, for example, in safety lockers or in gaining admittance to permission-restricted entrances.

4.5.3.1 Computer simulations

In our computer simulations two fingerprints of 256×256 pixels were taken from the same finger of the same person. Both fingerprints are illustrated in Figs. 4.11 and 4.12.

One of the prints was taken as the reference object while the other was used as the testing input. As the first stage in our simulated experiment, a conventional Fourier analysis was done. The input was Fourier transformed, multiplied by the conjugate of the Fourier transform of the reference object, and finally inverse Fourier transformed. The central line profile of the obtained correlation peak is shown in Fig. 4.13.

Note that, in order to obtain fair performance comparisons in each simulation, the MF was normalized by the energy of the reference pattern. As one can note, the quality of the correlation peak is not very good. For the second stage, a FRT analysis was done. The input pattern was fractionally Fourier transformed by the order of p, multiplied by the conjugate of the fractionally Fourier-transformed (also by fractional order of p) reference object. The multiplication was eventually fractionally Fourier transformed by the order of -1 (an inverse Fourier transform). This is according to the definition of the optimal FC that is given in Ref. 9. This procedure was done for different fractional orders. The fractional order that was optimal according to the correlation peak form was found to be $p = 0.9$. Figure 4.14 illustrates the central line profile of the output correlation peak obtained for $p = 0.9$.

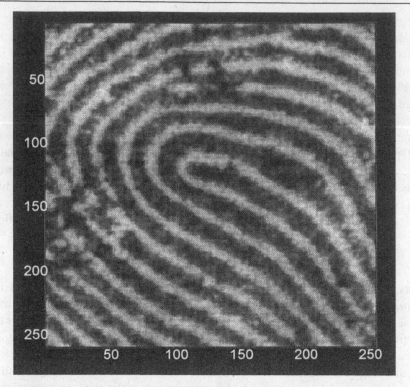

Fig. 4.11. Fingerprint used as the reference object.

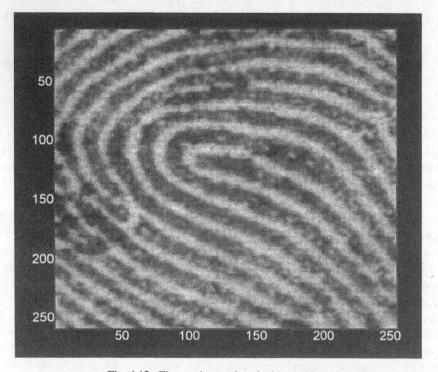

Fig. 4.12. Fingerprint used as the input object.

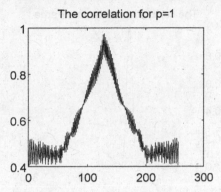

Fig. 4.13. Central line profile of the correlation peak when Fourier processing is used.

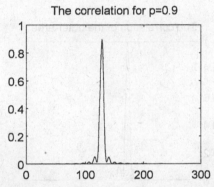

Fig. 4.14. Central line profile of the output correlation when FRT processing of $p = 0.9$ is used.

As one can note, the peak is much narrower than that of Fig. 4.13. Thus, for the fingerprint case, partial space-variant filtering obtains a much better performance than a regular Fourier analysis. For the third stage, a LFRT was applied. As mentioned above, the inner area of the print is less space variant than the outer area. Thus the fractional order for the inner area should be smaller than that for the outer area. The processing filter should be more space invariant at its outer region than at its inner ones. We divided the processing zone into two areas: the inner area, which was defined for the pixels $64 < x < 192$, $64 < y < 192$, and the outer area, which was all the rest. After some computer investigation we found that the optimal fractional order for the inner area is $p = 0.86$ and for the outer one is $p = 0.94$. Thus the constructed processor performed a FRT of $p = 0.86$ in the inner area and a FRT of $p = 0.94$ in outer one. The processor may be optically implemented as illustrated in Ref. 20 and in section 4.5.1.

The LFRT processor of Ref. 20 is multichannel. The two processing areas are separated by reference gratings. A suitable correlation peak is obtained in each processing channel. The processing over the inner area contains the performance of a FRT of $p = 0.86$ over the inner area, multiplying it by the inner area of the reference object that is fractionally Fourier transformed with $p = 0.86$ and obtaining a FRT of $p = -1$ over the result. The processing over the outer area is exactly the same procedure, but is performed with the fractional order of $p = 0.94$ instead of 0.86. The profile of the central line in the obtained correlation peak is shown in Fig. 4.15. Figure 4.16 illustrates the same but for the outer region.

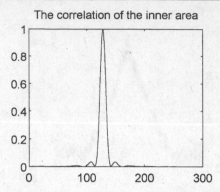

Fig. 4.15. Central line profile of the correlation peak obtained for the inner area of LFRT processing.

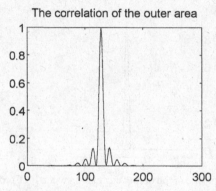

Fig. 4.16. Central line profile of the correlation peak obtained for the outer area of LFRT processing.

As one can note, both peaks are very sharp and narrow. The obtained peaks are ∼10% higher than the peaks of Fig. 4.14 or Fig. 4.13. Thus the performances of the LFRT are better compared with those of the conventional Fourier processor or the optimal uniform FRT processor. A complete LFRT would combine the two detected images to gain even more discrimination.

4.6 Anamorphic fractional Fourier transform for pattern recognition

4.6.1 Anamorphic fractional Fourier transform

The FRT concept can be extended to the anamorphic case [22, 23]. This modification permits the use of different fractional orders for two orthogonal axes of a two-dimensional image. The main potential of the anamorphic transform is its possibility of achieving different amounts of shift variance in the two different axes. A clear example for its advantage occurs when shift-invariant detection of objects along a row is needed. However, in the direction perpendicular to the row, there is no need to keep the shift-invariance property. Depending on the characteristics of the object to be detected, the decrease of shift invariance may result in a gain in the performance of the correlator, mainly in peak sharpness and the SNR [9, 33]. In addition, in special cases, the shift variance can help in locating the object. The detection peak will be produced only when the input object lies along the line where the shift invariance is kept.

The anamorphic FRT is defined as [34]

$$F^{p_x,p_y}(x', y') = \int_{-\infty}^{\infty} \int_{-\infty}^{\infty} f(x, y) \exp\left[i\pi\left(\frac{x^2 + x'^2}{T_x} + \frac{y^2 + y'^2}{T_y}\right)\right.$$

$$\left. - 2i\pi\left(\frac{xx'}{S_x} + \frac{yy'}{S_y}\right)\right] dx\, dy, \tag{4.62}$$

where

$$T_x = \lambda f_{1x} \tan\phi_x, \qquad S_x = \lambda f_{1x} \sin\phi_x, \qquad \phi_x = p_x(\pi/2),$$

$$T_y = \lambda f_{1y} \tan\phi_y \qquad S_y = \lambda f_{1y} \sin\phi_y, \qquad \phi_y = p_y(\pi/2) \tag{4.63}$$

where the subscripts x and y indicate the horizontal and the vertical directions of the system, respectively. Because here we are dealing with physical implementation as well, we use the physical coordinate set. The additivity property of the FRT allows us to implement the anamorphic FRT by cascading an amorphic setup, which performs the FRT with the order that is the lower between p_x and p_y and an anamorphic system that obtains a FRT in one axis and imaging in the other one [23]. Other setups providing a higher compactness or flexibility have also been proposed [22, 23].

An alternative approach for obtaining the anamorphic FRT is based on the setup described in Ref. 35. Instead of preparing a full setup containing two lenses and free space propagation, one illuminates the object with a converging beam. The convergence-phase factor, multiplying the object, may be changed by the displacement of the object along the optical axis. The matching between the distance object–filter and the convergence beam phase may produce any desired order and scaling factor for the FRT. This fact makes the last approach more convenient for the experimenter, as the exact sizes of the transparencies are often not precisely determined. Note that the FRT obtained in this method is inexact. It does not have the final quadratic-phase factor, and thus the correlation plane will be displaced along the optical axis. In the case of an anamorphic FRT the convergence of the beam at the output of the filter plane will be different in the two main axes. In order to focus the correlation, an anamorphic system will be needed.

The setup for performing the anamorphic fractional correlation is shown in Fig. 4.17.

The adjustment of distances a_x and a_y determines the proper order for the FRT. The scale factor of the FRT is variable, as a parameter independent of the order. For practical reasons one should try to reduce the number of cylindrical lenses in the constructed optical setup. In the chosen configuration, only three cylindrical lenses and one spherical lens are used.

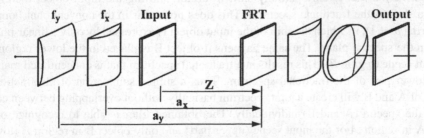

Fig. 4.17. Experimental setup for obtaining the anamorphic fractional correlation.

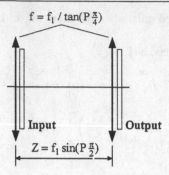

Fig. 4.18. Optical setup for performing a FRT operation.

The price to be paid for this simplification is the aspect ratio of the FRT (quotient between the x and that y scale ratios), which cannot be adjusted. Inserting an additional anamorphic image forming system (which provides different magnifications in both axes) stretches the image of the FRT plane and corrects the problem. The output of this imaging system is then taken as the input for the inverse-transforming subsystem. This additional complexity can be avoided in most practical cases. As shown in Figs. 4.17 and 4.18, because the Z distance is equal for both axes, one may write the following:

$$Z = Z_x = f_{1x} \sin\frac{p_x\pi}{2} = Z_y = f_{1y} \sin\frac{p_y\pi}{2}. \tag{4.64}$$

Thus the aspect ratio (AR) between the two axes is

$$AR = \frac{f_{1x}}{f_{1y}} = \frac{\sin\dfrac{p_y\pi}{2}}{\sin\dfrac{p_x\pi}{2}}. \tag{4.65}$$

4.6.2 Multiple fractional-Fourier-transform filters

One of the motivations of the anamorphic FRT approach is to design a composite filter that is able to recognize object A or a certain deformation of this object in region $\mathcal{A}$ and object B or a certain deformation of it in region $\mathcal{B}$. In order to obtain these capabilities, an anamorphic FC is used.

To implement the demands, we place object A in region $\mathcal{A}$, which is assumed to be in the upper part of the input scene. If a FRT of 0.5, for example, is performed over this input over the y axis, the obtained fractional spectrum will be concentrated mainly in the upper region of the output plane (region $\mathcal{A}$). Because the fractional order of 0.5 determines a transformation that is only partially shift invariant, shifting the input object therefore causes a shift of the fractional spectrum. This does not occur in the conventional Fourier transform ($p = 1$), in which the shift of the input object is expressed by only a linear-phase factor in the spectral plane. The same happens if object B is placed in the lower region of the input scene (region $\mathcal{B}$). This results in a fractional spectrum that is concentrated mainly in the lower part of the fractional spectrum. Thus a simple summation of the fractional spectra of A and B will create a joint spectrum basically without overlapping between each one of the spectra (A and B) individually. The obtained filter is able to recognize only object A in region $\mathcal{A}$ of the input scene (upper part) and only object B in region $\mathcal{B}$ (lower part).

Assuming that along axis x one wishes to obtain full shift invariance, i.e., when object A is located in region $\mathcal{A}$, object A can move along the x axis and yet be recognized, and the same property should be ensured for object B. Thus a FRT with a fractional order of 1 (conventional Fourier transform) over the x-direction should be performed.

We now summarize the procedure for preparing the filter for a certain example. We denote by p_x the fractional order performed in the x axis and by p_y the fractional order in the y axis. Object A is shifted to the center of region $\mathcal{A}$ and an anamorphic FRT is calculated with the fractional orders of $p_x = 1$ and $p_y = 0.5$. The complex conjugate of this distribution is taken. Now this filter itself will detect the presence of object A at the center of region $\mathcal{A}$ by producing a correlation peak located in the center of the output plane. Because it is more convenient if the peak is produced over the object and not in the center of the output, the filter is multiplied by a linear-phase factor that deviates the correlation peak to the object's location. This linear-phase factor is calculated according to the distance between the center of the input image and the position where the target has been displaced in the first step of the filter preparation (the center of region $\mathcal{A}$). The same process is repeated for object B, with the corresponding displacement to the center of region $\mathcal{B}$. The two distributions obtained in this way are added to construct the final filter. The resulting filter is placed in the appropriate fractional Fourier domain in the anamorphic correlator.

If, instead of detecting object A in region $\mathcal{A}$ and object B in region $\mathcal{B}$, one prefers to detect a certain deformation of object A in region $\mathcal{A}$ and a different deformation of object B in region $\mathcal{B}$, the same approach can be applied. Let us assume that a one-dimensional (1-D) x-direction scaling-invariance property of object A is required to be detected in region $\mathcal{A}$ and a 1-D y-direction scaling invariance of object B is required to be detected in region $\mathcal{B}$. One way of obtaining a 1-D scale invariance is to use the logarithmic harmonic decomposition [36]. Thus here object A was decomposed to the proper logarithmic harmonic (x scaling invariant), and the harmonic was shifted to the center of region $\mathcal{A}$. In the same manner, object B was decomposed to the proper logarithmic harmonic (y scaling invariant), and the harmonic was shifted to the center of region $\mathcal{B}$. Then a FRT with the fractional orders of $p_x = 1$, $p_y = 0.5$ is performed over the sum, and a complex conjugate of this function is obtained. The resultant filter is placed in the fractional domain of the anamorphic FC. Note that, when this approach is used, any invariant property can be detected in region $\mathcal{A}$ or $\mathcal{B}$ if the proper harmonic decompositions are used in each region.

4.6.3 Optical implementation

In the FRT processor the scale of a filter's distribution is crucial for obtaining the desired results. In a case in which the input object is recorded on photographic film and the filter is generated by a computer, the scales can be matched during the recording process. However, this requires high accuracy. Moreover, if instead of a computer-generated hologram, spatial light modulators are used for either the input transparency or the filter, the scale cannot be controlled. Special difficulties appear when the sizes of input and output do not match. A way to overcome this problem is to use an adjustable anamorphic FRT correlator [35]. It is based on the setup depicted in Fig. 4.18, in which the order can be varied by a change in the focal length of the lenses and the distance between input and output. Illuminating the input transparency with a nonparallel beam and displacing the object along the optical axis vary the convergence of the beam that illuminates the input. This is fully equivalent to changing the focal length of the first lens in the setup of Fig. 4.18. The separation between the input and the output must be varied accordingly so that the condition for being a FRT of the desired order is fulfilled. The second lens is removed from the setup. The result is a fractional transform with a variable scale, but with an additional quadratic-phase factor in the output plane. Because in a FC the anamorphic transformer is only the first stage of the complete correlator, the quadratic-phase factor will change only the position of the output correlation plane.

4.6.4 Results

4.6.4.1 Computer simulations

Several computer simulations were performed to demonstrate the performances of the proposed filter. For those simulations the input image illustrated in Fig. 4.19 was used.

The constructed filter is supposed to recognize an F-18 airplane in the upper part of the input and a Tornado airplane in the lower part. As mentioned above, for the construction of the filter, the F-18 airplane was shifted to the center of the upper part of the image and the Tornado was shifted to the center of the lower part. The above-outlined procedure was followed to obtain a filter with the fractional orders of $p_x = 1$, $p_y = 0.5$. The chosen fractional orders caused the filter to be fully shift invariant in the x direction and also to allow a small amount of shift invariance in the y axis. In the input scene illustrated in Fig. 4.19, the y positions of the centers of the F-18 airplanes in the upper part and the y positions of the Tornado airplanes' centers in the lower part were separated by few pixels. Figure 4.20 is the obtained output plane. One can note good correlation peaks that indicate the existence of the F-18 in the upper part and the Tornado in the lower part.

The possibility of recognizing different deformation properties in different parts of the image is demonstrated in Fig. 4.21.

In this case the purpose was to obtain an x scale-invariant recognition of F-18 airplanes in the upper part of the scene and a y scale-invariant recognition of the same target in the lower part. For the construction of the filter, an x scale-invariant logarithmic harmonic of the F-18 was calculated and shifted to the center of the upper region of the image. Then a

Fig. 4.19. Input image used for computer simulations and optical experiments.

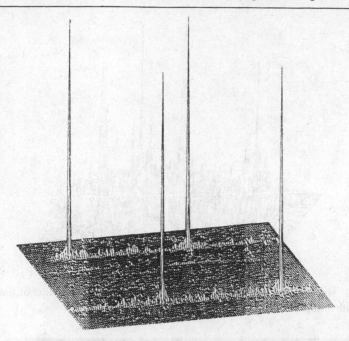

Fig. 4.20. Numerical calculation of the correlation that shows detection of the F-18 target in the upper part of the image and the Tornado in the lower part.

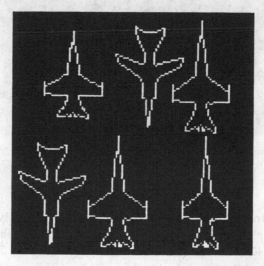

Fig. 4.21. Input image used for computer simulations.

y scale-invariant logarithmic harmonic of the F-18 was calculated and shifted to the center of the lower region of the image. The above-outlined procedure was performed to obtain a filter with $p_x = 1$, $p_y = 0.5$. Successful recognition is demonstrated by the distinct correlation peaks of Fig. 4.22.

A threshold of 35% of the maximum intensity value is enough to detect the true target peaks from the peaks that correspond to other objects and from the background.

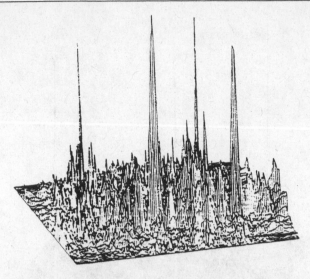

Fig. 4.22. Numerical calculation of the correlation that shows detection of the F-18 target with invariance to a 1-D scale in two axes.

Fig. 4.23. Experimental results for a multiple anamorphic FRT correlation with the image in Fig. 4.19.

4.6.4.2 Optical results

In order to test the performance of the suggested approach experimentally, a binary computer-generated mask was plotted. Then it was reduced by 20% with a high-resolution camera. The hologram was a 128 × 128 pixel Lohmann encoding mask [37]. The input scene size was 4.2 × 4.2 mm, and the filter size was 8.5 × 12 mm. This aspect ratio between the axes was calculated according to Eq. (4.65) [AR $= \sin(0.5\pi/2)/\sin(\pi/2) = 8.5/12$]. At the output plane, a CCD camera connected with a Matrox image LC framegrabber was used in order to grab the correlator output. The input scene is illustrated in Fig. 4.19. The experimentally obtained output is shown in Fig. 4.23.

These pictures show the intensity at the output correlation plane. The experimentally obtained results appearing in the first diffraction order match with the computer simulations illustrated in Fig. 4.20. In order to adjust the system a special plate was used, as explained in Refs. 24 and 35.

4.7 Fractional joint transform correlator

The JTC configuration is based on the simultaneous presentation of two patterns at the input plane, each laterally shifted from the center of the axis. Thus such an approach does not require the generation of a complex filter and aligning it with high accuracy in the Fourier plane. As Fig. 4.24 shows, a schematic configuration of the JTC contains a $4\text{-}f$ setup that at its Fourier plane includes a square-law converter device (a device that converts field distribution to amplitude distribution).

The output first diffraction orders of such a method are the correlation between the two input patterns. Several approaches for implementing the square-law conversion have been suggested, such as photofilm [3], a spatial light modulator [38], and liquid-crystal light valves [39]. When the shift invariance of the input plane is not important, a fractional JTC configuration may be used [25]. This configuration is first analyzed and then optically implemented in this section. The advantages of such a configuration in the optical pattern recognition field are similar to the advantages of the conventional JTC approach, but, in addition, the amount of the shift invariance can be controlled.

Because the mathematical analysis done in this section relies on the Wigner distribution function (WDF), a brief background is given [25, 40].

4.7.1 Wigner distribution function

The WDF of a signal $U(x)$ is defined as

$$W(x, v) = \int U(x + x'/2)U^*(x - x'/2)\exp(-2\pi i v x')\,dx',\qquad(4.66)$$

or alternatively, with $\tilde{U}(v)$, the spectrum of $U(x)$ $[\tilde{U}(v) = \int U(x)\exp(-2\pi i v x)\,dx]$:

$$W(x, v) = \int \tilde{U}(v + v'/2)\tilde{U}^*(v - v'/2)\exp(+2\pi i v' x)\,dv'.\qquad(4.67)$$

U may be reconstructed out of W, apart from a constant phase factor:

$$\int W(x/2, v)\exp(2\pi i v x)\,dv = U(x)U^*(0).\qquad(4.68)$$

The signal intensity $|U(x)|^2$ and the power spectrum $|\tilde{U}(v)|^2$ can be obtained from W by two orthogonal projection integrals:

$$\int W(x, v)\,dv = |U(x)|^2,\qquad(4.69)$$

$$\int W(x, v)\,dx = |\tilde{U}(v)|^2.\qquad(4.70)$$

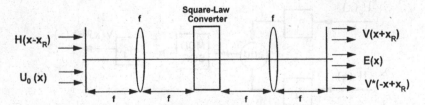

Fig. 4.24. Schematic optical setup for performing the joint transform correlation.

Shifting of the WDF occurs in the (x, v) domain if the signal $U(x)$ is shifted and is illuminated by a tilted plane wave, $\exp(2\pi i \bar{v} x)$:

$$U(x) \longrightarrow U(x - \bar{x}) \exp(2\pi i \bar{v} x), \tag{4.71}$$

$$W(x, v) \longrightarrow W(x - \bar{x}; v - \bar{v}). \tag{4.72}$$

Note that, as illustrated in Ref. 14, the FRT operation is defined as what occurs to U_0 while the WDF is rotated by an angle $\phi = p\pi/2$:

$$U_0(x) \longrightarrow U_p(x), \tag{4.73}$$

$$W_0(x, v) \longrightarrow W_0(x \cos\phi - v \sin\phi, v \cos\phi + x \sin\phi) = W_p(x, v), \tag{4.74}$$

where $U_p(x)$ is a FRT with the fractional order of p applied over $U_0(x)$, W_0 is the WDF of U_0, and W_p is the WDF of U_p.

4.7.2 Concept of the joint fractional correlator

Based on the FRT operation, we generalize the classical JTC (Fig. 4.25). To do so, the classical Fourier transforms are now replaced by FRT's (Fig. 4.26).

There is a hidden problem in this flow chart. We identify this problem by stepping into the Wigner domain.

Step I:

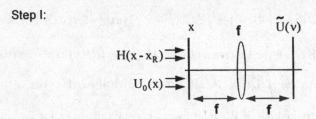

Step II:

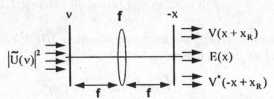

Fig. 4.25. JTC. U_0, input; H, reference signal; $\tilde{U}$, complex amplitude in the Fourier domain; $|\tilde{U}|^2$, transmission of the holographic filter; V, output; E, extraneous term.

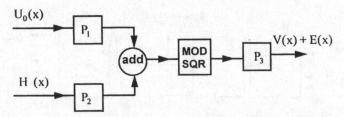

Fig. 4.26. Flow diagram of the joint fractional correlator. MOD SQR, modulus-square.

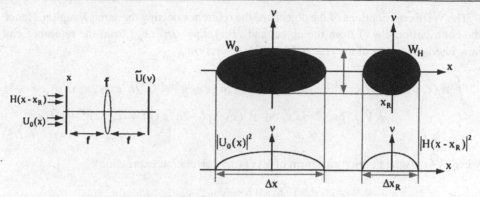

Fig. 4.27. WDF of the input to Fig. 4.25, step I, and the projection of that WDF, which is the input intensity to Fig 4.25, step II.

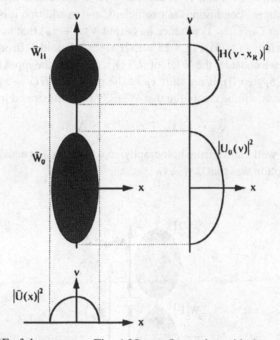

Fig. 4.28. WDF of the output to Fig. 4.25, step I, together with the two projections, intensity (x) and power spectrum (ν).

4.7.3 Removal of the extraneous terms

The need to remove some extraneous terms already occurred in the classical JTC system. However, in the classical JTC case the solution is simple and obvious. Thus we now identify that problem explicitly because it prepares us for solving the more general problem in the joint fractional correlator (JFC) system.

The WDF shown in Fig. 4.27 is the WDF of the two-part signal $U(x) = U_0(x) + H(x - x_R)$ shown in Fig. 4.25, step I.

The Fourier transform in Fig. 4.25, step I, corresponds to a 90° rotation in the Wigner domain (Fig. 4.28). (Note that a space variable x and a frequency variable ν are added in Fig. 4.28 because here we are using the normalized coordinates.)

The WDF contributions of the object and the reference occupy the same X region. Hence the contribution $\tilde{U}_0(x)$ from the object and $\tilde{H}(x)\exp(-2\pi ixx_R)$ from the reference can now interact as parts of the observable WDF projection:

$$\int \tilde{W}(x,v)\,dv = |\tilde{U}(x)|^2 = |\tilde{U}_0 + \tilde{H}\exp(-2\pi ixx_R)|^2 = \tilde{U}_0\tilde{H}^*\exp(2\pi ixx_R) + \cdots$$

$$= \tilde{V}(x)\exp(2\pi ixx_R) + \tilde{V}^*(x)\exp(-2\pi ixx_R) + |\tilde{U}_0(x)|^2 + |\tilde{H}(x)|^2,$$

$$(4.75)$$

where $\tilde{V}(x)$ is the Fourier transform of $V(x)$ (the correlation expression):

$$\tilde{V}(x) = \int \tilde{U}_0(v)\tilde{H}^*(v)\exp(2\pi ivx)\,dv. \qquad (4.76)$$

This overlapping between $\tilde{U}_0$ and $\tilde{H}$ is necessary for a successful operation of the JTC. However, the overlapping condition is not sufficient, and in addition it is also required that the wanted term with $\tilde{U}_0\tilde{H}^* = \tilde{V}$ produce an output $V(x + x_R)$ that is laterally separated from the other terms in Eq. (4.75) (Fig. 4.25, step II, right-hand side). In order to comprehend the second request, we consider the WDF of $|\tilde{U}(x)|^2$, which is the input for the second step of the JTC (Fig. 4.25, step II). If the shift x_R of the reference $H(x - x_R)$ is large enough, three separated islands will be created in that WDF. This is illustrated in Fig. 4.29:

$$x_R - (\Delta x_H + \Delta x_0/2) \geq \Delta x_0. \qquad (4.77)$$

This condition is well known from holography, in which Δx_H is usually close to zero. In our case the assumption was that $0 \leq \Delta x_H \leq \Delta x_0$.

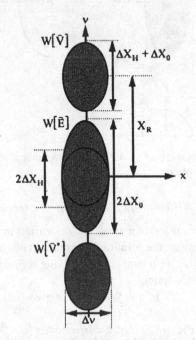

Fig. 4.29. WDF of the intensity output of Fig. 4.25, step I, which is the input for Fig. 4.25, step II.

The second stage of the JTC (Fig. 4.25, step II) is again a Fourier transform that corresponds to a 90° rotation (anticlockwise) of the WDF (see Fig. 4.29). The desired term $V(x + x_R)$ is clearly separated from the extraneous term $E(x)$ and $V^*(-x + x_R)$ if the condition of relation (4.77) is fulfilled. Thus the proper shift of x_R [relation (4.77)] of the reference pattern was necessary and sufficient for the proper JTC operation.

A similar consideration for the JFC yields the fact that the reference $H(x)$ has to be shifted and tilted by proper amounts:

$$H(x) \longrightarrow H(x - x_R)\exp(2\pi i x \nu_R). \tag{4.78}$$

In order to explain this point we translate the front part of the JFC flow chart (Fig. 4.26) into the Wigner domain. The WDF of the centered object $U_0(x)$ is rotated by an angle $\phi_1 = p_1\pi/2$ (Fig. 4.30).

The WDF of the reference signal is shifted at first (Fig. 4.31):

$$W_H(x, \nu) \longrightarrow W_H(x - x_R, \nu - \nu_R). \tag{4.79}$$

Then W_H is subsequently, rotated by an angle of $\phi_2 = p_2\pi/2$. This angle must be such that W_H will end up on the ν axis just above the WDF of U_0 (Fig. 4.32).

Expressing the shift in polar coordinates, we obtain

$$X_R = R\cos(\pi/2 - \phi_2) = R\sin\phi_2, \tag{4.80}$$

$$\nu_R = R\sin(\pi/2 - \phi_2) = R\cos\phi_2. \tag{4.81}$$

This colocation in the x axis is needed to satisfy the condition for letting W_0 and W_H interact when the modulus-square operation of Fig. 4.26 is applied.

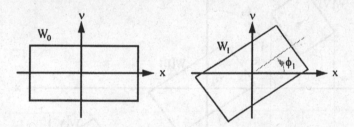

Fig. 4.30. FRT with degree p_1 rotates the WDF of the object by the angle $\phi_1 = p_1(\pi/2)$.

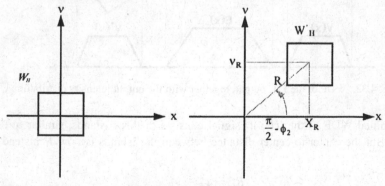

Fig. 4.31. Shift and tilt of the reference signal H cause a shift of the WDF of H. That shift can be expressed by a radius R and an angle $\pi/2 - \phi/2$.

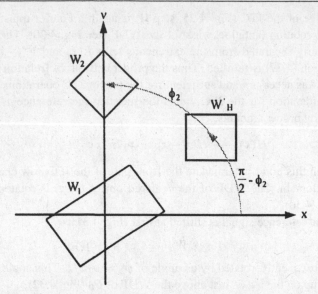

Fig. 4.32. FRT with degree p_2 rotates the WDF of the reference by an angle $\phi_2 = p_2(\pi/2)$, bringing it to the frequency axis on top of the object WDF.

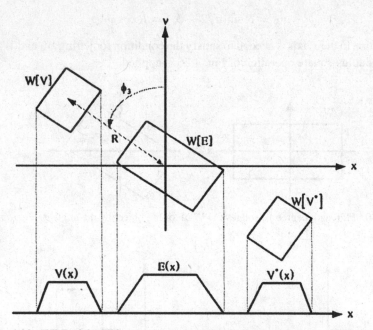

Fig. 4.33. WDF of the JTC output, together with the output intensity distribution.

The obtained WDF of the $|U(x)|^2$ signal consists of three islands, similar to those in Fig. 4.29. But the center-to-center distance between the islands is now R instead of x_R. Note that

$$R^2 = x_R^2 + v_R^2. \tag{4.82}$$

The final step is a rotation of the WDF by an angle $\phi_3 = p_3\pi/2$, as indicated by Fig. 4.26.

The final WDF is shown in Fig. 4.33. In this figure, the radial shift R was sufficiently large to create full separation between the terms $V(x)$ and $E(x)$.

For deriving the condition for R that ensures the separation, one has to calculate the locations of the rectangular WDF islands. The result will be a generalization of the classical condition of separation [relation (4.77)]. However, not only the sizes Δx_0 and Δx_H will appear in the inequality but also the bandwidths Δv_0 and Δv_H, together with cosine and sine factors, involving the three angles $\phi_1, \phi_2,$ and ϕ_3. The shortest R will be needed if $\phi_3 = \pi/2$, whereas a small value of ϕ_3 requires large R. If the angle ϕ_2 is small, the R is affected mainly by the tilt v_R. But with ϕ_2 close to $\pi/2$, the R depends mainly on the shift x_R [as seen from Fig. (4.32) and Eqs. (4.81) and (4.82)].

4.8 Concluding remarks

In this chapter we have attempted to describe the status of the field of fractional correlation that is based on the FRT. The described analysis started by analyzing the performances of the FC compared with those of conventional correlator configurations. Section 4.3 illustrated that the performances are object dependent. For white noise, according to the SNR criterion, the performances of the shift-variant correlator (the FC) are not worse than those of the conventional correlator. For other types of noise, the performances of the FC may be even better (for example, when chirp noise is involved). Then a configuration for obtaining a real-time FC was demonstrated in section 4.4. This configuration is a real-time one because, in order to change the amount of shift variance (the fractional order), one need change only the longitudinal distance of the filter. There is no need to change the distances between the lenses, change their focal length, or recompute the filter. In Section 4.5 the LFRT was defined. This FRT-based transform performs the FRT with different fractional orders in different spatial regions. An optical FC-based on this transform may be constructed. The main advantage of using a FC that is based on the LFRT is when nonstationary noises (with statistical properties that are varied over the space) or special pattern recognition aspects are involved. A pattern recognition example for which such a use is essential was demonstrated for fingerprint recognition. The fingerprint pattern has a spatially varied form. Its central region is more or less constant, whereas the outer region is changed from instant to instant because a person never presses a finger with equal force. Thus, in order to recognize or reconstruct those prints, an LFRT processor whose spatial shift variance is changed with space is required. In Section 4.6 another extension for the FC was illustrated. This time the suggested processor was an anamorphic one. A practical example for which object detection is needed along a row was demonstrated. For cases in which full shift invariance is needed in the one axis and only a limited amount of variance is needed in the other axis, such a processor can be very helpful. The demonstrated example included a case in which a certain first object (or certain deformation property) is needed to be detected in the first region and rejected in the second spatial region and vice versa for the second object (or deformation property) and the second spatial region. Section 4.7 dealt with the JTC configuration applied for the partially shift-variant case. The advantages of the conventional JTC (i.e., it does not require the generation of a complex filter and aligning it with high accuracy in the Fourier plane) were combined with the ability to control the amount of the shift-variance property and implemented optically. In sum, this chapter shows the potential of the FC in various optical signal processing fields. It indicates that the FC adds a degree of freedom that can increase the flexibility of the conventional optical covolver/correlator.

Acknowledgments

The authors acknowledge the fruitful discussions and the generous assistance from A. W. Lohmann and R. G. Dorsch from Erlangen University, C. Ferreira and J. Garcia from the University of Valencia, H. J. Caulfield from Alabama A & M University, Y. Bitran from Tel-Aviv University, and A. Kutay from Bilkent University.

References

[1] A. VanderLugt, "Signal detection by complex spatial filtering," IEEE Trans. Inf. Theory **IT-10**, 139–146 (1964).

[2] A. W. Lohmann and H. W. Werlich, "Incoherent matched filter with Fourier holograms," Appl. Opt. **7**, 561–563 (1968).

[3] C. S. Weaver and J. W. Goodman, "A technique for optically convolving two functions," Appl. Opt. **5**, 1248–1249 (1966).

[4] I. E. Rau, "Detection of differences in real distributions," J. Opt. Soc. Am. **56**, 1490–1494 (1966).

[5] H. O. Bartelt, "Applications of the tandem component: an element with optimum light efficiency," Appl. Opt. **24**, 3811–3816 (1985).

[6] J. R. Leger and S. H. Lee, "Hybrid optical processor for pattern recognition and classification using a generalized set of pattern recognition functions," Appl. Opt. **21**, 274–287 (1982).

[7] J. A. Davis, D. M. Cottrell, N. Nestorovic, and S. M. Highnote, "Space-variant Fresnel-transform optical correlator," Appl. Opt. **31**, 6889–6893 (1992).

[8] G. G. Mu, X. M. Wang, and Z. Q. Wang, "A new type of holographic encoding filter for correlation: a lensless intensity correlator," in *International Conference on Holographic Applications,* J. Ke and R. J. Pryputniewicz, eds., Proc. SPIE **673**, 546–549 (1986).

[9] D. Mendlovic, H. M. Ozaktas, and A. W. Lohmann, "Fractional correlation," Appl. Opt. **34**, 303–309 (1995).

[10] D. Mendlovic, Y. Bitran, R. G. Dorsch, and A. W. Lohmann, "Fractional correlation: experimental results," Appl. Opt. **34**, 1329–1335 (1995).

[11] D. Mendlovic and H. M. Ozaktas, "Fractional Fourier transformations and their optical implementation: part I," J. Opt. Soc. Am. A **10**, 1875–1881 (1993).

[12] R. G. Dorsch, A. W. Lohmann, Y. Bitran, D. Mendlovic, and H. M. Ozaktas, "Chirp filtering in the fractional Fourier domain," Appl. Opt. **33**, 7599–7602 (1994).

[13] H. M. Ozaktas, B. Barshan, D. Mendlovic, and L. Onural, "Convolution, filtering, and multiplexing in fractional Fourier domain and their relation to chirp and wavelet transforms," J. Opt. Soc. Am. A **11**, 547–559 (1994).

[14] A. W. Lohmann, "Image rotation, Wigner rotation, and the fractional Fourier-transform," J. Opt. Soc. Am. A **10**, 2181–2186 (1993).

[15] D. Mendlovic, M. Ozaktas, and A. W. Lohmann, "Graded-index fibers, Wigner-distribution functions, and the fractional Fourier-transform," Appl. Opt. **33**, 6188–6193 (1994).

[16] H. L. Van Trees, *Detection, Estimation and Modulation Theory, Part I* (Wiley, New York, 1968).

[17] B. V. K. Vijaya Kumar and L. Hassebrook, "Performance measures for correlation filters," Appl. Opt. **29**, 2997–3006 (1990).

[18] J. L. Horner, "Light utilization in optical correlators," Appl. Opt. **21**, 4511–4514 (1982).

[19] Z. Zalevsky, D. Mendlovic, and H. J. Caulfield, "Fractional correlator with real-time control of the space-invariance property," Appl. Opt. **36**, 2370–2375 (1997).

[20] D. Mendlovic, Z. Zalevsky, A. W. Lohmann, and R. G. Dorsch, "Signal spatial-filtering using the localized fractional Fourier-transform," Opt. Commun. **126**, 14–18 (1996).

[21] Z. Zalevsky, D. Mendlovic, and H. J. Caulfield, "Localized partially space-invariant filtering," Appl. Opt. **36**, 1086–1092 (1997).

[22] A. Sahin, H. M. Osaktas, and D. Mendlovic, "Optical implementation of the two-dimensional fractional Fourier-transform with different orders in two dimensions," J. Opt. Soc. Am. A **12**, 134–138 (1995).

[23] D. Mendlovic, Y. Bitran, R. G. Dorsch, C. Ferreira, J. Garćia, and H. M. Ozaktas, "Anamorphic fractional Fourier transforming: optical implementation and applications," Appl. Opt. **34**, 7451–7456 (1995).

[24] J. Garćia, D. Mendlovic, Z. Zalevsky, and A. Lohmann, "Space-variant simultaneous detection of several objects using multiple anamorphic fractional Fourier-transform filters," Appl. Opt. **35**, 3945–3952 (1996).

[25] A. W. Lohmann and D. Mendlovic, "Fractional joint transform correlator," Appl. Opt. accepted for publication.

[26] B. Javidi, "Nonlinear joint power spectrum based optical correlation," Appl. Opt. **28**, 2358–2367 (1989).

[27] A. Papoulis, *Probability, Random Variables, and Stochastic Processes* (McGraw-Hill, New York, 1984), pp. 215–283.

[28] J. L. Horner, "Clarification of Horner efficiency," Appl. Opt. **31**, 4629 (1992).

[29] J. L. Horner and P. D. Gianino, "Phase only matched filtering," Appl. Opt. **23**, 812–816 (1984).

[30] A. W. Lohmann, D. Mendlovic, and Z. Zalevsky, "Synthesis of pattern recognition filters for fractional Fourier processing," Opt. Commun. **128**, 199–204 (1996).

[31] D. Mendlovic, R. G. Dorsch, A. W. Lohmann, Z. Zalevsky, and C. Ferreira, "Optical illustration of a varied fractional Fourier-transform order and the Radon-Wigner display," Appl. Opt. **35**, 3925–3929 (1996).

[32] H. M. Ozaktas and D. Mendlovic, "Multistage optical implementation architecture with least possible growth of system size," Opt. Lett. **18**, 296–298 (1993).

[33] Y. Bitran, Z. Zalevsky, D. Mendlovic, and R. G. Dorsch, "Performance analysis of the fractional correlation operation," Appl. Opt. **35**, 297–303 (1996).

[34] A. Sahin, H. Ozaktas, and D. Mendlovic, "Optical implementation of the two-dimensional fractional Fourier-transform with different orders in the two dimensions," Opt. Commun. **120**, 134–138 (1995).

[35] J. Garcia, R. Dorsch, A. W. Lohmann, C. Ferreira, and Z. Zalevsky, "Flexible optical implementation of fractional Fourier transform processors. Applications to correlation and filtering," Opt. Commun. **133**, 393–400 (1997).

[36] D. Mendlovic, N. Konforti, and E. Marom, "Shift and projection invariant pattern recognition using logarithmic harmonics," Appl. Opt. **29**, 4784 (1990).

[37] A. W. Lohmann and D. P. Paris, "Binary Fraunhofer holograms, generated by computer," Appl. Opt. **6**, 1739–1748 (1967).

[38] F. T. S. Yu, S. Jutamulia, T. W. Lin, and D. A Gregory, "Adaptive real-time pattern recognition using a liquid crystal TV based joint transform correlator," Appl. Opt. **26**, 1370–1372 (1987).

[39] F. T. S. Yu and X. J. Lu, "A real-time programmable joint transform correlator," Opt. Commun. **52**, 10–16 (1984).

[40] A. W. Lohmann and B. H. Soffer, "Relationships between two transforms: Radon–Wigner and fractional Fourier," J. Opt. Soc. Am. A **11**, 1798–1801 (1994).

5

Optical implementation of mathematical morphology

Tien-Hsin Chao

5.1 Introduction

Mathematical morphology is a rapidly developing image processing tool that combines nonlinear processing with shape and size analysis. The basic idea behind mathematical morphology is to probe a given image with a sequence of elementary patterns, known as structuring elements, thereby extracting important shape and size information. Mathematical morphology rests on a solid mathematical foundation and leans on concepts from algebra, topology, and stochastic geometry [1–4].

Morphological processing is one of the excellent choices for image preprocessing (image editing) operations for many reasons. First, morphological operations have demonstrated the ability to remove noise and clutter [5]. In addition, more complex sequences of morphological operations have been demonstrated to perform image recognition based on the object of interest's size, shape, and relation with its local surroundings [6]. These examples demonstrate the power of using morphological processing for image editing and clutter and noise removal in the problem of interest. Figure 5.1 shows a schematic of the objective of morphological image editing. In this figure, the objects of interest are surrounded by both noise and objects that are both larger and longer than it. When information on the object's general shape and size is used, it is possible to process the image morphologically to leave only the objects of interest. This is advantageous, for example, in a correlation application, as correlations with the edited image will output stronger and less noisy correlation peaks. The second reason for choosing morphological processing is that it can be computed through a series of local neighborhood operations. This makes it suitable for parallel implementation in either electronics or optoelectronics that are capable of operating at high speeds.

Mathematical morphology was introduced as a tool for investigating the geometric structure in binary images. The fundamental morphological operations on which the entire subsequent development depends are briefly reviewed here [1–5].

Assume an image A in Euclidean space E. Geometric information is extracted from a set A when it is probed with another smaller set B, known as the structuring element. The important morphological operator known as erosion, which is denoted by $A \ominus B$, is given by

$$A \ominus B = \bigcap_{b \in B} A + b,$$

where $+$ denotes translation. In this operation, A is translated by every element b of B in R^2 and then the intersection is taken.

126

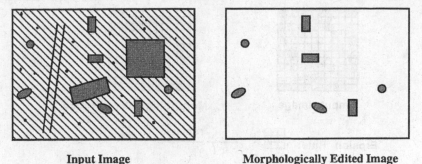

Input Image   **Morphologically Edited Image**

Fig. 5.1. Image editing by morphological processing.

The dual of erosion, known as dilation and denoted by $A \oplus B$, is given by

$$A \oplus B = \bigcup_{b \in B} A + b.$$

We construct $A \oplus B$ by translating A by each element of B and then taking the union of all the resulting translates.

Dilation expands a shape, whereas erosion shrinks it. The composition of these two operators gives two more operators, which are known as opening and closing and are denoted by $A \circ B$, and $A \bullet B$, respectively. They are given by

$$A \circ B = (A \ominus B) \oplus B,$$

$$A \bullet B = (A \oplus B) \ominus B.$$

It can be shown that $A \circ B$ is the union of all translated structuring elements B that are included in A. Therefore opening removes objects of smaller size than the structuring element and thus is capable of separating regions that barely touch each other. It also smoothes boundaries of high curvature. On the other hand, closing fills in the small gaps of sizes comparable with those of the structuring element and results in boundary smoothing.

5.1.1 Binary morphology

Figure 5.2 shows the effect of the erosion operation on a binary image [7]. In this example, the white pixels represent a 1 and the dark pixels a 0. The structuring element consists of a 3×3 cross. Hence the operations will utilize pixel information only from the input pixel and its four local neighbors to compute the morphological output. The erosion operation consists of sliding the structuring element across the input image and finding areas in the image that contain the structuring element. In areas that match the 3×3 cross, the single 1 pixel is placed on the output image. Areas that do not match have a 0 placed on the output image. The effect of this operation on an extended object is to remove the edges of the object (hence the analogy to an erosion process). When a different size and shape of the structuring element are chosen, the sizes and the orientations of objects that will be eroded can be controlled.

Figure 5.2 also shows the binary dilation operation. In this case any occurrence of a 1 pixel is substituted by the structuring element. The effect of this operation is to expand the boundaries of an object (hence the dilation analogy). Again, when the shape and the size of the structuring element are controlled, the amount and the orientation of the boundary expansion can be controlled.

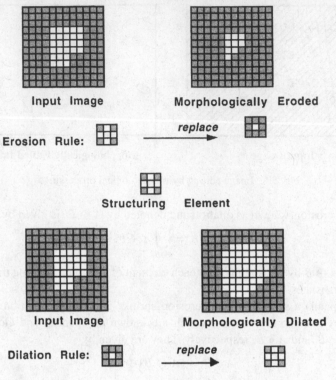

Fig. 5.2. Binary erosion and dilation operators.

The erosion and the dilation operators form the basic operations for morphological processing. Through combinations of these operators and choice of structuring elements, more complex morphological operators can be formed. Two important operations are the opening operator, which is erosion followed by dilation, and the closing operator, which is dilation followed by erosion. When pointwise image operations, such as binary arithmetic (AND, OR, complement) operations, and conditional statements are added, powerful image processing algorithms can be realized.

An example of morphological image processing is the removal of salt-and-pepper noise from an input image through a sequence of opening and closing operations. Salt-and-pepper noise refers to single-pixel bit errors on an image. Figure 5.3 shows a large bloblike shape that is our object of interest corrupted by salt-and-pepper noise. The procedure for cleaning up the salt-and-pepper noise is also illustrated in Fig. 5.3. The removal of salt noise (white pixels on a black background) is accomplished by the opening of the image with a 3 × 3 cross structuring element. The erosion stage removes the salt noise as well as the boundary of the object of interest. The dilation stage restores the boundary of the object of interest. A closing operation (dilation followed by erosion) is then used to remove the pepper noise.

Morphological operations can also be used to perform edge enhancement. The input image is first eroded and then subtracted from the original. The effect is to leave the boundary of the original image.

Aside from noise removal and edge enhancement, morphological processing can implement a large number of image processing algorithms. Among these are size and orientation determination, skeletonization, hit-or-miss, and object removal.

Opening Morphological Operator
Removes Salt Noise

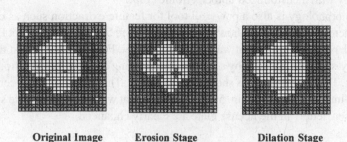

Original Image   Erosion Stage   Dilation Stage

Salt Noise Removal by Morphological Opening

Closing Morphological Operator
Removes Pepper Noise

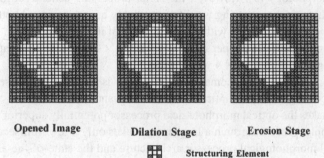

Opened Image   Dilation Stage   Erosion Stage

Structuring Element

Pepper Noise Removal by Morphological Closing

Fig. 5.3. Salt-and-pepper noise removal by morphological opening and closing.

5.1.2 Gray-scale morphology

Morphological operations on gray-level images are similar to those performed on binary images. Instead of a substitution rule, a minimum or a maximum operation is used. Erosion is performed by detecting the pixel values in the input image that overlaps the structuring element and substituting the center pixel with the minimum value detected. The dilation operation is performed by substituting the center pixel with the maximum pixel value. Mathematical expressions for the gray-level morphology operations are given below.

For an image $A(x, y)$ and a structuring element $B(x, y)$ of zero height, the morphological operators are

$$D(x, y) = \max_{i, j \in B}[A(x + i, y + j)], \qquad \text{dilation,}$$

$$E(x, y) = \min_{i, j \in B}[A(x + i, y + j)], \qquad \text{erosion,}$$

where i and j are taken over the pixels in the structuring element B.

Although it can be shown that gray-level morphology is suitable for image processing and editing, it is difficult to implement directly with an optical processor; the maximum and the minimum operators necessary for completing dilation and erosion operations can be achieved only with a customized optoelectronic chip.

Gray-scale morphology is also a powerful tool for feature extraction such as car license plate enhancement. An example is shown in Fig. 5.4. Figure 5.4(a) shows the Jet Propulsion Laboratory (JPL) commuter van that was used as the input. For processing, first the input is closed with gray-scale morphology by a rhombus structuring element (one dilation followed by one erosion). The closed image is shown in Fig. 5.4(b). The difference between the input and the opened image is shown in Fig. 5.4(c). It is seen that when an appropriate structuring element is chosen, letters in the license plate are clearly enhanced.

5.2 Optical morphological processor

Morphological processing has been demonstrated to be capable of removing noise and clutter as well as having the ability to edit an image, through edge enhancement, image subtraction, image shape representation, smoothing, recognition, skeletonization, and coding [8]. Current software and electronic hardware are impressively fast and able to perform mathematical morphology [9]. However, the serial processing scheme utilized in digital electronic morphological processing would unavoidably become a speed bottleneck when processing a large image template with a large number of iterations required. For example, removing the typical salt-and-pepper noise from a 256×256 image template takes up to 1-min processing time with a SUN 4 workstation.

Optical implementation of mathematical morphology is promising because of the inherent parallel processing nature of optics. The recent advancement in optical and optoelectronic devices makes the optical morphological processor potentially superior in processing speed. For example, processing such a large image takes only a few milliseconds with the proposed optical morphological processor architecture and the state-of-the-art optical and optoelectronic devices and components.

Many optical morphological processing architectures have been presented in recent years. O'Neill and Rhodes [10] introduced an incoherent processing technique that used a computer-generated holographic pupil transparency to produce the desired point-spread function for morphological processing. This technique relies on synthesized pupil function and tends to reduce the space–bandwidth product of the optical processor. Hereford and Rhodes later extended the binary morphological processing idea to gray-scale image processing by using the thresholding decomposition approach [11]. Ochoa *et al.* have also proposed the use of the thresholding decomposition idea to optical median filtering implementation [12]. Casasent and Botha [13] have also introduced a cascaded optical correlator system for symbolic-substitution-based morphological processing. In this system, a basic structuring element was constructed with a Fourier hologram filter. Because of the large dynamic range of the Fourier spectrum, this holographic filter synthesis suffers from nonuniformity and may cause errors in the thresholded output. More recently, Li *et al.* [14] introduced a partially coherent optical morphological processor in which a shadow-casting approach was used to obtain basic erosion and dilation processing. The limitation of this shadow-casting approach is that the image resolution is severely reduced as the detector is placed on a nonimaging plane.

As shown in Fig. 5.5, we compute the binary dilation operation by convoluting the binary input with a binary structuring element and then thresholding the output. We can also obtain this binary dilation by replacing each bright pixel of the input image with the structuring element and then thresholding the output.

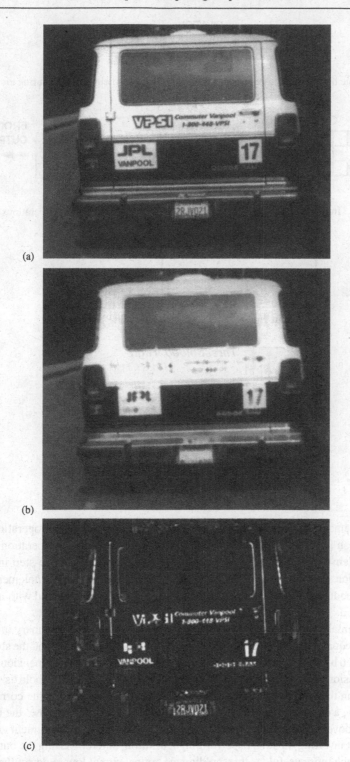

(a)

(b)

(c)

Fig. 5.4. Feature extraction by gray-scale morphology: (a) input image of the JPL commuter van, (b) morphologically closed image of the input, (c) difference between (a) and (b). The license plate data are sharply enhanced.

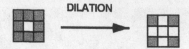

Fig. 5.5. Basic binary dilation operation with a rhombus four-neighbor structuring element.

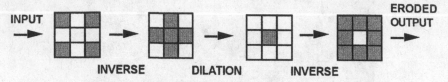

Fig. 5.6. Implementing optical binary erosion by inverse and dilation operations.

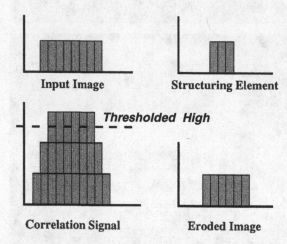

Fig. 5.7. Binary erosion by correlations.

We can implement the erosion processor by combining two inversion operations and a dilation operation [15, 16], as shown in Fig. 5.6. Note that the erosion operation converts a structuring element to a single pixel. This is accomplished by a three-step inversion–dilation–inversion sequential operation. Because inversion can be easily implemented with a spatial light modulator (SLM), the erosion operation can be implemented with an optical processor as well.

The fundamental morphological operations of erosion and dilation on binary images can be shown to be equivalent to a cross-correlation between the input image and the structuring element followed by a threshold operation [10]. Figure 5.7 shows a one-dimensional analog of a binary erosion operation in which a correlation is used. The threshold is set to be slightly less than the sum of all pixels in the structuring element. When the correlation is thresholded low, a dilation operation is implemented (Fig. 5.8). In this case, the threshold is set slightly above the zero value. It can be shown that we can also implement an erosion operator by first inverting the image, performing a dilation, and reinverting the output. This technique is much more useful, as thresholding an optical signal low to detect the absence of light is more difficult than thresholding high at a level that is dependent on the size of the structuring element.

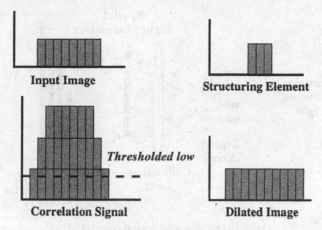

Fig. 5.8. Binary dilation by correlations.

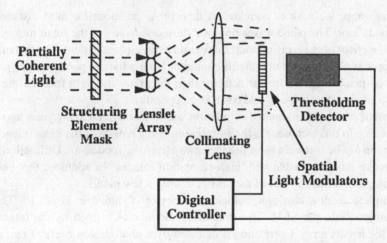

Fig. 5.9. Basic schematic of a shadow-cast optical morphological processor.

There are a number of methods of achieving the cross-correlation operation necessary to implement binary morphology. One method proposed by Botha and Casasent [17] utilized a holographic optical correlator. Another approach with a shadow-cast system [14] has also be demonstrated.

5.2.1 Shadow-cast optical morphological processor

The shadow-cast optical morphological processor is based on a correlator architecture. As opposed to the conventional optical correlator, in the shadow-cast system no Fourier transform occurs; rather, the correlation is formed by a shift-and-add operation. Figure 5.9 shows a typical shadow-cast correlator system. In this system, a collimated laser beam illuminates a lenslet array, forming a matrix of point sources. The pattern of these point sources corresponds to one of the images that we wish to correlate. The array of points is then collimated by a second lens. Each point produces a plane wave propagating at an angle proportional to the position of the lenslet. The array of plane waves illuminates a SLM that

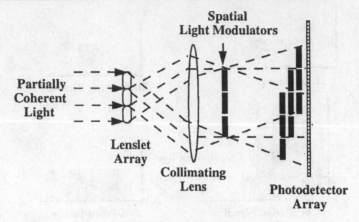

Fig. 5.10. Shadow-cast optical correlation.

contains the image we wish to correlate. A detector is positioned a short distance away from the modulator. The plane waves produce shifted versions of the input image. Figure 5.9 shows the effect of the array of point sources on a three-pixel pattern. The distance from the modulator is chosen such that the illumination from adjacent point sources produces exactly a one-pixel image shift. The detector then integrates the light from all the shifted images, completing the shift-and-add correlation operation.

This form of producing an optical correlation was demonstrated many years ago [18]. A major limitation to this technique is the assumption that a replica of the input image and its shifted version can be formed a short distance away from the modulator. Diffraction effects, however, make this impossible with high-resolution images. In addition, this geometric optics assumption limits the size of the lenslet to only a few pixels.

The complete shadow-cast morphological processor is shown in Fig. 5.10. The input image is entered into the SLM. The structuring element is formed by the transmission pattern of the lenslet array. Correlation is detected by a photodetector array (e.g., a CCD) and binary morphological operations formed by thresholding with a computer.

5.2.2 Reconfigurable optical morphological processor

Two reconfigurable optical morphological processors are introduced in this section. The primary advantage of this reconfigurable system is that the size and the shape of the structuring element can be arbitrarily changed with each morphological processing cycle. This reconfigurability is a major improvement over the conventional correlator-based or shadow-cast morphological processors in which fixed holographic filters or illumination light-source patterns are used as structuring elements [11–14]. Moreover, with the use of a two-dimensional (2-D) thresholding photodetector array (thresholding and feedback), an iterative morphological processor can be built to perform many complex morphological operations. Optical architecture and the principle of operation of two types of reconfigurable optical morphological processors are discussed in this section. The extension of this binary morphology technique to gray-scale image processing by a thresholding decomposition technique is also discussed.

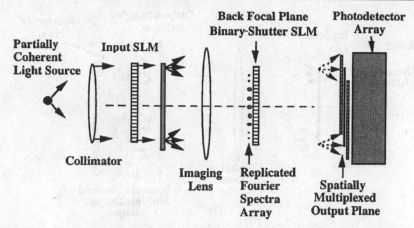

Fig. 5.11. Reconfigurable optical morphological processor with a diffraction grating and a Fourier-plane shutter SLM.

5.2.3 Optical morphological processor with a diffraction grating and a shutter spatial light modulator

An optical morphological processor system architecture is shown in Fig. 5.11 [15, 16]. A partially coherent light source (e.g., a filtered mercury arc lamp) is collimated to illuminate an input SLM. An imaging lens is used to image the input onto a photodetector array in the output plane with appropriate magnification. A diffraction grating with a uniform 2-D array of diffraction patterns (e.g., a cross Dammann grating or a custom-designed multiphase e-beam grating) is inserted between the input SLM and the imaging lens. A binary high-speed SLM is placed over the back focal plane (the Fourier plane) of the imaging lens. This diffraction grating will replicate an $N \times N$ array of Fourier spectra of the input image.

With the insertion of the grating, the system has been effectively converted into a multiple imaging system. An $N \times N$ array of replicated images will be present in the output plane. The spacing between two neighboring images can be varied when the distance between the grating and the input SLM is adjusted. When this spacing is adjusted to be equal to the pixel size of the output image, the system is effectively an optical morphological processor that performs binary dilation (after photodetection and thresholding) with a structuring element consisting of $N \times N$ pixels.

In order to achieve dynamic reconfigurability, a binary-shutter SLM is inserted into the Fourier-transform plane.

A spatial pattern that matches that of the desired structuring element shape can be down-loaded into the shutter SLM to allow the appropriate diffraction orders to pass through. This will enable optical dilation with an arbitrary structuring element to take place in the output plane.

For erosion, the input SLM is operated in its contrast reversal mode. The contrast reversal can be achieved by varying the bias of the SLM (for a Hughes liquid-crystal light valve) or varying the orientation of the output analyzer. As described above, a binary input will be inverted, dilated, and reinverted (after feedback into the SLM) to form its eroded counterpart.

This diffraction grating and shutter SLM-based optical morphological processor is capable of morphological dilation and erosion operations. By an iterative combination of these

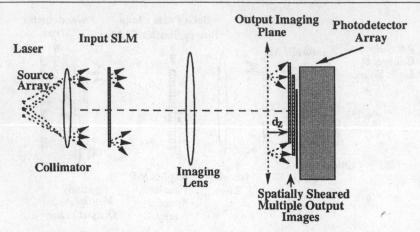

Fig. 5.12. Schematic diagram of a reconfigurable optical shadow morphological processor with a laser source array.

optical dilations and erosions, many more useful morphological operations, such as opening, closing, hit-or-miss, top hat, and salt-and-pepper noise removal, can also be accomplished.

The unique advantage of this system is that, unlike the shadow-cast system, the output plane is a precise imaging plane. The blurring caused by diffraction from the shadow-cast system can be avoided. The limitation of this system is that the use of a diffraction grating will introduce spectral smearing. A spectrally limited but spatially extended light source needs to be used to optimize image resolution and to avoid coherence noise.

5.2.4 Optical morphological processor with a laser source array

A modified shadow-cast technique for optical implementation of a morphological operations system has recently been reported [19]. As shown in Fig. 5.12, this shadow-casting system consists of a laser source array (a three-element array is shown for simplicity), a collimator, an input image, and an imaging lens that focuses the input into an output conjugate imaging plane with appropriate magnification. The input image, under illumination by collimated laser beams emanated from the three-element laser source array, will be imaged onto the conjugate output imaging plane. At a distance dz away from this conjugate imaging plane, the original single-output image will be separated into three translated images. The translation between each two neighboring images is Dx. Thus the system impulse response of this shadow-casting system is exactly the same shape as that of a scaled version of the laser source array. When the laser source array is configured into a desired structuring element (e.g., a rhombus), the output will be the convolution between the input and the structuring element. Thus binary morphological dilation can be achieved by binary thresholding the output. As described above, morphological opening and closing can be achieved by alternating dilation and erosion operations.

5.3 Miniature system architecture

A system architecture for a miniaturized optical shadow-casting morphological processor is shown in Fig. 5.13. A switchable 2-D laser source array is employed for its high-speed

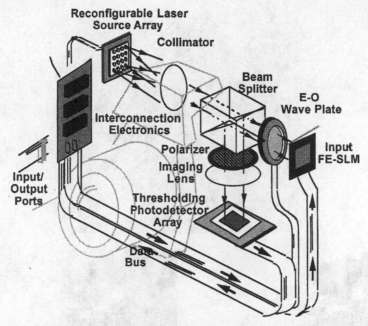

Fig. 5.13. Architecture of an optical morphological processor approximately the size of a 35-mm camera. E-O, electro-optical.

dynamic reconfigurability. By utilizing this laser source array, we can configure arbitrary structuring elements by turning on the desired laser source array elements. Because the switching time of this laser array is less than a microsecond, ultrafast system reconfiguration can be easily achieved.

A liquid-crystal ferroelectric spatial light modulator (FE SLM) can be used for incoherent-to-coherent input image conversion. Shadow casting is performed by an imaging lens that maps the input SLM to the output detector in the above-mentioned manner.

Because of the selection of state-of-the-art optical and optoelectronic devices and components such as an optoelectronic integrated-circuit laser diode array and the FE SLM, a miniature system package, approximately the size of a camera, is feasible, as shown in Fig. 5.13. The compatible high speeds of all the system devices, including the laser source (microsecond switching time), the FE SLM (1000–10,000 frame rate), and the high-speed photodetector array (1000 frame rate or higher), would enable ultrahigh-speed optical morphological processing that is not achievable with state-of-the-art electronics.

The salient features provided by this optically implemented morphological processor include

- High-speed parallel processing (2–3 orders of magnitude faster than a digital parallel processor), a processing speed that is invariable with increases in image and structuring set sizes.
- Analog input, analog output.
- Dynamic reconfigurability and iterative processing.
- Versatile morphological processing algorithm implementation capability.
- Miniature and user-friendly packaging.

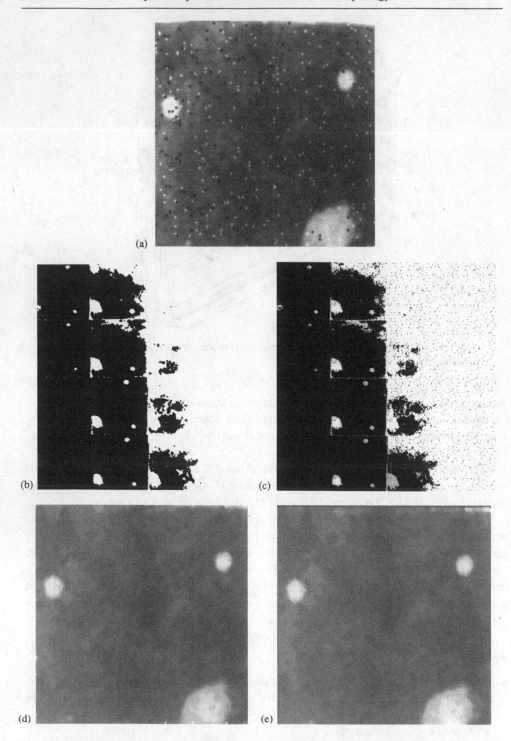

(a)

(b)   (c)

(d)   (e)

Fig. 5.14. Gray-scale morphological processing with gray-scale decomposition and binary morphological processing: (a) input image of a mine field with 3% added salt-and-pepper noise, (b) gray-scale decomposition of the input into 16 binary slices, (c) salt-and-pepper noise removal by binary opening and closing, (d) recombined gray-scale image from (c), (e) results of cleanup image by digital gray-scale morphological processing.

5.4 Gray-scale optical morphological processor

A gray-scale input image will first be decomposed into a series of binary images with a gray-scale decomposition technique. To perform a dilation, each of the binary images is convoluted with a structuring set. The output is then thresholded to obtain the dilated image. The multiply processed binary images are then superimposed to give the gray-scale dilated image. Erosion will be performed similarly with a complementing operation.

The binary morphological processing technique can be extended to process gray-scale images through the method of gray-scale decomposition [11, 12]. For example, a gray-scale image $I(x, y)$ that has a 2-D gray-scale distribution can be broken up into a number of binary slices. This process can be analytically represented as

$$f_i(x, y) = \begin{cases} 1 & I(x, y) \geq \text{Th}(i) \\ 0 & \text{otherwise} \end{cases},$$

where $I(x, y)$ is the input image intensity distribution, $f_i(x, y)$ is the ith binary slice, and Th(i) is the threshold value that corresponds to that slice. If each binary slice has a constant amplitude of a, the gray-scale image can be expressed as

$$I(x, y) = \sum_{i=1}^{K} a f_i(x, y).$$

Thus it can be seen that a gray-scale image can be represented as the sum of binarized slices obtained by means of thresholding.

For illustration, an example problem of gray-scale morphological processing by use of the gray-scale decomposition method has been studied as shown in Fig. 5.14. For comparison, a result obtained with direct gray-scale morphological processing is also provided.

Figure 5.14(a) shows a photo of a mine field. The two circular disks are surface mines. 3% salt-and-pepper noise has been added to the original image. Figure 5.14(b) shows the 16 slices of binary images cut out from the input. Each slice is evenly spaced by 16 gray levels. To remove the salt-and-pepper noise, binary opening and closing operations are performed on each of the 16 images. Results are shown in Fig. 5.14(c). The 16 binary images are then recombined into a single gray-scale image, as shown in Fig. 5.14(d). The salt-and-pepper noise is completely removed. The same input as that shown in Fig. 5.14(a) is then directly processed with digital gray-scale opening and closing. The result is shown in Fig. 5.14(e). Comparing Figs. 5.14(d) and 5.14(e) shows that results obtained with gray-scale decomposition and binary morphological processing are the same as those obtained with digital gray-scale morphological processing.

Therefore it is concluded that binary optical morphological processing can be extended to gray-scale image processing by means of gray-scale decomposition. The quality will be retained while speed is greatly increased.

Acknowledgments

The JPL work described in this chapter was performed by the Center for Space Microelectronics Technology, Jet Propulsion Laboratory, California Institute of Technology. Technical contributions and valuable discussions from J. Yu, B. Lau, and Y. Park were greatly appreciated.

References

[1] J. Serra, *Image Analysis and Mathematical Morphology* (Academic, London, 1982).

[2] C. R. Giardina and E. R. Dougherty, *Morphological Methods in Image and Signal Processing* (Prentice-Hall, Englewood Cliffs, NJ, 1988).

[3] H. J. A. M. Heijmans, *Morphological Image Operators* (Academic, Boston, 1994).

[4] A. Banerji and J. Goutsias, "Detection of mines and minelike targets using gray-scale morphological image reconstruction," in *Detection Technologies for Mines and Minelike Targets*, A. Dubey, I. Cindrich, J. M. Ralston, and K. Rigano, eds., Proc. SPIE **2496**, 836–849 (1995).

[5] R. Haralick, "Recognition methodology: algorithms and architecture," in *Image Pattern Recognition: Algorithm Implementations, Techniques, and Technology*, F. J. Corbett, ed., Proc. SPIE **755** (1987).

[6] F. Mok, "Volume holographic storage in lithium niobate photorefractive crystals," presented at the Image Processing Workshop, California Institute of Technology, Pasadena, CA, 17–18 May 1990.

[7] J. Yu, T.-H. Chao, P. Dumont, W. C. Fang, T. Glavich, and B. Lau, "Symbolic optical correlator," Final Rep. (California Institute of Technology, Pasadena, CA, 1991).

[8] P. Maragos, "Tutorial on advances in morphological image processing analysis," Opt. Eng. **26**, 623–632 (1987).

[9] K. S. Huang, *A Digital Optical Cellular Image Processor*, Vol. 24 of World Scientific Series in Computer Science (World Scientific, Singapore, 1990).

[10] K. S. O'Neill and W. T. Rhodes, "Morphological transformation by hybrid optical-electronic methods," in *Hybrid Imaging Processing*, D. Casasent and A. Tescher, eds., Proc. SPIE **638**, 41–44 (1986).

[11] J. M. Hereford and W. T. Rhodes, "Nonlinear optical image filtering by time-sequential threshold decomposition," Opt. Eng. **27**, 274–279 (1988).

[12] E. Ochoa, J. P. Allebach, and D. W. Sweeney, "Optical median filtering using threshold decomposition," Appl. Opt. **26**, 252–260 (1987).

[13] D. Casasent and E. Botha, "Optical symbolic substitution for morphological transformations," Appl. Opt. **27**, 3806–3809 (1988).

[14] Y. Li, A. Kostrzewski, D. H. Kim, and G. Eichmann, "Compact parallel real-time programmable optical morphological image processor," Opt. Lett. **14**, 981–983 (1989).

[15] T.-H. Chao, "Dynamically reconfigurable optical morphological processor and its applications," B. Javidi, ed., Proc. SPIE **1772**, 21–29 (1992).

[16] H. Langenbacher, T. H. Chao, T. Shaw, and J. Yu, "64 × 64 thresholding photodetector array for optical pattern recognition," D. Casasent, ed., Proc. SPIE **1959**, 350–358 (1993).

[17] E. C. Botha and D. Casasent, "Applications of optical morphological transformations," Opt. Eng. **28**, 501–505 (1989).

[18] J. Tanida and Y. Ichioka, "Optical logic array processor using shadowgrams," J. Opt. Soc. Am. **73**, 800–804 (1983).

[19] T.-H. Chao, "Optical morphological processor for high-speed image," JPL Task Plan 80-3805 (technical report submitted to the U.S. Office of Naval Research, Arlington, VA, 1995).

6

Nonlinear optical correlators with improved discrimination capability for object location and recognition

Leonid P. Yaroslavsky

6.1 Introduction: a review of the theory

In this chapter the synthesis of optical correlators for object location and recognition is described. For object (target) location, the correlator must determine the position (coordinates) of a small target object on an observed image that generally contains this object and is surrounded by a clutter of background objects and image details that may camouflage the target object. The localization device has to locate the target object as accurately as possible with the lowest possible probability of false identification of the target object with one of the background objects. In object recognition, the observed image of an object has to be identified with a certain image from a set of template images. The recognition device has to be able to make the identification with the lowest possible probability of misrecognition.

The capability of the localization and recognition devices to discriminate between the target object and false-background objects (in target localization) or wrong template images (in object recognition) is called their discrimination capability. It is this discrimination capability of localization and recognition devices that we are concerned with here.

As is well known, the gold standard for devices for localization or recognition of objects observed in a mixture with additive Gaussian noise is the combination of a matched filter and a unit for localizing the signal maximum at its output. However, for target location in images with a cluttered background or for recognition of images that cannot be regarded as copies of one and the same image with Gaussian noise added, this device is far from being optimal and has low discrimination capability.

This classical localization scheme with matched filtering can be regarded as a special case of localization and recognition devices that use, instead of the matched filter, a linear filter whose frequency response is optimized for a particular task. The function of the linear filter in localization (recognition) devices is to transform the signal space in a way that enables decision making on the basis of individual signal samples only, rather than on the basis of the entire signal. Because of separation into independent linear and pointwise nonlinear units, data analysis and implementation (both digital and optical) of such devices are much simplified. This motivates the use of this scheme in image processing and, in particular, for target location, provided the filter is optimized to guarantee the highest possible discrimination capability.

The synthesis of such an optimal filter was outlined in Ref. 1. It was shown that the requirement for the highest discrimination capability (or lowest probability of misrecognition) is approximately equivalent to the requirement that the filter has to provide the maximum ratio of the signal value in the point of the output plane, where the target object is located, to the standard deviation of the signal in the background part of the output plane or, for object recognition, tó the standard deviation of the filter output for the entire set of objects to be rejected. Because a number of random factors, such as sensor noise, unknown reference-object position, variability of its orientation, size, etc., are involved in localization (recognition), this requirement should be satisfied on average over these factors. For object location, the frequency response of the optimal filter was found in Ref. 1 to be

$$H_{opt}(\mathbf{f}) = \frac{AV_{ob}[TO^*(\mathbf{f})]}{AV_{imsys}AV_{x_0}(|B(\mathbf{f})|^2)}, \tag{6.1a}$$

where $TO^*(\mathbf{f})$ is the complex conjugate of the target-object Fourier spectrum, $|B(\mathbf{f})|^2$ is the power spectrum of the background component of the input image, and AV_{ob}, AV_{imsys}, and AV_{x_0} denote averaging of the corresponding variables over unknown parameters that define the variability of the target-object signal, averaging over realizations of imaging system noise, and averaging over unknown coordinates of the target, respectively. The term background-image component means the part of the observed image outside the area occupied by the target object. The filter of Eq. (6.1a), if realizable, is adaptive and provides the best (in the class of linear filters) possible performance for the given individual input image. Below it is referred to as the optimal adaptive correlator (OPAC).

One can show that, for object recognition, a similar set of filters

$$H_{opt}^{(i)}(\mathbf{f}) = \frac{AV_{ob}\left[TIM_i^*(\mathbf{f})\right]}{AV_{set}[|TIM_{k \neq i}(\mathbf{f})|^2]} \tag{6.1b}$$

is optimal, where $H_{opt}^{(i)}(\mathbf{f})$ is the optimal filter for the ith template image, $TIM_i^*(\mathbf{f})$ is the complex conjugate of the ith template image, and $AV_{set}[|TIM_{k \neq i}(\mathbf{f})|^2]$ denotes averaging spectra of the set of the template images, excluding the ith image. It is assumed here that the input image is subjected to filtering with the above filters, and the image identification is made by the selection of the template image that produces the highest output. Because of the obvious similarity between optimal filters for object location and recognition, the discussion here is confined to the former. The extension of the results to the latter is straightforward.

The design of the OPAC requires knowledge of the power spectrum of the background objects averaged over (random) coordinates of the target object. These data are not known before the target is located and have to be determined from the observed input image.

As a zero-order approximation of the background power spectrum, one may use the squared module of the entire observed image spectrum:

$$AV_{imsys}AV_{x_0}|B(\mathbf{f})|^2 \approx |IM(\mathbf{f})|^2. \tag{6.2}$$

This approximation is based on the assumption that the object size is much smaller than the size of the entire image (area of search).

For more accurate estimation of the background power spectrum, one can use the following two models for the representation of the background-image component:

$$b(\mathbf{x}) = w(\mathbf{x} - \mathbf{x}_0)\, im(\mathbf{x}), \tag{6.3}$$

$$b(\mathbf{x}) = im(\mathbf{x}) - to(\mathbf{x} - \mathbf{x}_0), \tag{6.4}$$

where $im(\mathbf{x})$ is the observed image, $b(\mathbf{x})$ is its background component, $w(\mathbf{x} - \mathbf{x}_0)$ is a window function,

$$w(\mathbf{x} - \mathbf{x}_0) = \begin{cases} 0 & \text{within the target object} \\ 1 & \text{elsewhere} \end{cases}, \qquad (6.5)$$

$\mathbf{x}_0$ is the target coordinate, and $to(\mathbf{x} - \mathbf{x}_0)$ is the target object.

For the model of Eq. (6.3), one can show that, with the assumption of uniform *a priori* distribution of the object coordinates $\mathbf{x}_0$ over the picture area S,

$$AV_{imsys}AV_{x_0}|B(\mathbf{f})|^2 \approx AV_{imsys}\{|IM(\mathbf{f})|^2 \bullet [|W(\mathbf{f})|^2/S]\}, \qquad (6.6)$$

where $\bullet$ denotes convolution and $|W(\mathbf{f})|^2$ is the squared magnitude of the window-function Fourier spectrum.

For the model of Eq. (6.4), one can show that, with the same assumption of uniform distribution of the object coordinates $\mathbf{x}_0$ over the picture area S,

$$AV_{imsis}AV_{x_0}|B(\mathbf{f})|^2 \approx AV_{imsis}[|IM(\mathbf{f})|^2 + |TO(\mathbf{f})|^2]. \qquad (6.7)$$

There are two options to account for the variability of the target object. First, one can subdivide the range of the variations into segments small enough to establish that the variations within each segment are negligible, and then one can design the filter for each individual version of the object. The price of this is an increase in computational costs.

Second, one can perform object averaging over the range of its variations and design a single filter for such an averaged object. Localization and recognition in this case are computationally much less costly, but the process obviously might have a lower discrimination capability (higher probability of misrecognition).

For object localization, there might also be one more cause for the object variability that has to be taken into account. This is the variability that is due to the background component of the image. In some applications, one may regard the object as being cut into the background such that the target-object signal component in the observed image can be described as

$$to(\mathbf{x} - \mathbf{x}_0) = to_0(\mathbf{x} - \mathbf{x}_0) - \bar{w}(\mathbf{x} - \mathbf{x}_0)\,im(\mathbf{x}), \qquad (6.8)$$

where $\bar{w}(\mathbf{x} - \mathbf{x}_0)$ is a window-function complement to the above function $w(\mathbf{x} - \mathbf{x}_0)$:

$$\bar{w}(\mathbf{x} - \mathbf{x}_0) = \begin{cases} 1 & \text{within the target object} \\ 0 & \text{elsewhere} \end{cases}, \qquad (6.9)$$

and $to_0(\mathbf{x})$ is the pure target-object signal that is cut into the image (it is assumed to be zero outside the target object). In this representation, the variations of the observed target object are defined by the term $\bar{w}(\mathbf{x} - \mathbf{x}_0)im(\mathbf{x})$. Therefore

$$AV_{ob}[to(\mathbf{x})] = AV_{x_0}[to(\mathbf{x})] = to_0(\mathbf{x}) - AV_{x_0}[\bar{w}(\mathbf{x})\,im(\mathbf{x} - \mathbf{x}_0)], \qquad (6.10)$$

or, in the spectral domain,

$$AV_{x_0}[TO(\mathbf{f})] = TO_0(\mathbf{f}) - AV_{x_0}\{[IM(\mathbf{f})\exp(i2\pi\mathbf{f}\mathbf{x}_0)] \bullet \bar{W}(\mathbf{f})\}$$

$$= TO_0(\mathbf{f}) - \bar{W}(\mathbf{f}) \bullet [IM(\mathbf{f}) \cdot CF(\mathbf{f})], \qquad (6.11)$$

where $CF(\mathbf{f})$ is the characteristic function (Fourier transform) of the distribution density of the object coordinate $\mathbf{x}_0$ and $\bullet$ designates convolution. If $\mathbf{x}_0$ is uniformly distributed over an area of search S that is much larger than the object size, its distribution characteristic function $CF(\mathbf{f})$ can be approximated by the delta function divided by S; therefore

$$AV_{x_0}[TO(\mathbf{f})] \approx TO_0(\mathbf{f}) - \bar{W}(\mathbf{f})IM(0)/S = TO_0(\mathbf{f}) - \bar{W}(\mathbf{f}) \cdot \overline{im}, \qquad (6.12)$$

where $\overline{im}$ is the arithmetic mean over the background-image component.

With estimations (6.6) and (6.7) of the background power spectrum, the optimal filter can be implemented either as

$$H_{opt}(\mathbf{f}) \propto \frac{AV_{ob}[TO^*(\mathbf{f})]}{AV_{imsys}|IM(\mathbf{f})|^2 \bullet |W(\mathbf{f})|^2} \qquad (6.13)$$

or as

$$H_{opt}(\mathbf{f}) \propto \frac{AV_{ob}[TO^*(\mathbf{f})]}{AV_{imsys}|IM(\mathbf{f})|^2 + |TO(\mathbf{f})|^2} \qquad (6.14)$$

with the appropriate selection of $AV_{ob}[TO^*(\mathbf{f})]$ as described above.

6.2 Nonlinear optical correlators

One can implement the described adaptive filters in nonlinear optical correlators. Two types of nonlinear optical correlator are now regarded as the most feasible: coherent optical correlators with a nonlinear light-sensitive optical light modulator installed in the correlator's Fourier plane [1] (Fig. 6.1) and joint transform correlators (JTC's) [2–6] with input images that are joint spectrum nonlinearly transformed in a computer or in an electronic amplifier before it modulates the output spatial light modulator (Fig. 6.2). In the setup of Fig. 6.1, a spatial light-sensitive nonlinear light modulator plays the role of a nonlinear element. Its transparency at each point is controlled by, and therefore depends on, the input-image power spectrum energy at this point.

In the setup of Fig. 6.2, an input image and a target object put side by side on the input plane are jointly Fourier transformed, and their joint power spectrum $|IM(\mathbf{f}) + TO(\mathbf{f})|^2$ is read out by a video camera and nonlinearly transformed pointwise in a nonlinear amplifier. The amplifier output is recorded on a spatial light modulator and Fourier transformed by the second lens to produce an output signal that represents the input-image-to-reference-object correlation signal in each one of its sidebands. The position of the reference object in the input image is indicated by the position of the correlation signal's highest peak.

The design and the implementation of such correlators require answering a number of practical questions such as

- What type of nonlinear transformation should one use in the above nonlinear correlators to ensure the highest discrimination capability?
- How sensitive is the correlator's discrimination capability to design factors such as the accuracy of realization of the nonlinear transformation, the limitation of the dynamic range of the nonlinear optical media and electronic components used, and the accuracy of optical alignment?

Several other modifications of the nonlinear optical correlators are also known. Among them, phase-only filters (POF's) and phase-only correlators (POC's) are the most popular, mainly because of their high light efficiency and relatively simple implementation [6–10].

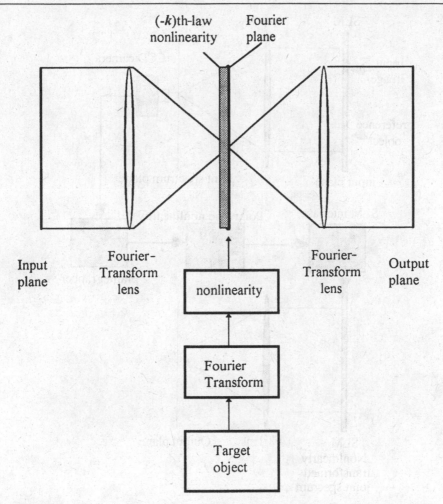

Fig. 6.1. Schematic diagram of the optical correlators with the $(-k)$th-law nonlinearity.

It is instructive to compare them with the above-mentioned implementations of the optimal correlator.

Here the recent experimental results that address the above-mentioned issues are reviewed. In Section 6.3 the results for the correlators of Fig. 6.1 with $(-k)$th-law nonlinearity in their Fourier plane are presented, and their discrimination capability is compared with that of the conventional matched filter, the POF, and the POC. In Section 6.4 the use of a logarithmic or $(1/k)$th-law nonlinearity in JTC's is justified, and corresponding experimental results are presented. In conclusion, the discussion is summarized.

6.3 Nonlinear optical correlators with $(-k)$th-law nonlinearity in the Fourier plane

Optical correlators with a nonlinearity applied to the input-image spectrum or to the target-object spectrum have attracted considerable attention from researchers looking for ways to improve the performance of optical correlators in pattern recognition and target location.

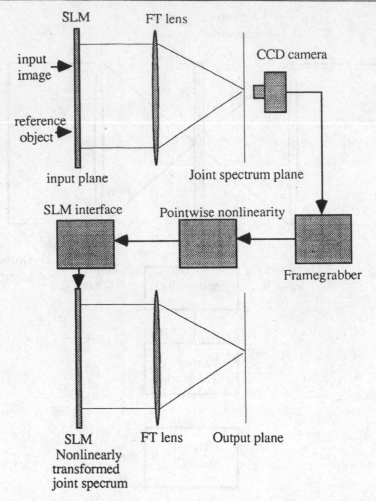

Fig. 6.2. Schematic diagram of the JTC with a nonlinear transformation of the joint spectrum. SLM, spatial light modulator; FT, Fourier-transform.

An important subclass of such correlators is an optical correlator with $(-k)$th-law nonlinearity, which is described by the relationship

$$\text{Out} = (|\text{In}|^2)^{-k}, \tag{6.15}$$

where In and Out are input and output signals, respectively, of the nonlinear transformation. According to the theory, one may expect that such correlators with nonlinearity in their Fourier plane, as shown in Fig. 6.1, and $k = 1$ will approximate the OPAC of Eqs. (6.1). Computer and optical experiments [11–13] have shown that such correlators do considerably outperform the classical matched filter in terms of discrimination capability. Here the results of an investigation into the correlator's sensitivity to the nonlinearity index k and the limitation of the nonlinearity's dynamic range and into the estimation method of the power spectrum of the background component of the input image are presented. The correlator's discrimination capability is compared with that of the POF's and the POC's.

Table 6.1. *Variety of correlators implemented in the computer model*

Correlator type	e	k	n	bl	lim
Matched filter	0	0	0	0	∞
POF	0	0	1/2	0	∞
POC	0	1/2	1/2	≤ 0	∞
Nonlinear	0 and 1	>0	0	>0	$<\infty$

The simulation was carried out in the computer model that has been designed to implement the following signal processing scheme:

$$c = \left| \text{IFT}\left[\text{IM LDR}\left(\{ \text{CONV}[h^{[\text{bl}]}, (|I|^2 + e|\text{TO}|^2)] \}^{-k} \right)(|\text{TO}|^2)^{-n}\text{TO}^* \right] \right|^2, \qquad (6.16)$$

where c is the correlator output signal, IM is the input-image Fourier spectrum, TO^* is the complex conjugate of the target-object Fourier-spectrum, IFT is the inverse Fourier-transform operator, $\text{CONV}[h^{[\text{bl}]}, (\cdot)]$ is a bl-fold linear convolution operator with point-spread function h, and LDR is a pointwise nonlinear operator:

$$\text{LDR}(x) = \begin{cases} x & x < \lim \\ \lim & \text{otherwise} \end{cases}. \qquad (6.17)$$

The model parameters bl, e, k, n, and lim define different processing modes. The parameter bl defines the degree of spectrum linear smoothing, or blur, that is assumed in the method of approximation (6.7) of background-image-component spectrum estimation. When bl $= 0$, no signal smoothing is performed. In the setup of Fig. 6.1, the convolution of approximation (6.6) can be associated with placement of the nonlinear spatial light modulator slightly out of the lens focus or with its low-resolution power. The parameter e takes one of two values, 0 or 1, and determines whether the target-object power spectrum is used to modify the observed image power spectrum [Eq. (6.8)]. The parameter k is the index of the nonlinearity. It defines the nonlinear transformation applied to the estimation of the observed input-image power spectrum. The parameter n defines nonlinear transformation applied to the target-object power spectrum. It was used to simulate the POF and the POC ($n = 1/2$). The parameter lim defines the degree of the dynamic range limitation.

The model covers a broad variety of nonlinear correlators. Some specific cases are shown in Table 6.1. In this table, the suboptimal correlators indicate the family of correlators that uses an estimation of the image-background-component power spectrum rather than the exact power spectrum, which has to be used in the optimal correlator design.

In the experiments, a range of values for the nonlinearity index k has been tested. For the exponent n, only those cases in which $n = 0$ (matched filter) and $n = 1/2$ (POF) were tested. As was mentioned above, the convolution operator was implemented as a bl-fold repetition (bl $= 0, 1, \ldots, 7$) of the elementary blur operator with a circular symmetric frequency response:

$$H(\mathbf{f}) = \exp(-f^2/0.1), \qquad (6.18)$$

where $f \in [0, 1]$ is spatial frequency normalized to the correlator's bandwidth. This function approximates the power spectrum of the target window function required by formula

Fig. 6.3. Image and its fragments used in the experiments.

(6.13) for the estimation of the image-background-component power spectrum according to approximation (6.6).

In the experiments with a limitation on the nonlinear material dynamic range, the parameter lim was selected as a $1/4^{lim}$ fraction of the signal maximum with lim $= 0, 1, \ldots, 7$. Thus the case in which lim $= 0$ corresponded to no dynamic range limitation, and the case in which lim $= 7$ corresponded to a limitation at the level of 1/16,384 of the signal maximal value.

The experiments were conducted with 16 128×128 pixel fragments of a 512×512 pixel satellite photograph of an urban area as test images (Fig. 6.3) and a small circular spot with a diameter of ~ 5 pixels as the target object embedded within the test images. The optical Fourier transform was approximated by the two-dimensional discrete Fourier transform. Two performance measures of the correlator's discrimination capability were computed in each experiment: the ratio of the object-signal maximum to the signal standard deviation over the background area in the correlation plane, or the signal-to-noise ratio (SNR), in terms of the background-signal variance (SNRV), and the ratio of the highest object-signal maximum to the maximum over the background area in the correlation plane, or the SNR, in terms of the background-signal maximum (SNRM). The former connects the experimental results with the analysis presented in Ref. 1, whereas the latter is a more adequate indicator of the correlator's ability to locate a target on a cluttered background reliably. In order to compare the discrimination capability of suboptimal correlators with the potential discrimination capability offered by the OPAC, the latter has also been implemented in the model. In this case, to allow an exact estimation of the image power spectra, the target object was not embedded within the test images.

6.3.1 Optimal adaptive correlator

The results for OPAC's designed individually for each image from the set of test images are represented in Fig. 6.4 along with those for the conventional correlator (matched filter). Figure 6.4(a) shows that the SNRV values at the output of the conventional correlator are almost uniformly distributed in the broad range from ~0.3 to 3 for the set of test images. This distribution allows the suggestion that the selected set of images is representative enough. Values for the SNRV and the SNRM as functions of the nonlinearity index k obtained for all the test images are plotted in Figs. 6.4(b) and 6.4(c), respectively. The plots clearly make it evident that

- The nonlinear correlators with the matched filter and with $(-k)$th nonlinearity in the Fourier plane can potentially outperform the conventional correlator considerably in terms of the discrimination capability. The observed gain is in the range from ~20 to more than 200 for the SNRV and from ~15 to more than 100 for the SNRM.
- Although the optimal value of k in terms of the SNRV values is equal to 1, the optimal value of k in terms of the SNRM tends to be slightly higher; on the average, it is approximately 10% higher.

It is remarkable that the SNRM decays relatively slowly for values of k higher than the optimal value. Observations show that for $k > 1$ the correlator's discrimination is basically due to a few frequency components of the signals, that is, the optimal filter acts as a bandpass filter with a very narrow bandwidth.

6.3.2 Suboptimal correlators with $(-k)$th-law nonlinearity and empirical estimation of the image power spectrum

As mentioned above, two estimation methods of estimating the background-image spectrum have been studied for correlators with $(-k)$th-law nonlinearity: linear smoothing (LS) method [$e = 0$ in approximation (6.6)] and the adding target spectrum (ATS) method [$e = 1$ in approximation (6.7)]. Experimental SNRV and SNRM values for these methods are plotted in Fig. 6.5 for a typical image from the test set. Averaged data for the LS method and two values of the smoothing parameter bl are plotted in Fig. 6.6 to illustrate the gain in the correlator's discrimination capability obtained by spectrum smoothing (bl $= 4$ corresponds to the highest average gain over the set of test images) compared with the direct use of the observed image spectrum as an estimation of the background-image power spectrum. Averaged data for the entire set of images and for the optimal value of the smoothing parameter bl $= 4$ are plotted in Fig. 6.7. These results suggest the following conclusions:

- Suboptimal nonlinear correlators provide a significant (on the average, ~20-fold for the SNRV and eightfold for the SNRM) improvement in the discrimination capability compared with that of the conventional matched filter (a vivid comparison of output signals is presented in Fig. 6.8). A twofold to fivefold gap still remains between the discrimination capability of the suboptimal correlators and that potentially achievable for an exactly known power spectrum of the background-image component. The estimation improves the nonlinear correlator's discrimination capability considerably.
- The LS method outperforms the ATS method of spectrum estimation if the degree of image spectrum smoothing is properly selected; the ATS method is much less sensitive

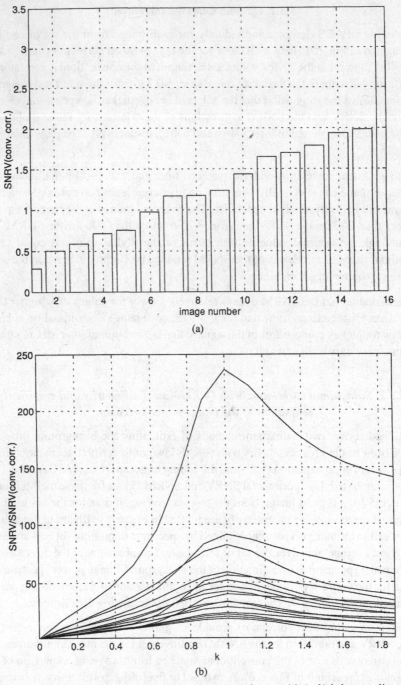

Fig. 6.4. Discrimination capability of nonlinear correlators with $(-k)$th-law nonlinearity and an exactly known image power spectrum as a function of the index k ($k=1$ corresponds to the optimal correlator): (a) sorted sequence of SNRV's at output of the conventional correlator ($k=0$) for the set of test images, (b) gain in SNRV values compared with that of the conventional correlator as a function of k for the set of test images, (c) gain in SNRM values compared with that of the conventional correlator as a function of k for the set of test images.

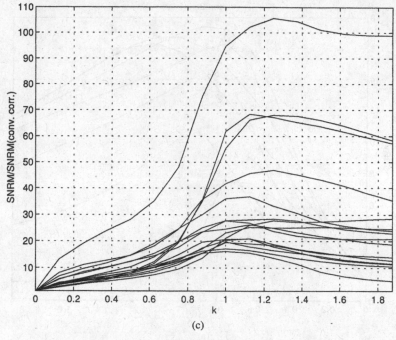

(c)

Fig. 6.4. (*cont.*)

to the degree of image spectrum smoothing and provides considerable improvement even without the spectrum smoothing.

- The optimal value of the nonlinearity index k is close to 1 for the correlators' performance evaluation in terms of both the SNRV and the SNRM; small deviations of k from its optimal value are not critical.
- A considerable degree of spectrum smoothing is required for achieving better discrimination capability; the optimal degree of smoothing corresponds to the resolution power of the nonlinear light modulator, $\sim$5–10 times lower than that required for the image. One can interpret this fact as an indication that the resolution power (or the square root of the number of degrees of freedom) of the nonlinear media should be of the order of magnitude of a $1/\sqrt{\text{TGTSIZE}}$ fraction of the required resolution power in the image domain, where TGTSIZE is the number of pixels (resolution cells) in the target object. Obviously this may considerably simplify the correlator's optical design and alignment.

Along with the nonlinearity's resolution power, the limitation of the nonlinearity's dynamic range is another important issue in the design of nonlinear correlators. For the optimal correlator, one would expect that the limitation of the nonlinearity's dynamic range deteriorates the correlator's discrimination capability. The simulation has shown that, while this deterioration does occur, it is not too severe (Fig. 6.9).

For the suboptimal correlator with the LS method of spectrum estimation, the experiments have shown that the dynamic range limitation does not necessarily deteriorate the correlator's discrimination capability. Moreover, a sort of trade-off is possible between the degree of the dynamic range limitation and the nonlinearity index k: the higher the limitation, the higher the nonlinearity index required for better discrimination capability. This is illustrated in Fig. 6.10, which shows that an appropriate choice of the nonlinearity index k

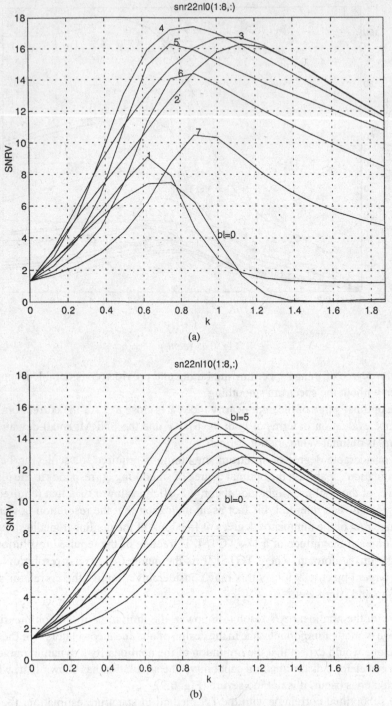

Fig. 6.5. Discrimination capability of the suboptimal correlators with $(-k)$th-law non-linearity as a function of the index k and the degree of spectrum smoothing for a typical image from the test set: (a) SNRV for the LS method of spectrum estimation, (b) SNRV for the ATS method of spectrum estimation, (c) SNRM for the LS method of spectrum estimation, (d) SNRM for the ATS method of spectrum estimation.

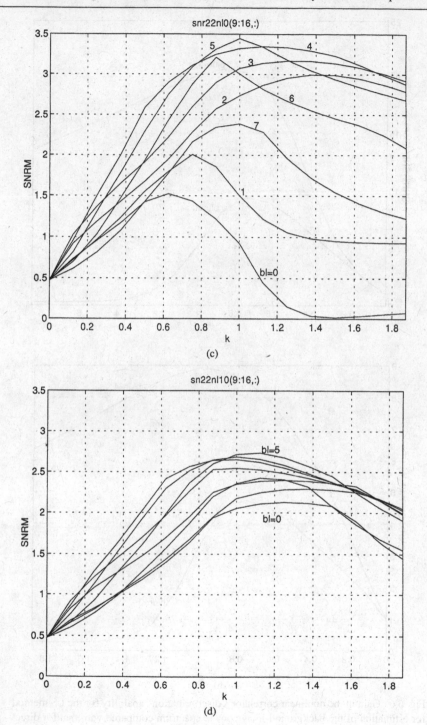

Fig. 6.5. (*cont.*)

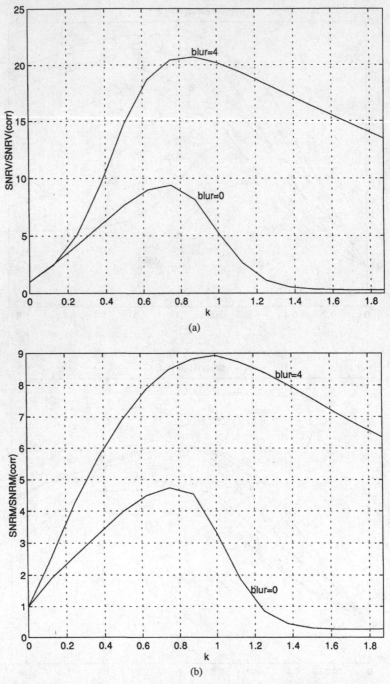

Fig. 6.6. Gain in the nonlinear correlator's discrimination capability for the LS method for estimation of the background-image power spectrum compared with that for direct use of the observed image power spectrum: (a) gain in the SNRV in relation to the conventional correlator matched filter, (b) the same as (a), but for the SNRM.

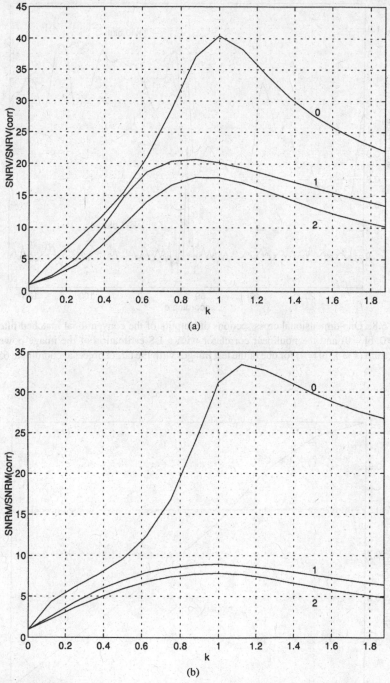

Fig. 6.7. Comparison of the optimal correlator (curve 0) with the correlators with the LS (curve 1) and the ATS (curve 2) methods for spectrum estimation with the optimized smoothing index bl: (a) gain in the SNRV in relation to the conventional correlator matched filter, (b) the same as (a), but for the SNRM.

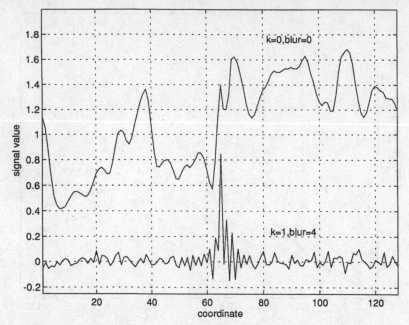

Fig. 6.8. One-dimensional cross sections of outputs of the conventional matched filter ($k = 0$, bl $= 0$) and the nonlinear correlator with a LS estimation of the image power spectrum ($k = 1$, bl $= 4$) for one of the test images with the target object at coordinate 65.

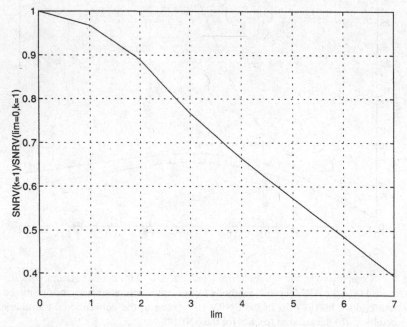

Fig. 6.9. Losses in the discrimination capability of the optimal correlator as functions of the dynamic range limitation degree.

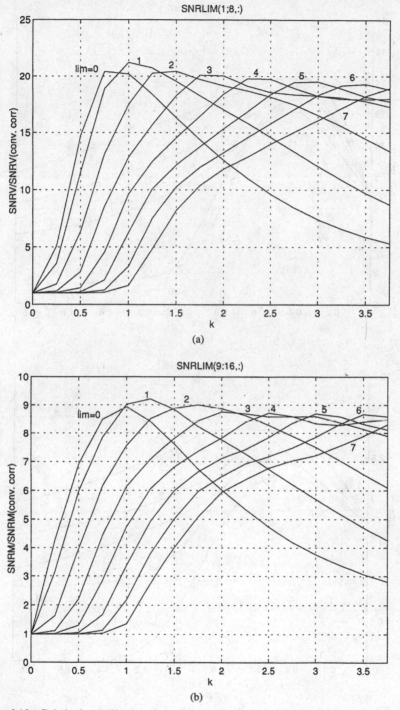

Fig. 6.10. Gain in the nonlinear correlator's discrimination capability for the LS method of spectrum estimation and limitation of the nonlinear media's dynamic range: (a) gain in the SNRV in relation to the conventional correlator matched filter, (b) the same as (a), but for the SNRM.

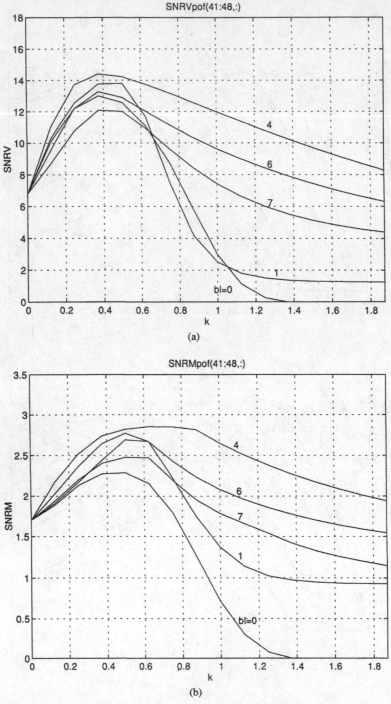

Fig. 6.11. Discrimination capability of the correlators with the POF and the $(-k)$th-law nonlinearity as a function of the index k and the degree of spectrum smoothing for a typical image from the test set: (a) the SNRV for the LS method of spectrum estimation, (b) the SNRM for the LS method of spectrum estimation.

provides practically the same discrimination capability as that of the nonlinear correlator with an unlimited dynamic range. It is remarkable that the dynamic range limitation may even improve, although slightly, the discrimination capability of the correlator without the dynamic range limitation. This can be attributed to the fact that power spectrum estimation by spectrum linear smoothing is not perfect enough.

Correlators with a strongly restricted dynamic range have a higher light efficiency than do correlators with no restrictions, because, according to Eq. (6.17), the nonlinear material can be made entirely transparent in the spectral plane wherever the signal that modulates it exceeds the limitation threshold. Therefore it might be advisable to use the observed trade-off between the nonlinearity index and the degree of dynamic range limitation to improve the correlator's light efficiency while preserving its discrimination capability.

6.3.3 Phase-only filters and phase-only correlators

The use of POF's instead of matched filters in optical correlators is advocated by the high light efficiency of the POF's. However, the discrimination capability of the correlators with POF's is significantly lower than that of the optimal and suboptimal nonlinear correlators [14, 15]. The corresponding graphs for the SNRV and the SNRM for the nonlinear correlators, with the POF representing the reference object, are given in Fig. 6.11 for the same test image as that of Fig. 6.5. One can see from Fig. 6.11 that the optimal nonlinearity index k in this case is approximately $k = 0.5$, which corresponds to the POC. The graphs also show that spectrum smoothing performed before its nonlinear transformation considerably improves the discrimination capability of the POC's, although their discrimination capability remains lower than that of the suboptimal nonlinear correlators with a matched filter that represents the target object. A comprehensive comparison of all the correlators described is presented in Fig. 6.12. These plots demonstrate a hierarchy of correlators with $(-k)$th-low nonlinearity in terms of their discrimination capability.

6.4 Nonlinear joint transform correlators

Nonlinear JTC's (NLJTC's) have been proposed as a tool for real-time pattern recognition, and different modifications of NLJTC's have been studied [2–6]. Here the issues of optimization of the nonlinearity of NLJTC's and the sensitivity of their discrimination capability to optical misalignments and to the dynamic range and resolution power limitations of nonlinear spatial light modulators that can be used in NLJTC's are addressed. It is shown that NLJTC's with logarithmic nonlinearity approximate the OPAC. By means of computer simulation, it is also shown that the NLJTC's with a $(1/k)$th-law nonlinearity within a limited dynamic range and the binary JTC may exhibit similar discrimination capabilities provided an appropriate selection of the nonlinearity index k, dynamic range limitation threshold, and the binarization threshold is made.

6.4.1 Logarithmic joint transform correlators

In this subsection, the use of logarithmic nonlinearity in NLJTC's is justified. With estimate (6.7), signal filtering in a suboptimal adaptive correlator is described by the formula

$$\text{OUT}(\mathbf{f}) = \frac{\text{IM}(\mathbf{f}) \cdot \text{TO}^*(\mathbf{f})}{\text{AV}_{\text{imsys}}|\text{IM}(\mathbf{f})|^2 + |\text{TO}(\mathbf{f})|^2} \tag{6.19}$$

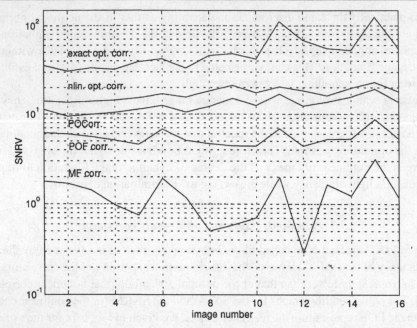

Fig. 6.12. Comparison of the discrimination capability of the optimal correlator (exact opt. corr.), suboptimal nonlinear optical correlators with the LS method of power spectrum estimation (nlin. opt. corr.), the POC (POCorr), the POF (POF corr), and a matched filter (MF corr) for the set of test images in terms of the SNRV.

(similar filtering schemes were also discussed, from different assumptions, in Refs. 16–18). Let $\Phi(\cdot)$ be a pointwise nonlinear transformation. If

$$\Phi(\cdot) = \log(\cdot), \tag{6.20}$$

the transformed joint power spectrum $\text{OUT}_{\text{NLJTC}}$ at the output of this nonlinear device can be written as

$$\text{OUT}_{\text{NLJTC}}(\mathbf{f}) = \log |\text{IM}(\mathbf{f}) + \text{TO}(\mathbf{f})|^2$$
$$= \log[|\text{IM}(\mathbf{f})|^2 + |\text{TO}(\mathbf{f})|^2 + \text{IM}(\mathbf{f}) \cdot \text{TO}^*(\mathbf{f}) + \text{IM}^*(\mathbf{f}) \cdot \text{TO}(\mathbf{f})]. \tag{6.21}$$

In target location, the size of the reference object is usually much smaller than the size of the input image. Therefore, for the majority of the spectral components,

$$|\text{IM}(\mathbf{f})|^2 + |\text{TO}(\mathbf{f})|^2 \gg |\text{IM}(\mathbf{f})| \cdot |\text{TO}(\mathbf{f})|. \tag{6.22}$$

With this assumption, $\text{OUT}_{\text{NLJTC}}$ is approximately equal to

$$\text{OUT}_{\text{NLJTC}}(\mathbf{f}) \approx \log[|\text{IM}(\mathbf{f})|^2 + |\text{RO}(\mathbf{f})|^2]$$
$$+ \frac{\text{IM}^*(\mathbf{f}) \cdot \text{TO}(\mathbf{f})}{|\text{IM}(\mathbf{f})|^2 + |\text{TO}(\mathbf{f})|^2} + \frac{\text{IM}(\mathbf{f}) \cdot \text{TO}^*(\mathbf{f})}{|\text{IM}(\mathbf{f})|^2 + |\text{TO}(\mathbf{f})|^2}. \tag{6.23}$$

In a JTC configuration, the two last terms displaying the correlation function are readily separated. The last term of this expression reproduces expression (6.19), which corresponds to the OPAC with an estimation of the background-image-component power spectrum by Eq. (6.7) but without averaging AV_{imsys}. Therefore one can conclude that the logarithmic

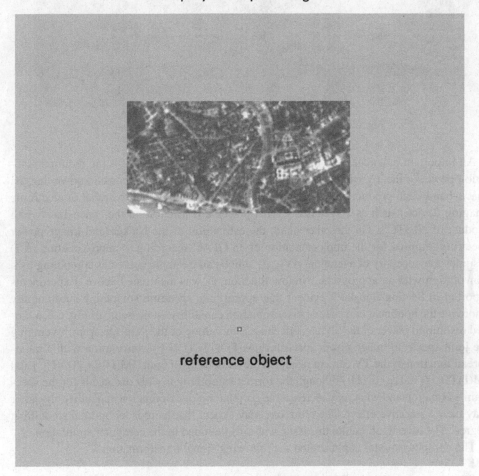

Fig. 6.13. Example of the input image of the JTC.

nonlinearity in the NLJTC promises a reasonable approximation to the OPAC and there-fore promises improved discrimination capability. Note that the averaging $\mathrm{AV}_{\mathrm{imsys}}$ can be implemented by a kind of smoothing (for example, by a linear blur) of the involved signal.

In order to verify this conclusion, computer simulation experiments were conducted with the above-mentioned set of test images and the target object. Arrangement of the input images and the target object that was used as an input for the JTC is shown in Fig. 6.13. A pairwise arrangement of test images allowed the performance of experiments with two images of the set made in parallel. In order to reduce boundary effects, the input images and the test image were inscribed into a uniform background with the gray level equal to the average gray level of the images. Experiments with this set of test images were aimed at investigation of the discrimination capability of adaptive nonlinear correlators in images with the same target but a different background. In order to verify the results obtained, an experiment was also performed with stereoscopic images (Fig. 6.14). In this case, the target object was a 21×21 pixel fragment of one image (shown in the box in Fig. 6.14), and the test image in which this fragment was to be located was the second image.

Left and right images with a reference object marked by box

Fig. 6.14. Stereoscopic images used in the experiments.

An important practical issue in the design of the JTC's of Fig. 6.2 is the required reso-lution power of the TV camera that reads out the joint spectrum. It is well known that the space–bandwidth product of optical system is usually much higher than that of electronic imaging devices such as TV cameras. This places a restriction on the space–bandwidth product of NLJTC's. On the other hand, the estimation of the background-image power spectrum required for the implementation of an OPAC according to approximation (6.7) assumes the necessity of averaging AV_{imsys}, which can be implemented as smoothing by a convolution with an appropriate window function, as was mentioned above. Experiments reported in Section 3 make it evident that appropriate spectrum smoothing substantially improves the nonlinear correlators' discrimination capability. In the setup of Fig. 6.2, a lim-ited resolution power of the TV camera causes smoothing of the joint spectrum. Averaging the joint spectrum image power spectrum $|IM(f) + TO(f)|^2$ by convolution with a point-spread function of the TV camera results in the smoothing of both $[IM(f)|^2 + |TO(f)|^2]$ and $IM(f)TO^*(f)$ in Eq. (6.11). Although the former smoothing is what one needs for the spec-trum estimation and what may increase the correlator's discrimination capability, the latter may have a negative effect. Therefore one may expect that there is an optimal smoothing degree. The investigation into this issue was also included in the computer simulation.

The computer model implemented the following signal transformation:

$$\text{coroutput} = \left| \text{IFT}\left[\text{CONV}\left(h^{[bl]}, \text{LDR}\{\log[|\text{FT}(\text{corinput})|^2]\}\right)\right]\right|^2 \qquad (6.24)$$

Here, coroutput and corinput are the output and the input images, respectively, of the correlator, FT and IFT are direct and inverse discrete Fourier transforms, respectively, that were used as approximations to an optical Fourier transform, LDR is the dynamic range limitation transformation of Eq. (6.17), and $\text{CONV}[h^{[bl]}, \bullet]$ is a bl-fold convolution operator with a point-spread function h. Frequency responses of the convolution operator for the parameter $bl = 0, 1, \ldots, 11$ are shown in Fig. 6.15.

Parameter lim of the dynamic range limitation was selected as a 0.1^{lim-1} fraction of the signal maximum with $lim = 1, 2, \ldots, 12$. Thus the case in which $lim = 1$ corresponded to no dynamic range limitation, and the case in which $lim = 12$ corresponded to limitation at the level of 10^{-11} of the signal maximal value.

The same two performance measurements of the correlator's discrimination capability as for the above correlators with the $(-k)$th-law nonlinearity, SNRV and SNRM, were computed in each experiment.

The experimental results are presented in Figs. 6.16 and 6.17. Plots in Fig. 6.16 represent average values of the SNRV and the SNRM for the logarithmic JTC as functions of the dynamic range limitation parameter lim; averaging has been made over the set of test images.

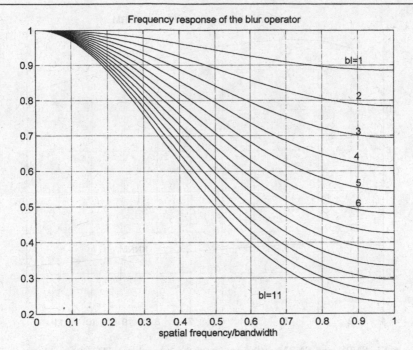

Fig. 6.15. Frequency response of the blur operator used in the experiments.

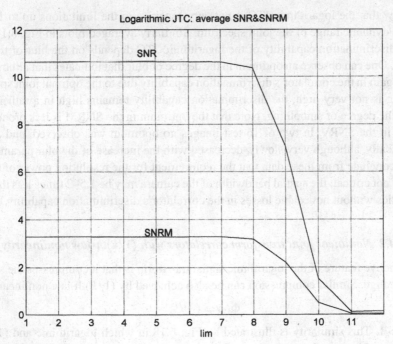

Fig. 6.16. SNRV and SNRM at the output of the logarithmic JTC versus limitation threshold.

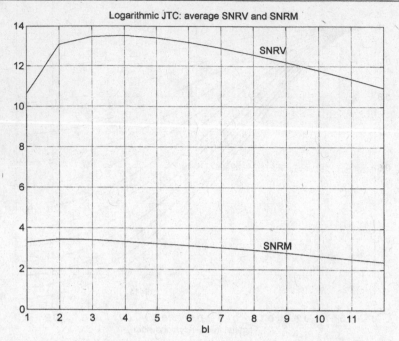

Fig. 6.17. SNRV and SNRM at the output of the logarithmic JTC versus blur parameter bl.

They show that the logarithmic JTC is not very sensitive to the limitations up to 10^{-7} of the entire dynamic range of the joint spectrum. Similarly averaged plots in Fig. 6.17 show how the discrimination capability of the logarithmic JTC depends on the blur of the joint spectrum. One can observe an optimum in the degree of blur that indicates that although the expected gain in the correlator's discrimination capability due to the optimal joint spectrum smoothing is not very high, the discrimination capability remains high in a rather broad range of the degree of smoothing. Note that the optimum in the SNRM is less pronounced than that in the SNRV. In two of 16 test images no optimum was observed, and SNRM monotonically, although very slowly, decreased with the increase of the blur parameter bl. One can conclude from these data that the requirement for the resolution power of the TV camera is not critical: the spatial bandwidth of the camera may be 1.5–2 times less than that of the optics without noticeable losses in the correlator's discrimination capability.

6.4.2 Nonlinear joint transform correlators with (1/k)th-law nonlinearity

The distinctive feature of the logarithmic signal transform is that it compresses the signal's dynamic range. Similar compression can be also achieved by $(1/k)$th-law nonlinearity,

$$\Phi(\cdot) = (\cdot)^{1/k}, \tag{6.25}$$

when $k \gg 1$. This similarity is illustrated by Fig. 6.18 in which logarithmic and $(1/k)$th-law nonlinearities are plotted together after a corresponding normalization by a constant. Therefore one can expect that nonlinear JTC's with $(1/k)$th-law nonlinearity [Eq. (6.25)] and $k \gg 1$ will perform nearly as well as the logarithmic JTC.

The simulation experiments with NLJTC's with $(1/k)$th-law nonlinearity were carried out with the same computer model as that for the logarithmic nonlinearity, except the logarithmic

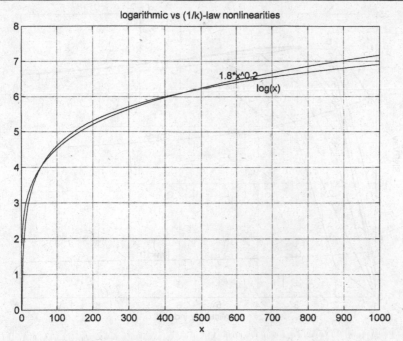

Fig. 6.18. Illustration of the similarity between logarithmic and $(1/k)$th-law nonlinearities.

transformation was substituted by the $(1/k)$th-law transformation and spectrum smoothing was not applied:

$$coroutput = |IFT(LDR\{[|FT(corinput)|^2]^{-k}\})|^2. \tag{6.26}$$

The corresponding averaged experimental data are plotted in Figs. 6.19(a) and 6.19(b) for the SNRV and the SNRM, respectively, as functions of the nonlinearity index k for the dynamic range limitation parameter lim $= 1, \ldots, 10$. Similar results were obtained for stereoscopic test images [Fig. 6.19(c)]. They show that

- NLJTC's with $(1/k)$th-law nonlinearity may have considerably improved discrimination capability compared with that of the JTC without nonlinear transformation of the joint spectrum (case $k = 1$). In terms of the SNRM, the averaged gain exceeds 3 times. A comparison of the corresponding data for the NLJTC's with $(1/k)$th-law nonlinearity and the logarithmic JTC shows that, with an appropriate selection of the parameters k and lim, the former performs slightly better.
- As for the logarithmic JTC, the discrimination capability of the NLJTC with $(1/k)$th-law nonlinearity is not very sensitive to the limitation threshold, provided proper selection of the nonlinearity index $k \gg 1$ is made.
- A trade-off exists between the nonlinearity index k and the dynamic range limitation parameter lim: a higher degree of the dynamic range limitation requires lower values of k. With this trade-off, the discrimination capability remains practically the same.
- The discrimination capability of the NLJTC with $(1/k)$th-law nonlinearity does not depend noticeably on k, provided k exceeds a minimal value determined by the dynamic range limitation level.

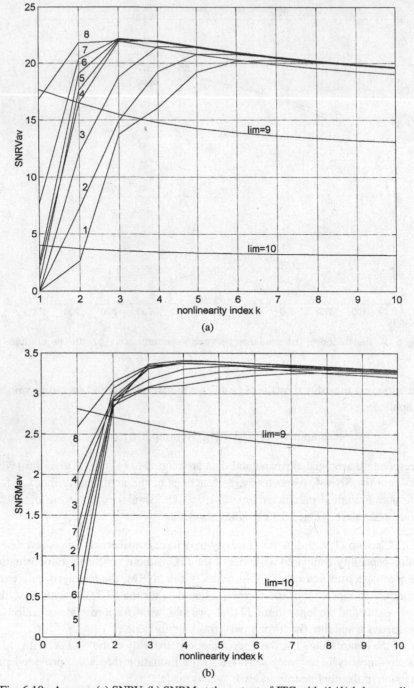

Fig. 6.19. Average (a) SNRV, (b) SNRM at the output of JTC with $(1/k)$th-law nonlinearity versus nonlinearity index k for the set of test images; (c) plot of the SNRV for stereoscopic images.

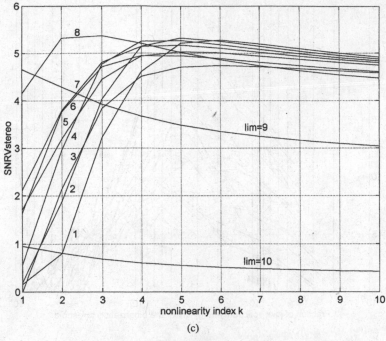

Fig. 6.19. (*cont.*)

6.4.3 Binary joint transform correlators

Many nonlinear media are binary, that is, they can be in only two states: transparent or opaque. This feature can be described as hard limiting (binarization):

$$\text{LDR}_{hlim}(x) = \begin{cases} 0 & x < \text{lim} \\ 1 & \text{otherwise} \end{cases}. \tag{6.27}$$

From the experiments with the $(1/k)$th-law nonlinearity, one can conclude that even a simple dynamic range limitation alone may substantially improve the NLJTC discrimination capability, provided the limitation threshold is properly chosen. This fact allows us to assume that binary JTC's with the hard limitation according to Eq. (6.27) may also have sufficiently high discrimination capability. Experiments reported in the literature [5, 19–21] also support this assumption. The simulation results confirmed this conjecture. The simulation was carried out with the same set of test images according to the model

$$\text{coroutput} = |\text{IFT}(\text{LDR}_{hlim}\{[|\text{FT}(\text{corinput})|^2]\})|^2. \tag{6.28}$$

The simulation results are plotted in Fig. 6.20 for all images of the test set as functions of the fraction of the joint spectrum energy under the binarization threshold. They clearly show that, for all the test images, there is an optimal value of the binarization threshold for which the correlator's discrimination capability reaches the levels close to those achievable for the logarithmic and $(1/k)$th-law NLJTC's. This optimal value corresponds to the binarization threshold in the range from approximately a 3×10^{-4} to an 8×10^{-4} fraction of the entire energy of the joint spectrum. This range is of the same order of magnitude as that of the ratio 7.5×10^{-4} of the area occupied by the target object (5×5 pixels) to the area of the input image (128×256 pixels) as one would expect it to be. One can also see that the

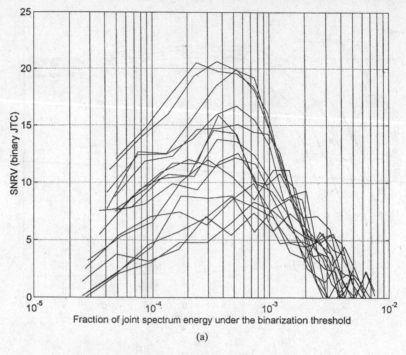

(a)

SNRMbin for the set of test images

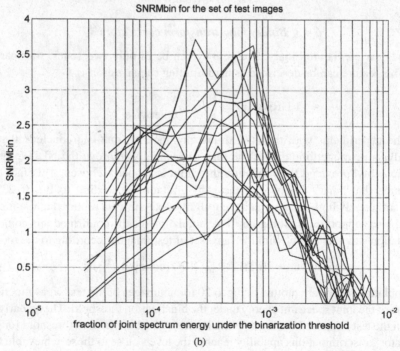

(b)

Fig. 6.20. (a) SNRV, (b) SNRM at the output of the binary JTC versus the fraction of the joint spectrum energy under the binarization threshold for the set of test images.

discrimination capability of the binary JTC is relatively tolerant to reasonable deviations of the binarization threshold from its optimal value. The achievable SNRV and SNRM for binary JTC's are lower than in the optimal NLJTC but are still much higher than those for the matched filter [$\lim = 1$, $k = 1$ in Fig. 6.19(a)].

6.5 Conclusion

The discrimination capability of several types of nonlinear optical correlators for target location and recognition has been investigated by computer simulation: ideal OPAC's, suboptimal correlators with $(-k)$th-law nonlinearity and a matched filter in their Fourier domain, POF's and POC's, and NLJTC with logarithmic and $(1/k)$th-law nonlinearities applied to joint spectrum binary JTC's. In the simulations, design factors such as limitation of the nonlinear media and electronic component dynamic range and resolution power have been taken into consideration. The experiments were conducted with a representative set of test images, and the results were evaluated both statistically by averaging over the test set and individually for each of the test images in the set.

The general conclusion is that, with an appropriate selection of the nonlinearity parameters, the dynamic range limitation, and the binarization thresholds, the nonlinear correlators exhibit the substantially improved discrimination capability, and the technical requirements for the nonlinear spatial light modulators required for the implementation of the nonlinear correlators and for their electronic components and optical alignment are surprisingly low.

It is also worth mentioning that the experiments have supported the basic assumption on which the theory of optimal adaptive correlators [1] was based: The discrimination capability of the correlators (defined as the SNRM of the correlator's response to the target object to the highest peak of the correlator's response to the objects to be rejected) is directly associated with the SNRV of the correlator's response to the target object to the standard deviation of the correlator's output signal measured over the area not occupied by the target object. In all the experiments, optimal values of the nonlinear correlators' parameters in terms of SNRV and in terms of SNRM were found, to all practical purposes, to be identical.

Acknowledgment

The author thanks P. Chavel, the Group of Physics of Images of the Institute of Theoretic and Applied Optics and the Centre National de la Recherche Scientifique, France, for the support of this research during the author's fellowship at the Centre National.

References

[1] L. P. Yaroslavsky, "The theory of optimal methods for localization of objects in pictures," in *Progress in Optics*, E. Wolf, ed. (Elsevier, Amsterdam, 1993), Vol. XXXII, pp. 147–201.

[2] C. S. Weaver and J. W. Goodman, "A technique for optically convoluting two functions," Appl. Opt. **5**, 1248–1249 (1966).

[3] F. T. S. Yu and X. J. Lu, "A real-time programmable joint transform correlator," Opt. Commun. **52**, 47 (1984).

[4] F. T. S. Yu, S. Jutamulia, T. W. Lin, and D. A. Gregory, "Adaptive real-time pattern recognition using a liquid crystal TV based joint transform correlator," Appl. Opt. **26**, 1370–1372 (1987).

[5] F. T. S. Yu and T. Nagata, "Binary phase-only joint transform correlator," Microwave Opt. Technol. Lett. **2**, 15–17 (1989).

[6] F. T. S. Yu and D. Gregory, "Optical pattern recognition: architectures and techniques," Proc. IEEE **84**, 733–752 (1966).

[7] H. Bartelt, "Unconventional correlators," in *Optical Signal Processing*, J. L. Horner, ed. (Academic, San Diego, 1987), pp. 97–128.

[8] T. Szoplik and K. Chalasinska-Macukow, "Towards nonlinear optical processing," in *International Trends in Optics*, J. W. Goodman, ed. (Academic, Boston, San Diego, New York, 1991), pp. 451–464.

[9] D. L. Flannery and J. L. Horner, "Fourier optical signal processors," Proc. IEEE **77**, 1511–1527 (1989).

[10] K. Chalasinska-Macukow, "Generalized matched spatial filters with optimum light efficiency," in *Optical Processing and Computing*, H. H. Arsenault, T. Szoplik, and B. Macukow, eds. (Academic, Boston, 1989), pp. 31–45.

[11] L. P. Yaroslavsky, "Optical correlators with $(-k)$th-law nonlinearity: optimal and suboptimal solutions," Appl. Opt. **34**, 3924–3932 (1995).

[12] V. N. Dudinov, V. A. Krishtal, and L. P. Yaroslavsky, "Localization of objects in images by the coherent optics techniques," Geod. Cartogr. 42–46 (1977).

[13] L. P. Yaroslavsky, *Digital Picture Processing. An Introduction* (Springer-Verlag, Berlin, Heidelberg, 1985), Chap. 9.

[14] J. Campos, F. Turon, L. P. Yaroslavsky, and M. J. Yzuel, "Some filters for reliable recognition and localization of objects by optical correlators: a comparison," Int. J. Opt. Comput. **2**, 341–366 (1993).

[15] L. P. Yaroslavsky, "Is the phase-only filter and its modifications optimal in terms of the discrimination capability in pattern recognition?" Appl. Opt. **31**, 1677–1679 (1992).

[16] P. Refregier, V. Laude, and B. Javidi, "Nonlinear joint-transform correlation: an optimal solution for adaptive image discrimination and input noise robustness," Opt. Lett. **19**, 405–407 (1994).

[17] H. Inbar, N. Konforti, and E. Marom, "Modified joint transform correlator binarized by error diffusion. I. Spatially constant noise-dependent range limit," Appl. Opt. **33**, 4434–4441 (1994).

[18] H. Inbar and E. Marom, "Modified joint transform correlator binarized by error diffusion. II. Spatially variant range limit," Appl. Opt. **33**, 4444–4451 (1994).

[19] F. T. S. You, Q. W. Song, Y. Suzuki, and M. W, "Hard clipping joint transform correlator using a microchannel spatial light modulator," Microwave Opt. Technol. Lett. **1**, 323–326 (1988).

[20] F. T. S. Yu, F. Cheng, T. Nagata, and D. A. Gregory, "Effects of fringe binarization of multiobject joint transform correlation," Appl. Opt. **28**, 2988–2990 (1989).

[21] A. Tanone, C.-M. Uang, F. T. S. Yu, E. C. Tam, and D. A. Gregory, "Effects of thresholding in joint transform correlation," Appl. Opt. **31**, 4816–4822 (1992).

7

Distortion-invariant quadratic filters

Gregory Gheen

7.1 Introduction

In this chapter, quadratic filters are developed for invariant pattern recognition, and their implementation in coherent optical processors is discussed. There are a number of reasons for implementing quadratic filters for optical pattern recognition systems. First, quadratic filters can be mapped onto a variety of optical processor architectures [1]. This provides flexibility in developing optical pattern recognition systems in which, for example, a change in the optical processor architecture can be used to avoid specialized optical components that may be physically unrealizable or too expensive to develop. Second, quadratic filters offer advantages over linear filters in various pattern recognition problems. Quadratic filters are the Bayes optimum for separating two Gaussian-distributed signals [2]. They have been used to classify sonar signals to mitigate factors such as multipath arrival and phase distortion caused by fluctuations in the propagation medium [3]. Furthermore, the structure of a quadratic filter can model the distortions caused by linear transformation groups such as rotation, scaling, shearing, etc. [4]. Finally, quadratic filters provide an avenue to exploit the high throughput potential of optical processors. In general, quadratic filters are computationally intensive and this has limited their use in many applications. However, optical processors offer the potential for very large throughputs and can provide real-time implementation of quadratic filters. Hence the attributes of quadratic filters and optical processors are well matched to each other.

This chapter is divided into three sections. Section 7.2 serves as a technical introduction. The topics discussed in Section 7.2 include the mathematical notation used in the chapter, the Bayes theory for pattern recognition, quadratic filters and their optical implementation, and the role of normalization in classification. In Section 7.3, distortion-invariant quadratic filters are developed. Two approaches to distortion-invariant filter design are described. First, invariant quadratic filters are developed with the requirement that the filter generate the same normalized response to all the targets in a training class. Next, invariant quadratic filters are developed from a group theoretical approach in which the target class is defined by a single exemplar signal under the action of a linear transformation group. Although the basic approach of these two filter designs is different, they converge to the same filter if the underlying distortion is described by a linear transformation group. It is shown that this result provides a new interpretation for the correlation matrix and gives new meaning to the role of principal component analysis in pattern recognition. In Section 7.4, the performance of invariant quadratic filters is characterized by the Fisher ratio as a performance metric. The Fisher ratio is derived as a function of various factors including the second-order statistics of the target, the clutter probability density functions (pdf's), and the signal-to-noise ratio (SNR). Comparing the Fisher ratio of a quadratic filter with a linear filter provides a basis for assessing the relative performance gain offered by quadratic filters.

7.2 Technical background

7.2.1 Notation

In this section, signals are defined as vectors that belong to either a finite-dimensional or an infinite-dimensional signal space. Finite-dimensional signals are denoted by boldfaced, lowercase letters (e.g., $\mathbf{f}$). Infinite-dimensional signals are denoted by the Dirac bracket notation (e.g., $|\mathbf{f}\rangle$ or its dual $\langle\mathbf{f}|$). In general, signals in the real world are finite-dimensional (i.e., contain a finite number of measured values) and are obtained by the sampling of a bandlimited signal over some interval. For our purposes, infinite-dimensional signals are bandlimited signals given over a finite interval. They are infinite dimensional because they are continuous and contain an infinite number of values. The main reason for introducing infinite-dimensional vectors is to avoid the issue of interpolation when switching between different coordinate systems (e.g., rectangular to polar).

An inner product or correlation of two N-dimensional signals is denoted by $\mathbf{h}^t\mathbf{f}$ and has the typical meaning

$$\mathbf{h}^t\mathbf{f} = \sum_{n=1}^{N} h_n f_n = \|\mathbf{h}\|\,\|\mathbf{f}\|\cos\theta, \tag{7.1}$$

where t denotes the conjugate transpose of $\mathbf{h}$, f_n are the nth elements in vector $\mathbf{f}$, $\|\mathbf{f}\|$ is the magnitude or length of vector $\mathbf{f}$, and θ is the angle between vectors $\mathbf{f}$ and $\mathbf{h}$. Likewise, the inner product between two infinite-dimensional signals is denoted by $\langle\mathbf{h}\,|\,\mathbf{f}\rangle$ and has the meaning

$$\langle\mathbf{h}\,|\,\mathbf{f}\rangle = \int_{a}^{b} h^*(t)f(t)\,\mathrm{d}t = \|\mathbf{f}\|\,\|\mathbf{h}\|\cos\theta, \tag{7.2}$$

where * denotes the complex conjugate, the integral is over the interval $[a, b]$ in t, $\|\mathbf{f}\| = \langle\mathbf{f}\,|\,\mathbf{f}\rangle^{1/2}\|\mathbf{f}\| = \langle\mathbf{f}\,|\,\mathbf{f}\rangle^{1/2}$ is the magnitude of $|\mathbf{f}\rangle$, and θ is the angle between vectors $|\mathbf{f}\rangle$ and $|\mathbf{h}\rangle$. For images, the integral in Eq. (7.2) is over a two-dimensional (2-D) interval defined by the image support.

7.2.2 Bayes decision theory and discriminant functions

The Bayes decision theory is the foundation of statistical pattern recognition and serves as the starting point for this subsection. We assume a two-class problem, in which one class is the target class (i.e., the set of signals to be detected) and the other class is the clutter class (i.e., any nontarget signal). The objective is to design a classifier that discriminates between the target class and the clutter class. The two-class problem can be extended to a multiclass problem by the solution of a series of two-class problems. Inherent in the Bayes decision theory is the assumption that the target and the clutter statistics are known. However, in reality, the associated statistical distribution can be extremely hard to model because of the variability that occurs in the signal collection process. For example, sonar signals can be affected by noise, superimposed clutter, time-varying Doppler shifts, and multipath arrivals.

Images can be affected by noise, variations in perspective, changes in lighting conditions, variations in the imaging environment, occlusions, etc. This wide variability in the target class makes it difficult to enumerate or analytically model a particular target class and is the principal reason why the pattern recognition problem is so difficult to solve.

The Bayes decision theory employs a risk function to perform classification [2]. The risk associated with classifying the input signal $\mathbf{f}$ as the target is denoted by $R(t\,|\,\mathbf{f})$ (i.e., the risk of deciding t, given $\mathbf{f}$). The risk associated with classifying $\mathbf{f}$ as clutter is denoted by $R(c\,|\,\mathbf{f})$. Classification is performed according to the Bayes decision rule:

$$\text{classify } \mathbf{f} \text{ as target if} \quad R(t\,|\,\mathbf{f}) \leq R(c\,|\,\mathbf{f}),$$

$$\text{classify } \mathbf{f} \text{ as clutter if} \quad R(t\,|\,\mathbf{f}) > R(c\,|\,\mathbf{f}). \tag{7.3}$$

The risk functions are given by

$$R(t\,|\,\mathbf{f}) = \lambda_{tt} P(t\,|\,\mathbf{f}) + \lambda_{tc} P(c\,|\,\mathbf{f}),$$

$$R(c\,|\,\mathbf{f}) = \lambda_{ct} P(t\,|\,\mathbf{f}) + \lambda_{cc} P(c\,|\,\mathbf{f}), \tag{7.4}$$

where λ_{tt} is the loss associated with classifying $\mathbf{f}$ as a target when it is a target, λ_{tc} is the loss associated with classifying $\mathbf{f}$ as a target when it is clutter, $P(t\,|\,\mathbf{f})$ is the *a posteriori* probability (i.e., the probability of a target being present given $\mathbf{f}$), and so on. When the Bayes rule is used, the *a posteriori* probability can be written as

$$P(t\,|\,\mathbf{f}) = \frac{p(\mathbf{f}\,|\,t)P(t)}{p(\mathbf{f})}, \tag{7.5}$$

where $p(\mathbf{f}\,|\,t)$ is the state-conditional probability density (i.e., the density function for the target class), $P(t)$ is the *a priori* probability (i.e., the probability of a target occurring), and $p(\mathbf{f})$ is the pdf for $\mathbf{f}$ [i.e., $p(\mathbf{f}) = p(\mathbf{f}\,|\,t) + p(\mathbf{f}\,|\,c)$]. Substituting Eqs. (7.4) and (7.5) into expressions (7.3) gives the Bayes Decision rule:

$$\text{classify } \mathbf{f} \text{ as target if} \quad \frac{(\lambda_{ct} - \lambda_{tt})p(\mathbf{f}\,|\,t)P(t)}{(\lambda_{tc} - \lambda_{cc})p(\mathbf{f}\,|\,c)P(c)} \geq 1,$$

$$\text{classify } \mathbf{f} \text{ as clutter} \quad \text{otherwise.} \tag{7.6}$$

In general, the values for the loss coefficients are ambiguous. To remove this difficulty, the error rate is frequently minimized by the imposition of a unit loss for incorrect classifications and no loss for correct classifications. Substituting these loss values into expressions (7.6) gives the decision rule for the minimum-error rate classifier as

$$\text{classify } \mathbf{f} \text{ as target if} \quad \frac{p(\mathbf{f}\,|\,t)P(t)}{p(\mathbf{f}\,|\,c)P(c)} \geq 1,$$

$$\text{classify } \mathbf{f} \text{ as clutter} \quad \text{otherwise.} \tag{7.7}$$

The minimum-error classifier has the general form $g(\mathbf{f}) \geq T$, where $g(\mathbf{f})$ is a discriminant function that operates on input signal $\mathbf{f}$ and T is a threshold value. The discriminant function divides the signal space into two regions: One region is associated with the target class and the other region is associated with the clutter class. Solving a pattern recognition problem is equivalent to finding a discriminant function that can distinguish between the target region(s) and the clutter region(s).

Discriminant functions and the decision boundaries that they implement provide a general framework for describing classifiers and the pattern recognition problem. In this chapter the primary focus is on quadratic discriminant functions (i.e., quadratic filters). Below it is shown how quadratic filters arise from the Bayes decision theory under certain simplifying assumptions.

The primary difficulty with the Bayes decision theory is that it requires knowledge of the target and the clutter pdf's. This presents a problem as these are usually unknown. One approach that is used to overcome this problem is to assume a particular functional form for the target pdf and the clutter pdf. The available data are then used to determine the parameters of the assumed functional form. This approach is called the parameteric approach to pattern classification. In the parameteric approach, a Gaussian pdf is frequently assumed. This may be motivated by fundamental considerations such as the central limit theorem; however, more times than not, a Gaussian distribution is adopted because it is completely described by its first- and second-order statistics (i.e., the mean and the covariance matrix of a training set).

The pdf's for Gaussian-distributed target and clutter classes are given by

$$p(\mathbf{f}\,|\,t) = (2\pi)^{-N/2}|C_t|^{-1/2}\exp\left[-(\mathbf{f}-\mathbf{m}_t)^t C_t^{-1}(\mathbf{f}-\mathbf{m}_t)/2\right],$$
$$p(\mathbf{f}\,|\,c) = (2\pi)^{-N/2}|C_c|^{-1/2}\exp\left[-(\mathbf{f}-\mathbf{m}_c)^t C_c^{-1}(\mathbf{f}-\mathbf{m}_c)/2\right], \qquad (7.8)$$

respectively, where N is the dimension of the signal space, the vectors $\mathbf{m}_t$ and $\mathbf{m}_c$ are the means of the target and the clutter pdf's, respectively, C_t and C_c are the target and the clutter covariance matrices, respectively, and $|C_t|$ and $|C_c|$ denote their determinants. Substituting these distributions into the expression for the minimum error classifier in expression (7.7) and taking the logarithm of both sides gives

$$\tfrac{1}{2}\mathbf{f}^t\left(C_c^{-1} - C_t^{-1}\right)\mathbf{f} + \left(\mathbf{m}_t^t C_t^{-1} - \mathbf{m}_c^t C_c^{-1}\right)\mathbf{f} > T, \qquad (7.9)$$

where $T = \tfrac{1}{2}(\mathbf{m}_t^t C_t^{-1}\mathbf{m}_t - \mathbf{m}_c^t C_c^{-1}\mathbf{m}_c + \ln\{|C_c|\} - \ln\{|C_t|\}) - \ln\{P(t)\} + \ln\{P(c)\}$. Inequality (7.9) is a linear-quadratic discriminant function. This result shows that a quadratic filter can provide Bayes optimum separation of two Gaussian-distributed signal distributions.

7.2.3 Quadratic filters and their optical implementation

In general, a quadratic filter has the form

$$Q(\mathbf{f}) = \mathbf{f}^t \mathcal{M}\mathbf{f} + \mathbf{h}^t\mathbf{f} + c, \qquad (7.10)$$

where $\mathbf{f}$ is an N-dimensional vector representing the input signal, $\mathcal{M}$ is an $N \times N$ symmetric matrix, $\mathbf{h}$ is an N-dimensional vector, and c represents a constant bias. Equation (7.10) defines a second-order polynomial function on an N-dimensional vector space. Equation (7.10) can be interpreted as a truncated polynomial series approximating a nonlinear function $g(\mathbf{f})$ over some region of the N-dimensional signal space. The complete expansion of the function is described by a Volterra series expansion of the function $g(\mathbf{f})$ [5].

Here we are interested primarily in quadratic filters of the form

$$Q(\mathbf{f}) = \mathbf{f}^t \mathcal{M}\mathbf{f}, \qquad (7.11)$$

as this form is easily implemented in coherent optical processors. In order to distinguish between Eqs. (7.10) and (7.11), we refer to Eq. (7.10) as a linear-quadratic filter and to Eq. (7.11) as a pure quadratic filter or simply a quadratic filter.

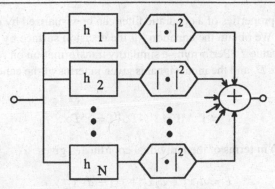

Fig. 7.1. Quadratic filter is equivalent to a bank of linear filters followed by a square-law detector and summation.

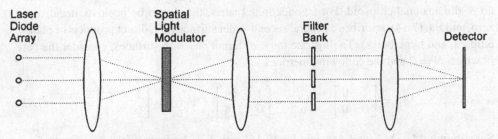

Fig. 7.2. Implementation of a quadratic filter using a multichannel correlator with mutually incoherent illumination.

To demonstrate how quadratic filters can be implemented in a coherent optical processor, we decompose the quadratic filter by using a similarity transformation. This gives

$$Q(\mathbf{f}) = \mathbf{f}^t \mathcal{M} \mathbf{f} = \mathbf{f}^t \mathcal{A} \mathcal{D} \mathcal{A}^t \mathbf{f}$$

$$= \mathbf{f}^t \left[\lambda_1 \psi_1 \psi_1^t + \lambda_2 \psi_2 \psi_2^t + \cdots + \lambda_N \psi_N \psi_N^t \right] \mathbf{f}$$

$$= \lambda_1 \left| \psi_1^t \mathbf{f} \right|^2 + \lambda_2 \left| \psi_2^t \mathbf{f} \right|^2 + \cdots + \lambda_2 \left| \psi_N^t \mathbf{f} \right|^2, \tag{7.12}$$

where ψ_i, with $i = 1, \dots, N$, are the eigenvectors of matrix $\mathcal{M}$ and λ_i are the eigenvalues. Equation (7.12) can be interpreted as a weighted sum of the squared magnitudes of N correlations where the input signal is $\mathbf{f}$ and the filter has the impulse response ψ_i. According to Eq. (7.12), a quadratic filter is equivalent to the bank of N linear filters, as shown in Fig. 7.1.

For optical implementation, we note that a coherent optical correlator performs the correlation operation in complex amplitude and detects the result in intensity [6]. Thus the magnitude-squared operation in Eq. (7.12) is automatically obtained by the detection process. A quadratic filter can thus be achieved in one of two ways. One approach would be to serially integrate the response of N coherent correlations on the output detector. This has the advantage of being simple and reducing the frame rate of the output detector. The second approach is to implement all correlation in parallel with a multichannel optical correlator architecture, as shown in Fig. 7.2. In this approach, a laser diode light source is used in each channel in order to achieve incoherent addition of the coherent correlations. In addition to these two basic strategies for optically implementing quadratic filters, a hybrid approach could be used that includes a combination of both strategies.

The classification properties of a quadratic filter can be visualized by examination of its decision surface [7]. We obtain the equation for the decision surface by setting Eq. (7.11) equal to a threshold value T. Performing a similarity transformation on matrix $\mathcal{M}$ converts it to a diagonal matrix $\mathcal{D}$, and the input signal is given in terms of the canonical coordinates **x**. This gives

$$T = \mathbf{f}^t \mathcal{M} \mathbf{f} = \mathbf{f}^t \mathcal{A} \mathcal{D} \mathcal{A}^t \mathbf{f} = \mathbf{x}^t \mathcal{D} \mathbf{x}. \tag{7.13}$$

Expanding Eq. (7.13) in terms of the individual coordinates gives

$$T = d_1 x_1^2 + d_2 x_2^2 + \cdots + d_N x_N^2, \tag{7.14}$$

where x_i denotes the ith components of vector **x** and d_i denotes the ith diagonal elements of matrix $\mathcal{D}$. If $\mathcal{M}$ is a positive definite matrix, all the d_i are positive and Eq. (7.14) defines an N-dimensional ellipsoid. For a nondefinite matrix, the d_i can be positive, negative, or zero, and Eq. (7.14) describes a general second-order surface (i.e., direct products of planes, ellipses, and hyperbolas). To illustrate these different decision surfaces, consider the case in which $N = 2$ and the following matrices:

$$\mathcal{M}_L = \begin{bmatrix} 1 & 0 \\ 0 & 0 \end{bmatrix}, \quad \mathcal{M}_E = \begin{bmatrix} 1 & 0 \\ 0 & 1/4 \end{bmatrix}, \quad \mathcal{M}_H = \begin{bmatrix} 1 & 0 \\ 0 & -1 \end{bmatrix}. \tag{7.15}$$

Substituting $\mathcal{M}_L$, $\mathcal{M}_E$, and $\mathcal{M}_H$ into Eq. (7.14) with $T = 1$ generates the decision surfaces shown in Figs. 7.3(a), 7.3(b), and 7.3(c), respectively. For nondiagonal matrices, the principal axis of the figures would be rotated by some angle. Quadratic filters can implement different combinations of these three basic decision surfaces. In contrast, a linear filter implements a single hyperplane decision boundary. The quadratic filter with matrix rank of unity [shown in Fig. 7.3(a)] is similar to a linear filter; however, it contains an additional hyperplane decision surface placed symmetrically with respect to the origin. The presence of this additional decision boundary means that a sign change in the signal has no effect on classification. In general, this added symmetry is beneficial as it is likely that the negative of a signal indicates the presence of the same thing as the signal itself.

Returning to the linear-quadratic filter given in Eq. (7.10), we want to address the following question: What is the effect of the linear and constant term on the decision boundary? In order to answer this question, we rewrite Eq. (7.10) by using the vector equivalent of

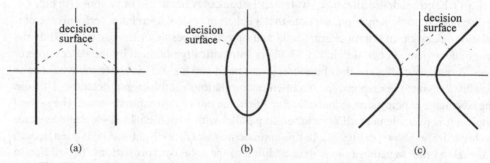

Fig. 7.3. Decision surfaces implemented by the three matrices in Eqs. (7.15): (a) $\mathcal{M}_L$, (b) $\mathcal{M}_E$, (c) $\mathcal{M}_H$.

completing the squares. This gives

$$Q(\mathbf{f}) = \mathbf{f}^t \mathcal{M} \mathbf{f} + \mathbf{h}^t \mathbf{f} + c$$
$$= \mathbf{f}^t \mathcal{M} \mathbf{f} + \mathbf{h}^t \mathcal{M}^{-1} \mathcal{M} \mathbf{f} + \tfrac{1}{4} \mathbf{h}^t \mathcal{M}^{-1} \mathbf{h} - \tfrac{1}{4} \mathbf{h}^t \mathcal{M}^{-1} \mathbf{h} + c$$
$$= (\mathbf{f} + \tfrac{1}{2} \mathcal{M}^{-1} \mathbf{h})^t \mathcal{M} (\mathbf{f} + \tfrac{1}{2} \mathcal{M}^{-1} \mathbf{h}) - \tfrac{1}{4} \mathbf{h}^t \mathcal{M}^{-1} \mathbf{h} + c, \qquad (7.16)$$

where $\mathcal{M}^{-1}$ denotes the inverse of $\mathcal{M}$. The constant term $-\tfrac{1}{4} \mathbf{h}^t \mathcal{M} \mathbf{h} + c$ in Eq. (7.16) changes the value of the effective threshold. Because the setting of the threshold value is under the designer's control, the constant term can be accounted for by a different threshold level. From Eq. (7.16) we see that the presence of a linear term in a linear-quadratic filter causes a translation of the decision boundary. In a pure quadratic filter (i.e., one with no linear terms), the decision surface is centered at the origin of the signal space. Adding a linear term translates the quadratic decision surface away from the origin by the amount $\tfrac{1}{2} \mathcal{M}^{-1} \mathbf{h}$.

7.2.4 Normalization of input signals

Scaling the intensity of an input signal scales the response produced by a quadratic filter. For example, doubling the intensity of a signal quadruples the output response of a quadratic filter. This is undesirable when one is performing threshold detection, as a change in the intensity of a signal could produce a false classification. To remove this difficulty, the filter's output response should be normalized with respect to the intensity of the input signal [8]. Alternatively one could adjust the threshold value based on the intensity of the input signal.

For a linear filter, the decision surface specified by threshold detecting the normalized response is given by

$$\frac{\mathbf{h}^t \mathbf{f}}{(\mathbf{h}^t \mathbf{h})^{1/2} (\mathbf{f}^t \mathbf{f})^{1/2}} = T. \qquad (7.17)$$

Squaring both sides of Eq. (7.17) leaves the basic character of the decision surface unaffected (except for adding a symmetric decision boundary with respect to the origin) and gives

$$\mathbf{f}^t (\mathbf{h} \mathbf{h}^t) \mathbf{f} = T^2 (\mathbf{f}^t \mathbf{f})(\mathbf{h}^t \mathbf{h}),$$
$$\mathbf{f}^t [T^2 (\mathbf{h}^t \mathbf{h}) \mathcal{I} - \mathbf{h} \mathbf{h}^t] \mathbf{f} = 0,$$
$$\mathbf{f}^t \mathcal{M} \mathbf{f} = 0, \qquad (7.18)$$

where $\mathcal{M} = T^2 (\mathbf{h}^t \mathbf{h}) \mathcal{I} - \mathbf{h} \mathbf{h}^t$ and $\mathcal{I}$ denotes the identity matrix. Equations (7.18) show that a normalized linear filter is equivalent to a quadratic filter. The shape of the decision surface can be determined with the help of Eq. (7.1). The numerator in Eq. (7.17) can be written as $\|\mathbf{f}\| \|\mathbf{h}\| \cos \theta$, where θ is the angle between $\mathbf{f}$ and $\mathbf{h}$. Noting that the magnitudes in the numerator and the denominator cancel, we find that the decision surface is given by a hypercone that makes an angle $\theta = \cos^{-1}(T)$ with the vector $\mathbf{h}$. This hypercone decision boundary is illustrated in Fig. 7.4 for the simple case of $N = 3$. The hypercone decision boundary can be interpreted as a nearest-neighbor classifier in angle space. Signals that are within a certain angular distance from filter vector $\mathbf{h}$ are classified as targets.

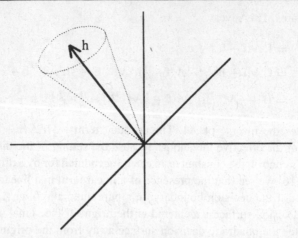

Fig. 7.4. Hypercone decision surface implemented by a normalized linear filter in three dimensions.

Two different procedures are used to normalize the response of the quadratic filter. The first approach is similar to the approach described above for the linear filter. Although this approach removes the effects of changing the signal intensity, it does not lead to a nearest-neighbor interpretation of classification as in the case of a linear filter. The second approach is a slight modification of the first approach, which removes the affects of changing the signal intensity and allows a nearest-neighbor classification result to be achieved.

The first normalization procedure for the quadratic filter generates a decision boundary given by

$$\frac{\mathbf{f}^t \mathcal{M} \mathbf{f}}{\|M\| \|\mathbf{f}\|^2} = \frac{\mathbf{f}^t \mathcal{M} \mathbf{f}}{[\mathrm{TR}(\mathcal{M}^t \mathcal{M})]^{1/2} \mathbf{f}^t \mathbf{f}} = T, \tag{7.19}$$

where $\|M\|$ is the norm of matrix $\mathcal{M}$, $\|\mathbf{f}\|$ is the norm of $\mathbf{f}$, and $\mathrm{TR}(\ )$ denotes the matrix trace operation. Matrix $\mathcal{M}$ is a symmetric matrix and can be diagonalized by an appropriate similarity transform (i.e., $\mathcal{M} = \mathcal{A}\mathcal{D}\mathcal{A}^t$, where $\mathcal{D}$ is a diagonal matrix with the eigenvalues of $\mathcal{M}$ along its diagonal and $\mathcal{A}$ is a matrix with the eigenvectors of $\mathcal{M}$ as its column vectors). Equation (7.19) can then be rewritten in canonical form as

$$\frac{\mathbf{f}^t \mathcal{A}\mathcal{D}\mathcal{A}^t \mathbf{f}}{[\mathrm{TR}(\mathcal{D}^2)]^{1/2} \mathbf{f}^t \mathcal{A}\mathcal{A}^t \mathbf{f}} = \frac{\mathbf{x}^t \mathcal{D} \mathbf{x}}{[\mathrm{TR}(\mathcal{D}^2)]^{1/2} \mathbf{x}^t \mathbf{x}} = \frac{\sum_n \lambda_n |x_n|^2}{\left(\sum_n \lambda_n^2\right)^{1/2} \sum_n |x_n|^2} = T, \tag{7.20}$$

where λ_n are the eigenvalues of $\mathcal{M}$ and x_n are the canonical coordinates of $\mathbf{f}$. The numerator in Eq. (7.20) can be interpreted as an inner product between two vectors, one with components λ_n, the other with components $|x_n|^2$. The two expressions in the denominator can be interpreted as the Euclidean norm (or l^2 norm) of the vector containing components λ_n and the l^1 norm of the vector with components $|x_n|^2$. We cannot interpret Eq. (7.20) as defining a nearest-neighbor classifier in angle space as was done for the linear filter above because of the presence of the l^1 norm.

A slight modification of the normalization procedure in Eq. (7.20) allows a nearest-neighbor interpretation for the quadratic filter. Instead of normalizing the quadratic filter with

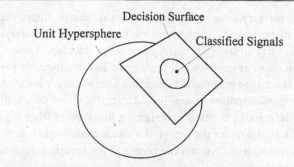

Fig. 7.5. Alternative interpretation of a normalized linear filter in which the input signals are constrained to the unit hypersphere and the hyperplane decision surface slices off a portion of the hypersphere to perform minimum distance classification.

the squared magnitude of the input signal, we use the following normalization procedure:

$$\frac{\sum_n \lambda_n |x_n|^2}{\left(\sum_n \lambda_n^2\right)^{1/2}\left(\sum_n |x_n|^4\right)^{1/2}} = T. \tag{7.21}$$

This modified expression cancels the effect of any scale change in the input signal and can be interpreted as a nearest-neighbor classifier in angle space as in the case of the linear filter. In this case, a hypercone decision boundary is implemented in the feature space defined by the components $|x_n|^2$ and the axis of the hypercone is defined by the components λ_n.

At this point, one may wonder what has been gained by using a quadratic filter. In order to answer this question, we restrict our attention to the input signal that has a unit norm. For a normalized linear filter, the intersection of the hypercone decision boundary with the unit hypersphere slices out a circular region that defines the target region on the unit hypersphere. This is illustrated in Fig. 7.5. This forms a minimum distance classifier on the unit hypersphere, that is, any point within a set distance from the central point of this region is classified as a target. For the quadratic filter, classification is based on the distance from a manifold (i.e., a hypersurface) defined by the set of equations $\lambda_n = |x_n|^2$ for $n = 1, 2, \ldots, N$ (where $|x_1|^2 + |x_2|^2 + \cdots + |x_N|^2 = 1$). If x_n is always positive real, then this set of N equation defines a single point defined by $\lambda_n = x_n^2$. In this case, nothing is gained by using a quadratic filter. However, if x_n is a complex quantity, then each equation $\lambda_n = |x_n|^2$ defines a circle and the set of N equations defines an N-dimensional torus. Hence the quadratic filter performs a nearest-neighbor classification with respect to an N-dimensional torus. Below we see in Section 7.3 that when a signal is transformed through the action of a linear transformation group, the quantities x_n are indeed complex. In this case, classification with respect to a torus manifold makes invariant pattern recognition possible.

7.3 Invariant quadratic filters

In pattern recognition, a mathematical model, or set of models, is used to describe the target set. In general, a signal corresponding to a particular target varies because of various factors in the data-collection process. For example, an image of an automobile depends on perspective (range, aspect, and rotation), lighting conditions (direction of illumination, direct or diffuse illumination, shadows), etc. The ability to model or compensate for one or more of these factors is an important part of the pattern recognition process.

In this section, our primary interest is in developing quadratic filters that can compensate for some range of perspective changes in the target-sensor geometry. Two different approaches are taken. First an invariant quadratic filter is developed when a filter is constrained to provide an equal response to all the signals in a training set [9]. Next a group theoretical approach is adapted to construct quadratic filters that are invariant to the actions of a particular linear transformation group. It is shown that these two approaches lead to the same filter when the training set used to design the first type of filter is generated by the uniform sampling of a signal under the action of a linear transformation group. This leads to insight into the correlation matrix and the potential for invariant feature extraction by principal component analysis.

7.3.1 Quadratic filters invariant to a training set

In this subsection, invariant quadratic filters are developed from a set of exemplar training signals. The quadratic filter is designed according to two basic criteria. First it is constrained to generate an equal normalized response (ENR) to every signal in the training set. The ENR criterion forces the filter to be invariant to the training set and is similar to the equal correlation peak constraint used to construct linear synthetic discriminant functions. The second criterion is to maximize the normalized response for the exemplar training signals. This serves as the pattern recognition metric for optimizing the quadratic filter's performance.

In Section 7.2, two different normalization procedures were presented. Here we use the normalization procedure given in Eq. (7.19), despite the fact that it does not provide a nearest-neighbor interpretation. The reason for doing this is twofold. First, and most important, an analytical solution for the quadratic filter was found with the normalization procedure given in Eq. (7.19), whereas no analytical solution was found with the normalization procedure in Eq. (7.21). (This is not to imply that one does not exist; rather it is to admit the author's inability to find the solution.) Second, it is shown that, in the case in which the training set can be modeled by the action of a linear transformation group, the two normalization procedures produce the same filter.

The targets signals to be detected are represented by a training set S given by

$$S = \{s_1, s_2, \ldots, s_N\}, \tag{7.22}$$

where the column vectors s_i correspond to individual training signals. Finding the quadratic filter $\mathcal{M}$ that satisfies the two criteria described above is a constrained optimization problem that can be solved with the method of Lagrange multipliers. This leads to the following functional form to be maximized:

$$\Psi = \sum_{n=1}^{N} \frac{\left(s_n^t \mathcal{M} s_n\right)^2}{\text{TR}(\mathcal{M}^t \mathcal{M})\left(s_n^t s_n\right)^2} + \sum_{n=1}^{N} \lambda_n \left(\frac{s_n^t \mathcal{M} s_n}{s_n^t s_n} - 1\right). \tag{7.23}$$

The first term on the right-hand side of Eq. (7.23) is the sum of the quadratic filter's squared normalized response to the training set. The second term contains the Lagrange multipliers λ_n for $n = 1, 2, \ldots, N$ and requires the quadratic filter to produce a unit response for a normalized signal from the training set, i.e.,

$$\frac{s_n^t \mathcal{M} s_n}{s_n^t s_n} = 1. \tag{7.24}$$

This corresponds to the ENR constraint in which the value of unity is selected arbitrarily, as any value could be obtained when $\mathcal{M}$ is scaled accordingly. Differentiating Eq. (7.23) with respect to $\mathcal{M}$ and setting the result to zero give

$$\frac{d\Psi}{d\mathcal{M}} = 2\sum_{n=1}^{N} \frac{\text{TR}(\mathcal{M}^t\mathcal{M})(\mathbf{s}_n^t\mathcal{M}\mathbf{s}_n)\mathbf{s}_n\mathbf{s}_n^t - (\mathbf{s}_n^t\mathcal{M}\mathbf{s}_n)^2\mathcal{M}}{\text{TR}^2(\mathcal{M}^t\mathcal{M})(\mathbf{s}_n^t\mathbf{s}_n)^2} + \sum_{n=1}^{N} \lambda_n \frac{\mathbf{s}_n\mathbf{s}_n^t}{\mathbf{s}_n^t\mathbf{s}_n} = 0. \quad (7.25)$$

Rearranging Eq. (7.25) and using the relationship in Eq. (7.24) give $\mathcal{M}$ as

$$\mathcal{M} = \frac{\text{TR}(\mathcal{M}^t\mathcal{M})^2}{N} \sum_{n=1}^{N} \left[\frac{\lambda_n}{2} + \frac{1}{\text{TR}(\mathcal{M}^t\mathcal{M})}\right]\frac{\mathbf{s}_n\mathbf{s}_n^t}{\mathbf{s}_n^t\mathbf{s}_n}. \quad (7.26)$$

Next the value of the Lagrange multiplier needs to be determined. Writing out Eq. (7.24) while substituting Eq. (7.26) for $\mathcal{M}$ gives

$$\frac{\mathbf{s}_k^t\mathcal{M}\mathbf{s}_k}{\mathbf{s}_k^t\mathbf{s}_k} = \frac{\text{TR}(\mathcal{M}^t\mathcal{M})^2}{N} \sum_{n=1}^{N} \left[\frac{\lambda_n}{2} + \frac{1}{\text{TR}(\mathcal{M}^t\mathcal{M})}\right]\frac{\mathbf{s}_k^t\mathbf{s}_n\mathbf{s}_n^t\mathbf{s}_k}{\mathbf{s}_n^t\mathbf{s}_n\mathbf{s}_k^t\mathbf{s}_k} = 1. \quad (7.27)$$

Let r_{nk} denote the normalized cross correlation between $\mathbf{s}_n$ and $\mathbf{s}_k$. Then Eq. (7.27) can be rewritten as

$$\frac{\text{TR}(\mathcal{M}^t\mathcal{M})^2}{N} \sum_{n=1}^{N} \left(\frac{\lambda_n}{2} + \frac{1}{\text{TR}(\mathcal{M}^t\mathcal{M})}\right)r_{nk}^2 = 1. \quad (7.28)$$

The expression in Eq. (7.28) corresponds to a vector–matrix equation in which

$$\frac{\lambda_n}{2} + \frac{1}{\text{TR}(\mathcal{M}^t\mathcal{M})}$$

are elements of an N-dimensional vector and r_{nk}^2 are elements of an $N \times N$ symmetric matrix. We can solve this vector–matrix equation to get

$$\frac{\lambda_n}{2} + \frac{1}{\text{TR}(\mathcal{M}^t\mathcal{M})} = \frac{N}{\text{TR}(\mathcal{M}^t\mathcal{M})^2}[\Lambda^{-1}\mathbf{u}]_n, \quad (7.29)$$

where Λ^{-1} is the inverse of the matrix with elements r_{nk}^2, $\mathbf{u}$ is an N-dimensional vector with all its elements equal to unity, and the square bracket subscripted by n means to take the nth element of the resulting vector contained in the bracket. Substituting Eq. (7.29) back into Eq. (7.26) gives the optimized quadratic filter as

$$\mathcal{M} = \sum_{n=1}^{N} [\Lambda^{-1}\mathbf{u}]_n \frac{\mathbf{s}_n\mathbf{s}_n^t}{\mathbf{s}_n^t\mathbf{s}_n}. \quad (7.30)$$

Examining Eq. (7.30), we see that, except for the constants $[\Lambda^{-1}\mathbf{u}]_n$, the optimum quadratic filter is the correlation matrix of the normalized training set. This result is consistent with the quadratic classifiers discussed in the communication literature, except that the addition of the ENR constraint causes the presence of the term $[\Lambda^{-1}\mathbf{u}]_n$.

To verify that the quadratic filter in Eq. (7.30) produces the same normalized response for any signal in the training set, we compute the normalized response of the quadratic filter given in Eq. (7.30) with the ith signal in the training set. The result is given by

$$\frac{s_i^t \mathcal{M} s_i}{\mathrm{TR}(\mathcal{M}^t \mathcal{M}) s_i^t s_i} = \frac{1}{\mathrm{TR}(\mathcal{M}^t \mathcal{M})} \sum_{n=1}^{N} [\Lambda^{-1} \mathbf{u}]_n \frac{s_i^t s_n s_n^t s_i}{s_n^t s_n s_i^t s_i}$$

$$= \frac{1}{\mathrm{TR}(\mathcal{M}^t \mathcal{M})} \sum_{n=1}^{N} [\Lambda^{-1} \mathbf{u}]_n r_{ni}^2 = \frac{1}{\mathrm{TR}(\mathcal{M}^t \mathcal{M})}. \qquad (7.31)$$

From Eq. (7.31) it is evident that the ENR constraint is achieved. In addition, Eq. (7.31) provides insight into the role played by the coefficient $[\Lambda^{-1} \mathbf{u}]_n$ in achieving the ENR constraint.

7.3.2 Quadratic filters invariant to a linear transformation group

To develop a deeper insight into invariant filter design, it is necessary to have some knowledge of group theory. Here we use ideas from group theory to formulate an optimal quadratic filter that is invariant to linear transformations of the input signal. We begin by discussing the basic properties of a group and defining the action of a group of the signal space by the group representation. Next the notion of an invariant subspace is introduced. This allows the signal space to be decomposed into a set of orthogonal subspace [5]. For a one-parameter group, these invariant subspaces correspond to an eigendecomposition of the linear transformation group. This provides an invariant basis that can be used to develop invariant quadratic filters.

From an abstract perspective, a group consists of a set of objects $G = \{g_1, g_2, \dots,\}$ and a binary operation, $\bullet$, that satisfies the following properties [10]:

(1)   Closure: Given $g_i \in G$, $g_j \in G$, and $g_i \bullet g_j = g_k$, then $g_k \in G$.
(2)   Associativity: $(g_i \bullet g_j) \bullet g_k = g_i \bullet (g_j \bullet g_k)$ for all $g_i \in G$.
(3)   Identity element: $\exists e \in G$ such that $g_i \bullet e = e \bullet g_i = g_i$.
(4)   Inverse: For each $g_i \in G \exists g_i^{-1} \in G$ such that $g_i \bullet g_i^{-1} = g_i^{-1} \bullet g_i = e$.

Here we are interested in linear transformation groups that operate on the coordinate system of the signal's support. Some examples of linear transformation groups include the rotation group and the translation group. It is easy to verify that each of these linear transformations satisfies the properties given above.

The effects of a group action on a vector space are described by the group representation. A group representation is a mapping of a group G onto a group T of $\mathcal{D} \times \mathcal{D}$ nonsingular matrices that preserve the group operation, i.e.,

$$\text{if } g_i \bullet g_j = g_k, \qquad \text{then } \mathcal{T}_{g_i} \mathcal{T}_{g_j} = \mathcal{T}_{g_k}, \qquad (7.32)$$

where g_i are elements of group G and $\mathcal{T}_{g_i}$ are $\mathcal{D} \times \mathcal{D}$ matrices that are elements of the group representation T. The requirement that the matrices $\mathcal{T}_{g_i}$ be nonsingular follows from the need for an inverse element in the group T. In short, a group representation is a group whose elements are matrices. The matrices in a representation are assumed to be unitary with no loss in generality (see Ref. 10, p. 74).

A group representation defines the action of a group on a particular vector space. For example, the real numbers ϕ defined on the interval $[0, 2\pi]$ form a group under modular addition mod$[2\pi]$. This group can be mapped onto the 2-D matrices SO(2) whose elements are the special orthogonal matrices,

$$T_\phi = \begin{bmatrix} \cos(\phi) & \sin(\phi) \\ -\sin(\phi) & \cos(\phi) \end{bmatrix}. \tag{7.33}$$

Note that the group SO(2) operates on a 2-D vector space (e.g., the x–y coordinate system), whereas the mod[2π] addition operates on a one-dimensional (1-D) space.

In this chapter, the primary interest is in group representations that operate on vectors in a signal space. For example, consider an $\mathcal{N} \times \mathcal{N}$ image and a linear transformation group that translates this image in the x direction. We assume that the image support has a torus topology (i.e., the right and the left edges are connected and the top and bottom are connected) so that as the image is shifted to the right, the part that falls off the right-hand side of the image support enters the left-hand side of the image support. The following two actions have the same effect:

(1)   Fix the coordinate system of the image support and translate the image x_o pixels to the right, or
(2)   Fix the image and translate the coordinate system x_o pixels to the left.

The equivalence of these two operations is expressed mathematically as

$$T_{x_o}\{s(x, y)\} = s(g_{x_o}[x], y) = s(x - x_o, y), \tag{7.34}$$

where T_{x_o} is a matrix operator that translates the image x_o pixels to the right and $g_{x_o}[\,]$ is a linear operator that translates the coordinate system x_o pixels to the left. Both actions produce the same result; however, the two operators are fundamentally different. T_{x_o} operates on the image that is an $N \times N$ dimensional vector, whereas $g_{x_o}[\,]$ operates on the x coordinate, which is 1-D. This is the key distinction between a group and its group representation.

The concept of invariant subspace plays an important role in our development. A signal space can be decomposed into a set of orthogonal invariant subspaces that is a function of the particular linear transformation group under consideration. An invariant subspace has the property that all the vectors in this subspace are mapped back into the same subspace when operated on by a particular linear transformation group. This concept is closely related to the eigenvector decomposition of a matrix in which the matrix transforms an eigenvector into a scaled version of itself. However, there are two exceptions to this analogy. First, an invariant subspace may have a dimension greater than unity, whereas an eigenvector is 1-D. Second, all the matrix operators that belong to a given linear transformation group share the same set of invariant subspaces. Thus, in order to satisfy the definition of an invariant subspace, a vector needs to be the eigenvector for all the matrices in the group representation. In general, this is not possible, and this leads to the more general notion of an invariant subspace. However, when a group is commutative (i.e., $T_{g_i}T_{g_j} = T_{g_j}T_{g_i}$ for all g_i and g_j), the group representations share common eigenvectors and the invariant subspaces are the eigenvectors of the group representation.

In this chapter, only commutative groups are considered. Hence the concept of an invariant subspace can be replaced with the more familiar notion of a subspace of degenerate eigenvectors. All commutative groups can be represented as direct products of a set of one-parameter groups. A one-parameter group is parameterized by a single variable. Examples of one-parameter groups include the 2-D rotation group and translation in one dimension. Examples of a two-parameter commutative group include translations in the x–y plane and rotation and scaling with respect to the same point in a plane. In analyzing cummutative groups, it is convenient to consider the constituent one-parameter group separately and combine the results of the individual analyses at a later time.

The eigenvalues of a transformation group play an important role in determining the invariant features of a signal. Therefore we seek an expression for the eigenvalues of a general one-parameter transformation group. We obtain the general form of the eigenvalues by noting that all one-parameter groups defined over a finite interval can be mapped continuously onto the group SO(2) given in Eq. (7.33). The eigenvalues of representation of SO(2) are easily derived as follows. Because the representation is assumed to be unitary, the eigenvalues must have the general form $\lambda_n(\phi) = \exp[if(\phi)]$, where $i = \sqrt{-1}$, f is a real-valued function, and ϕ is the group parameter. Furthermore, the representation must obey the rules of group multiplication {i.e., $\lambda_n(\phi)\lambda_n(\theta) = \lambda_n(\phi + \theta)$ for all $\phi \in [0, 2\pi]$ and $\theta \in [0, 2\pi]$}. This gives the relationship $f(\phi) + f(\theta) = f(\phi + \theta)$ from which we can conclude that $f(\phi) = a\phi + b$. Boundary conditions force a to be an integer and b is set to zero with no loss in generality. Thus the irreducible representations of SO(2) are

$$\lambda_n(\phi) = \exp(in\phi) \qquad \text{for } n = -\infty\text{–}\infty. \tag{7.35}$$

In general, a one-parameter group $g(t)$ will be parameterized over some finite interval $t \in [a, b]$. Furthermore, there will be a continuous, one-to-one function $\alpha(t)$ that maps t onto $[0, 2\pi]$, the parameter of the group SO(2). Thus the eigenvalues for an arbitrary one-parameter group have the general form

$$\lambda_n[g(t)] = \exp[in\alpha(t)], \tag{7.36}$$

where $t \in [a, b]$ and n is some integer. The eigenvalues for a two-parameter commutative group $g(s, t)$ can be written as the direct product of two one-parameter eigenvalues, i.e.,

$$\lambda_{nm}[g(s, t)] = \exp[in\alpha(s)]\exp[im\beta(t)], \tag{7.37}$$

where n and m are integers.

An eigendecomposition of a general one-parameter group representation has the form

$$\mathcal{T}_{g(t)} = \sum_j \sum_n \lambda_n[g(t)]\Psi_{n,j}\Psi_{n,j}^t, \tag{7.38}$$

where $\lambda_n[g(t)] = \exp[in\alpha(t)]$ is the nth eigenvalue of the group representation, $\Psi_{n,j}$ denotes the eigenvector of the nth eigenvalue, where j signifies a possible degeneracy in the eigenvalue, and t denotes the conjugate transpose. According to Eq. (7.38) the group representation can be decomposed into a set of eigenvectors that do not depend on the group parameter t; only the eigenvalues depend on the group parameter.

We now use the eigenvalues of a group to find the invariant features of a signal (i.e., an eigendecomposition of the signal). The key concept employed is that the eigenvalues of a group form an orthogonal set of functions over the group parameter space. For example, for a one-parameter group $g(t)$ the orthogonal relationship of the eigenvalues is given by

$$\int_{t=a}^b \lambda_n[g(t)]\lambda_m^*[g(t)]\,dg(t) = \frac{1}{2\pi}\int_{t=a}^b \exp[in\alpha(t)]\exp[-im\alpha(t)]\,|\alpha'(t)|\,dt = \delta_{nm}, \tag{7.39}$$

where δ_{nm} is the Kronecker delta function. We easily verify Eq. (7.39) by making a change of variables to produce the standard orthogonal relationship between the Fourier kernel functions. The orthogonal relationship given in Eq. (7.39) can be used to construct a projection operator that projects a signal onto the subspace spanned by the eigenvectors of a given eigenvalue. The projection of the signal $\mathbf{s}$ onto the subspace spanned by the eigenvectors of eigenvalue $\lambda_n[g(t)]$ is given by

$$P_n[\mathbf{s}] = \int_{t=a}^{b} \lambda_n^*[g(t)] T_{g(t)}[\mathbf{s}] \, dg(t). \tag{7.40}$$

Substituting Eq. (7.38) into Eq. (7.40) and using the orthogonality condition given in Eq. (7.39) give the projection operator as

$$P_n[\mathbf{s}] = \sum_j \Psi_{n,j} \Psi_{n,j}^t \mathbf{s}. \tag{7.41}$$

According to Eq. (7.41) the invariant features generated by the formula in Eq. (7.40) are the orthogonal projections of the signal $\mathbf{s}$ onto the subspace generated by the eigenvectors of $\lambda_n[g(t)]$. This proves that Eq. (7.40) is the desired projection operator.

The eigenvectors of the linear transformation group, $\Psi_{n,j}$, correspond to the possible set of distortion-invariant linear features. These invariant linear features also serve as the building blocks for an invariant quadratic filter. Because invariance is achieved automatically by use of the eigenvectors of the linear transformation group, the only criterion that needs to be determined in the quadratic filter's design is how to optimize its performance. We write the generic, unoptimized quadratic filter in the following form:

$$Q[\mathbf{f}] = \mathbf{f}^t \sum_{n,j,k} c_{n,j,k} \Psi_{n,j} \Psi_{n,k}^t \mathbf{f}. \tag{7.42}$$

We begin by optimizing the performance of the quadratic filter in Eq. (7.42) with respect to the normalized response given in Eq. (7.19). Substituting Eq. (7.42) into Eq. (7.19) gives

$$\frac{\mathbf{s}^t \sum_{n,j,k} c_{n,j,k} \Psi_{n,j} \Psi_{n,k}^t \mathbf{s}}{\left(\sum_{n,j,k} |c_{n,j,k}|^2 \right)^{1/2} \mathbf{s}^t \mathbf{s}} = \frac{\sum_{n,j,k} c_{n,j,k} s_{n,j}^* s_{n,k}}{\left(\sum_{n,j,k} |c_{n,j,k}|^2 \right)^{1/2} \mathbf{s}^t \mathbf{s}}, \tag{7.43}$$

where $s_{n,j}$ is the projection of $\mathbf{s}$ onto the eigenvector $\Psi_{n,j}$. The expression in Eq. (7.43) is optimized when $c_{n,j,k} = s_{n,j} s_{n,k}^*$. If we substitute Eq. (7.42) into the other normalization procedure given by Eq. (7.21), we obtain the same result, that is, the optimum quadratic filter that is invariant to a one-parameter linear transformation group and that maximizes the normalized response to the signal $\mathbf{s}$ is given by

$$Q[\mathbf{f}] = \mathbf{f}^t \sum_{n,j,k} s_{n,j} s_{n,k}^* \Psi_{n,j} \Psi_{n,k}^t \mathbf{f}. \tag{7.44}$$

Examining Eq. (7.41), we see that the optimum quadratic filter can also be written as

$$Q[\mathbf{f}] = \mathbf{f}^t \sum_{n,j} P_n[\mathbf{s}] P_n^t[\mathbf{s}] \mathbf{f}. \tag{7.45}$$

Equation (7.45) provides a simple formula for generating the optimum invariant quadratic filter given in Eq. (7.42).

7.3.3 Principal component analysis and invariant feature extraction

In this subsection, the results from the previous two subsections are combined to achieve invariant feature extraction. In particular, it is shown that when the training set is generated by uniform sampling a signal that undergoes a linear transformation, the filter design given in Subsection 7.3.1 converges to the filter design given in Subsection 7.3.2. Besides showing that the two filter designs are consistent, this result has further implications. In particular, it provides insight into principal component analysis that is significantly different from finding the principal axes of a Gaussian-distributed target distribution. In this case the principal components correspond to invariant features of the training set.

Assume that a training set S is generated by a one-parameter, linear transformation group operating on a signal $\mathbf{s}$. The training set is represented as $S = \{\mathbf{s}_1, \mathbf{s}_2, \ldots, \mathbf{s}_N\}$, where $\mathbf{s}_\alpha = T_{g(\alpha)}[\mathbf{s}]$. When this training set is used, the quadratic filter formulated in Eq. (7.30) is given by

$$\mathcal{M} = \sum_\alpha [\Lambda^{-1}\mathbf{u}]_\alpha \frac{\mathbf{s}_\alpha \mathbf{s}_\alpha^t}{\mathbf{s}_\alpha^t \mathbf{s}_\alpha} = \frac{1}{\mathbf{s}^t \mathbf{s}} \sum_\alpha [\Lambda^{-1}\mathbf{u}]_\alpha \mathbf{s}_\alpha \mathbf{s}_\alpha^t,$$

$$\mathcal{M} = \frac{1}{\mathbf{s}^t \mathbf{s}} \sum_\alpha [\Lambda^{-1}\mathbf{u}]_\alpha T_{g(\alpha)}[\mathbf{s}] T_{g(\alpha)}^*[\mathbf{s}^t], \qquad (7.46)$$

where the denominator is independent of α, as the group representation $T_{g(\alpha)}$ is unitary. Substituting the expression for $T_{g(\alpha)}$ given in Eq. (7.38) into Eq. (7.46) give the quadratic filter as

$$\mathcal{M} = \frac{1}{\mathbf{s}^t \mathbf{s}} \sum_\alpha [\Lambda^{-1}\mathbf{u}]_\alpha \sum_{n,j} \lambda_n[g(\alpha)] s_{n,j} \Psi_{n,j} \sum_{m,k} \lambda_m^*[g(\alpha)] s_{m,k}^* \Psi_{m,k}^t,$$

$$\mathcal{M} = \frac{1}{\mathbf{s}^t \mathbf{s}} \sum_{n,j} \sum_{m,k} \left\{ \sum_\alpha [\Lambda^{-1}\mathbf{u}]_\alpha \lambda_n[g(\alpha)] \lambda_m^*[g(\alpha)] \right\} s_{n,j} s_{m,k}^* \Psi_{n,j} \Psi_{m,k}. \qquad (7.47)$$

In order for the quadratic filter in Eq. (7.47) to converge to the quadratic filter generated from group theoretical consideration, the term in the braces must equal the Kronecker delta function, i.e.,

$$\sum_\alpha [\Lambda^{-1}\mathbf{u}]_\alpha \lambda_n[g(\alpha)] \lambda_m^*[g(\alpha)] = \delta_{nm}. \qquad (7.48)$$

Comparing Eq. (7.48) with Eq. (7.39), we see that Eq. (7.48) is a discretized version of the orthogonality condition given by Eq. (7.39). As a result of sampling the group to form a training set, the eigenvalues do not necessarily form an orthogonal basis. In general, they form a skew basis. However, if the sampling is performed uniformly over the group manifold, the sampled eigenvalues $\lambda_n[g(\alpha)]$ are the kernel functions of a discrete Fourier transform that form an orthogonal basis. In addition, the matrix Λ is circulant Toplez so the values for $[\Lambda^{-1}\mathbf{u}]_\alpha$ are constant (i.e., independent of α). Thus, for a uniformly sampled training set, Eq. (7.47) reduces to

$$\mathcal{M} = \sum_\alpha \frac{\mathbf{s}_\alpha \mathbf{s}_\alpha^t}{\mathbf{s}_\alpha^t \mathbf{s}_\alpha} = c \sum_{n,j,k} s_{n,j} s_{n,k}^* \Psi_{n,j} \Psi_{n,k}^t, \qquad (7.49)$$

where c is some constant. Equation (7.49) provides a simple formula for generating invariant quadratic filters and for performing invariant feature extraction. According to Eq. (7.49), the correlation matrix of the normalized training is the optimum invariant quadratic filter, and the eigenvectors of the matrix are the invariant features of the target's training set.

7.4 Performance analysis of invariant quadratic filters

In this section, we attempt to quantify the performance of invariant quadratic filters. The basic purpose of an invariant quadratic filter is to classify an input signal into one of two classes: a target class or a clutter class. The target class is represented by a prespecified signal and all the signals that can be generated from this exemplar by the action of a linear transformation group. The clutter represents all nontarget signals and is described by some pdf. The performance of the filter will depend on the particular choice of the target, the type of linear transformation, and the clutter pdf. The metric used to evaluate the performance of the classifier is also an important consideration. Ideally a metric such as the false classification rate is preferred; however, this leads to some intractable calculations. Instead, we adopt the Fisher ratio, as it can be solved analytically and provides a good indication of how well the filter separates the two classes. The Fisher ratio is given by

$$\frac{|E\{p(t)\} - E\{p(c)\}|^2}{\mathrm{Var}\{p(t)\} + \mathrm{Var}\{p(c)\}},\tag{7.50}$$

where $p(t)$ and $p(c)$ indicate the probability density of the filter's output to the target class and the clutter class, respectively, $E\{\}$ is the expectation operator, and $\mathrm{Var}\{\}$ is the variance operator. This provides an indication of how well the filter separates the two classes, as a large Fisher ratio requires both a significant separation in the average response and a small relative variance.

The performance of the invariant quadratic filter is assessed by a comparison of its Fisher ratio with the Fisher ratio of an invariant linear filter and a matched filter. Comparison with a matched filter is used to illustrate the performance degradation that results from making a filter invariant to a linear transformation group. Comparison with the invariant linear filter is used to illustrate the performance gains achieved by use of a quadratic filter over a linear filter for invariant classification.

7.4.1 Assumptions and models for target and clutter

In order to develop an expression for the Fisher ratio, it is necessary to adopt a specific model for the target class and the clutter class. The target model is defined by an image $f(j, k)$, which is defined on a $(2N+1)\times(2N+1)$ support. We generate the target class by operating on an exemplar target image with a one-parameter translation group. Although this may appear restrictive, it really is not, as all one-parameter linear transformation groups are equivalent to the one-parameter translation group. Thus in-plane rotations scaling, foreshortening, rotation–scaling, and scaling–foreshortening can all be considered translations. Our primary motivation for using the translation group is to avoid the complications associated with interpolating on a discrete image support.

The Fourier kernel functions form the invariant basis for the translation group. Thus the input signal can be written as

$$f(j, k) = \frac{1}{(2N + 1)^{1/2}} \sum_{n=-N}^{N} f_n(k) \exp\left(i\frac{2\pi}{2N + 1}nj\right),\tag{7.51}$$

where

$$f_n(k) = \frac{1}{(2N + 1)^{1/2}} \sum_{j=-N}^{N} f(j, k) \exp\left(-i\frac{2\pi}{2N + 1}nj\right)\tag{7.52}$$

gives the nth-order harmonic of the target image. The distribution of energy among the different harmonic components plays an important role in the filter's performance. Here we assume a uniform distribution of energy among the harmonic components. We assume that the energy in the first m harmonics is equal and that all other harmonics contain zero energy, i.e.,

$$E_n = \sum_{k=-N}^{N} |f_n(k)|^2 = \alpha \qquad \text{for } |n| \leq m,$$

$$E_n = 0 \qquad\qquad\qquad \text{otherwise}, \qquad\qquad (7.53)$$

where α is some constant.

The clutter pdf is not known and must be assumed. The clutter can be characterized by its projection onto the invariant basis vectors of the signal space. We assume that the distribution of the clutter when projected onto an invariant basis vector approaches a zero-mean Gaussian random variable. This assumption can be partially justified with the central limit theorem [11]. When a high-dimensional distribution is projected onto a low-dimensional space, the resulting distribution tends toward a Gaussian if the joint pdf is a product of many uncorrelated lower-dimensional pdf's. In this case, the act of projection simply sums together a set of uncorrelated random variables that approach a Gaussian distribution by means of the central limit theorem. The zero-mean assumption is reasonable for all harmonics except for $n = 0$. Because images are positively valued, the zeroth-order harmonic usually contains a large dc bias. However, images can be demeaned to remove this large dc bias and this type of preprocessing is assumed here. The marginal pdf's of the projected clutter on the nth harmonic are then given by

$$p_n(x, y) = \frac{1}{2\pi\sigma_n^2} \exp\left(-\frac{x^2 + y^2}{2\sigma_n^2}\right), \qquad\qquad (7.54)$$

where x and y denote the real and the imaginary components of the response, respectively. A uniform energy distribution is also assumed for the clutter model. For this case, the variance of the Gaussian clutter is given by

$$\sigma_n^2 = \frac{\gamma}{2} \qquad \text{for } |n| \leq p,$$

$$\sigma_n^2 = 0 \qquad \text{otherwise}, \qquad\qquad (7.55)$$

where γ is a constant.

We make the following observations about clutter. In general, the pdf of the clutter is invariant to the transformation being considered. This is justified because the clutter images should experience the same transformation as the target images. Without any *a priori* knowledge, it is reasonable to assume that all transformations occur with equal probability. This invariance of the clutter pdf to the linear transformation implies that the clutter's marginal pdf's obtained when the clutter is projected onto the invariant bases of the linear transformation group are uncorrelated. For translation invariance, this implies that the clutter distribution is stationary.

The noise is modeled as additive white, zero-mean Gaussian noise. The variance of the noise is denoted by σ_N^2. Noise is added to only the target class; however, it is assumed that noise is already incorporated into the clutter model.

7.4.2 Fisher ratio of filters

The Fisher ratio is computed and compared for three different types of filters: a matched filter, an invariant linear filter, and an invariant quadratic filter. The Fisher ratio of the matched filter depends on its alignment (i.e., translation) with respect to the filter template and is thus not invariant. The response of the matched filter when properly aligned serves as a reference point for the comparison.

The derivation of the Fisher ratio for the three different filters is tedious but straightforward. The final results are simply presented. The Fisher ratios for the matched filter, the invariant linear filter, and the invariant quadratic filter are given by

$$\text{FR}_M = \frac{\left|\left[\sum_{n=-N}^{N} E_n\right]^2 + \sigma_N^2 \sum_{n=-N}^{N} E_n - \sum_{n=-N}^{N} \sigma_n^2 E_n\right|^2}{4\sigma_N^2\left[\sum_{n=-N}^{N} E_n\right]^3 + 2\sigma_N^4\left[\sum_{n=-N}^{N} E_n\right]^2 + 2\left[\sum_{n=-N}^{N} \sigma_n^2 E_n\right]^2}, \tag{7.56}$$

$$\text{FR}_L = \frac{\left|E_n^2 + \sigma_N^2 E_n - \sigma_n^2 E_n\right|^2}{4\sigma_N^2 E_n^3 + 2\sigma_N^4 E_n^2 + 2\sigma_n^4 E_n^2}, \tag{7.57}$$

$$\text{FR}_Q = \frac{\left|\sum_{n=-N}^{N}\left(E_n^2 + \sigma_N^2 E_n - \sigma_n^2 E_n\right)\right|^2}{\sum_{n=-N}^{N}\left(4\sigma_N^2 E_n^3 + 2\sigma_N^4 E_n^2 + 2\sigma_n^4 E_n^2\right)}, \tag{7.58}$$

respectively, where E_n denotes the energy in the nth harmonic of the target, σ_N^2 is the variance of the noise, and σ_n^2 is the variance of the clutter projected along the nth harmonic of the target. In Eq. (7.56), the expression for the Fisher ratio was formed from the magnitude-squared response of the matched filter in order to compensate for the magnitude-squared operations present in the other two filters. In general, using the magnitude-squared response of a matched filter does not affect its performance; however, it does affect the Fisher ratio, adversely. We do this in order to get all three filters to converge to the Fisher ratio under the condition that causes the three filters to be equivalent.

7.4.3 Relationship between filter performance and key parameters

According to Eqs. (7.56)–(7.58), there are three key factors that affect the Fisher ratio of the filters: (1) E_n, the distribution of target energy among the different harmonic components, (2) σ_n^2, the relative distribution of clutter energy along the target's harmonic components, and (3) σ_N^2, the noise level. In order to explore the effect of each of these components further, we examine the case of uniformly distributed target and clutter.

If the target and the clutter distributions are known, the relative magnitude can be determined as follows. The input signals are normalized to achieve intensity invariance. We assume that the target signal and clutter signal have a unit energy with no loss in generality. For a one-parameter group acting on a 2-D signal, the invariant filters provide a processing gain [i.e., the component $f_n(k)$ is target specific]. The clutter is assumed to be uniformly distributed across the degenerate eigenvectors of a given harmonic order. This gives a processing gain of $(2N+1)^{1/2}$ (i.e., the square root of the degrees of freedom in each harmonic component). This leads to the following relationship between the energy components of the target and the clutter:

$$\left(1 + \frac{1}{\text{SNR}}\right) \sum_{n=-N}^{N} E_n = 1 = (2N+1) \sum_{n=-N}^{N} \sigma_n^2. \tag{7.59}$$

By using Eq. (7.59) we can relate the noise variance σ_N^2 to the SNR by the expression

$$\text{SNR} = \frac{\sum_{n=-N}^{N} E_n}{\sigma_N^2 (2N+1)^2} = \frac{\text{SNR}/(\text{SNR}+1)}{\sigma_N^2 (2N+1)^2}. \tag{7.60}$$

According to Eq. (7.60), the variance of the noise projected onto a given harmonic component is small (i.e., $\sigma_N^2 \ll 1$) for a large space–bandwidth product (i.e., large N).

We assumed that the target energy is uniformly distributed among D harmonic terms and that the clutter is uniformly distributed across all harmonic components. A unit energy is assumed for the target and the clutter images. Thus the nonzero-energy components of the target are given by $E_n = 1/D$, and the clutter variance is given by $\sigma_n^2 = (2N+1)^{-2}$. With these assumptions, the Fisher ratios of the matched filter, the invariant linear filter, and the invariant quadratic filter become

$$\text{FR}_M = \frac{[(2N+1)^2\text{SNR} - \text{SNR}]^2}{4(2N+1)^2\text{SNR} + 2 + 2(\text{SNR}+1)^2} \approx \frac{(2N+1)^2\text{SNR}}{4}, \tag{7.61}$$

$$\text{FR}_L = \frac{[(2N+1)^2\text{SNR} - D \bullet \text{SNR}]^2}{4D(2N+1)^2\text{SNR} + 2D^2 + 2D^2(\text{SNR}+1)^2}, \tag{7.62}$$

$$\text{FR}_Q = \frac{[(2N+1)^2\text{SNR} - D \bullet \text{SNR}]^2}{4(2N+1)^2\text{SNR} + 2D + 2D(\text{SNR}+1)^2}, \tag{7.63}$$

respectively. The Fisher ratios of the three filters are plotted in Fig. 7.6 for different values of SNR, N, and D. From these plots, the following relationships are evident. When all the energy of the target is contained in a single harmonic component (i.e., when $D = 1$), the Fisher ratios of all the filters converge to the same value, as expected. As the energy in the target gets distributed over more harmonic components, the performance of the matched filter remains unchanged, while the performance of the invariant linear filter drops off quickly. The performance of the invariant quadratic filter also drops off, but more slowly than that of the invariant linear filter. Another aspect that is evident from the plots and the equations is that the Fisher ratio is approximately proportional to the space–bandwidth product of the signal [i.e., $(2N+1)^2$]. This is similar to a result obtained by Kumar and Pochapsky [12]. The Fisher ratio is also approximately proportional to the SNR. From the plots in Fig. 7.6, the dependence of the quadratic filter's Fisher ratio on D appears to be magnified with an increase in the SNR. Examining Eq. (7.63), we find that the increased sensitivity to D due to SNR is caused by the increased influence of clutter, that is, at low SNR's, the variance of the noise tends to dominate the denominator of the Fisher ratio. However, as the SNR becomes large, the variance of the clutter acquires more influence. This influence is given by the last term in the denominator of Eq. (7.63). Thus when the SNR is large, the influence of D is more pronounced.

The examples given above illustrate the relationship between the filter's performance and the distribution of the energy among the harmonic orders of the target. If most of the target energy is concentrated in a single harmonic order, the invariant linear filter performs well. As the energy in the target becomes spread over more harmonic orders, the performance of the invariant filter drops and the relative advantage of the quadratic filter increases. This is a simple yet important result, as it provides a qualitative measure for assessing the relative merit of invariant linear and quadratic filters.

A relationship that was not explored above was the effect that different relative clutter distributions have on performance. If we assume a large SNR, we can ignore the effects of

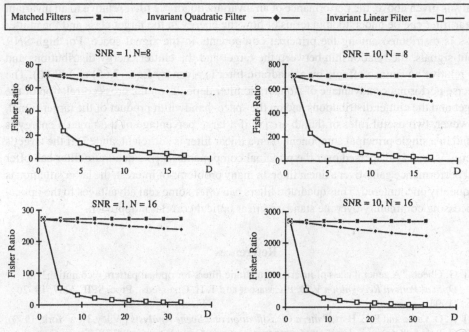

Fig. 7.6. Fisher ratios of the matched filter, the invariant linear filter, and the invariant quadratic filter as functions of D for different values of SNR and N.

noise. In this case, the Fisher ratio of the matched filter, the invariant linear filter, and the invariant quadratic filter can be written as

$$FR_M = \frac{1}{2}\left| \frac{\left[\sum_{n=-N}^{N} E_n\right]^2}{\sum_{n=-N}^{N} \sigma_n^2 E_n} - 1 \right|^2, \tag{7.64}$$

$$FR_L = \frac{1}{2}\left| \frac{E_n}{\sigma_n^2} - 1 \right|^2, \tag{7.65}$$

$$FR_Q = \frac{\left|\sum_{n=-N}^{N}\left(E_n^2 - \sigma_n^2 E_n\right)\right|^2}{2\sum_{n=-N}^{N}\left(\sigma_n^4 E_n^2\right)}, \tag{7.66}$$

respectively. According to Eq. (7.64), the Fisher ratio of the matched filter is related to the total energy in the target divided by the cross correlation between the energy in the harmonic of the target and the clutter. Thus a large Fisher ratio is achieved when the harmonic energy components of the target and the clutter are uncorrelated. For an invariant linear filter, the maximum Fisher ratio is achieved by the choice of the harmonic component that maximizes the target-to-clutter ratio. For an invariant quadratic filter, the expression is a little more complicated; however, a large Fisher ratio is achieved by minimization of the cross correlation of the squared harmonic energy of the target and the clutter.

Although the results presented above were obtained for a particular type of distortion (i.e., one-parameter translation) the results are valid for any one-parameter group and can be extended to the invariant quadratic filters discussed in Subsection 7.3.1. For example, the distribution of target energy among the harmonic components can be replaced with the distribution of energy among the principal components of a quadratic filter. According to the analysis given above, the performance of an invariant quadratic filter relative to an invariant linear filter can be understood in terms of how the energy in the target class and the clutter class is distributed among the principal components in the signal space. For high-SNR input signals, the relationship between the target and the clutter energy distributions and the relative advantage offered by a quadratic filter is given by Eqs. (7.65) and (7.66). The precise performance advantage of a quadratic filter depends on the cross correlation of the target and the clutter distributions and on the space–bandwidth product of the target signal. However, two useful rules of thumb are (1) if a large percentage of the target's energy is found in a single principal component, then a linear filter is sufficient, and (2) if the target's energy is widely dispersed over the principal components, then a quadratic filter can offer real performance gains over a linear filter. In many problems of interest, the latter situation is frequently encountered. Thus quadratic filters can offer some real advantages to the optical processing community over the standard linear optical correlator approach.

References

[1] G. Gheen, "A general class of invariant quadratic filters for optical pattern recognition," in *Optical Pattern Recognition V*, D. P. Casasent and T. H. Chao, eds., Proc. SPIE **2237**, 19–26 (1994).

[2] R. O. Duda and P. E. Hart, *Pattern Classification and Scene Analysis* (Wiley, New York, 1973).

[3] C. Baker, "Optimum detection of a random vector in Gaussian noise," IEEE Trans. Commun. **14**, 802–805 (1966).

[4] G. Gheen, "Quadratic filters invariant to linear transformation groups," in *Optical Pattern Recognition III*, D. P. Casasent and T. Chao, eds., Proc. SPIE **1701**, 198–209 (1992).

[5] M. Schetzen, "Nonlinear modeling based on the Wiener theory," Proc. IEEE **69**, 1557–1573 (1981).

[6] F. T. S. Yu, *Optical Information Processing* (Wiley, New York, 1983).

[7] N. Nilsson, *Learning Machines* (McGraw-Hill, New York, 1965).

[8] F. M. Dickey and L. A. Romero, "Normalized correlation for pattern recognition," Opt. Lett. **16**, 1186–1188 (1991).

[9] B. V. K. Kumar, "Tutorial survey of composite filter design for optical correlators," Appl. Opt. **31**, 4773–4801 (1992).

[10] E. Wigner, *Group Theory and its Applications to Quantum Mechanics of Atomic Spectra* (Academic, New York, 1959).

[11] A. Papoulis, *Probability, Random Variables, and Stochastic Processes*, 2nd ed. (McGraw-Hill, New York, 1984).

[12] B. V. K. Kumar and E. Pochapsky, "Signal-to-noise ratio considerations in modified matched spatial filters," J. Opt. Soc. Am. A **3**, 777–784 (1986).

8

Composite filter synthesis as applied to pattern recognition

Shizhuo Yin and Guowen Lu

8.1 Introduction

Because classical matched filters [1] are sensitive to distortions such as scale and rotation, numerous techniques have been proposed to develop distortion-invariant matched filters for pattern recognition [2–15]. Among them, the composite filter is one of the most effective filters against distortions, particularly for three-dimensional rotation variance. Because there are already some thorough review papers on the general composite filters [16, 17], to save the pages we do not repeat the detailed discussions on the general composite filters. Instead, we just provide a brief summary of those works. We concentrate our discussion on a special type of composite filter that was recently developed at Penn State with the simulated annealing technique [18–20].

The concept of the composite filter can be traced back to Caulfield and Maloney's work in the late 1960's [21]. They suggested that using a linear combination of the correlation output improves the discrimination ability of the classical VanderLugt matched filter. Later, in 1980, Hester and Casasent proposed the concept of the synthetic discriminant function (SDF) [22]. The key feature of this technique is that the composite filter image is a linear combination of all target training-set images. Since then, a variety of techniques have been proposed to improve the performance of the SDF. In 1984, Casasent proposed a modified SDF technique by considering not only the correlation peak point but also the surrounding area of the correlation [23] for which multiobject shift-invariant and distortion-invariant pattern recognition can be achieved. To improve the noise performance of the SDF, several methods have been employed. For example, Kumar proposed the minimum-variance SDF [24], which can have deterministic constraints and minimized output variance at the same time. Mahalanobis *et al.* developed the minimum average correlation energy filters so that sharper correlation peaks could be obtained [25]. Kallman also proposed a low-noise composite filter design technique by optimizing the cross correlation of target and nontarget training sets [26]. However, the above types of filter usually have a large dynamic range, which makes them difficult to implement with the currently available spatial light modulators (SLM's). To alleviate this problem, several methods have been proposed [27–30]. Each approach has its advantages and limitations.

Recently, a powerful mathematical tool, simulated annealing, has been used for the bipolar phase filter design, which can significantly improve the performance of the conventional bipolar phase filter [30–32].

More recently, at Penn State we have applied the simulated annealing technique to the out-of-plane rotation-invariant composite filter design [18–20]. The major advantages of this filter are both the limited dynamic range and the distortion invariance. Furthermore, this technique can also be extended to multiclass pattern recognition. A detailed discussion about this technique is provided below.

8.2 Bipolar composite filter synthesis by simulated annealing

8.2.1 Simulated annealing algorithm

The simulated annealing algorithm is a computational optimization process in which an energy function should be established based on the optimization criteria. In other words, the state variables of a system are adjusted until a global minimum energy is achieved. When the system variable u_i is randomly perturbed by Δu_i, the change of system energy can be calculated as

$$\Delta E = E^{\text{new}} - E^{\text{old}}, \tag{8.1}$$

where $E^{\text{new}} = E(u_i + \Delta u_i)$ and $E^{\text{old}} = E(u_i)$. Note that if $\Delta E < 0$, the perturbation Δu_i is unconditionally accepted. Otherwise, the acceptance of Δu_i is based on the Boltzmann probability distribution $p(\Delta E)$, that is [30],

$$p(\Delta E) = \frac{1}{1 + \exp\left(\dfrac{\Delta E}{kT}\right)}, \tag{8.2}$$

where T is the temperature of the system used in the simulated annealing algorithm and k is the Boltzmann constant. The process is then repeated by random perturbation of each of the state variables and a slow decrease in the system temperature T, which can prevent the system from being trapped in a local minimum energy state. In other words, by continually adjusting the state variables of the system and slowly decreasing the system temperature T, we can find a global minimum energy state of the system. This is known as the simulated annealing process [31].

8.2.2 Bipolar composite filter synthesis

Consider the use of a simulated annealing algorithm to synthesize a bipolar composite filter (BCF), which optimizes detection in the presence of similar (antitarget) objects. To do so, we use two sets of training images, namely the target and the antitarget sets for the construction of the BCF, as denoted by

$$\text{Target set} = \{t_m(x, y)\},$$

$$\text{Antitarget set} = \{a_m(x, y)\}, \tag{8.3}$$

where $m = 1, 2, \ldots, M$, M is the number of training images, and (x, y) is the spatial-domain coordinate system. In addition, we define $W^t(x, y)$ and $W^a(x, y)$ as the desired correlation distributions for the target and the antitarget images, respectively. The mean-square error between the desired and the actual correlation distributions is defined as the energy function E of the system. The configuration of the BCF can be viewed as the state variable of the system. Then the simulated annealing algorithm can be used. The remaining task is to optimize the state variable of the system (i.e., the configuration of the BCF) so

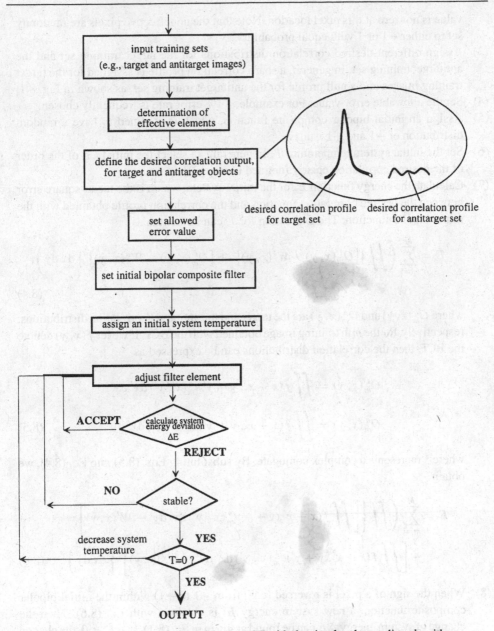

Fig. 8.1. Flow chart for synthesizing the BCF with the simulated annealing algorithm.

that a global minimum energy state can be achieved. A flow chart that shows the simulated annealing procedure for synthesizing a BCF is shown in Fig. 8.1; this flow chart can be described by the following:

(1) Read the input data for the target and the antitarget training sets, which consist of different out-of-plane-rotation images.

(2) Determine the effective pixels of the BCF in the spatial domain. The pixel is considered to be an effective pixel if there is at least one training target or antitarget image whose

value is nonzero at this pixel location. Note that the noneffective pixels are randomly set to either -1 or 1 with equal probability.

(3) Assign different desired correlation distributions for the target training set and the antitarget training set. In general, a sharp correlation profile is assigned for the target training images and a null profile for the antitarget training set, as shown in Fig. 8.1.

(4) Set the allowable error value. For example, a 1% error rate is frequently chosen.

(5) Assign an initial bipolar composite function, which is assumed to have a random distribution of -1 and $+1$ values.

(6) Set the initial system temperature T. In general, we select the initial kT of the order of the desired correlation energy [defined in step (3)].

(7) Calculate the energy function E for the initial system, which is the mean-square error between the desired correlation function and the correlation profile obtained with the initial bipolar function. The expression for E can be written as

$$E = \sum_{m=1}^{M} \left(\iint \left\{ [O_m^t(x, y) - W^t(x, y)]^2 + [O_m^a(x, y) - W^a(x, y)]^2 \right\} dx\, dy \right),$$

(8.4)

where $O_m^t(x, y)$ and $O_m^a(x, y)$ are the target and the antitarget correlation distributions, respectively, for the mth training image obtained with the BCF. If we let $f(x, y)$ denote the BCF, then the correlation distributions can be expressed as

$$O_m^t(x, y) = \iint f(x + x', y + y') t_m^*(x', y')\, dx'\, dy',$$

$$O_m^a(x, y) = \iint f(x + x', y + y') a_m^*(x', y')\, dx'\, dy',$$

(8.5)

where $*$ represents a complex conjugate. By substituting Eqs. (8.5) into Eq. (8.4), we obtain

$$E = \sum_{m=1}^{M} \left(\iint \left\{ \left[\iint f(x + x', y + y') t_m^*(x', y')\, dx'\, dy' - W^t(x, y) \right]^2 \right. \right.$$
$$\left. \left. + \left[\iint f(x + x', y + y') a_m^*(x', y')\, dx'\, dy' - W^a(x, y) \right]^2 \right\} dx\, dy \right).$$

(8.6)

(8) When the sign of a pixel is reversed (e.g., from -1 to $+1$) within the initial bipolar composite function, a new system energy E' is obtained with Eq. (8.6). Then the change of system energy ΔE can be found as given in Eq. (8.1). If $\Delta E < 0$, the change is unconditionally accepted. Otherwise, it will be accepted based on the Boltzmann probability distribution as given in Eq. (8.2). Because $f(x, y)$ is a bipolar function, the iteration equation for $f(x, y)$ in the above reversing process can be shown to be

$$f^{(n+1)}(x, y) = -f^{(n)}(x, y) \qquad\qquad \text{if } \Delta E < 0,$$

$$f^{(n+1)}(x, y) = f^{(n)}(x, y)\,\mathrm{sgn}\{\mathrm{ran}[p(\Delta E)] - p(\Delta E)\} \qquad \text{if } \Delta E \geq 0,$$

(8.7)

where the superscript (n) represents the nth iteration, $\mathrm{ran}[\cdot]$ represents a uniformly distributed random function in the range $[0, 1]$, sgn represents a signum function, and $p(\Delta E)$ is the Boltzmann distribution function.

(9) Step (8) is repeated for each pixel until the system energy E is stable. Then a reduced temperature T is assigned to the system and step (8) is again repeated.

(10) Step (8) and step (9) are continuously repeated until a global minimum E is found.

In practice, the system will reach a stable state if the number of iterations is larger than $4 \times N$, where N is the total number of pixel elements within the BCF [31–33].

One of the important aspects of the BCF design is the synthesis of a spatial-domain impulse response function by which we can easily select effective pixels [defined in step (2)] and set noneffective pixels to either -1 or 1 with equal probability. Thus the correlation between input objects and noneffective pixels would not contribute to the correlation peak intensity (CPI) of the system. To optimize the noise performance of the BCF further, a bright background pattern can be added to the antitarget training set.

8.2.3 Target detection with a bipolar composite filter

Two training sets are used for the synthesis of the BCF's: one is used to detect an American M60 tank and the other is for the detection of a Russian T72 tank. These training images were provided by the U.S. Army Missile Command and were obtained with a CCD camera viewing at a tilt angle of 45° with respect to the rotation axis of the targets. Each training set, as shown in Fig. 8.2, consists of 14 out-of-plane-rotation images, in which each image corresponds to a 5° increment rotation in the perspective view. Because the training images are confined within a 128×128 pixel array of 256 gray levels, the BCF's would also be limited by a 128×128 array pixel. We assign a Gaussian profile to the desired target correlation distribution and set the antitarget correlation distribution to zero. By setting the width of this Gaussian function within 5 pixels in an array of a 256×256 output plane, the correlation amplitude is of the order of average image energy. Thus when the simulated annealing procedure described above is used, two BCF's (for M60 and T72 tanks) are synthesized, as shown in Fig. 8.3. When these two BCF's are correlated with the training images of Fig. 8.2, we obtain CPI's as functions of out-of-plane rotation, which are plotted in Fig. 8.4, in which we can see that the peak intensity for the target set is at least threefold higher than that for the antitarget set.

8.2.4 Pattern discrimination capability of a bipolar composite filter

In order to test the discrimination capability of the designed BCF's, the following discrimination ratio (DR) is chosen as the figure of merit:

$$DR \triangleq \frac{(CPI)_{target}}{(CPI)_{antitarget}}. \tag{8.8}$$

In view of Fig. 8.4(a), the DR for the BCF's to detect the target objects is shown in Fig. 8.5, in which the average DR is measured to be $\sim$5.5 units. For comparison, the DR's of SDF filters and bipolar filters (obtained by directly binarizing the SDF's) are also plotted in the same figure, from which the BCF offers the highest DR among the three types of filters. The calculated DR of the BCF is $\sim$3.4 times larger than that of the SDF filters and 7 times larger than that of the bipolar SDF filter. We have also seen that the DR of the bipolar SDF filter is smaller than unity, which indicates that the filter is not capable of distinguishing the target and the antitarget objects. Thus it is obviously not a feasible method to obtain a bipolar SDF filter by simply binarizing the SDF.

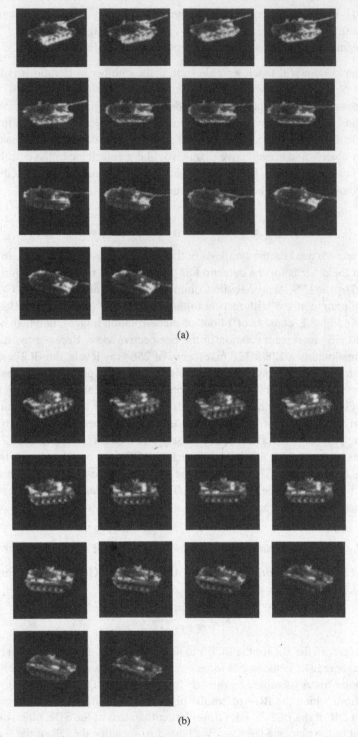

Fig. 8.2. Two sets of out-of-plane-rotation training images. Each image is rotated by 5°: (a) T72 (Russian) tank images, (b) M60 (American) tank images.

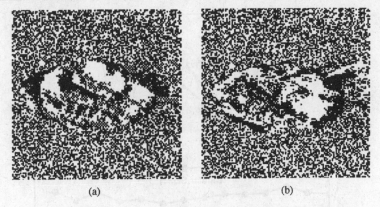

Fig. 8.3. Bipolar composite references: (a) BCF_{M60}, (b) BCF_{T72}.

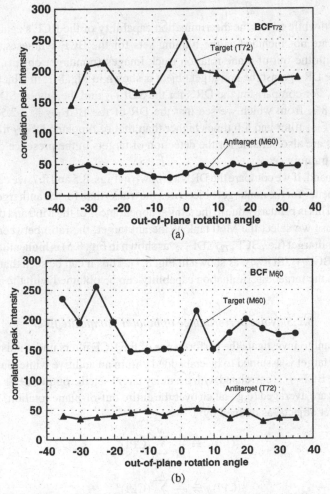

Fig. 8.4. Computed CPI's as functions of out-of-plane rotation.

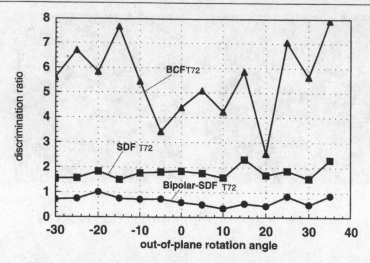

Fig. 8.5. DR between the target T72 and the antitarget M60 tanks.

It is also worthwhile to test the discrimination capability of the BCF's against nontarget objects, which are not included as the training sets for the BCF syntheses. The BCF_{T72} is correlated with the out-of-plane-rotation truck images (nontarget object), as shown in Fig. 8.6, and the DR against the nontarget object is shown in Fig. 8.7, in which the average DR is ~7 units. For comparison, the DR's for the SDF_{T72} and the bipolar SDF_{T72} are also plotted in Fig. 8.7, from which we see that the DR of the BCF_{T72} is ~2.5 times better than that of SDF_{T72} filter and 4.6 times higher than that of bipolar SDF_{T72} filter. Thus the designed BCF's are also suitable for the detection of targets in the presence of objects not included in the training sets.

On the other hand, if we compare the DR's as shown in Figs. 8.5 and 8.7, we can see that the BCF performs better for the nontarget object (i.e., the truck) than for the antitarget object (i.e., the M60 tank). This is primarily due to the similar appearances of the M60 and the T72 tanks. However, because we select the M60 tank as the antitarget, the ratio between DR_{BCF} and DR_{SDF} for the antitarget (i.e., BCF_{T72}/SDF_{T72}, as shown in Fig. 8.5) is higher than that for the nontarget (i.e., BCF_{T72}/SDF_{T72}, as shown in Fig. 8.7). Thus it can be seen that a significant improvement in the target discrimination capability can be obtained with the antitarget set.

8.2.5 Noise performance of bipolar composite filter

It is also important to investigate the performance of the BCF filters under noisy conditions. In this case, the target is assumed to be embedded within an additive white Gaussian noise. When the above BCF_{T72} is correlated with the noisy target set, the output CPI's are obtained. When the CPI's are averaged (e.g., taken over the entire out-of-plane rotation) for the target and the antitarget objects, as defined by

$$\langle CPI \rangle_t \triangleq \frac{1}{M} \sum_{m=1}^{M} (CPI)_m^t,$$

$$\langle CPI \rangle_a \triangleq \frac{1}{M} \sum_{m=1}^{M} (CPI)_m^a, \tag{8.9}$$

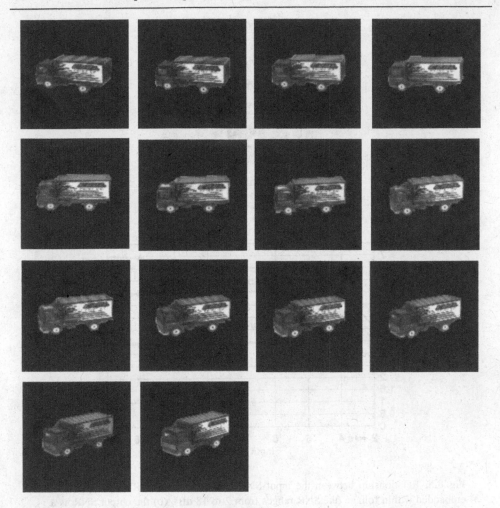

Fig. 8.6. Set of out-of-plane-rotation truck images. Each out-of-plane image is rotated by 5°.

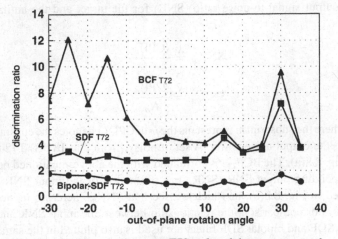

Fig. 8.7. DR between the target T72 tank and the nontarget truck.

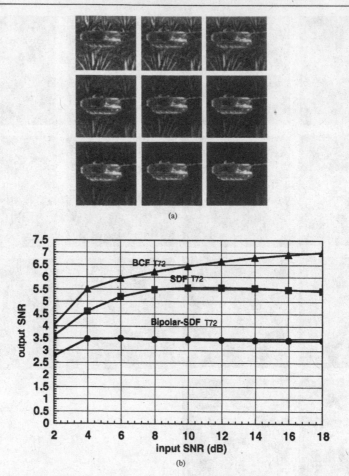

(b)

Fig. 8.8. Relationship between the input SNR and the output SNR: (a) a T72 tank embedded within foliage (the SNR ranges from 2 to 18 dB), (b) the output SNR as a function of the input SNR.

the (average) output signal-to-noise ratio (SNR) for the target and the antitarget can be written as

$$\mathrm{SNR}_{ot} \triangleq \frac{\langle \mathrm{CPI} \rangle_t}{N_o},$$

$$\mathrm{SNR}_{oa} \triangleq \frac{\langle \mathrm{CPI} \rangle_a}{N_o}, \qquad (8.10)$$

respectively, where the subscript m represents the mth CPI, $\langle \rangle$ denotes the ensemble average, and N_o denotes the output noise power. A set of noisy (from 2 to 18 dB) T72 tank images are shown in Fig. 8.8(a). The $\mathrm{BCF_{T72}}$ is correlated with the above-mentioned noisy images, and the corresponding output SNR (SNR_{ot}) as a function of the input SNR (denoted as SNR_i) is plotted in Fig. 8.8(b), in which we see that SNR_{ot} increases monotonically as SNR_i increases. For comparison, the relationship between the input SNR and the output SNR when the SDF and bipolar SDF filters are used is also plotted in the same figure. We note that the BCF offers the highest output SNR among them. The DR as a function of

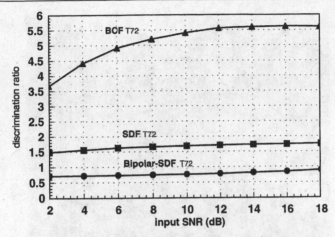

Fig. 8.9. DR as a function of the input SNR between the target T72 tank and the antitarget M60 tank.

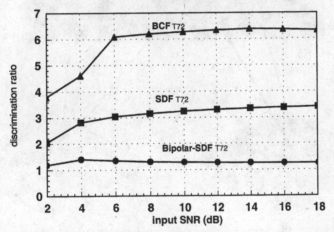

Fig. 8.10. DR as a function of the input SNR between the target T72 tank and the nontarget truck.

SNR_i for the target (i.e., T72 tank) is also plotted in Fig. 8.9, in which the DR increases monotonically as the SNR_i increases and saturates at DR $= \sim 5.52$. The DR for the SDF_{T72} and the bipolar SDF_{T72} filters are also plotted in the same figure, in which we see that the DR of these two types of SDF filter performs more poorly than that of the BCF_{T72}. For example, the DR for the bipolar SDF is lower than unity for the whole SNR_i range, which does not make it a feasible method for pattern recognition.

We also calculated the DR by using the BCF_{T72} against the nontarget (i.e., the truck, as shown in Fig. 8.6) at different SNR_i, as plotted in Fig. 8.10, in which the DR obtained with the SDF filter and the bipolar SDF filter are also plotted in the same figure. From this figure, we see the superior performance of BCF filter over that of the SDF and the bipolar SDF filters, in which the BCF maintains the higher discrimination capability against the nontarget object, even under noisy conditions.

To test the performance of the BCF within the noisy background further, we selected a special input pattern, as shown in Fig. 8.11(a), which consists of a set of out-of-plane-rotation

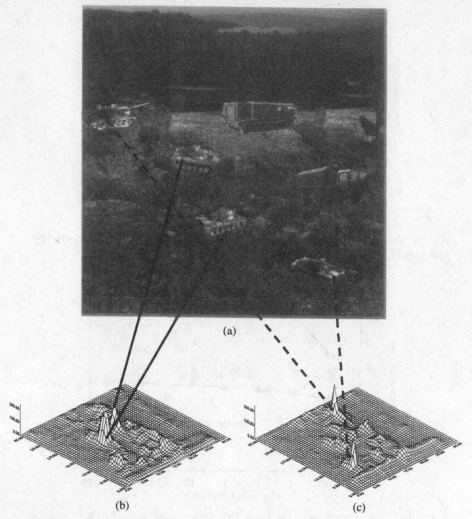

Fig. 8.11. (a) Targets embedded within the background noise, (b) correlation output of a noisy input scene with the BCF_{M60}, (c) correlation output of a noisy input scene with the BCF_{T72}.

objects, including two M60 tanks, two T72 tanks, and a couple of military vehicles. The CPI's obtained with the BCF's are plotted in Figs. 8.11(b) and 8.11(c). We see that the CPI's for target objects are ~3 times higher than those for the antitarget and the nontarget objects.

8.3 Multilevel composite filter synthesis by simulated annealing

To improve the performance of the BCF, we have developed a multilevel composite filter (MLCF). This filter can be used to discriminate highly similar targets with high discrimination capability. The MLCF is also synthesized by the simulated annealing algorithm. The filter fabrication process is similar to the one described in Subsection 8.2.2. However, multiple levels are used to replace two levels $(+1, -1)$. For the purpose of convenience, we select an odd number as the total number of levels, that is $[-n, n]$, $n = 0, 1, 2, \ldots, N$,

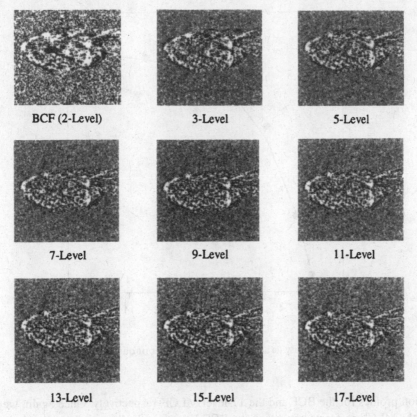

BCF (2-Level)	3-Level	5-Level
7-Level	9-Level	11-Level
13-Level	15-Level	17-Level

Fig. 8.12. Set of nine MLCF's.

and $2N + 1$ is the number of levels. For example, if a MLCF has five ($N = 2$) levels, the values of the levels are $+2$, $+1$, 0, -1, and -2. During the filter synthesis process, each time we change only one level (e.g., from $+2$ to $+1$). Thus, for each pixel, it requires $2N$ times of iterations instead of one time, as used in the BCF case.

To make the comparison easier, we use the same set of objects (as shown in Fig. 8.2) as the training set, in which the T72 tank and the M60 tank are selected as the target set and the antitarget set, respectively. Figure 8.12 shows a set of nine MLCF's with a number of levels from 2 to 17. Note that, because these filters have a very limited number of levels (less than 20), they are still relatively easy to implement with the current commercially available SLM's. For example, currently a liquid-crystal TV (LCTV) has a contrast ratio larger than 100, which makes it easy to display 20 quantization levels.

To test the performance of the MLCF, the ratio between the central correlation peak and the maximum sidelobe is defined as the detectability of the filter. Figure 8.13 shows the detectability as a function of the number of levels. From this figure, we can see that the detectability of the filter improves as the number of levels increases. However, the detectability reaches a saturation limit as the number of levels goes beyond 17. The filter that has seven or nine levels may be a good choice in terms of trade-off between the detectability and the number of levels.

To test the discrimination capability of the MLCF in a noisy environment, Fig. 8.11(a) is again used as the input scene. Figures 8.14(b) and 8.14(c) show the three-dimensional

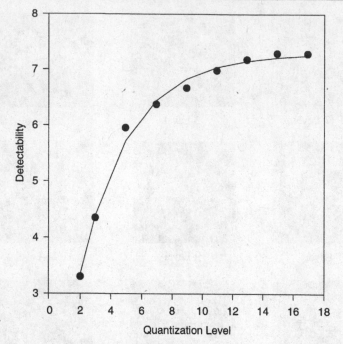

Fig. 8.13. Detectability as a function of the number of quantization levels.

correlation profiles for the BCF and the 17-level MLCF, respectively. Once again, we can see that the MLCF performs better than the BCF. For example, the ratio between the target correlation peaks and the nontarget correlation peaks is ~3 for the BCF whereas it is ~6 for the 17-level MLCF.

8.4 Multitarget composite filter synthesis

Because the correlation peak profile can be arbitrarily selected during the composite filter synthesis process, as described in Subsection 8.2.2, filters can be designed to have specific correlation peak signatures. Each signature can represent one type of target. Thus one filter can be used to recognize multiple targets at the same time.

These different shapes for optical correlation signals may be easily detected from spurious background noise or from each other in the correlation plane by use of morphological processing techniques [34]. For example, the target and the antitarget sets shown in Fig. 8.2 are also used to design this multitarget composite filter (MTCF) that yields two Gaussian-shaped peaks but with a different orientation for each set of images. For the M60 tank, the two peaks are in the off-diagonal direction, whereas for the T72 tank, the two peaks are in the diagonal direction. Nine levels (i.e., $-4, -3, -2, -1, 0, 1, 2, 3, 4$) are selected for the filter synthesis. The definition of level is the same as the one described in Section 8.3. Figure 8.15 shows the synthesized MTCF. The calculated cross correlations of this MTCF against the training-set images, as shown in Fig. 8.2, are shown in Figs. 8.16(a) and 8.16(b). The calculation result is consistent with the theoretical prediction. Therefore, indeed, we can design a MTCF by using the simulated annealing technique.

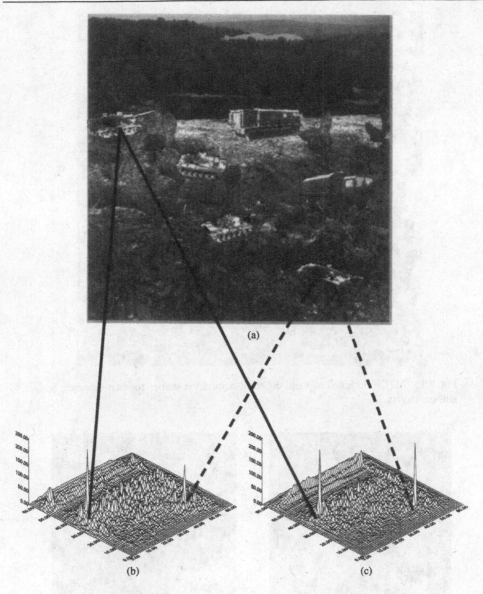

Fig. 8.14. (a) Targets embedded within the background noise, (b) correlation output with the BCF_{T72}, (c) correlation output with the 17-level $MLCF_{T72}$.

8.5 Optical implementation of a bipolar composite filter by a photorefractive crystal hologram

The optical setup for implementing the designed BCF filters in a VanderLugt-type correlator is shown in Fig. 8.17, in which a photorefractive crystal is used for recording the wavelength-multiplexed BCF filters [35, 36]. The photorefractive crystal is a 1 cm × 1 cm × 1 cm cube of Ce:Fe-doped $LiNbO_3$ crystal (in the y-cut direction). An Ar laser is used as the light source. The laser light is divided into two paths. One is used to illuminate the BCF, which is displayed on a phase-modulated LCTV, and the other one is used as the reference beam.

Fig. 8.15. MTCF designed to yield different correlation shapes for two separate, but similar, targets.

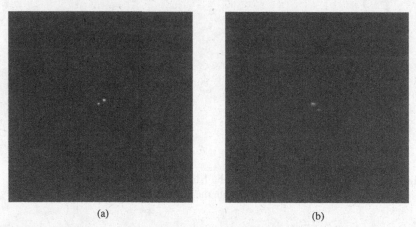

(a)   (b)

Fig. 8.16. Calculated correlation responses of the MTCF, shown in Fig. 8.15, with representative (a) M60 tank, (b) T72 tank images.

Then a set of wavelength-multiplexed BCFs are recorded in the crystal, in which one is recorded with $\lambda = 488$ nm and the other one is constructed with $\lambda = 515$ nm. The exposure time for each hologram is approximately 5 s.

During the correlation detection process, the reference beam is blocked by a shutter, and the LCTV is now operating in the amplitude-modulation mode driven by the input scene provided by a 256 gray-level image board. When the output wavelength of the Ar laser is scanned (i.e., from 488 to 515 nm), the output correlation results are obtained, as shown

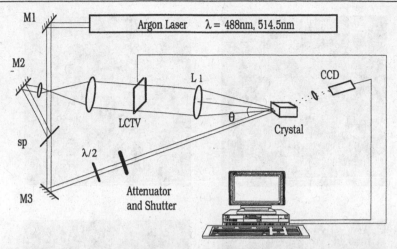

Fig. 8.17. Optical implementation. M's, mirrors; SP, splitter; L, lens.

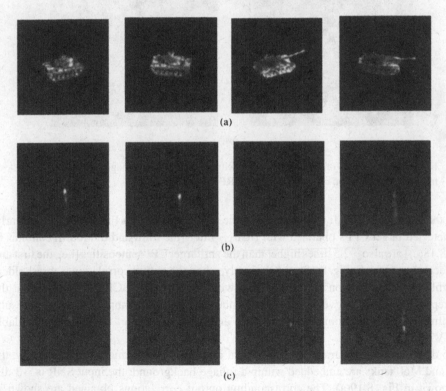

Fig. 8.18. Correlation outputs: (a) oriented input objects, (b) correlation outputs with the BCRF$_{M60}$, (c) correlation outputs with the BCRF$_{T72}$.

in Fig. 8.18. Figure 8.18(a) shows the input out-of-plane-rotation images; Figs. 8.18(b) and 8.18(c) are the corresponding correlation outputs obtained with the two BCRF filters. The out-of-plane-rotation angle between the two M60 tanks [i.e., the first and the second columns in Fig. 8.18(a)] is ∼50°, and the angle between the two T72 tanks [i.e., the third and the fourth columns in Fig. 8.18(a)] is also ∼50°. The detected target CPI's obtained with BCF$_{M60}$ filter [i.e., the first and the second columns in Fig. 8.18(b)] are ∼2.5 times higher

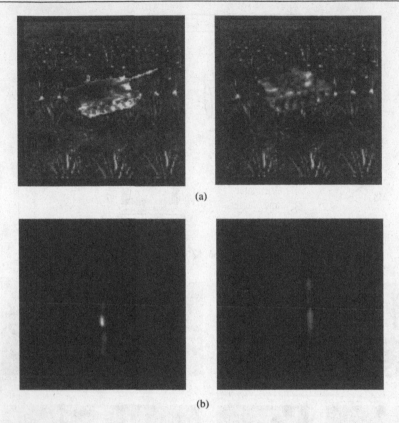

Fig. 8.19. Detection of target embedded within background foliage: (a) input noisy targets, (b) correlation outputs with the BCRF$_{T72}$.

than the antitarget CPI's [i.e., the third and the fourth columns in Fig. 8.18(b)]. Similarly, the detected target CPI's obtained with BCF$_{T72}$ filter [the third and the fourth columns in Fig. 8.18(c)] are also ∼2.5 times higher than the antitarget CPI's intensities [i.e., the first and the second columns in Fig. 8.18(c)]. Obviously, when the output correlation is thresholded, a highly reliable detection can be obtained with the proposed BCF filters. Note that the discrepancy (∼30%) between the experimental data and the computer-simulated results is primarily due to the limited contrast ratio and the resolution of the currently available LCTV.

To demonstrate the extraction of the signal in a noisy environment experimentally, the T72 and M60 tanks are embedded within a foliage background; the input SNR is ∼3 dB, as shown in Fig. 8.19(a). The corresponding output correlations obtained are shown in Fig. 8.19(b), in which the CPI's (i.e., the first column) are measured to be ∼2 times higher than the antitarget peak intensities. Thus it can be seen that the cluttering background has little adverse effect on the BCF target extraction.

8.6 Implementation in a joint transform correlator

Because the composite filters discussed in this chapter are synthesized in the spatial domain, they also can be easily implemented in a joint transform correlator (JTC) system

[37, 38]. The JTC has certain advantages over the VanderLugt correlator, even though the JTC usually has lower detection efficiencies compared with those of a VanderLugt correlator [1, 39]. For example, the JTC does not have the stringent filter alignment problem; therefore it is more suitable for real-time implementations and is robust to environmental disturbances. However, to realize the complex-valued spatial-domain composite filters in the JTC, complex function implementations are demanded. A full complex modulation may be obtained by the combination of an amplitude-modulated SLM and a phase-modulated SLM [40], but it makes the system very complicated. In this section, a position-encoding JTC is introduced [41–43], with which complex-valued references for the JTC can be implemented with an amplitude-modulated SLM.

8.6.1 Position encoding

It has been known that, as demonstrated in Figs. 8.20(a) and 8.20(b), a real-valued quantity can be decomposed into $c_1\phi_0 + c_2\phi_1$, and a complex-valued quantity can be decomposed into $c_1\phi_0 + c_2\phi_{2/3} + c_3\phi_{4/3}$, where c_1, c_2, and c_3 are nonnegative real coefficients, and $\phi_k = \exp(j2\pi k)$ represents an elementary phase vector [44, 45]. As shown by Lee [44], these decompositions can be optically realized with the position-encoding method, as illustrated

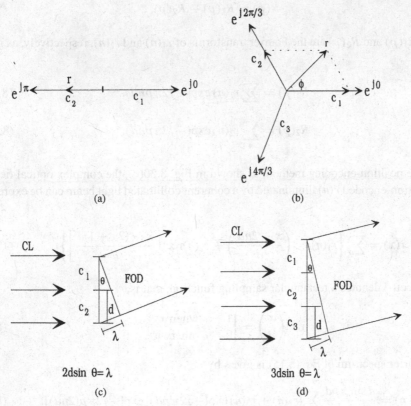

(a)

(b)

$2d\sin\,\theta=\lambda$

(c)

$3d\sin\,\theta=\lambda$

(d)

Fig. 8.20. Position encoding for real-valued or complex-valued quantities: (a) decomposition for a real-valued quantity, (b) decomposition for a complex-valued quantity, (c) position encoding for a real-valued quantity, (d) position encoding for a complex-valued quantity. CL, collimated light; FOD, first-order diffraction.

in Figs. 8.20(c) and 8.20(d). Thus, when the above position-encoding method is used with an amplitude-modulated SLM, real- or complex-valued correlations can be implemented in the JTC. For simplicity, one-dimensional notation is used for the analysis below. We also assume that the functions discussed are discrete and that the pixel sizes of the SLM's are $d \times d$.

We assume that the reference $r(n)$ is a real-valued discrete function, as the spatial-domain composite filters discussed in this chapter are real-valued functions. As can be seen from Fig. 8.20(a), $r(n)$ can be decomposed into

$$r(n) = r_1(n) - r_2(n), \tag{8.11}$$

where r_1 and r_2 are given by

$$r_1(n) = r(n), \qquad r_2(n) = 0, \qquad \text{if } r(n) > 0,$$
$$r_1(n) = 0, \qquad r_2(n) = -r(n), \qquad \text{if } r(n) \leq 0. \tag{8.12}$$

It can be easily shown that the Fourier transform $R(p)$ of $r(n)$ can be obtained from $R_1(p)$ and $R_2(p)$, as given by

$$R(p) = R_1(p) - R_2(p), \tag{8.13}$$

where $R_1(p)$ and $R_2(p)$ are the Fourier transforms of $r_1(n)$ and $r_2(n)$, respectively, as given by

$$R_1(p) = \sum_n r_1(n) \exp(-j2\pi pnd), \tag{8.14a}$$

$$R_2(p) = \sum_n r_2(n) \exp(-j2\pi pnd). \tag{8.14b}$$

With the position-encoding method, as shown in Fig. 8.20(c), the complex optical field of the position-encoded $r(n)$ illuminated by a coherent collimated light beam can be expressed as

$$i(x) = \sum_n \left\{ r_1(n)\text{rect}\left(\frac{x - 2nd}{d}\right) + r_2(n)\text{rect}\left[\frac{x - (2n+1)d}{d}\right] \right\}, \tag{8.15}$$

where $\text{rect}(\cdot)$ denotes a rectangular sampling function, that is,

$$\text{rect}\left(\frac{x}{d}\right) = \begin{cases} 1 & \text{when } x \leq d \\ 0 & \text{otherwise} \end{cases}. \tag{8.16}$$

The Fourier spectrum of Eq. (8.15) is given by

$$I(p) = \frac{d \sin \pi pd}{\pi pd} \sum_n [r_1(n) + r_2(n) \exp(-j2\pi pd)] \exp[-j2\pi p(2nd)], \tag{8.17}$$

where $p = \mu/\lambda f$, μ is the spatial frequency coordinate, λ is the wavelength of the light source, and f is the focal length of the transform lens. As illustrated in Fig. 8.20(c), the

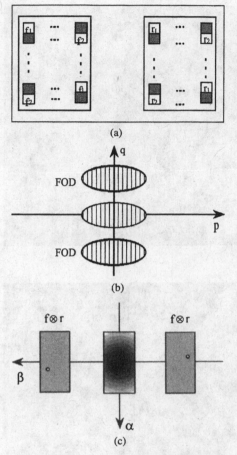

Fig. 8.21. Position encoding for the JTC: (a) position-encoding input, (b) JTPS, (c) output correlation of a real-valued function.

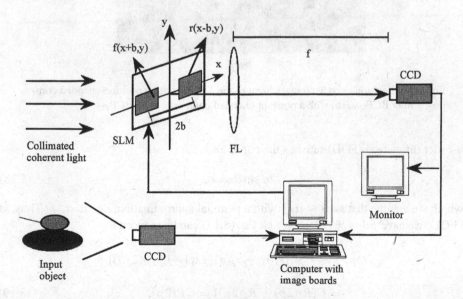

Fig. 8.22. Single-SLM JTC. FL, Fourier lens.

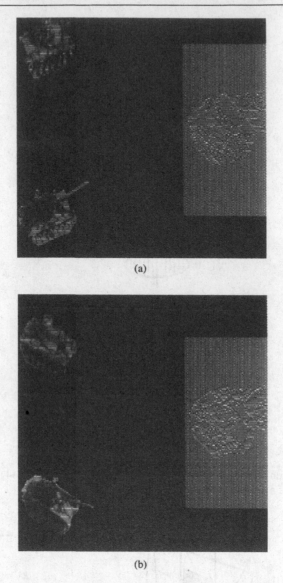

Fig. 8.23. Position-encoded images input to the JTC (a) with a position-encoded composite filter BCF_{M60}, (b) with a position-encoded composite filter BCF_{T72}.

first-order diffraction (FOD) satisfies the condition

$$2d \sin \theta = \lambda, \tag{8.18}$$

in which we assume that $\sin \theta \approx \mu / f$ with a paraxial approximation, i.e., $\mu \ll f$. Thus, at the FOD, we have $pd \approx \pm 1/2$, by which Eq. (8.17) can be reduced to

$$I(p) = \frac{2d}{\pi} \sum_n [r_1(n) - r_2(n)] \exp[-j2\pi p(2nd)]$$

$$= C[R_1(2p) - R_2(2p)] = C R(2p), \tag{8.19}$$

Fig. 8.24. JTPS of Fig. 8.23(a).

where C is a proportionality constant. Therefore a scaled Fourier spectrum of the real-valued function $r(n)$ can be obtained at the FOD at the off-axis angle $\theta = \lambda/2d$ (rad). In the same manner, the Fourier spectrum of the input function $f(n)$ can also be obtained at the FOD by the decomposition of $f(x)$ into $f_1(n)\phi_0 + f_2(n)\phi_1$.

8.6.2 Position-encoding joint transform correlator

Now we are ready to implement correlation or convolution between real-valued functions on the JTC with an amplitude-modulated SLM based on the position-encoding technique discussed above. Even though we have discussed only the one-dimensional case, we can easily extend it to a two-dimensional case by applying position encoding along only one of two dimensions, that is, either the x axis or the y axis. Let the input function and the reference be $f(x, y)$ and $r(x, y)$, respectively, and the input to the JTC be $f(x + b, y) + r(x - b, y)$. Then the JTC with real-valued inputs can be performed as follows.

According to Eqs. (8.11) and (8.12), the real-valued inputs can be decomposed into

$$f(x, y) = f_1(x, y) - f_2(x, y),$$
$$r(x, y) = r_1(x, y) - r_2(x, y).$$

If the position encoding is applied along the y axis, as shown in Fig. 8.21(a), then the joint transform power spectrum (JTPS) at one of the FOD's as illustrated in Fig. 8.21(b), can be written as

$$\text{JTPS} = |R(p, 2q)|^2 + |F(p, 2q)|^2$$
$$+ 2|R(p, 2q)F(p, 2q)| \cos(2pb + \phi_R - \phi_F), \qquad (8.20)$$

where ϕ_R and ϕ_F are the phase components of $R(2p, q)$ and $F(2p, q)$, respectively. After an inverse Fourier transform, the output correlation distribution, as shown in Fig. 8.21(c),

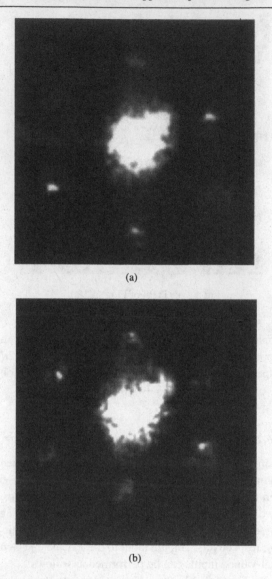

(a)

(b)

Fig. 8.25. Output light distributions obtained with (a) the position-encoded BCF_{M60}, (b) the position-encoded BCF_{T72}.

can be obtained:

$$C(\alpha, \beta) = r(\alpha, \beta/2) \otimes r(\alpha, \beta/2) + f(\alpha, \beta/2) \otimes f(\alpha, \beta/2) + r(\alpha - 2b, \beta/2)$$
$$\otimes f(\alpha, \beta/2) + r(-\alpha - 2b, -\beta/2) \otimes f(-\alpha, -\beta/2). \qquad (8.21)$$

Obviously the correlation plane is elongated twice along the β axis. Nevertheless, we can eliminate the elongation by magnifying the JTPS twice along the q axis. The situation with position encoding along the y axis is similar to the above case and is not discussed further here.

Notice that the price we pay for using position encoding is that two pixels are used for real functional values. As for a complex functional value, three pixels are required.

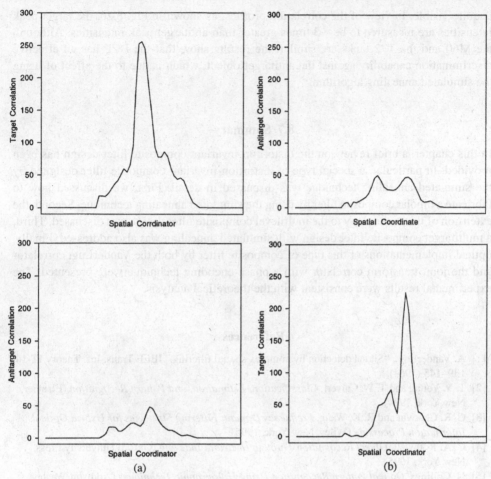

Fig. 8.26. Target and antitarget correlation profiles obtained with (a) the position-encoded BCF_{M60}, (b) the position-encoded BCF_{T72}.

Nevertheless, the major advantage of using position encoding for this specific application is that the bipolar nature of the BCF can be implemented in the amplitude-modulated SLM. Because the zero-order JTPS is not used, the dc component caused by low contrast and the dead zones of the SLM can be avoided.

8.6.3 Experimental demonstration

Figure 8.22 shows a single-SLM-based JTC system [46], in which a LCTV operated in the amplitude-modulation mode is used as the input SLM. For experimental demonstration, the filters BCF_{M60} and BCF_{T72} shown in Figs. 8.3(a) and 8.3(b) are used to detect the tanks shown in Fig. 8.2. The position-encoding method is used to obtain the position-encoded BCF's and the input objects along the y direction that are shown in Figs. 8.23(a) and 8.23(b), respectively. The JTPS of Fig. 8.23(a) captured by the CCD is shown in Fig. 8.24, in which one of the first-order Fourier spectra ($JTPS_1$) is used for the second-half cycle operation. The $JTPS_1$ is displayed on the LCTV, and the output correlation distributions obtained are shown in Figs. 8.25(a) and 8.25(b), respectively. Note that the extraction of the target objects is quite visible. In view of the correlation profiles, as shown in Fig. 8.26, the target peak intensities are measured to be ~3 times greater than antitarget peak intensities. Although the M60 and the T72 tanks are similar, the results show that the BCF has an effective discrimination capability against the antitarget object, which is due to the effect of using the simulated annealing algorithm.

8.7 Summary

In this chapter, a brief review on the distortion-invariant composite filter design has been provided. In particular, a special type of distortion-invariant composite filter designed by the simulated annealing technique was discussed in detail. First, we discussed how to fabricate a bipolar composite filter by using the simulated annealing technique. Second, the extension of this technology to the multilevel composite filter design was discussed. Third, a multitarget composite filter design with simulated annealing was also addressed. Finally, optical implementations of this type of composite filter by both the VanderLugt correlator and the joint transform correlator with a phase-encoding technique were presented. The experimental results were consistent with the theoretical analysis.

References

[1] A. VanderLugt, "Signal detection by complex spatial filtering," IEEE Trans. Inf. Theory **IT-10**, 139–145 (1964).

[2] T. Y. Young and T. W. Calvert, *Classification, Estimation, and Pattern Recognition* (Elsevier, New York, 1974).

[3] C. R. Chatwin and R. K. Wang, *Frequency Domain Filtering Strategies for Hybrid Optical Information Processing* (Wiley, New York, 1995).

[4] C. A. Rothwell, *Object Recognition Through Invariant Indexing* (Oxford University Press, New York, 1995).

[5] N. Collings, *Optical Pattern Recognition Using Holographic Techniques* (Addison-Wesley, Reading, MA, 1988).

[6] S. Yin, L. Cheng, and G. Mu, "3-D target recognition by serial-code-filters," Optik **82**, 129–131 (1989).

[7] J. R. Ullmann, *Pattern Recognition Techniques* (Russak, New York, 1973).

[8] T. H. Reiss, *Recognizing Planar Objects Using Invariant Image Features* (Springer-Verlag, Berlin, 1993).

[9] F. T. S. Yu and S. Yin, eds., *Coherent Optical Processing*, Vol. MS52 of SPIE Milestone Series (Society of Photo-Optical Instrumentation Engineers, Bellingham, WA, 1992).

[10] H. Nasr, ed., *Model-Based Vision*, Vol. MS72 of SPIE Milestone Series (Society of Photo-Optical Instrumentation Engineers, Bellingham, WA, 1993).

[11] S. Jutamulia, ed., *Optical Correlators*, Vol. MS76 of SPIE Milestone Series (Society of Photo-Optical Instrumentation Engineers, Bellingham, WA, 1993).

[12] C. Y. Suen and P. S. Wang, eds., *Thinning Methodologies for Pattern Recognition* (World Scientific, River Edge, Singapore, 1994).

[13] M. Suk and S. M. Bhandarker, *Three-Dimensional Object Recognition from Range Images* (Springer-Verlag, New York, 1992).

[14] A. K. Jain and P. J. Flynn, *Three-Dimensional Object Recognition Systems* (Elsevier, New York, 1993).

[15] D. P. Casasent and T.-H. Chao, eds., *Optical Pattern Recognition VI*, Proc. SPIE **2490** (1995).

[16] B. V. K. Vijaya Kumar, "Tutorial survey of composite filter designs for optical correlators," Appl. Opt. **31**, 4773–4801 (1992).

[17] B. V. K. Vijaya Kumar and A. Mahalanobis, "Recent advances in distortion-invariant correlation filter design," in Ref. 15, pp. 2–13.

[18] S. Yin, M. Lu, C. M. Chen, F. T. S. Yu, T. D. Hudson, and D. McMillen, "Design of a bipolar composite filter using a simulated annealing algorithm," Opt. Lett. **20**, 1409–1411 (1995).

[19] F. T. S. Yu, M. Lu, G. Lu, S. Yin, T. Hudson, and D. McMillen, "Optimum target detection usign a spatial-domain bipolar composite filter with a joint transform correlator," Opt. Eng. **34**, 3200–3207 (1997).

[20] T. Hudson, D. McMillen, A. Kransteuber, S. Yin, M. Lu, and F. T. S. Yu, "Rotation-invariant bipolar filter design for object classification," in *Photorefractive Fiber and Crystal Devices: Materials, Optical Properties, and Applications*, F. T. S. Yu, ed., Proc. SPIE **2529**, 274–284 (1995).

[21] H. J. Caulfield and W. T. Maloney, "Improved discrimination in optical character recognition," Appl. Opt. **8**, 2354–2356 (1969).

[22] C. F. Hester and D. Casasent, "Multivariant technique for multiclass pattern recognition," Appl. Opt. **19**, 1758–1761 (1980).

[23] D. Casasent, "Unified synthetic discriminant function computing formulation," Appl. Opt. **23**, 1620–1627 (1984).

[24] B. V. K. Vijaya Kumar, "Minimum variance synthetic discriminant functions," J. Opt. Soc. Am. A **3**, 1579–1584 (1986).

[25] A. Mahalanobis, B. K. K. Vijaya Kumar, and D. Casasent, "Minimum average correlation energy filters," Appl. Opt. **26**, 3633–3640 (1987).

[26] R. Kallman, "The construction of low noise optical correlators," Appl. Opt. **25**, 1032–1033 (1986).

[27] L. Horner and P. D. Gianino, "Applying the phase-only filter concept to synthetic discriminant function correlation filter," Appl. Opt. **24**, 851–855 (1985).

[28] D. A. Jared and D. Ennis, "Inclusion of filter modulation in synthetic discriminant function construction," Appl. Opt, **28**, 232–239 (1989).

[29] B. Reid, P. W. Ma, J. D. Downie, and E. Ochoa, "Experimental verification of modified synthetic discriminant function filters for rotation invariance," Appl. Opt. **29**, 1209–1214 (1990).

[30] S. Kirkpatrick, C. D. Gelatt Jr., and M. P. Vecchi, "Optimization by simulated annealing," Science **220**, 671–680 (1983).

[31] S. Kim, M. R. Feldman, and C. C. Guest, "Optimum encoding of binary phase-only filters with a simulated annealing algorithm," Opt. Lett. **14**, 545–547 (1989).

[32] M. S. Kim and C. C. Guest, "Simulated annealing algorithm for binary phase only filter in pattern recognition," Appl. Opt. **29**, 1203–1208 (1990).

[33] N. Yoshikawa and T. Yatagai, "Phase optimization of a kinoform by simulated annealing," Appl. Opt. **33**, 863–868 (1994).

[34] W. M. Growe, "Optical morphological processing of optical correlator signals," in *Photonics for Processors, Neural Networks, and Memories*, J. L. Horner, B. Javidi, S. T. Kowel, and W. J. Miceli, eds., Proc. SPIE **2026**, 297–301 (1993).

[35] P. Yeh, *Introduction to Photorefractive Nonlinear Optics* (Wiley, New York, 1993).

[36] F. T. S. Yu, S. Yin, and A. S. Bhalla, "Wavelength-multiplexed holographic construction using a Ce:Fe: doped PR fiber with a tunable visible-light laser diode," IEEE Photon. Technol. Lett. **5**, 1230–1233 (1993).

[37] C. S. Weaver and J. W. Goodman, "Technique for optically convolving two functions," Appl. Opt. **5**, 1248–1249 (1966).

[38] F. T. S. Yu and X. J. Lu, "A real-time programmable joint transform correlator," Opt. Commun. **52**, 10–16 (1984).

[39] F. T. S. Yu, Q. W. Song, Y. S. Cheng, and D. A. Gregory, "Comparison of detection efficiencies for VanderLugt and joint transform correlators," Appl. Opt. **29**, 225–232 (1990).

[40] D. A. Gregory, J. C. Kirsch, and E. C. Tam, "Full complex modulation using liquid crystal television," Appl. Opt. **31**, 163–165 (1992).

[41] F. T. S. Yu, G. Lu, D. Zhao, and F. Cheng, "Implementation of complex gray leveled functions for a joint transform processor by using a decomposition method," in *OSA Annual Meeting*, Vol. 6 of 1993 OSA Technical Digest Series (Optical Society of America, Washington, D.C., 1993), p. 46.

[42] F. T. S. Yu, G. Lu, M. Lu, and D. Zhao, "Application of position-encoding to a complex joint transform correlator," Appl. Opt. **34**, 1386–1388 (1995).

[43] G. Lu, "Study of phase-encoding techniques for joint transform correlator as applied to pattern recognition and classification," Ph.D. dissertation (Pennsylvania State University, University Park, PA, 1996).

[44] W. H. Lee, "Sampled Fourier transform hologram generated by computer," Appl. Opt. **9**, 639–643 (1970).

[45] C. B. Burckhardt, "A simplification of Lee's method of generating holograms by computer," Appl. Opt. **9**, 1949 (1970).

[46] F. T. S. Yu, S. Jutamulia, T. W. Lin, and D. A. Gregory, "Adaptive real-time pattern recognition using a liquid crystal TV based joint transform correlator," Appl. Opt. **26**, 1370–1372 (1987).

9

Iterative procedures in electro-optical pattern recognition

Joseph Shamir

9.1 Introduction

Interest in Fourier-transform- (FT-) based optical correlators started to evolve from VanderLugt's first introduction of a holographic matched filter [1, 2] into coherent optical processors. Although accepted as a significant success, the idea of the matched filter is rooted in the communication theory, in which the main issue is the detection of a signal immersed in noise. For this application the matched filter was found optimal, provided the noise was additive and had a white Gaussian distribution. The conditions for optimality of the matched filter are, in general, not fulfilled in optical pattern recognition systems, particularly if the objective is to discriminate among different objects. Moreover, noise in optical systems does not have a Gaussian distribution and is usually not additive but, rather, is signal dependent.

A significant progress in the field of optical pattern recognition was marked by the introduction of spatial filters based on linear combinations of simpler filters [3, 4] that could be implemented as computer-generated holograms (CGH's) [5]. This approach already required the marriage of optics and electronics, which led to improved performance but still maintained the intrinsic drawbacks of the matched filter. After this fact was realized it became common practice to design the appropriate filters in a highly nonlinear fashion, significantly deviating from the matched-filter approach [6]. Although a classical matched filter can be recorded in a holographic recording system, these synthetic filters (SF's) require sophisticated numerical processing.

After a filter is designed it must be implemented and aligned properly in the optical system. Both of these steps are tedious and time consuming, making the whole process difficult to use in practical applications. The problem of alignment can be mitigated by use of the joint transform correlator (JTC) [7–9] architecture, in which a reference function (RF) is placed beside the input pattern instead of a filter in the FT plane. Of course, the benefits of the JTC architecture are penalized by some disadvantages, such as a reduction in the available space–bandwidth product.

In view of the rapid progress of digital signal processors, one should ask the question: Is there any need to continue the research in optical signal processing? The objective of this chapter is to review developments toward a positive answer and to demonstrate that the best strategy is to exploit the attributes of both electrons and photons in a hybrid system. One should look for computing architectures in which photonic processing complements (rather than competes with) electronic computing.

The main reason for the incredible advance in electronic computing technology is the strict control one has on electrons by electrons because of the strong interaction among the electrons themselves. The strong electron–electron interaction is exploited in conventional digital computing to implement switching circuits that are the basis for digital operations. Such strong interaction does not exist among photons, making them unsuitable for digital operations. However, the same lack of interaction provides a property that is ideal for an information transfer for which interchannel interference must be minimal. A light wave can perform a large number of calculations in parallel with no cross talk among the individual channels. The calculation that is best suited for an optical implementation is the solution of the wave equation, in which the input information is presented as a set of boundary conditions. In principle, an infinite amount of information contained in the boundary conditions can be processed in a finite time, although in practice some limitations exist [10, 11].

Because photons practically do not interact with each other, the solution of the wave equation is probably the only thing they can do without some interaction with a mediator. Thus there is little hope for the realization of a photonic-only computer to replace the electronic computer. However, a relatively strong interaction does exist between photons and electrons (or rather, electronic states). Such an interaction was practiced in the early photographic process in which photons interact with electronic states in the photosensitive material, and, after development (chemical amplification), the photographic material modulates the light illuminating it. Thus, in principle, the marriage of optics and electronics dates back to the invention of the photographic process. Unfortunately the photographic process is rather slow and not practical for real-time applications. Even the so-called real-time media, such as photorefractives, depend on the diffusion of charge carriers, which is a rather slow process. To increase the operating speed, we need the intervention of some accelerating force. Such forces exist when the excited electrons are subject to electric fields, as is the case in electronic photodetection. Accordingly, to make a practical signal processor, we must add to the optical part a detector (photodiode, CCD, etc.) that has a controllable response and can be coupled to the outside world. The generated information can now be reintroduced into the system or transferred to another stage of the processor. The combined technology of photodetection and spatial light modulation constitutes the heart of a hybrid system that complements the attributes of electronics with those of optics.

The iterative methods treated in this chapter can be also described as optimization procedures. Motivations for using these optimization algorithms in pattern recognition are addressed in Section 9.2. This is followed by a brief overview of iterative methods with an emphasis on two families of algorithms, genetic algorithms (GA's) and projection-onto-constraint-sets (POCS) algorithms, which were found to be especially efficient for pattern recognition applications. One possible mathematical definition of the pattern recognition problem, discussed in Section 9.4, is used for the implementation of hybrid electro-optical pattern recognition systems in Section 9.5. The design of spatial filters for linear and nonlinear pattern recognition procedures is executed with the help of POCS algorithms with the various examples presented in Section 9.6. To extend the distortion-invariance capabilities, POCS algorithms are also used in the implementation of the adaptive pattern recognition systems presented in Section 9.7.

To keep the scope of this chapter within reasonable limits, the discussion is restricted to correlator-based pattern recognition methods implemented in two general architectures, the $4f$ correlator and the JTC.

9.2 Pattern recognition and optimization

The implementation of optical pattern recognition by correlation methods requires the design of a SF in the $4f$ architecture or a RF in the JTC architecture. From a mathematical point of view, the two approaches are identical and, to be specific, we usually address the $4f$ correlator that contains a SF. Because, basically, a $4f$ correlator is uniquely defined, the main issue is the design of a proper SF. Before designing such a SF, we must define an objective and a criterion that determine whether that objective has been achieved. The objective may be one or more tasks such as known target detection, target identification, discrimination among several target classes, target location, target tracking, etc. A processing system designed for one task is not necessarily optimal for a different task. Most conventional procedures for generating a SF assume the task of target recognition and define the criterion for recognition as the appearance of a strong cross-correlation peak at the origin of the output plane if the desired object is present at the origin. However, solving a problem for one point in the output plane does not guarantee that large sidelobes would not plague other regions of the output plane. Solutions that exist mathematically are not always practical for implementation on real physical systems. In particular, if the required end result or some constraints are not linear, such as binary filters or phase-only filters (POF's) [12], these procedures may fail quite frequently. Moreover, even if the processing system containing this filter operates as designed, we may still not know the exact position of the object because the relation of the correlation peak to the actual position of the object may be not well defined [13–15].

Advanced procedures started to deviate from strictly linear methods by supplementing them with optimization processes. These include the minimum average correlation energy filter [16], the minimum-variance filter [17], and the generalized SF [18], which contains most of the previous types as special cases. Mathematically, all these SF's have an analytic representation, but nevertheless, numerical methods are usually needed for their actual evaluation.

Alternative approaches rely completely on nonlinear procedures [19–26] in which complex weighting coefficients were allowed and response was controlled over the entire output plane in parallel. These latter procedures can seldom be put into a closed analytic form suitable for straightforward calculations. Therefore, in most cases, numerical methods must be employed.

Iterative methods are preferred for solving nonlinear and noisy problems as well as signal reconstruction from incomplete information. There are also several additional incentives to motivate the applications of iterative processes that can, in principle, accommodate various environmental situations. Some of those motivations are outlined here.

Animal eyes appear to work, at least partly, in an iterative fashion. The optical equipment (cornea, lens, retina, etc.) comprises more or less unchangeable givens. Some eyes are optically superior to others. All eyes are unique. Seeing involves not just the optical equipment but also the neural equipment (optic nerve, brain, etc.) that operates in conjunction with the optical equipment. The neural network contains numerous readily adjustable variables (connections) that evolve as the individual learns to see. This arrangement is desirable because it can readily accommodate rather gross defects in the optical system. Good performance does not necessarily require good optics, a situation that may be quite advantageous for many applications. Of course, in this subject we encounter an optimization process that is performed within the complete system and not an off-line procedure to design just one system component (the SF, in our case).

In any mass production of items, there are two ways to ensure satisfactory performance. First, we can build, assemble, and maintain all components and the relationships between them to within close predetermined tolerances. Most, but not all, technologically mass-produced equipment is so designed. Second, we can assume that some errors will inevitably creep in and that therefore adaptive fixes should be built in to allow use of imperfect hardware. Many integrated circuits and computers are designed for such error tolerance. Naturally this increases yield and reliability. Optical correlators made of inferior components imperfectly assembled might be much less expensive than perfect correlators. It would be advantageous if they could work almost as well.

Biological systems and some computers show a remarkable ability to program around injuries or failed parts, that is, they achieve reliability through adaptivity. Optical correlators, too, can undergo partial component failures. In addition, goals (and hence the definition of optimality) may change. The optimization techniques originally employed could also be used for later adaptation.

Taking into consideration all the aspects discussed above indicates that a feasible procedure for the implementation of an efficient pattern recognition system in its general sense is actually an optimization problem. This optimization must take into account the well-defined tasks, the required criteria, and all the system and problem parameters. An important conclusion of this section is that not only is a good filter design an optimization process, but performing this optimization within the complete optical system has many advantages. In Section 9.3 a brief review of some optimization procedures that were found useful for optical pattern recognition procedures is provided. Some of them are applicable for real-time hybrid electro-optical implementation whereas others are better suited for off-line filter design.

9.3 Iterative optimization algorithms: an overview

The implementation of an optimization algorithm starts by the definition of a goal. In our case the goal is to design a SF (or a RF) that performs, in a given optical system, the desired task under a given set of criteria. We represent the transfer function of the filter by a function h that may be an actual complex transfer function or its FT. The optimization is implemented by minimizing some distance function to this goal, $d(h)$, which is a functional of the filter function h. Other terms used for this distance function include cost function, figure of merit, energy function, and fitness. These terms are used loosely, in correspondence with the historically accepted ones for the various algorithms. The variables of h are usually the coordinates x, y over a plane perpendicular to the general propagation direction of the light in the optical system. For numerical processing it is customary to use pixelated coordinates; thus the function h is given by its sampled form $h(i, j)$. The distance function, which must be carefully selected for a given application, represents some generalized distance from the present iteration of $h(i, j)$ to the desired final solution of the problem or to a previous iteration. Thus the optimization is carried out by manipulation of the elements of the function $h(i, j)$ in such a way as to optimize the distance function with respect to a prescribed rule. Below some of the most frequently encountered optimization procedures are described.

9.3.1 Gradient-descent algorithm

One of the best-known algorithms for optimization is the gradient-descent (GD) algorithm [27]. We update the function h for the next iteration by using the relation

$$h^{(t+1)}(i, j) = h^{(t)}(i, j) - \eta \left[\nabla d^{(t)} \right]_{i, j},$$

(9.1)

where η is a step size and $[\nabla d^{(t)}]_{i,j}$ is the (i, j)th element of the $(N \times N)$-dimensional gradient of the distance function derived for the tth iteration of h.

In principle, this is a convenient algorithm, but it tends to get stuck in local minima and it can be destructively affected by noise. It is also based on the assumption that a gradient of the distance function always exists, which is not necessarily the case.

9.3.2 Hill-climbing procedure

The hill-climbing (HC) optimization procedure is also known as a direct binary search. After the distance function of the tth iteration is calculated, a change is induced over one element (sample or pixel) of $h(i, j)$ to obtain the $(t + 1)$th iteration. This modification changes $d^{(t)}$ by an amount

$$\Delta d^{(t+1)} = d\left[h^{(t+1)}(i, j)\right] - d\left[h^{(t)}(i, j)\right]. \tag{9.2}$$

The new function $h^{(t+1)}(i, j)$ is accepted if $\Delta d^{(t+1)} \leq 0$; otherwise it is rejected. The procedure is now repeated for the next element of h and so on. Eventually the procedure is recycled until a desired minimum is achieved.

While in the GD algorithm, we need the calculation of gradients that requires the knowledge of the functional behavior of various parameters; the HC algorithm is a trial-and-error procedure. The functional relationships are not strictly necessary but the algorithm is extremely slow.

9.3.3 Simulated annealing

We seek a global energy minimum in a piece of metal by annealing. The metal is heated to a high temperature at which patches within it are quite mobile. It is then slowly cooled to allow it to settle into its lowest energy state. Simulated annealing (SA) [28] uses the same approach. In analogy to the natural process, the distance function here is a nonnegative energy, the energy function, that is to be minimized. The energy depends on the set of free variables that are to be manipulated. We start with any random set of parameters and then stochastically perturb the variable set by a large amount (energy measure) at high temperature and by lower amounts at lower temperatures. If the perturbed variables lower the energy, we accept the perturbation. If the perturbed variables increase the energy, we may or may not accept the perturbation. That choice is determined stochastically from a probability distribution function governed by the temperature. At high temperature, the probability of accepting an energy-increasing perturbation is high, and it decreases with the lowering of the temperature. The temperature is slowly decreased until a steady-state minimum is achieved. This is a simplified overview of what is not really a method but a family of methods.

The SA algorithm is, in principle, capable of reaching a global optimum if it is run for an infinite time. Using the previous notation, at the tth iteration we induce a random change in the elements of $h^{(t)}$ to obtain the $(t + 1)$th iteration of the function h. This changes the energy function $d^{(t)}$ by an amount $\Delta d^{(t+1)}$, similar to Eq. (9.2). The new function h is accepted if $\Delta d^{(t+1)} < 0$. In this algorithm, however, for $\Delta d^{(t+1)} \geq 0$ the iteration may be also accepted. This acceptance is made conditionally, based on the acceptance probability

$$\text{Pr}_{\text{accept}} = \exp\left[\frac{-\Delta\, d^{(t+1)}}{T}\right],\tag{9.3}$$

where T is the temperature parameter.

The procedure is now repeated, starting from the new function $h^{(t+1)}$ and decreasing the temperature slowly as the process continues. The cooling rate and the steps of the random perturbation are important parameters that depend on the specific process implemented. Because the achievement of the global minimum is guaranteed after only an infinite number of iterations, the process is terminated when an adequately low energy is obtained.

9.3.4 Genetic algorithms

In biological evolution, it is not the individual but a population (species) that evolves. The success of the individual (phenotype) gives it an improved chance of breeding. In breeding, the genetic structure (genotype) of the offspring is made up of genotypic contributions from both parents. In addition, errors (mutations) occasionally occur. The offspring then competes for the right to reproduce in the next generation. Thus a gene pool evolves that not only governs future generations but also bears within its memory a knowledge of where it has been.

Optimization procedures based on the above ideas are called genetic algorithms (GA's) [29]. In a GA, a genome or vector is specified as a way to describe the system. It contains (usually) all the information needed to describe the system. A figure of merit is then evaluated for each member of a pool of genomes. Winners are selected for genetic exchange (usually called crossover) and mutation. Losers are usually dropped from the pool to keep the pool size constant.

To use a GA we should have the following features:

(1)   A chromosomal representation of solutions to the problem, usually binary.
(2)   An evaluation function that gives the fitness of the population. This is, in principle, again a distance function.
(3)   Combination rules (genetic operators) to produce new structures from old ones: reproduction, crossover, and mutation.

There are several variants of the GA. One possible algorithm that was used for generating spatial filters for pattern recognition in a hybrid electro-optical system [24, 30] is summarized as follows:

(1)   Start: Select at random a population of m members (binary functions) $\{h_1, h_2, \ldots, h_m\}$. In our case these functions represent possible solutions for the filter functions. Evaluate the values of the corresponding cost functions d_i $\{i = 1, 2, \ldots, m\}$. Compute the average value of the cost function

$$\theta = \frac{1}{m}\sum_{i=1}^{m} d_i.\tag{9.4}$$

Set a discrete time parameter t to zero. Define a probability P for a mutation to occur and set it to some P_{max}.

(2)   Crossover/mutate: Select the function h_l that corresponds to the minimal cost function d_l. Pick from the population a function h_j at random. The two functions h_l and h_j are the parents to be used for generating an offspring function. Select a random integer k between 0 and n, where n is the dimension of the vectors h. Create the offspring function h_c by taking the first k elements from one of the parents randomly and the remaining $n - k$ elements from the other parent. Induce a mutation (inverting the sign of the elements 1 to 0 or 0 to 1) with probability P on each element of the offspring vector h_c. Evaluate the offspring cost function d_c.

(3)   Reproduce: Pick at random a function h_d from the population subject to the constraint $d_d \geq \theta$. Replace h_d in the population with the new offspring h_c and update the average value of the cost function:

$$\theta \to \theta + \frac{1}{m}(d_c - d_d). \tag{9.5}$$

(4)   Setting parameters: Set the new parameters $t \to t + 1$ and $P \to P_{\max}(1/t)^r$. Terminate the procedure when adequate discrimination (a predetermined value of the cost function d) is achieved. (Alternatively, the process can be terminated after a predetermined number of iterations. The latter criterion is useful if there is no *a priori* knowledge about the expected behavior of the cost function.) If $P > P_{\min}$ go to 2; otherwise go to 1. Selection of the parameters r, $P_{\min}$, and $P_{\max}$ depends on the particular problem at hand.

9.3.5 Projections-onto-constraint-sets algorithms

Projection-based algorithms constitute an extremely powerful class of optimization algorithms. In the POCS algorithm [31] the distance function is measured from the function h^t obtained at the tth iteration to sets of various constraints defined in the problem. These constraints represent a set of conditions that must be fulfilled by the solution. The constraints are usually determined by the results of some measurements, the physical characteristics of the experimental system, and some demands on the required solution. In some specific applications, the constraints include the explicit form of the desired solution, and then some system parameters are the variables to be determined. For the POCS algorithm to work, some functional relationships are necessary and the physical constraints must be precisely known and defined.

The POCS algorithm is an iterative process that basically transfers a function from one domain to another (for example, the FT domain and the image domain), and in every domain it is projected onto one or several constraint sets. The procedure is repeated in a cyclic way until the solution converges to a function h that satisfies all the constraints simultaneously. If all the constraint sets are closed and convex and they have at least one common domain, then the process converges weakly. In our context it is difficult, and may sometimes be impossible, to fulfill these conditions, so the convergence is not always guaranteed. Nevertheless, in modified versions of the POCS algorithm, generalized distance functions can be employed that will not increase from one iteration to the next. In any case, an exact solution is inaccessible in a finite time, and the algorithm is terminated when the distance function attains a value that is defined to be adequately small. Thus a good solution can be obtained even if the various constraints, are inconsistent with each other. The solution obtained will have the smallest possible distance to all constraints, although no one of them may be completely satisfied.

To design a spatial filter for pattern recognition in an optical correlator, the main constraint is defined by a desired intensity distribution over the correlation plane. Additional
constraints will include some physical characteristics and limitations of the recording media
(transmittance cannot exceed unity, resolution is limited to pixel or grain size, etc.). After
the set of constraints is set up, the distance function $d(h')$ can be defined as some weighted,
generalized distance between the actually detected correlation distribution and the other
characteristics of the filter function to the predefined constraints.

The traditional POCS algorithms are based on a serial approach [31–34]. However, the
range of applicability of the serial algorithm is limited and for many problems the more
recently introduced parallel projection method [35, 36], based on Ref. 37, must be used.
This is the case, in particular, in which not all the constraints are convex and some of the
constraints are inconsistent [38].

9.3.5.1 Serial projections

Given a Hilbert space $\mathcal{H}$, a distance function d on $\mathcal{H}$, and a closed convex set C in $\mathcal{H}$,
projection from $\mathcal{H}$ onto C with respect to the distance function d is an operation P that
associates to every element $h \in \mathcal{H}$ the (unique) element h' in C closest to h, where close is
measured by d:

$$P\{h\} = h' \qquad \text{if and only if} \qquad h' \in C, \qquad \inf_{y \in C} d(y, h) = d(h', h) \qquad (9.6)$$

(projected vectors are henceforth marked by a prime). Usually d is derived from the pre-
vailing Hilbert space structure,

$$d(h, h') = \|h - h'\|_{\mathcal{H}} := \int |h(x) - h'(x)|^2 \, dx. \qquad (9.7)$$

This Euclidean distance function is convenient for these projection algorithms because
of its simplicity and fast calculability. At the expense of computational complexity, more
complicated cost functions can be used as well.

If the sets C_i are closed with respect to d and convex, the projection element exists and
is unique. If the sets are not convex, the situation is more complicated but, under certain
conditions, procedures still exist for determining a unique projection [39].

Sometimes the projection operation is modified to admit relaxation that may enhance
the convergence properties. For example, P may be replaced by the relaxed operator P_λ,
defined by

$$P_\lambda\{h\} = P\{h\} + \lambda\,(P\{h\} - h), \qquad (9.8)$$

where λ is a real relaxation parameter with $|\lambda| < 1$.

Given N closed convex sets $C_i, i = 1, \ldots, N, C_i \subset \mathcal{H}$, with a nonempty intersection
$C_0 = \bigcap_{i=1}^{N} C_i$, we can associate a separate relaxed projection P_{i,λ_i} with each set C_i and
corresponding projection P_i. To obtain an element in C_0 we may iterate the composed
operator T, which is defined by

$$T = P_{N,\lambda_N} P_{N-1,\lambda_{N-1}} \cdots P_{1,\lambda_1}, \qquad (9.9)$$

by using the following algorithm:

Algorithm 1, a serial POCS algorithm: Given an arbitrary initial function $h^0(x)$,

$$h^{k+1} = T(h^k), \qquad k \geq 0. \tag{9.10}$$

For any arbitrary initial function h^0 we are assured that the infinite sequence $\{h^0, h^1, h^2, \ldots, \}$ generated by algorithm 1 converges weakly [31] to an element in C_0, provided all projections are performed with respect to the same distance function [40, 41] and all the N sets are closed and convex (in finite dimension, e.g., $\mathcal{H} = \mathbb{C}$, weak and strong convergence are the same). If some of the N sets are not convex, we are assured of a monotonic nonincrease of some error function along the iterates, provided that $N \leq 2$. If $N > 2$ this is not guaranteed even if only one set is not convex.

9.3.5.2 Parallel projections

We start by defining generalized weighted L^2 norm-squared distance functions with weights W_i:

$$d_i(h_1, h_2) := \|H_1 - H_2\|_{W_i}^2 := \int_{-\infty}^{\infty} |H_1(u) - H_2(u)|^2 W_i(u)\, du, \qquad h_1, h_2 \in \mathcal{H}, \tag{9.11}$$

where $W_i(u)$ is an essentially positive and essentially bounded weighting function and upper case letters denote the FT of the lower case functions, e.g., $H_i(u) := \mathcal{F}\{h_i(x)\}$. We also define a cost functional:

$$\hat{J}(h)^2 := \sum_{i=1}^{N} \beta_i d_i \left[P_{C_i}^{d_i}(h), h \right] = \sum_{i=1}^{N} \beta_i \left\| \mathcal{F}\{P_{C_i}^{d_i}(h)\} - \mathcal{F}\{h\} \right\|_{W_i}^2, \tag{9.12}$$

where $\beta_i > 0$ attributes an *importance* to the projection and $P_{C_i}^{d_i}(h)$ denotes the projection of h onto the set C_i with respect to the distance function d_i, i.e.,

$$P_{C_i}^{d_i}(h) = h' \qquad \text{if and only if} \qquad \inf_{h_1 \in C_i} d_i(h_1, h) = d_i(h', h), \qquad h' \in C_i. \tag{9.13}$$

We also denote by P_{i,λ_i} the relaxed projections, as above [for which we omit the superscript $(\cdot)^{d_i}$ for brevity]. If the sets C_i are closed with respect to d_i and convex, the projection element exists and is unique. If the sets are not convex, procedures exist for determining the (unique) projection [39]. With these definitions we may state the parallel projection algorithm in its space (time) representation, generating the sequence of successive estimates $\{h^0, h^1, \ldots, \}$. Although the algorithm operates in an infinite-dimensional Hilbert space as well [35, 38], we assume finite dimension (as it is implemented on a digital computer).

Algorithm 2, a parallel POCS algorithm: Given an arbitrary initial function $h^0(x)$, calculate

$$v_i^{k+1}(x) := P_{i,\lambda}[h^k(x)], \quad \text{for all } i = 1, 2, \ldots, N,$$

$$h^{k+1}(x) = \mathcal{F}^{-1} \left\{ \frac{\sum_{i=1}^{N} \beta_i W_i(u) \mathcal{F}\{v_i^{k+1}\}(u)}{\sum_{i=1}^{N} \beta_i W_i(u)} \right\}, \tag{9.14}$$

where $\mathcal{F}$ and $\mathcal{F}^{-1}$ denote the FT and its inverse, respectively. A similar algorithm can be implemented in the spatial frequency domain.

A detailed mathematical justification of this algorithm is provided in Ref. 38. Here we note only that iterates generated by this parallel algorithm converge weakly to C_0, provided that all sets are closed and convex and $\lambda \in (-1, 1)$. The individual projections may be defined with respect to different distance functions, in contrast to the serial algorithm. If some, or all, of the sets are nonconvex, the cost function $\hat{J}$ is still nonincreasing along the iterates, provided that $\lambda \in (0, 1)$, ensuring us of improved estimates along the iterates. This holds for an arbitrary number of sets, as opposed to the serial algorithm given in Subsection 9.3.5.1.

9.3.6 Discussion

Some of the optimization procedures assume that all physical parameters involved are known. These algorithms can be implemented on high-precision digital computers. In their present form these algorithms cannot be implemented if the processors have only a limited accuracy, if there are unknown parameters in the system, or if nonnegligible noise exists in the processor. In an actual optical system, all these factors are present. It has space–bandwidth limitations, it has conventional optical aberrations and also unknown aberration parameters, there are distortions and dead zones in the spatial light modulators (SLM's), and, in general, no physical element is ideal. In the presence of unknown parameters or parameters that are difficult to quantize, it is impossible to evaluate the distance from various constraints exactly. Under these circumstances, we cannot have a unique, deterministic update rule for the function at each iteration. Therefore, if implementation on a hybrid electro-optical architecture is desired, we must use procedures in which unknown parameters may be also present. Stochastic algorithms, such as SA and the GA, are of this nature. These algorithms can function with unknown parameters, naturally. During the design process, the filter function is updated in a random fashion, and an update is accepted or rejected in view of its actual performance. For this reason, these stochastic algorithms are also immune to system aberrations and other distortions, such as those introduced by a nonideal SLM. Naturally the results of these designs must be used in the same environment and with the same distortions as during the iterative design, that is, the design algorithm must be performed within the same system in which the results are to be used.

While implementing a filter synthesis algorithm on a real physical system, we assume that the function h is written on a SLM whose pixel settings are the convenient variables. All other system parameters are treated as fixed, even if some of them are unknown. The incident optical beam quality, the input SLM performance, the behavior of the lenses, and the pixel-by-pixel detector performance are examples of these unknowns that we seek not to determine but to accommodate. The principle behind this attitude is reminiscent of the living eye, which gives quite a poor optical performance. However, its combination with the processor (the brain) provides adaptation and correction by an extensive learning (or optimization) procedure.

In general, functions of many variables will exhibit many local extrema. This is usually also the case for our distance function. We use mathematical procedures designed to optimize the distance function in the sense of finding the global (not just the local) optimum.

In a practical implementation of these algorithms there are many elements of art as well as science in how to select or determine items such as

- the energy function and the cooling schedule for simulated annealing,
- the genome structure, the genetic exchange procedure, figure of merit, population size, and the mutation rate in the GA, or
- the proper distance function in conventional POCS algorithms and a set of distance functions and weights in parallel POCS algorithms.

Occasionally any of the methods can get stuck in some nonoptimum position or simply fail to converge at all. The behavior depends on not just the method but also on the problem landscape. All methods assume that there is some sense in saying that a given solution is nearer to the optimum than others. If the landscape is without this property, neither method helps and, if the problem has a solution at all, different approaches should be investigated.

A certain sensitivity to starting solutions also exists in the various methods. If we start SA near the right answer, it may converge rapidly. However, if we start near a local optimum, the algorithm may get stuck around this point for an almost infinite time. Therefore, if the region of the desired solution is not known, it is preferred to start with a random solution. A similar discussion is valid for the GA as well. However, in the large population of solutions in the GA, instead of a single trial solution as in SA or most other algorithms, it is possible to include much more information in the starting stage. Thus it is a good practice to start with much genetic diversity and little prejudice (which might predispose the solution in a way to get stuck in a suboptimum solution), so we usually choose the starting population randomly and attempt to maintain some diversity throughout the iterative procedure.

The projection-based algorithms appear to be the most powerful and to have the fastest convergence among the algorithms reviewed in this section. These algorithms are best suited when all the constraint sets can be rigorously defined, such as in a digital computer design of spatial filters, even if the constraints are complicated. As indicated above, for applications in which the constraints cannot be exactly defined or there are unknown parameters, stochastic algorithms, such as SA and the GA, are more suitable. In Section 9.4 one possible mathematical approach toward the design of optimal spatial filters in the information theoretical sense is outlined. Although applicable for any optimization algorithms, this mathematical foundation is particularly useful for the implementation of the stochastic algorithms.

9.4 Detection criteria: information theoretical approach

In Section 9.3 it was indicated that an optimization procedure must be based on a measure that describes the quality of the solution. This measure, the cost function, in turn, can be defined only through the goal of the optimization. In our case the goal is the generation of a filter function that satisfies certain detection criteria. Of course, the actual procedure is in the opposite direction:

- Define the goal (detection of a known pattern immersed in noise, discrimination between several classes of patterns, location of a target, or any other possible task).
- Define the criteria that determine the achievement of the goal (a steep correlation peak for a pattern detection, minimum error in a tracking system, or the establishment of interconnection between selected communication channels in an interconnection network).
- Define the cost function that properly depends on the distance from the fulfillment of the criteria.
- Optimize the system by using the cost function as a quality measure.

When all the possible aspects of a specific problem are taken into account, the above steps can lead to entirely different routes for different problems. Below the mathematical and the algorithmic aspects of a possible direction that proved efficient for various applications are analyzed.

Our present objective is to derive a cost function for optimizing a filter in a $4f$ optical correlator or a reference function for a JTC architecture. The procedure is similar, and, to be specific, we assume a $4f$ optical correlator. The spatial filter, having a point-spread function $h^*(-x, -y)$, is placed in the FT plane. Inserting a function $f(x, y)$ at the input plane leads to a complex amplitude distribution

$$c(x_0, y_0) = \int_{-\infty}^{\infty} \int_{-\infty}^{\infty} f(x, y)h^*(x - x_0, y - y_0) \, dx \, dy \qquad (9.15)$$

over the output plane.

Assuming a set of possible input patterns $\{f_n(x, y)\}$ we define our goal as the detection of the presence of patterns out of a subset $\{f_n^D(x, y)\}$ while rejecting all other patterns denoted by the subset $\{f_n^R(x, y)\}$. Our idealized criterion for detection is the appearance of a strong and narrow peak for a match between the input and the filter function as contrasted with a uniform distribution for a pattern to be rejected. An important aspect of this approach is that the exact position of the detected peak is not important as long as it always appears at the same position for a given input. The advantage of this criterion is that it provides an additional degree of freedom for the optimization procedure. Although it appears that the localization information of the object is lost, this is not the case, as we may redefine the origin of the coordinate system after the filter is implemented. Besides, the actual position of an object with respect to the correlation peak with other procedures is also ambiguous and must be defined. For example, the correlation peak with a matched filter is obtained at the origin if the object is placed at the position where it was during the recording of the matched filter, regardless of where this position exactly is. The reason is that the linear phase that contains the position information is incorporated into the filter function and it is canceled during the correlation process.

The discrete form of $c(x_0, y_0)$, the amplitude distribution over the output plane, can be written as

$$c(m, n) = \sum_{i=1}^{N} \sum_{j=1}^{N} f(i, j)h^*(i - m, j - n); \qquad m, n = 1, 2, \ldots, (2N - 1), \qquad (9.16)$$

where $f(i, j)$ and $h(i, j)$ are the sampled representations of $f(x, y)$ and $h(x, y)$, respectively, on an $N \times N$ array.

The complex amplitude distribution over the output plane is usually detected electronically; thus the phase is lost in the process. In a more general sense we may apply a nonlinear operator $\mathcal{L}$ to generate a new, nonnegative function on (m, n):

$$\varphi(m, n) = \mathcal{L}[c(m, n)]; \qquad \varphi(m, n) \geq 0, \qquad \forall(m, n). \qquad (9.17)$$

The normalized form of φ,

$$\Phi(m, n) = \frac{\varphi(m, n)}{\sum_{j=1}^{2N-1} \sum_{l=1}^{2N-1} \varphi(j, l)}, \qquad (9.18)$$

has the properties of a probability density, i.e.,

$$0 \leq \Phi(m, n) \leq 1, \qquad \forall(m, n), \qquad \sum_{m, n} \Phi(m, n) = 1. \qquad (9.19)$$

In the special case in which $\mathcal{L}$ performs the operation of a square-law detector, the quantity $\Phi(m, n)\Delta_m \Delta_n$ represents the probability that a given photon detection event will occur in an area $\Delta_m \Delta_n$ centered at (m, n).

Let Ψ be a nonlinear functional that satisfies

$$|\Psi[\Phi(m, n)]| < \infty, \qquad \forall \Phi(m, n) \in [0, 1], \qquad \Psi \in \mathcal{C}^1, \tag{9.20}$$

where $\mathcal{C}^1$ is the space of continuous functions that have at least first derivatives. Let Ψ also be a functional through which we can define a generalized entropy given by

$$S = \sum_{m=1}^{2N-1} \sum_{n=1}^{2N-1} \Psi[\Phi(m, n)] \tag{9.21}$$

such that

$$S = S_{\min} \quad \text{for } \Phi(m, n) = \text{const.} \qquad \forall(m, n), \tag{9.22}$$

$$S = S_{\max} \quad \text{for } \Phi(m, n) = \begin{cases} 1 & m = k, n = l \\ 0 & \text{otherwise} \end{cases}. \tag{9.23}$$

For a given point represented by the pair kl, it turns out that strictly convex functions [30, 42] satisfy the above requirements. It should be noted that sometimes the negative of the above-defined generalized entropy is used, and then the minimum and the maximum values must be interchanged.

Our detection criterion states that for a rejected pattern, a representative of the R subset, we would like to obtain a uniform distribution over the whole output plane, whereas if a pattern from the D subset is presented the result should be a strong and narrow peak somewhere in the output plane. A quantitative description of these distributions is given by Eqs. (9.21)–(9.23).

Given a $4f$ correlator with a specific SF, the generalized entropy can be evaluated for each input pattern. Denote by S_k^D and S_k^R the entropies that correspond to patterns from the class to be detected and the class to be rejected, respectively. We then have

$$S_k^D = \sum_{m=1}^{2N-1} \sum_{n=1}^{2N-1} \Psi[\Phi_k^D(m, n)], \tag{9.24}$$

$$S_k^R = \sum_{m=1}^{2N-1} \sum_{n=1}^{2N-1} \Psi[\Phi_k^R(m, n)], \tag{9.25}$$

where $\Phi_k^D(m, n)$ corresponds to the distribution over the output plane for the kth pattern of the detected set D and $\Phi_k^R(m, n)$ is derived from the output distribution for the kth pattern of the set to be rejected R.

We define a generalized cost function M by the relation

$$M_h = \sum_{\{k \in R\}} S_k^R - \sum_{\{k \in D\}} S_k^D \tag{9.26}$$

if convex functions are used, or by

$$M_h = \sum_{\{k \in D\}} S_k^D - \sum_{\{k \in R\}} S_k^R \tag{9.27}$$

if the functions employed are concave. The subscript h indicates that these cost functions depend on the filter function h. Referring to Eq. (9.26), for a given filter $h(i, j)$, we calculate S_k for each member of the training set $f_n(i, j)$, add the values obtained for the subset R, and subtract the values obtained for the subset D. We minimize the resultant generalized cost function by varying the components of $h(i, j)$.

An ideal filter function h would generate a steep peak for $f^D(i, j)$, represented by a distribution of the form given in Eq. (9.23), and a uniform distribution for $f^R(i, j)$, as given by Eq. (9.22). An ideal minimization procedure performed on the generalized cost function should lead to this ideal filter, which satisfies

$$M_{\text{ideal}} = S_{\text{min}}^R - S_{\text{max}}^D. \tag{9.28}$$

9.4.1 Generalized information function

As shown in the previous subsection, in principle any convex or concave function may be used for the quantification of the detection criterion and the process of generating a highly discriminating filter function. The choice of a specific function depends on several aspects that must be considered. For example, choosing a convex function that leads to an analytic solution for the cost function of Eq. (9.26) or one that provides a quick minimization process is beneficial.

An interesting convex function known from communication theory is the entropy function [43], which provides a good quantitative description for our detection criterion [21, 22]. A maximum entropy is attained for the distribution given in Eq. (9.22), and a minimum entropy is obtained for the distribution of Eq. (9.23). This entropy function is a special case of a large family of functions known as generalized information functionals [44]. The general form of these information functionals can be expressed in terms of our notation by the relation

$$\mathcal{H}^\Psi = \sum_m \sum_n \Phi(m, n) \Psi\left[\frac{1}{\Phi(m, n)}\right], \tag{9.29}$$

where Ψ is a convex function that satisfies the limiting constraint

$$\lim_{\alpha \to 0} \alpha \Psi(1/\alpha) = 0, \tag{9.30}$$

with

$$\sum_m \sum_n \Phi(m, n) = 1, \qquad \Phi(m, n) \geq 0, \qquad \forall(m, n). \tag{9.31}$$

The functional $\mathcal{H}^\Psi$ in relation (9.29) is called the Ψ entropy, with the conventional entropy function being a special case in which

$$\Psi(\alpha) = \log(\alpha). \tag{9.32}$$

Thus

$$\mathcal{H}^{\log} = -\sum_m \sum_n \Phi(m, n) \log \Phi(m, n). \tag{9.33}$$

In Refs. 21 and 22 this entropy function was used for generating a highly selective spatial filter for pattern recognition: the entropy optimized filter. The approach adopted in Ref. 19

used a cost function that was also based on a convex function and led to quite good discrimination. These filters are shift invariant in the same way as matched filters and they can also be made robust with respect to some distortions if distorted images are included in the training set.

In Subsection 9.4.2 this generalized approach is demonstrated by a test of the procedure with several convex (concave) functions.

9.4.2 Performance comparison of different cost functions

Simulation and laboratory experiments were performed with several convex and concave functions. The filters for the laboratory experiments were designed numerically and implemented as CGH's. The following are samples of these:

$$\Psi_1(\alpha) = \log(\alpha), \tag{9.34}$$

$$\Psi_2(\alpha) = \sqrt{\alpha}, \tag{9.35}$$

$$\Psi_3(\alpha) = [(1/\alpha) - 1], \tag{9.36}$$

$$\Psi_4(\alpha) = \exp(-\alpha), \tag{9.37}$$

$$\Psi_5(\alpha) = \alpha^p, \qquad p > 1, \tag{9.38}$$

$$\Psi_6(\alpha) = [\max(\alpha)]^2. \tag{9.39}$$

When photoelectric detection methods are used, the operator $\mathcal{L}$ in Eq. (9.17) may be any polynomial P over the absolute value of $c(m, n)$ such that $\varphi(m, n) \geq 0$:

$$\mathcal{L}\{c(m, n)\} = P\{|c(m, n)|\} \tag{9.40}$$

In Refs. 19, 21, and 22, square-law detectors were assumed, and therefore P was chosen to be a quadratic polynomial that gives the intensity distribution over the output correlation plane. In our present investigation we take only the linear term,

$$\varphi(m, n) = \mathcal{L}\{c(m, n)\} = |c(m, n)|. \tag{9.41}$$

For example, for Ψ_1, Ψ_2, Ψ_3, and Ψ_4 we use the general form of the information functional given in Eq. (9.29), whereas for Ψ_5 and Ψ_6, Eqs. (9.24) and (9.25) were used. From a performance point of view, there is not much difference between the two choices of the cost function, but we should note that Ψ_1 and Ψ_3 are not defined for $\alpha = 0$ and therefore Eq. (9.29) must be used. A cost function was calculated as defined in Eq. (9.26) for each of the convex functions, whereas for concave functions Eq. (9.27) was used. Thus, when searching for the best filter $h(i, j)$ for each of the Ψ-entropy functions, we use a minimization procedure for minimizing the corresponding cost function:

$$M_h^{\Psi_1} = \sum_{m,n} \Phi^R(m, n) \log \Phi^R(m, n) - \sum_{m,n} \Phi^D(m, n) \log \Phi^D(m, n), \tag{9.42}$$

$$M_h^{\Psi_2} = \sum_{m,n} [\Phi^D(m, n)]^{1/2} - \sum_{m,n} [\Phi^R(m, n)]^{1/2}, \tag{9.43}$$

$$M_h^{\Psi_3} = \sum_{m,n} \{[\Phi^R(m, n)]^2 - \Phi^R(m, n)\} - \sum_{m,n} \{[\Phi^D(m, n)]^2 - \Phi^D(m, n)\}, \tag{9.44}$$

$$M_h^{\Psi_4} = \sum_{m,n} \Phi^R(m, n) \exp\left[-\frac{1}{\Phi^R(m, n)}\right] - \sum_{m,n} \Phi^D(m, n) \exp\left[-\frac{1}{\Phi^D(m, n)}\right], \quad (9.45)$$

$$M_h^{\Psi_5} = \sum_{m,n} [\Phi^R(m, n)]^p - \sum_{m,n} [\Phi^D(m, n)]^p \qquad \text{where } p = 2, \quad (9.46)$$

$$M_h^{\Psi_6} = \max^2 \Phi^R(m, n) - \max^2 \Phi^D(m, n). \quad (9.47)$$

In a demonstrative set of experiments the training set was represented on a matrix with $N \times N = 8 \times 8$ pixels padded with zeros. The total input and output planes were on a matrix of 32×32 pixels, and the filter was represented on a 16×16 matrix. For one of the experiments the training set contained only two letters, **P** and **F**, as shown in Fig. 9.1. The filter was generated to detect **P** and reject **F**. These two letters differ from each other by only four pixels; thus they are quite difficult to distinguish. This is indicated in the intensity distribution of Fig. 9.2, which was obtained by correlation of the input with a conventional matched filter for **P**.

In principle, any of the optimization algorithms described in Section 9.3 could be adapted to minimize M_h. However, the statistical nature of the criteria defined for detection and the cost functions based on the generalized information function are better suited for

Fig. 9.1. Training set with the detected pattern **P** and the pattern to be rejected **F**.

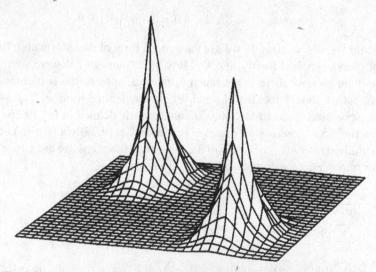

Fig. 9.2. Cross correlation of Fig. 9.1 with a conventional matched filter for **P**.

implementation with stochastic optimization algorithms, rather than algorithms of the POCS family.

For one set of experiments, SA was chosen according to the following algorithm: At the lth iteration, we have $M_l = M(\mathbf{h}_l)$. Inducing a small, random perturbation over the pixels in a limited region of the filter function, we obtain the $(l + 1)$th iteration of the filter that changes M by an amount

$$\Delta M = M^\Psi(\mathbf{h}_{l+1}) - M^\Psi(\mathbf{h}_l). \tag{9.48}$$

The iteration is accepted if $\Delta M \leq 0$; otherwise it is rejected or conditionally accepted, based on the acceptance probability given by Eq. (9.3). The procedure is now repeated, starting from the new filter. Decreasing the temperature slowly as the process continues causes M to approach its optimal value.

Figure 9.3 illustrates the correlation peaks for the input pattern of Fig. 9.1 with each of the filters that were generated by minimization of the cost functions based on the corresponding Ψ functional. The plots of Fig. 9.3 are results of simulation experiments, but similar results were obtained experimentally with the filters implemented as CGH's. The minimization process was stopped when a ratio of 4:1 was achieved between the correlation peak value for the letter $\mathbf{P}$ to that of the letter $\mathbf{F}$. It is interesting to compare these results with the

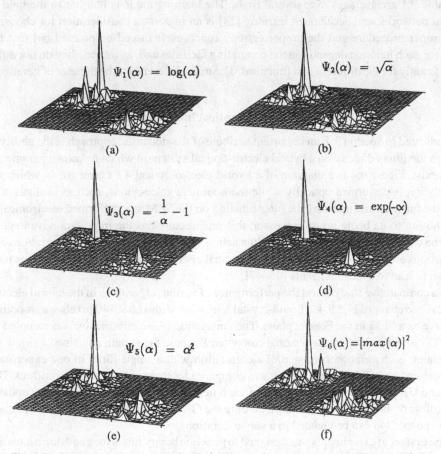

Fig. 9.3. Output correlation intensity for different filters generated with the noted Ψ function to detect $\mathbf{P}$ and reject $\mathbf{F}$. For each SF the discrimination ratio is 4:1.

Table 9.1. *Comparison of the convergence rate*
for various Ψ functions

Function	Source of $S^{D/R}$	Iterations	Time/iterations (s)
Ψ_1 [Eq. (9.34)]	Eq. (9.29)	13,654	1.59
Ψ_2 [Eq. (9.35)]	Eq. (9.29)	9643	1.49
Ψ_3 [Eq. (9.36)]	Eq. (9.29)	9278	1.47
Ψ_4 [Eq. (9.37)]	Eq. (9.29)	8951	1.53
Ψ_5 [Eq. (9.38)]	Eq. (9.24)	1538	1.48
Ψ_6 [Eq. (9.39)]	Eq. (9.24)	1567	1.47

performance of the conventional matched filter in Fig. 9.2 with the wide correlation peaks and practically no discrimination between the objects.

When iterative processes are used for optimization, an important parameter is the number of iterations required for achieving a result. Table 9.1 lists the number of iterations that were required for reaching the discrimination ratio of 4:1 for each convex or concave function. To take into account the statistical nature of the optimization procedure, the results shown in Table 9.1 are averages over several trials. The learning mode is familiar in the field of neural networks, and accelerated learning [28] is an important consideration for choosing the proper procedure and the proper entropy function. It should be pointed out that the time for each iteration depends on the computing facilities and, as we see, they do not differ significantly for the different cost functions. The main difference is the number of iterations.

9.5 Hybrid electro-optical implementation

As indicated in Section 9.4, an important attribute of the stochastic approach is the ability to design the filters directly on a hybrid electro-optical system in which unknown parameters may exist. Figure 9.4 is a diagram of a hybrid electro-optical $4f$ correlator in which the correlation is performed optically while the nonlinear calculations, such as evaluation of the cost function and updating the filter function on the SLM, are performed electronically. In addition to its being a fast processor, this architecture has the important advantage in that the computations take into account the actual system parameters. Thus aberrations and distortions are automatically corrected. The actual operations of such adaptive systems were shown to lead to excellent results [45–47].

In a comparative study to test the performance of various algorithms in the hybrid electro-optical correlator of Fig. 9.4, a liquid-crystal television with 166×140 pixels was modified to serve as a SLM in the Fourier plane. The correlation-plane information was sampled by a CCD camera and fed into a personal computer. Because the system contained a single $4f$ correlator, each iteration consisted of a correlation with a single filter. In one experiment the performance of the hybrid system was compared for three optimization algorithms, HC, SA, and GA. For the sake of comparison, each iteration consisted of a single filter updated regardless of the specific algorithm (including the GA, for which, usually, several members of the population can be updated in a single iteration).

The first set of experiments was designed to generate binary filters for good discrimination between the letter **T** (to be detected) and the letter **L** (to be rejected), as shown in Fig. 9.5. The objective was to generate a binary filter with a matrix size of 64×64. From a computational

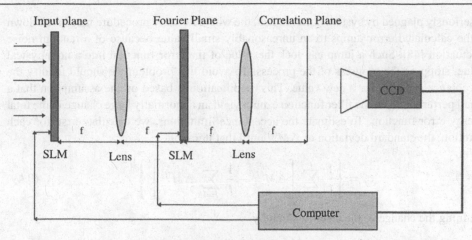

Fig. 9.4. Hybrid electro-optical 4f correlator. [This figure was included in *Trends in Optics*, A. Consortini, ed., Chap. 13 (Academic Press, London, 1996).]

Fig. 9.5. Training set for comparing various algorithms on the hybrid correlator of Fig. 9.4.

point of view this is a problem of large dimensions. An array of this size has, theoretically, 2^{4096} possible states, although some of them are physically meaningless and therefore never encountered. In the second set of experiments a matrix of 100×100 was used, that is, $2^{10,000}$ possible states. Increasing the size of the matrix generally increases the number of function evaluations required for finding a solution. This size increase is extremely bothersome for conventional digital operations but not too significant as long as the whole array is processed in parallel. In the electro-optical architecture this is the case.

The massive parallelism in an optical system is the main attribute contributed by the optical part in a hybrid electro-optical architecture. An additional advantage, as indicated above, is the self-correcting capability if a correlation filter is designed within the system in which it is to be used. Nevertheless, in an actual physical system, problems that do not exist in digital computer simulations may be encountered. These problems depend on the specific algorithm, the system architecture, and the quality of the electronic and optical components. For example, the implementation of the HC algorithm in a real-life environment is seriously plagued by system noise. In fact, the whole iterative procedure may break down if the calculated error jumps to an unreasonably small value because of a random noise fluctuation [48]. Such a jump can lock the value of the error function into a nonphysical value, stopping the progress of the process. To avoid this problem we should modify the acceptance criterion for a new value. This modification is based on the assumption that a small perturbation on the filter function cannot yield an abnormally large change in the total energy error function. To estimate the acceptance limitations, we calculate first, for each iteration, the standard deviation of $\Delta M^{(t)}$ until that iteration

$$\sigma_t = \left\{ \frac{1}{t} \sum_{j=1}^{t} \left[\Delta M^{(j)} - \frac{1}{t} \sum_{i=1}^{t} \Delta M^{(i)} \right]^2 \right\}^{1/2}. \tag{9.49}$$

Defining the change of the error function by

$$\Delta e^{(t+1)} = \Delta M^{(t+1)} - \Delta M^{(t)}, \tag{9.50}$$

we consider iterations for acceptance only if

$$\left| \Delta e^{(t+1)} \right| < \sigma_t, \qquad \Delta e^{(t+1)} < 0. \tag{9.51}$$

Experimental results of the HC algorithm are shown in Fig. 9.6, illustrating the convergence process. The three-dimensional representation of the output correlation-plane distributions is shown in Fig. 9.6(a) for the beginning of the process. Through the intermediary stages [Figs. 9.6(b) and 9.6(c)], the final result is shown in Fig. 9.6(d).

The final results obtained by the SA algorithm and the GA were similar. With each algorithm, a discrimination ratio of 1:3.5 was obtained before the procedure was stopped. Because all algorithms used the same cost function and the same physical system, the sharpness of the correlation peak obtained was also similar but its location varied within the correlation region because of the flexibility imposed by the detection criteria. The main

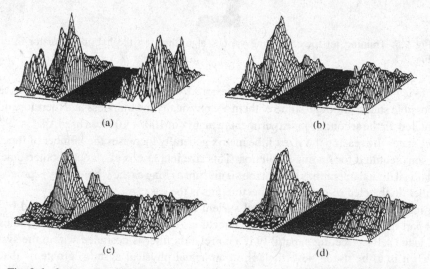

(a)   (b)

(c)   (d)

Fig. 9.6. Output-plane distribution during the convergence of the HC algorithm for the training set of Fig. 9.5. The various frames were picked out as the algorithm converged, starting from (a) and ending with (d).

difference was in the convergence rate with the GA, which was ∼50 times faster than that with the HC algorithm.

When SA is applied, several practical problems are faced. The first is the choice of a perturbation size for changing the configuration for each iteration. An unsuitable perturbation may cause a large deviation of the error function that may trap the solution in a nonphysical state. The cooling rate [the reduction of the temperature parameter T in Eq. (9.3)] must also be carefully chosen. The proper cooling rate must be determined separately for each specific problem [28]. In the experiments described here the perturbation was induced on randomly selected 10% of the filter elements. A graded cooling profile rate was found to lead to the best results among several alternatives investigated. The noise problem mentioned with regard to the HC procedure affects SA too and must be handled in a similar way.

A well-known problem in the GA is premature convergence. This problem is similar to the problem that arises in the SA algorithm because of an unmatched cooling rate [29]. In the hybrid system this problem is not as important as the proper choice of initial population. The whole process is based on sampling a noisy output correlation plane. When the signal is smaller than the noise level, no convergence occurs. Hence we need an initial population that bears enough energy to exceed the noise level. In the experiments described the population contained 100 filter functions. 60% of the initial filter population were one-dimensional grids with different spatial frequencies, 20% were high-pass filters with different cutoff frequencies, and the rest were randomly chosen binary filters. The proper mutation rate may also be important for satisfactory convergence, but the experiments indicated only a marginal influence of this parameter.

9.5.1 Performance comparison for various algorithms

A representative comparison was implemented with the various algorithms with the same cost function under the same laboratory conditions. Each process was performed 25 times, and the results given below represent their average. The performance measure is the convergence rate of the cost function, which can be represented by its value after a certain number of iterations.

The convergence profiles are compared in Fig. 9.7. The solid curves represent the convergence of the GA, whereas the dashed curves in Figs. 9.7(a) and 9.7(b) represent the convergence of the HC and SA, respectively. It is obvious that the GA drastically speeds up the convergence compared with HC and SA. It should be pointed out here that this superior performance was observed, although in the GA only one member in the population was updated for each iteration instead of working simultaneously with the whole population. Compared with the GA, the HC and the SA appear almost stationary in Fig. 9.7. Plotted with higher resolution, the two latter algorithms would exhibit a slow tendency of convergence, superposed on large fluctuations. Such a noisy behavior is absent for the GA, in which we operate on the average of the cost function for the whole population. Hence in the GA no special measures had to be taken to deal with the disturbance due to noise.

In another experiment the matrix size was increased to 100×100 pixels. Several faces such as those shown in the upper part of Fig. 9.8 were used; the GA was employed to instruct a hybrid correlator to distinguish among the faces. The SF was prepared to detect the right-hand face and reject the others. The intensity distribution across the correlation peaks, marked by the solid scanning line in the middle frame, indicates good discrimination. Shift invariance of the input is demonstrated in the lower frame, which is the output

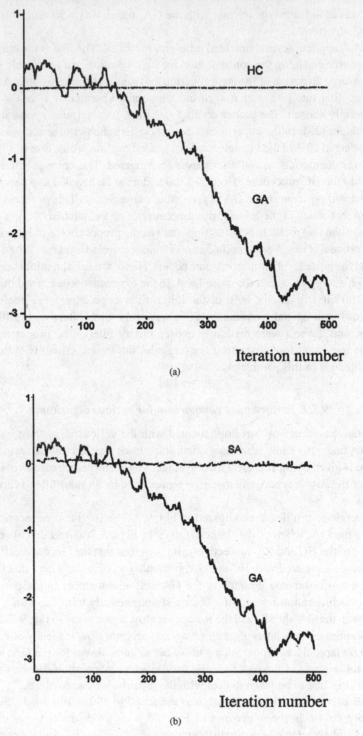

Fig. 9.7. Convergence profile for (a) the GA compared with the HC algorithm, (b) the GA compared with the SA algorithm.

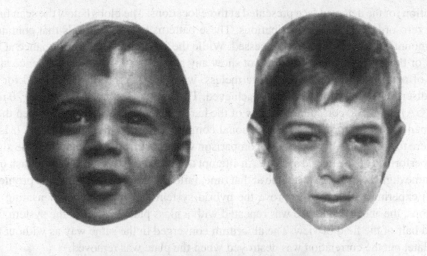

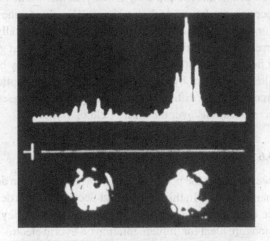

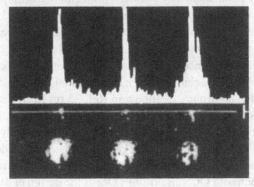

Fig. 9.8. Training set (top) for hybrid correlator and discrimination results obtained by the GA (middle). Shift invariance is demonstrated by the output plane distribution when the detected pattern was presented at three positions (bottom). The blobs below the scan lines are the zero-order diffraction distributions. [This figure was included in *Trends in Optics*, A. Consortini, ed., Chap. 13 (Academic Press, London, 1996).]

distribution for the detected face presented at three locations. The blobs below the scan lines are the zero-order diffraction distributions. These patterns have gray levels that contain a large amount of information to be processed. While the GD algorithm had no chance at all, the HC or the SA algorithms also did not show any significant signs of convergence after 3 days of optimization processing. Nevertheless, when a GA procedure was used for 20 min, a discrimination ratio of 1:4.5 was achieved. The time scales indicated above (20 min for the GA) should be considered in view of the hardware available at the time when these experiments were performed (a slow personal computer, slow framegrabber, etc.). It is in order here to emphasize that the above comparison was implemented with each of the algorithms performed on the hybrid system. An attempt to execute the GA for the above task on a mainframe digital computer, available at that time, failed because of the size of the problem.

In an experiment designed to prove the hybrid system capability of compensating for distortions, the above procedure was repeated with a glass plate located in the system that covered half of the field of view. The algorithm converged in the same way as without the glass plate, but the correlation was destroyed when the plate was removed.

All the procedures discussed in this section and, actually, in this whole chapter, consider shift-invariant pattern recognition. In an ideal situation this is always the case and, as is also shown in Fig. 9.8, fair shift invariance was also obtained experimentally. Nevertheless, with practical optical correlators, various problems may be encountered if good correlation is desired over the entire field of view of the input scene. Difficulties may be caused by the physical nature of the filter [49–51] or by an input device that is not optically perfect. A possible solution for overcoming these difficulties is to enlarge the training set by introducing the same pattern at different locations over the entire field of view.

9.6 Applications of projection algorithms

In the previous sections we have seen the power of statistical procedures in designing spatial filters with prescribed characteristics. In this section and Section 9.7 it is demonstrated that projection algorithms are even more powerful and flexible whenever they are applicable. Although statistical methods are ideal for implementation on hybrid electro-optical learning systems in which unknown parameters may exist, the projection algorithms are better suited for applications in which a set of constraints can be exactly defined [32, 34]. In this section various applications are described in which projection algorithms were employed to design filters for linear as well as nonlinear optical correlators. In the design process the constraints are basically composed of discrimination and peak energy (amplitude) constraints. Noise constraints, e.g., noise robustness, can be easily incorporated into the design process as well, at least for the parallel algorithm, as shown in Ref. 35 (in Ref. 35 the noise is taken into account for image restoration purposes and the idea is similar for pattern recognition). Moreover, it was shown in Ref. 50 that the presence of noise may actually assist in the case of multiple-object inputs. Thus, for brevity, noise problems are not considered further here; nor is shift invariance, which was demonstrated in Refs. 50 and 52. The rest of this section is devoted to two applications of POCS algorithms. One is the design of filters for a conventional linear correlator, and the second deals with basically nonlinear correlators in which the pattern recognition process itself is implemented in a nonlinear fashion. Section 9.7 takes us one step further, into the domain of adaptive correlators. In adaptive correlators the filter and correlator parameters vary during operation to adapt to the input function. These correlators require filters with special characteristics that can also be designed with the help of the POCS algorithms.

9.6.1 Class discrimination by a linear correlator

For a class discrimination problem we define a training set consisting of two classes. The class to be detected is placed in a region of space R_1 while the class to be rejected is situated in a region denoted by R_2. If the complete training set is presented simultaneously over the input plane, then R_1 corresponds to regions in the correlation plane that correspond to the positions of objects to be detected while the regions R_2 represent the location of objects to be rejected and also the empty regions surrounding the correlation peaks in R_1. The task is to design a filter h such that

(1) Its correlation with a given input function f satisfies some correlation constraint C_1, namely, in the detection region R_1 the correlation peak will be larger than some pre-determined value T_1, whereas in the rejection region, R_2, the correlation will be lower than some predetermined value T_2. During the learning stage the correlation peak is assumed to be contained in a single pixel. Because this is physically not possible, some of the peak energy will leak out into neighboring pixels constituting the background that should be below T_2. Also, T_1 and T_2 are appropriately chosen threshold values to provide sufficient discrimination (at least T_1/T_2) as well as sufficient energy in the peak (high absolute value of T_1). The appropriate values will depend on the specific application and the level of similarity between the two classes.
(2) Its FT, $\mathcal{F}\{h\}$, corresponds to a passive element (C_2).
(3) It has finite support, say $[-a, a]$ (C_3).

Any filter h that satisfies all the above three constraints is considered a solution. Mathematically the constraints are given by the following definitions:

$$C_1 := \{h \mid [h * f](j) \in \hat{C}_1, \forall j\},$$
$$\hat{C}_1 := \{\Phi(j) \mid |\Phi(j)| \le T_2 \quad \text{for } j \in R_2; \quad \Phi_{re}(j) \ge T_1, \quad \Phi_{im}(j) = 0 \quad \text{for } j \in R_1\},$$
$$C_2 := \{h \mid |\mathcal{F}\{h(j)\}| \le 1\},$$
$$C_3 := \{h \mid h(j) = 0 \quad \text{for } j \notin [-a, a]; \quad a > 0\}, \tag{9.52}$$

where

$$\Phi(x) = h(x) * f(x) \tag{9.53}$$

is the correlation function that has the sampled representation

$$\Phi(j) := [h * f](j), \qquad \Phi(j) := \Phi_{re}(j) + i\,\Phi_{im}(j),$$

where

$$\Phi_{re}(j) = \text{Re}\{\Phi(j)\}, \qquad \Phi_{im}(j) = \text{Im}\{\Phi(j)\}.$$

Actually the measured quantity is $|\Phi|^2$ and not its imaginary or real values. However, the constraint, $|\Phi(j)|^2 \ge \text{const.}$ is not a convex constraint set and convergence is then not guaranteed. Thus, with our choice, $C_1, \hat{C}_1, C_2$, and C_3 are closed convex sets.

Projections onto C_2, C_3 with respect to the distance function given by Eq. (9.11) with unity weighting [$W_i(u) = 1, i = 2, 3$], i.e., the Euclidean norm, are simple and are given by

$$P_{C_2}^{d_2}\{h(x)\} = \mathcal{F}^{-1}\{H'(u)\},$$

where

$$H'(u) = \begin{cases} H(u) & \text{if } |H(u)| \leq 1 \\ \exp[i\varphi_H(u)] & \text{otherwise} \end{cases},$$

$$P_{C_3}^{d_3}\{h(x)\} = \begin{cases} h(x) & \text{if } x \in [-a, a] \\ 0 & \text{otherwise} \end{cases}, \tag{9.54}$$

where $H(u) = \mathcal{F}\{h(x)\} \equiv |H(u)| \exp[i\varphi_H(u)]$. Unfortunately, projection onto C_1 with respect to the Euclidean norm is complicated and is a typical constrained deconvolution problem in itself [40, 53].

To perform the projection onto C_1 easily we follow the idea proposed in Ref. 40. We perform the projection onto C_1, with respect to the distance function induced by a weighted norm squared, with the appropriate weighting given by $W_1(m) := |\mathcal{F}\{f(j)\}|^2 = |F(m)|^2$:

$$d_1(H_1, H_2) = \sum_m W_1(m)|H_1(m) - H_2(m)|^2$$

$$= \sum_j |\mathcal{F}^{-1}\{[W_1(m)]^{1/2}\}(j) * [h_1(j) - h_2(j)]|^2. \tag{9.55}$$

This careful choice of the weighting function results in a simple projection, viz.,

$$(\mathcal{F}^{-1}\{V_1\} =) v_1 := P_{C_1}^{d_1}(h), \qquad \text{where } V_1(m) = \frac{\mathcal{F}\{\Phi'(j)\}(m)}{F(m)},$$

$$\Phi'(j) = \begin{cases} T_2 \exp[i\varphi_\Phi(j)] & \text{if } j \in R_2 \text{ and } |\Phi(x)| > T_2 \\ \Phi(j) & \text{if } j \in R_2 \text{ and } |\Phi(j)| \leq T_2 \\ T_1 & \text{if } j \in R_1 \text{ and } \Phi_{re}(j) < T_1 \\ \Phi_{re}(j) & \text{if } j \in R_1 \text{ and } C_{re}(j) \geq T_1 \\ \Phi(j) & \text{otherwise} \end{cases},$$

where

$$\Phi(j) = |\Phi(j)| \exp[+i\varphi_\Phi(j)] = \mathcal{F}^{-1}\{H(m)F(m)\} = h(j) * f(j). \tag{9.56}$$

For details see Ref. 53.

Algorithm 2 allows projections with respect to several different distance functions, and therefore the projection of different quantities into domains in which both the constraint set and the distance function are simple is possible [40]. Hence it is used for this filter synthesis task. The sequence $\{h^k\}_{k=0}^\infty$, generated by algorithm 2, converges to a function in C_0 that satisfies all constraints and is given by [see Eqs. (9.14)] $h^{k+1}(j) = \mathcal{F}^{-1}\{H^{k+1}(m)\}$, where

$$H^{k+1}(m) =$$

$$\frac{\mathcal{F}\{P_{C_1}^{d_1}(\mathcal{F}^{-1}\{H^k\})\}(m)W_1(m) + \mathcal{F}\{P_{C_2}^{d_2}(\mathcal{F}^{-1}\{H^k\})\}(m) + \mathcal{F}\{P_{C_3}^{d_3}(\mathcal{F}^{-1}\{H^k\})\}(m)}{W_1(m) + 1 + 1}, \tag{9.57}$$

d_1 is given by Eq. (9.55), the zero relaxation parameter ($\lambda = 0$) is used, and $d_2(h_1, h_2) = d_3(h_1, h_2) = \|h_1 - h_2\|$ (the Euclidean norm).

One experiment started from a filter h such that $h \notin C_1$, $h \in C_2$, and $h \notin C_3$. Figure 9.9 shows the input distribution. The task was to detect the letter **F** and reject all others. Figure 9.10(a) shows the correlation distribution with a conventional POF [12] matched to the letter **F**. Figure 9.10(b) shows the correlation distribution with the filter generated by algorithm 2. The improvement in both recognition and discrimination is obvious.

Fig. 9.9. Training set for a class discrimination experiment.

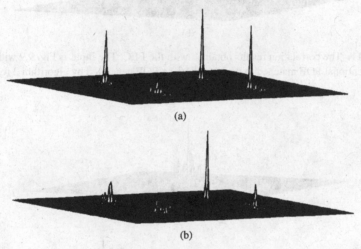

Fig. 9.10. Correlation results for the input of Fig. 9.9 with (a) a conventional POF matched to the letter **F**, (b) a filter generated by algorithm 2.

9.6.2 Class discrimination by the phase-extraction correlator

In Subsection 9.6.1 the procedure implemented a SF that was used in an essentially linear correlator. Several research results indicated that nonlinear correlators perform, under various circumstances, better than linear correlators [54–59]. One of these correlators uses an essentially POF, but instead of correlating this filter with the original input function, the input function is preprocessed by the extraction of only its phase and correlation of this phase with the filter. As a result, this phase-extraction correlator (PEC) [50, 52, 60] performs like a correlator with an ideal inverse filter.

In this subsection we use a projection algorithm to optimize the operation of a PEC with respect to class discrimination. The correlation function in the case of the PEC is given by

$$\Phi(x) = h_p(x) * f_p(x), \qquad (9.58)$$

where the subscript p refers to the phase part of the function involved. Accordingly, the convex constraint set C_2 in Eq. (9.52) must be replaced by the nonconvex constraint set C_{2-nc}:

$$C_{2-nc} := \{h \mid |\mathcal{F}\{h(j)\}| = 1\}. \qquad (9.59)$$

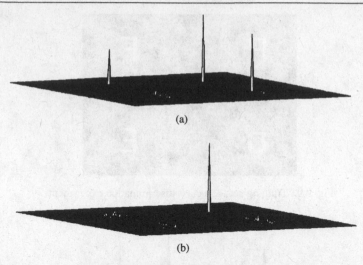

Fig. 9.11. The correlation results obtained with the PEC. The input is Fig. 9.9 with (a) a conventional POF matched to the letter **F**, (b) a POF generated by algorithm 2 for the PEC.

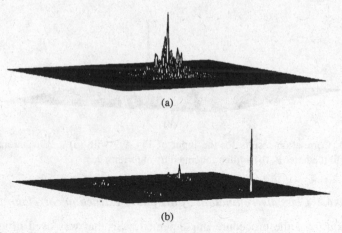

Fig. 9.12. (a) Impulse response of the modified filter, (b) same as Fig. 9.11(b), but with the letters of the input (from Fig. 9.9) **F** and **E** interchanged.

In this case h is a solution if $h \in C_1 \cap C_{2-\mathrm{nc}} \cap C_3$, i.e., $h = h_p$. Thus, in order to design a suitable POF h_p, it is necessary to iterate the operator $T := P_{C_1} P_{C_{2-\mathrm{nc}}} P_{C_3}$. Unfortunately, because one of the sets is nonconvex and we have more than two sets to project onto, we are not assured of any monotonic behavior of the iterates of algorithm 1 [39]. However, the parallel algorithm, algorithm 2, may be used, with ensured monotonic reduction of the cost function $\hat{J}$ of Eq. (9.12).

Figure 9.11(b) shows the result of the PEC, according to Eq. (9.58), in which the input is given by Fig. 9.9 and the filter is the POF matched to the letter **F**. This is again compared with the conventional POF of Fig. 9.11(a), and the improvement is obvious. Also, the correlation peaks are sharper in the PEC, compared with those in the linear correlator. This is due to the intrinsic high-frequency amplification of the PEC. However, as noted above, there may be some shift variance. To minimize this, we confined the impulse response of the filter to be narrow in the space domain (constraint C_3). The impulse response of the filter is shown in Fig. 9.12(a). Indeed, when taking the input shown in Fig. 9.9 and interchanging the positions of the letters **F** and **E**, we obtain the correlation function shown in Fig. 9.12(b), which is similar to Fig. 9.11(b) (note the interchange of letters), which demonstrates approximate shift invariance.

9.7 Adaptive procedures for distortion invariance

Pattern recognition by coherent optical correlators has many advantages, including high operating speed and parallelism. Although intrinsically shift invariant, optical correlators have a significant disadvantage: their high sensitivity to distortions such as rotation and scale. Several approaches toward rotation and scale invariance were investigated during the past few years, in particular, the use of circular harmonic component (CHC) filters [61–64] that are invariant to rotation and shift, radial harmonic filters [65, 66] that are scale and shift invariant, and coordinate transformations [67] such as the Melin transform that yield rotation and scale invariance but not shift invariance. In general, it is relatively easy to treat invariance with respect to two out of the three major distortion parameters: shift, in-plane rotation, and scale. The problem is much more difficult if invariance to more than two of these standard distortion parameters is desired and even more so when additional distortions may be present (out-of-plane rotations and actual shape variations).

The difficulty in obtaining filters with a high degree of robustness against various distortion parameters stems from the contradictory constraints imposed on a single filter by different distortion parameters. Moreover, any distortion invariance incorporated into a filter design necessarily reduces the discriminating power of the filter.

The purpose of this section is to demonstrate that the problems indicated above can be alleviated by use of an adaptive approach. This approach is based on the fact that some distortion parameters can be estimated before the actual pattern recognition is made with a filter that is adapted to the estimated parameters. Essentially this whole approach is an extenstion of iterative processes implemented in several stages. Iterative methods can be implemented for the distortion parameter estimations and also for the actual pattern recognition process. Because special filter designs are necessary for the latter, iterative methods may be used in the system preparatory stages. As in the applications discussed in Section 9.6, here too the POCS-based algorithms were found to be especially efficient.

A general adaptive architecture is composed of several parallel channels, two of which are shown explicitly in the diagram of Fig. 9.13. The input object is presented simultaneously to all channels. The distortion parameters are estimated in the lower channels, and the upper channel performs the final recognition after adaptation to the measured parameters.

As a case study, a pattern recognition system is presented here that is invariant to rotation, scale, and shift [68]. In this example the distortion parameter estimated is the scale of the object. Thus, in one channel, the scale of the input object is estimated by a fast and efficient optical measuring system, and the second channel is a $4f$ rotation- and shift-invariant adaptive optical correlator.

In Subsection 9.7.1, a possible optical procedure for estimating the scale of an input object, regardless of its position, orientation, and illumination intensity, is described. Of course, with such a task the operation must be limited to object classes that have some features

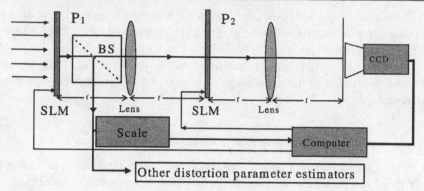

Fig. 9.13. Schematic diagram of a multichannel, adaptive pattern recognition system. The lower channels are the scale estimation channel and other optional parameter estimators, and the upper channel is an adaptive (controlled filter) $4f$ correlator. BS, beam splitter.

in common. Because the scale cannot be estimated precisely, the rest of the system must have a certain robustness to scale changes, in addition to the other distortion parameters. A design procedure for appropriate spatial filters, presented in Subsection 9.7.2, is based on POCS algorithms.

9.7.1 Scale measurement procedure

Several scale estimation procedures can be adapted. For the present application, a method based on the properties of the radial intensity distribution in the FT domain was found to perform well. Let $f(r, \theta)$ be the input function and $F(\rho, \phi)$ be its FT (in polar coordinates). When the scale of the input is extended by a factor a, the energy, when Parseval's theorem is used, becomes

$$E_a = \int_0^{2\pi} \int_0^\infty |a^2 F(a\rho, \phi)|^2 \rho \, d\rho \, d\phi = a^2 E_0, \qquad (9.60)$$

where E_0 is the energy of the original function $f(r, \theta)$. Let $E_a(\hat{\rho})$ be the energy of the object in the radial spatial-frequency interval $0 < \rho < \hat{\rho}$:

$$E_a(\hat{\rho}) = \int_0^{2\pi} \int_0^{\hat{\rho}} |a^2 F(a\rho, \phi)|^2 \rho \, d\rho \, d\phi = a^2 \int_0^{2\pi} \int_0^{a\hat{\rho}} |F(\rho, \phi)|^2 \rho \, d\rho \, d\phi. \qquad (9.61)$$

We define the ratio between $E_a(\hat{\rho})$ and the total energy E_a by

$$T(a, \hat{\rho}) = \frac{E_a(\hat{\rho})}{E_a} = \frac{\int_0^{2\pi} \int_0^{a\hat{\rho}} |F(\rho, \phi)|^2 \rho \, d\rho \, d\phi}{E_0}. \qquad (9.62)$$

Clearly $T(a, \hat{\rho})$ depends on $\hat{\rho}$ only through the integration limit; hence, in order to get a specific value of T independently of a, we must keep the integration limit fixed, i.e.,

$$a\hat{\rho} = C = \text{const.} \qquad (9.63)$$

The scale factor a can now be easily found by the following procedure: First we choose some reference value of $T(a, \hat{\rho})$ and define it as T_0; then we find, for an object with a known scale a_0, the frequency $\hat{\rho}_0$ that yields the predefined T_0. Now every new scale factor a can be derived from the simple relation

$$a = a_0(\hat{\rho}_0/\hat{\rho}). \tag{9.64}$$

This procedure has several important advantages:

- Because the processing is done in the Fourier plane, the measurement does not depend on object position (shift invariance).
- The scale is estimated by $T(a, \hat{\rho})$, which does not depend on the rotation angle ϕ, and therefore the method is also rotation invariant. Furthermore, because of the angular integration, the behavior of T is approximately the same for a large group of similar objects.
- The measurement does not depend on the light intensity or the object's energy, as $T(a, \hat{\rho})$ is a normalized quantity.

Obviously this procedure is accurate for a single object and can be repeated for each object of a whole group. Unfortunately, it is usually not known *a priori* which of the members of a class are going to appear on the input plane. Nevertheless, if the behavior of $T(a, \hat{\rho})$ is similar for all the objects in the group, the scales of all the objects can be estimated with Eq. (9.64) with a_0 and $\hat{\rho}_0$ of a single reference object properly selected from the group. It should be pointed out that, because of the intrinsic high discrimination of the correlation channel, when the objects are not similar the selectivity of the filters can be reduced together with the needed accuracy of the scale estimation.

In a few demonstrative experiments, two classes of objects were chosen. Each class contained several objects at different scales. For each class one object was chosen as the reference object and the scales of all the other objects were calculated according to Eq. (9.64). One such class contained the letters **F**, **H**, **P**, and **E**, each of them at 18 different scales. The four letters in the two extreme sizes are shown in Fig. 9.14(a). The plots of $T(a, \hat{\rho})$ for the eight letters [Fig. 9.14(b)] demonstrate their similarity for a given scale factor a. Figure 9.15 shows the errors that were obtained in the scale measurement of the four letters in all 18 scales. This was done for two cases: In Fig. 9.15(a) the reference object was the smallest **F**; hence a scale change of up to $(20/3) = 666\%$ was encountered. To show the insignificance of the exact reference chosen, in Fig. 9.15(b) the reference object was one of the largest **F** (linewidth of 18 pixels). In both cases the maximum error was under 7%, and it was even smaller for similar objects such as **F** and **P**. Experiments performed with different classes of objects, including full gray-level images, led to similar results that demonstrated the efficiency of this scale estimation procedure.

In the laboratory system the scale estimation can be implemented in two stages. In the first stage the input function $f(x, y)$ is written on the SLM, and then the amplitude of its FT $F(f_x, f_y)$ is recorded by the CCD camera. At the second stage this FT is written on the SLM, superposed by a circular mask of radius $\hat{\rho}$, and the integrated transmitted power is measured. The result is proportional to

$$\int_0^{2\pi} \int_0^{\hat{\rho}} |F(\rho, \phi)|^2 \, \rho \, d\rho \, d\phi,$$

which is $E_a(\hat{\rho})$ as defined in Eq. (9.61). E_0 can be found by making the mask radius large enough to transmit the entire energy of $F(f_x, f_y)$. From these values we can derive $T(a, \hat{\rho})$ at each radial frequency $\hat{\rho}$. As noted above, we wish to find the special value $\hat{\rho}$ that corresponds to the desired T_0. Employing an iterative process, we substitute the new radius

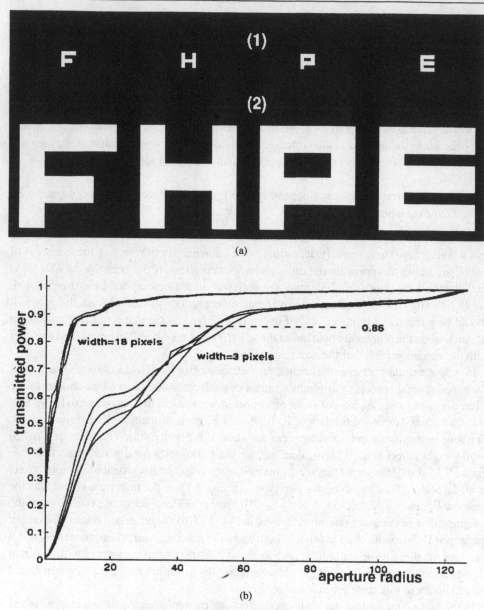

Fig. 9.14. (a) Letters **F**, **H**, **E**, and **P** that were used for testing the scale estimation procedure in the two extreme scales: (1) linewidth of 3 pixels, (2) linewidth of 20 pixels; (b) normalized radial intensity [$T(a, \hat{\rho})$ of Eq. (9.62)] for the letters shown in (a) as a function of aperture radius. [This figure is reproduced from Appl. Opt. **34**, 1891–1900 (1995).]

for the radius of the mask at every iteration:

$$
\rho_{i+1} = \begin{cases} \rho_i + \rho_t & T_i < T_0 \\ \rho_i - \rho_t & T_i > T_0 \end{cases},
\tag{9.65}
$$

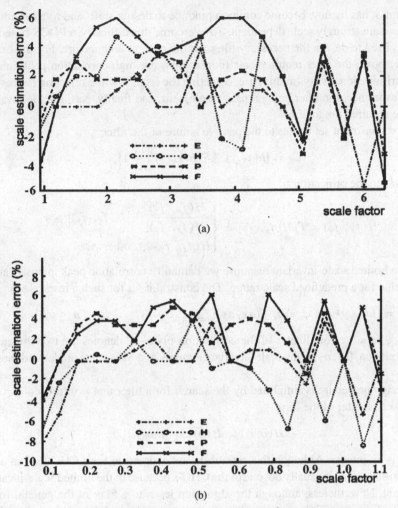

Fig. 9.15. Scale estimation error. The reference functions were (a) the letter F with a linewidth of 3 pixels, (b) the letter F with a linewidth of 18 pixels [see Fig. 9.14(a)]. [This figure is reproduced from Appl. Opt. **34**, 1891–1900 (1995).]

where ρ_t is initially the largest radius and is being changed at each step to $\rho_t = (\rho_t/2)$. In this way, for an input object of $N \times N$ pixels, the scale is found after not more than $\log_2(N)$ steps. An alternative approach is to put a computer-controlled iris behind the SLM instead of writing masks onto the SLM. In a detailed study to estimate the accuracy of the scale estimation [68] it was found that the error did not exceed 6%, in accordance with the theory.

9.7.2 Filter design

As indicated above, after the scale of the input object in the first channel is estimated, the second channel of the system requires a spatial filter that is adaptable to the scale and is invariant to the position and the orientation of the object. Moreover, this spatial filter must also be scale invariant for a scale change within ±6%, the accuracy of the scale estimation.

Although it has by now become common practice to design shift- and rotation-invariant filters, these are strongly scale dependent. To overcome this difficulty, a POCS-based algorithm was used to design the necessary filter for the present architecture. For the example discussed above, the filter requirements (constraints) are high correlation peak intensity, good discrimination ability (in this case, detecting the letter **F** while rejecting the rest), full rotation invariance, the filter being a passive element, and finally, partial scale invariance within the required range of 6%.

The first constraint set relates to the passive nature of the filter:

$$C_1 = \{h(x, y): |\mathcal{F}\{h(x, y)\}| \le 1\}, \tag{9.66}$$

which leads to the projection

$$\bar{H}(f_x, f_y) = P_1 H(f_x, f_y) = \begin{cases} \dfrac{H(f_x, f_y)}{|H(f_x, f_y)|} & |H(f_x, f_y)| > 1 \\ H(f_x, f_y) & \text{otherwise} \end{cases} \tag{9.67}$$

To get a limited scale-invariant response we demand a correlation peak intensity above a certain value for a predefined scale range. The constraint set for such a response is

$$C_2 = [h(x, y): |h(x, y) \otimes f(ax, ay)|_{\eta=0, \xi=0} \ge S_1, \qquad a_1 \le a \le a_2], \tag{9.68}$$

where (η, ξ) are the coordinates of the correlation plane, $\otimes$ denotes the two-dimensional cross correlation, $[a_1, a_2]$ is the scale-invariance range, and S_1 is some predefined threshold value.

The design procedure is simplified by the search for a filter that is intrinsically rotation invariant, i.e., a filter of the form

$$H(\rho, \phi) = H(\rho) \exp(jm\phi), \tag{9.69}$$

where m is an integer. Although this resembles the expression for a CHC [61] in a CHC decomposition, $H(\rho)$ is usually not one of the CHC's because of the limited scale-invariance requirement. Nevertheless, although the algorithm leads to a filter of the general form of Eq. (9.69), choosing an integer m results in a correlation peak at the origin determined by the correlation of the filter $H(\rho, \phi)$ with the mth CHC of the input. This is shown in the following analysis.

Let $f(r, \theta)$ be the input function and $F(\rho, \phi)$ be its FT. When this function is rotated by α and scaled by $a[f(r/a, \theta + \alpha)]$, the correlation distribution at the origin, with the filter of Eq. (9.69), becomes

$$c^{FH}(a, \alpha) = \int_0^{2\pi} \int_0^\infty F(a\rho, \phi + \alpha) H^*(\rho, \phi) \rho \, d\rho \, d\phi = 2\pi \exp(jm\alpha) C_m^{FH}(a), \tag{9.70}$$

where H^* is the complex conjugate of H. It is clear that the correlation peak intensity is rotation invariant. $C_m^{FH}(a)$ is defined by

$$C_m^{FH}(a) = \int_0^\infty F_m(a\rho) H^*(\rho) \rho \, d\rho, \tag{9.71}$$

where $F_m(\rho)$ is the mth CHC of $F(\rho, \phi)$:

$$F_m(\rho) = \frac{1}{2\pi} \int_0^{2\pi} F(\rho, \phi) \exp(-jm\phi) \, d\phi. \tag{9.72}$$

We note that, by using the rotation-invariant filter, we may consider $C_m^{FH}(a)$ instead of the more general two-dimensional expression for the correlation distribution in the output plane. The requirement to meet the appropriate constraints with this choice of filter reduces the computation complexity, as the calculations are performed in one dimension instead of two. As a result, a significant increase in the speed of the design procedure is achieved.

The problem with the correlation peak intensity constraint mentioned above, constraint C_2 of Eq. (9.68), is that forcing the intensity to be above a certain value is not a convex constraint and thus the convergence of the serial POCS algorithm is not guaranteed. We could use the parallel POCS algorithm but another possible solution is to break this constraint into two separate constraints, one for the real and the other for the imaginary parts of $C_m^{FH}(a)$, as was done in Eq. (9.52). The combination of these constraints will lead to the desired intensity response.

The new constraint set becomes

$$C_2 = \{h(x, y): h(x, y) \otimes f(ax, ay)_{\eta=0, \xi=0} \in B\}, \tag{9.73}$$

where the set B is defined by

$$B = \left\{ C_m: \begin{cases} C_m^R = \mathrm{Re}\{C_m^{FH}(a)\} & \geq S_1 \\ |C_m^I| = |\mathrm{Im}\{C_m^{FH}(a)\}| & \leq S_2 \end{cases} \quad a_1 \leq a \leq a_2 \right\}, \tag{9.74}$$

where S_2 is some predefined upper bound for the imaginary part.

Further details of these projections are elaborated in Ref. 68.

To improve the discrimination ability of the filter, a special discrimination constraint may be included in the design process. Experimental study indicated that adequate discrimination was obtained without such a constraint but, if needed, this can be added. The constraint set for this requirement is given by

$$C_3(\eta, \xi) = \{h(x, y): |h(x, y) \otimes g(x, y)|_{\eta, \xi} \leq S_{\mathrm{rej}}(\eta, \xi)\}, \tag{9.75}$$

where $g(x, y)$ is the object to be rejected. This set is convex and there is no need to break it into two separate constraints as was done with C_2. The only problem is that for a rejection constraint we must control the whole output plane to avoid false correlation peaks, although for recognition it is sufficient to obtain a single correlation peak with adequate energy.

The laboratory implementation of the optical correlator is based on SLM's, in both the input and the Fourier planes. This architecture dictates several requirements from the filter, including high light efficiency, a high correlation peak, and the ability to record the filter on the SLM that has a limited number of pixels. A phase-only version of the designed filter is a good candidate for implementation, but this requirement cannot be added to the design procedure because it is not a convex constraint. Although this problem could be treated with parallel projection algorithms, satisfactory results were obtained by designing according to the other constraints and then applying a nonlinear transformation on the filter [32, 34, 68]. The resulting filter remains rotation invariant and has sharp autocorrelation peaks and good discrimination capability [62–64].

In the experiments described here, the generally complex filter function $H(f_x, f_y)$ was encoded into a binary hologram. The power of the POCS algorithm was again demonstrated by the design of an efficient filter that could be displayed on a low-resolution SLM (400×400 pixels, in this case) in which a conventional CGH would not work. A unipolar

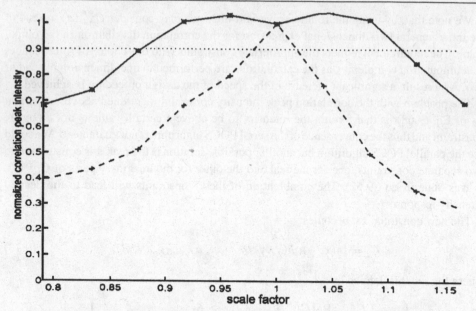

Fig. 9.16. Scale sensitivity of the POF designed with the POCS algorithm (solid curve) compared with that of the CHC POF (dashed curve). [This figure is reproduced from Appl. Opt. **34**, 1891–1900 (1995).]

binary POF representation [64],

$$H_B(f_x, f_y) = \begin{cases} 1 & \cos(\eta) > 0 \\ 0 & \text{otherwise} \end{cases}, \tag{9.76}$$

was used, where $H_P(f_x, f_y) = \exp[j\Phi(f_x, f_y)]$ is the continuous POF, $\eta = \Phi(f_x, f_y) + \beta(f_x, f_y)$, and β is the carrier frequency. When an appropriate carrier frequency is chosen, the desired correlation distribution between the input and the filter is obtained in the first diffraction order on the correlation plane.

In the experiments, the range of scale invariance was chosen to be ±8%, which fully compensated for the uncertainty of the scale estimation. The performance of the filter obtained is compared with the CHC POF in Fig. 9.16 with respect to the correlation peak intensity as a function of the input function's scale. The intensities are normalized so that for the original image both peaks become equal to 1. It is obvious from the plot that the correlation peak stays within 95% of its maximum value through a relative scale range of ±10%. The other filter is much more sensitive to scale and therefore is not suitable for the present application. The discrimination ability of the filter is shown in Fig. 9.17; Fig. 9.18 demonstrates full rotation invariance.

In the complete optical system of Fig. 9.13, the input object was displayed on the input SLM and its scale was measured. After the measurement was completed, the input image was digitally rescaled to match the predefined filter that was designed for optimal utilization of the SLM capacity. The final correlation was executed by the optical correlator between the rescaled input and the stored filter. The resulting correlation function indicated the presence (or absence) of the specific pattern and determined its location. Of course, the image rescaling procedure could be done optically by a computer-controlled zoom lens, which may be preferable for both speed and accuracy considerations. Nevertheless, the

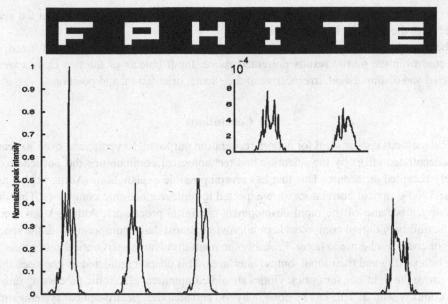

Fig. 9.17. Discrimination characteristics of the filter. The input (top) is correlated with a filter made for **F**. The correlation peak intensities (bottom) indicate good discrimination. The inset is an enlargement by a factor of 10^4 to show also the cross section of the correlations for **I** and **T**. [This figure is reproduced from Appl. Opt. **34**, 1891–1900 (1995).]

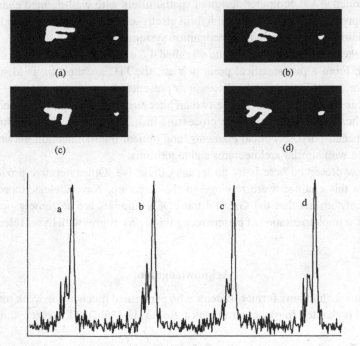

Fig. 9.18. Rotation-invariance characteristics of the binary filter. The intensity distribution over the correlation plane for several rotation angles of the input (including the zero-order term) and the corresponding intensity cross sections. [This figure is reproduced from Appl. Opt. **34**, 1891–1900 (1995).]

more flexible approach is the adaptation of the filter, which will be practical in the near future with the progress in SLM technology.

The experimental performance of the complete system was tested and performed as expected from the partial results presented above. Input objects of the two classes were detected and distinguished, irrespective of their scale, orientation, and position.

9.8 Conclusions

Optical methods appear ideal for pattern recognition purposes. Nevertheless, over 30 years of concentrated effort by the scientific and technological communities did not lead to a widely accepted procedure. This fact has several possible explanations. At the beginning, in the 1960's, optical correlators were expected to replace electronic computers. This did not happen because of the rapid development of digital processors. Although the speed and flexibility of digital computers kept improving almost daily, computing with the speed of light could not be made faster. Probably the main disadvantage of optical processors is their inflexibility and their input–output interfaces. This chapter indicated an approach that may prove the right one for optics: Optics should complement electronic processing rather than compete with it. This can be done in hybrid and adaptive electro-optical systems such as those presented here. With this statement accepted as a given fact, this chapter focused on the algorithmic aspect of such hybrid systems. In this context the power of iterative methods was demonstrated with strong indications that these methods may become the preferred procedure.

A major advantage of iterative processes is their intrinsic flexibility. Iterative algorithms can be exploited in the computer design of spatial filters with well-defined characteristics, in learning processes implemented on hybrid electro-optical architectures, and during the actual operation of adaptive pattern recognition systems.

For the sake of compactness only the so-called $4f$ optical correlators were discussed in this chapter. From a mathematical point of view, the JTC architecture is identical to the $4f$ correlator, and therefore all the discussions presented here are also relevant to the JTC. The technical differences between the two architectures can be accommodated with only slight modifications. Moreover, other processing tasks, such as image reconstruction, the design of special-purpose optical elements, and optical interconnection networks can be implemented with similar architectures and algorithms.

The review presented here is by no means exhaustive. Other iterative procedures, not discussed in this chapter, were reported in the literature. Nevertheless, there are many indications of the fact that the GA and the POCS families are extremely powerful and valuable to the implementation of pattern recognition procedures with hybrid electro-optical architectures.

Acknowledgment

It is a pleasure to thank my former students who performed much of the work that served as a basis for this chapter. In particular I thank J. Rosen, U. Mahlab, T. Kotzer, and E. Silvera.

References

[1] A. B. VanderLugt, "Signal detection by complex spatial filtering," IEEE Trans. Inf. Theory **IT-10**, 139–145 (1964).

[2] See, for example, S. P. Almeida and G. Indebetouw, "Pattern recognition via complex spatial filtering," in *Applications of Optical Fourier Transforms*, H. Stark, ed. (Academic, Orlando, FL, 1982).

[3] H. J. Caulfield and W. T. Maloney, "Improved discrimination in optical character recognition," Appl. Opt. **8**, 2354–2356 (1969).

[4] B. Braunecker, R. Hauch, and A. W. Lohmann, "Optical character recognition based on nonredundant correlation measurements," Appl. Opt. **18**, 2746–2753 (1979).

[5] O. Bryngdahl and F. Wyrowski, "Digital holography – computer-generated holograms," in *Progress in Optics*, E. Wolf, ed. (North-Holland, Amsterdam, 1990), Vol. 28, pp. 1–86.

[6] B. V. K. Vijaya Kumar, "Tutorial survey of composite filter designs for optical correlators," Appl. Opt. **31**, 4773–4801 (1992).

[7] C. S. Weaver and J. W. Goodman, "A technique for optically convolving two functions," Appl. Opt. **5**, 1248–1249 (1966).

[8] F. T. S. Yu and X. J. Lu, "A real-time programmable joint transform correlator," Opt. Commun. **52**, 10–16 (1984).

[9] F. T. S. Yu, S. Jutamulia, T. W. Lin, and D. A. Gregory, "Adaptive real-time pattern recognition using aligned crystal T.V. based joint transform correlator," Appl. Opt. **25**, 1370–1372 (1987).

[10] D. Gabor, "Light and Information," in *Progress in Optics*, E. Wolf, ed. (North-Holland, Amsterdam, 1961), Vol. 1, pp. 111–152.

[11] J. Shamir, H. J. Caulfield, and R. B. Johnson, "Massive holographic interconnections and their limitations," Appl. Opt. **28**, 311–324 (1989).

[12] J. L. Horner and P. D. Gianino, "Phase only matched filtering," Appl. Opt. **23**, 812–816 (1984).

[13] U. Mahlab, M. Fleisher, and J. Shamir, "Error probability in optical pattern recognition," Opt. Commun. **77**, 415–422 (1990).

[14] M. Fleisher, U. Mahlab, and J. Shamir, "Target location measurement by optical correlator: a performance criterion," Appl. Opt. **31**, 230–235 (1992).

[15] B. V. K. Vijaya Kumar, F. M. Dickey, and J. M. DeLaurentis, "Correlation filters minimizing peak location errors," J. Opt. Soc. Am. A **9**, 678–682 (1992).

[16] A. Mahalanobis, B. V. K. Vijaya Kumar, and D. Casasent, "Minimum average correlation energy filters," Appl. Opt. **25**, 3633–3640 (1987).

[17] B. V. K. Vijaya Kumar, "Minimum variance SDFs," J. Opt. Soc. Am. A **3**, 1579–1584 (1986).

[18] Z. Bahri and B. V. K. Vijaya Kumar, "Generalized synthetic discriminant functions," J. Opt. Soc. Am. A **5**, 562–571 (1988).

[19] R. R. Kallman, "Construction of low noise optical correlations filters," Appl. Opt. **25**, 1032–1033 (1986).

[20] R. D. Juday and B. J. Daiuto, "Relaxation method of compensation in an optical correlator," Opt. Eng. **26**, 1094–1101 (1987).

[21] M. Fleisher, U. Mahlab, and J. Shamir, "Entropy optimized filter for pattern recognition," Appl. Opt. **29**, 2091–2098 (1990).

[22] U. Mahlab and J. Shamir, "Phase only entropy optimized filter by simulated annealing," Opt. Lett. **14**, 146–148 (1989).

[23] G. Zalman and J. Shamir, "Maximum discrimination filter," J. Opt. Soc. Am. A **8**, 814–821 (1991).

[24] U. Mahlab and J. Shamir, "Optical pattern recognition based on convex functions," J. Opt. Soc. Am. A **8**, 1233–1239 (1991).

[25] G. Zalman and J. Shamir, "Reduced noise-sensitive optical pattern recognition," J. Opt. Soc. Am. A **8**, 1866–1873 (1991).

[26] J. P. Ding, M. Itoh, and T. Yatagai, "Iterative design of distortion-invariant phase-only filters with high Horner efficiency," Opt. Eng. **33**, 4037–4044 (1994).

[27] M. Avriel, *Nonlinear Programming: Analysis and Methods* (Prentice-Hall, Englewood Cliffs, NJ, 1976).

[28] P. J. M. van Luarhoven and E. H. L. Aarts, *Simulated Annealing: Theory and Applications* (Reidel, Dordrecht, The Netherlands, 1987).

[29] D. Lawrence, *Genetic Algorithm and Simulated Annealing* (Kaufmann, Los Altos, CA, 1987).

[30] U. Mahlab and J. Shamir, "Comparison of iterative optimization algorithms for filter generation in optical correlators," Appl. Opt. **31**, 1117–1125 (1992).

[31] D. C. Youla and H. Webb, "Image restoration by the method of convex projections: part 1 – theory," IEEE Trans. Med. Imaging **TMI-1**, 81–94 (1982).

[32] J. Rosen and J. Shamir, "Application of the projection-onto-constraint-sets algorithm for optical pattern recognition," Opt. Lett. **16**, 752–754 (1991).

[33] J. Rosen, "Learning in correlators based on projections onto constraint sets," Opt. Lett. **18**, 1183–1185 (1993).

[34] D. W. Sweeney, E. Ochoa, and G. F. Schills, "Experimental use of iteratively designed rotation invariant correlation filters," Appl. Opt. **26**, 3458–3465 (1987).

[35] T. Kotzer, N. Cohen, and J. Shamir, "Image reconstruction by a novel parallel projection onto constraint set method," Opt. Lett. **20**, 1172–1174 (1995).

[36] T. Kotzer, N. Cohen, J. Shamir, and Y. Censor, "Multi-distance, multi-projection parallel projection method," The International Conference on Optical Computing, OC'94, Edinburgh, 22–25 August 1994, Inst. Phys. Conf. Ser. No. 139, Part III, 301–304 (1995).

[37] Y. Censor and T. Elfving, "A multiprojection algorithm using Bregman projections in a product space," Num. Algorithms **8**, 221–239 (1994).

[38] T. Kotzer, N. Cohen, and J. Shamir, "A projection algorithm for consistent and inconsistent constraints," SIAM J. Optim. **7**, 527–546 (1997).

[39] A. Levi and H. Stark, "Image restoration by the method of generalized projections with application to restoration from magnitude," J. Opt. Soc. Am. A **1**, 932–943 (1984).

[40] T. Kotzer, N. Cohen, and J. Shamir, "Extended and alternative projections onto convex constraint sets: theory and applications," EE Pub. 900 (Technion, Israel Institute of Technology, Haifa, Israel, 1993); Proc. IEEE Internat. Conf. on Pattern Recognition, 77–81, 9–13 Oct. 1994, Jerusalem.

[41] T. Kotzer, J. Rosen, and J. Shamir, "Application of serial and parallel projection methods to correlation filter design," Appl. Opt. **34**, 3883–3895 (1995).

[42] A. W. Robert and D. E. Veberg, *Convex Function* (Academic, New York, 1973).

[43] C. E. Shannon and W. Weaver, *The Mathematical Theory of Communication* (University of Illinois, Urbana, IL, 1949).

[44] J. Ziv and M. Zakai, "On functionals satisfying a data-processing theorem," IEEE Trans. Inf. Theory **IT-19**, 275–283 (1973).

[45] J. Rosen, U. Mahlab, and J. Shamir, "Adaptive learning with joint transform correlator," Opt. Eng. **29**, 1101–1106 (1990).

[46] U. Mahlab, J. Rosen, and J. Shamir, "Iterative generation of complex RDF in joint transform correlators," Opt. Lett. **15**, 556–558 (1990).

[47] U. Mahlab, J. Shamir, and H. J. Caulfield, "Genetic algorithm for optical pattern recognition," Opt. Lett. **16**, 648–650 (1991).

[48] U. Mahlab, J. Rosen, and J. Shamir, "Iterative generation of holograms on spatial light modulators," Opt. Lett. **15**, 556–558 (1990).

[49] N. Douklias and J. Shamir, "Relation between object position and auto-correlation spots in the VanderLugt filtering process. 2: influence of the volume nature of the photographic emulsion," Appl. Opt. **12**, 364–367 (1973).

[50] T. Kotzer, J. Rosen, and J. Shamir, "Multiple-object input in nonlinear correlation," Appl. Opt. **32**, 1919–1932 (1993).

[51] H. Y. Zhou, F. Zhao, and F. T. S. Yu, "Angle-dependent diffraction efficiency in a thick photorefractive hologram," Appl. Opt. **34**, 1303–1309 (1995).

[52] T. Kotzer, J. Rosen, and J. Shamir, "Phase extraction pattern recognition," Appl. Opt. **31**, 1126–1137 (1992).

[53] T. Kotzer, N. Cohen, J. Shamir, and Y. Censor, "Summed distance error reduction of simultaneous multiprojections and applications," EE Pub. 909 (Department of Electrical Engineering, Technion, Israel Institute of Technology, Haifa, Israel, 1994).

[54] O. K. Ersoy and M. Zeng, "Nonlinear matched filtering," J. Opt. Soc. Am. A **6**, 636–648 (1989).

[55] B. Javidi, "Nonlinear joint power spectrum based optical correlation," Appl. Opt. **28**, 2358–2367 (1989).

[56] O. K. Ersoy, Y. Yoon, N. Keshava, and D. Zimmerman, "Nonlinear matched filtering 2," Opt. Eng. **29**, 1002–1012 (1990).

[57] B. Javidi and J. Wang, "Binary nonlinear joint transform correlation with median and subset thresholding," Appl. Opt. **30**, 967–976 (1991).

[58] W. B. Hahn and D. L. Flannery, "Design elements of binary joint transform correlation and selected optimization techniques," Opt. Eng. **31**, 896–905 (1992).

[59] H. Y. Zhou, F. Zhao, F. T. S. Yu, and T. H. Chao, "Improved interclass multiobject discrimination with phase-difference prewhitening technique," Opt. Eng. **32**, 2720–2725 (1993).

[60] J. Rosen, T. Kotzer, and J. Shamir, "Optical implementation of phase extraction pattern recognition," Opt. Commun. **83**, 10–14 (1991).

[61] Y. Hsu and H. Arsenault, "Optical pattern recognition using circular harmonic expansion," Appl. Opt. **21**, 4016–4019 (1982).

[62] J. Rosen and J. Shamir, "Circular harmonic phase filters for efficient rotation-invariant pattern recognition," Appl. Opt. **27**, 2895–2899 (1988).

[63] L. Leclerc, Y. Sheng, and H. Arsenault, "Rotation invariant phase-only and binary phase-only correlation," Appl. Opt. **28**, 1251–1256 (1989).

[64] L. Leclerc, Y. Sheng, and H. H. Arsenault, "Optical binary phase-only filters for circular harmonic correlations," Appl. Opt. **30**, 4643–4648 (1991).

[65] J. Rosen and J. Shamir, "Scale invariant pattern recognition with logarithmic radial harmonic filters," Appl. Opt. **28**, 240–244 (1989).

[66] D. Mendlovic, E. Marom, and N. Konforti, "Shift and scale invariant pattern recognition using Mellin radial harmonic," Opt. Commun. **67**, 172–176 (1988).

[67] D. Casasent and D. Psaltis, "New optical transforms for pattern recognition," Proc. IEEE **65**, 77–84 (1977).

[68] E. Silvera, T. Kotzer, and J. Shamir, "Adaptive pattern recognition with rotation, scale and shift invariance," Appl. Opt. **34**, 1891–1900 (1995).

10

Optoelectronic hybrid system for three-dimensional object pattern recognition

Guoguang Mu, Mingzhe Lu, Ying Sun, and Hongchen Zhai

10.1 Introduction

Pattern recognition is of considerable interest in signal and data processing, it is important in many practical applications, such as robotic vision and other intelligent systems. Because of the real-time and parallel processing features of optical information processors such as Fourier-transform and correlation operations inherent in these systems, optical pattern recognition is thus considered a promising candidate for pattern recognition technique. The possibility of optically producing a correlation of two-dimensional (2-D) functions is the basis of a coherent optical pattern recognition technique. Signal detection or pattern recognition by complex spatial matched filtering, which was first reported by VanderLugt [1], has been applied to various fields, such as the study of blood cells, malignant cancer tissues, and signal detecting from random noise. Many successful results have been achieved. However, there are some disadvantages in pattern recognition with a matched filter; for example, the matched filter must be reset at an exact position in the Fourier plane of a coherent optical system and this causes trouble in applications.

When a square-law detector and a spatial light modulator are used, the real-time optical pattern recognition system can be implemented with a joint transform correlator. Historically, many efforts have been devoted to increasing the discrimination capacity of the matched filter and to finding the means of providing a real-time operation. Basing their work on the relative role played by phase and amplitude in the Fourier domain for image reformation, Horner and Gianino [2] presented a phase-only matched-filtering technique and showed its potential advantages for applications in pattern recognition, such as its much better ability to discriminate in the absence of sidelobes. To remove amplitude mismatching in the phase-only filter, Mu et al. [3] proposed amplitude-compensated matched filtering for pattern recognition.

Compared with the amplitude correlation filter, the intensity correlators have some advantages [4], especially the correlation results that are invariant to the filter shift. Mu et al. [5] proposed a new technique with a holographic encoding filter for lensless intensity correlation; for this technique, the in-plane shift of the correlation filter does not influence the correlation result, and the allowable out-of-plane shift extends some millimeters, much longer than the allowable shift by means of the matched-spatial-filter technique. The transformation from incoherent light to coherent light is not necessary in an optical operation, because of its intensity correlation characteristics. These offer a convenient application to real-time pattern recognition.

It is well known that the matched-filter output will degrade severely in the presence of input-image distortions. Although the optical matched filter was regarded as a method for optical pattern recognition for many years, it suffers from an inability to recognize an object that has undergone rotation or a scale change. However, in many applications it is especially desirable to recognize the targets within different orientations and at different distances. Much research has therefore been directed toward developing a system in which image recognition is invariant to object rotation and scale. Hsu and Arsenault [6] achieved rotation-invariant correlation by utilizing the circular harmonic expansion of the reference object. With the expression of an object in another feature space, a Mellin transform enables the correlator to be scale invariant. An important component for optical information processing is the wedge–ring detector [7], which can detect the radial power spectrum and edges and the edge–angle correlation; the ring detector is rotation invariant whereas the wedge scale is invariant. The optical correlator with a wedge–ring detector is perhaps the simplest optical pattern recognition system that is rotation invariant and scale invariant simultaneously.

However, a more important but difficult problem in optical pattern recognition is recognizing a certain three-dimensional (3-D) target by projection of the object in any perspective angle. Several techniques, in which a single filter is used to recognize a large group of perspective views, have been developed [8, 9]. In these techniques, all sampling pictures are required for synthesizing the filter. Unfortunately, such a single filter loses its specificity to an image group. In fact, it is shown that the specificity of the synthetic discriminant function (SDF) filter will decrease when the number of object views to be encoded in the filter increases. Therefore one filter is not sufficient to solve the 3-D target recognition problem because there is such a large amount of information to be handled. Using a small set of filters may be a better way to solve this problem. Schils and Sweeney [10] presented the technique of a lock-and-tumble (LAT) filter for 3-D target recognition, in which a spectral iterative method is necessary for the design of this filter. Yin *et al.* [11] proposed a new type of a bank of filters, called serial-code filters (SCF's), for recognizing a 3-D object from arbitrary perspective views. The procedure for the design of these filters is much easier than that for LAT filters, and, because these filters are completely independent of each other, the number of required filters can also be much fewer than that of the sampling images.

In this chapter we present the principle of the Fresnel holographic filter (FHF) and some of its applications in optical pattern recognition, such as a lensless intensity correlator [12], a white-light intensity correlator [13], an optical intensity correlator with high discrimination [14], and a multichannel intensity correlator [15]. Based on the SCF architecture, a three-layer optical pattern recognition system is built up to recognize 3-D objects from an arbitrary perspective view.

10.2 Fresnel holographic filter

10.2.1 Principle of the Fresnel holographic filter

10.2.1.1 Synthesis of the Fresnel holographic filter

The principal scheme of the holographic filter recording system, shown in Fig. 10.1, is analogous to that of a common holographic recording system, except for the converging reference beam. An object $f(x, y)$ is placed in plane P_1, and the converging reference beam

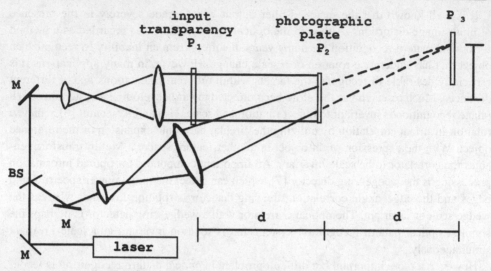

Fig. 10.1. Scheme of FHF recording system. M's, mirrors; BS, beam splitter.

is focused onto point B in-plane P_3. The amplitude distribution of the reference beam is

$$R(p, q) = K_1 \exp\left\{d' + p\cos\theta + \frac{1}{2d'}[(p\sin\theta)^2 + q^2]\right\}, \tag{10.1}$$

and the amplitude of the object beam is

$$O(p, q) = K_2 \int W(x, y) f(x, y) \exp\left(ik\left\{d + p\cos\theta\right.\right.$$
$$\left.\left. + \frac{1}{2d}[(x - p\sin\theta)^2 + (y - q)^2]\right\}\right) dx\, dy, \tag{10.2}$$

where k is the wave vector, d' is the distance between O and B, and $W(x, y)$ is the illuminating light distribution, which can be described as $W(x, y) = W_0 \exp[i\phi(x, y)]$; $\phi(x, y)$ is an arbitrary phase.

After suitable exposure and developing, the FHF is created. One of the terms of its transmittance function that we are interested in is given by

$$R(p, q)O^*(p, q) = K_3 \exp\left(-ik\left\{p\cos\theta + \frac{1}{2d'}[(p\sin\theta)^2 + q^2]\right\}\right)$$
$$\times \int W^*(x, y) f^*(x, y) \exp\left(-ik\left\{p\cos\theta + \frac{1}{2d}[(x - p\sin\theta)^2\right.\right.$$
$$\left.\left. + (y - q)^2]\right\}\right) dx\, dy. \tag{10.3}$$

10.2.1.2 Correlation process

The scheme of correlation system is shown in Fig. 10.2. The whole setup contains a filter as well as an input plane and an output plane. The object $f'(x_1, y_1)$ is placed in P_1, with a contact moving diffuser in front of it. Assume that the filter is replaced not exactly at its

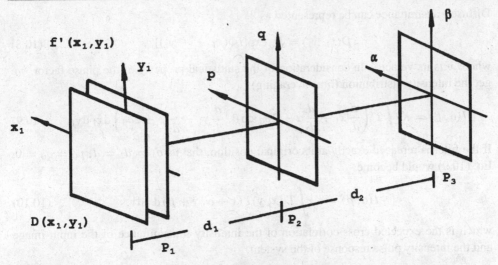

Fig. 10.2. FHF correlation system.

original position, but with an out-of-plane shift of $d_1 - d$ and an in-plane shift of $(-p_0, q_0)$. Then the complex field just behind the filter would be

$$C(p, q) = K_4 R(p + p_0, q + q_0) \int f'(x_1, y_1) D(x_1, y_1, t)$$

$$\times \exp\left(ik\left\{p\cos\theta + \frac{1}{2d_1}[(p\sin\theta - x_1)^2 + (q - y_1)^2]\right\}\right) dx_1\, dy_1 \quad (10.4)$$

and the complex field in the output plane P_3 is

$$E(\alpha, \beta) = K_5 \int D(x_1, y_1) f'(x_1, y_1)\, dx_1\, dy_1 \int W^*(x, y) f^*(x, y)\, dx\, dy$$

$$\times \exp\left\{\frac{ik}{2}[(x_1^2 + y_1^2)/d_1 + (\alpha^2 + \beta^2)/d_2 - (x - p_0\sin\theta)^2 + (y - q_0)^2]/d\right\}$$

$$\times \int \exp\left\{\frac{ik}{2}\left[(p^2\sin^2\theta + q^2)\left(\frac{1}{d_1} + \frac{1}{d_2} - \frac{1}{d} - \frac{1}{d'}\right)\right.\right.$$

$$\left. - 2p\sin\theta\left(\frac{x_1}{d_1} + \frac{\alpha}{d_2} - \frac{x - p_0\sin\theta}{d}\right) + p_0\sin\frac{\theta}{d'}\right.$$

$$\left.\left. - 2q\left(\frac{y_1}{d_1} + \frac{\beta}{d_2} - \frac{y - q_0}{d} + \frac{q_0}{d'}\right)\right]\right\} dp\, dq. \quad (10.5)$$

When the condition

$$\frac{1}{d} + \frac{1}{d'} - \frac{1}{d_1} - \frac{1}{d_2} = 0 \quad (10.6)$$

is satisfied, Eq. (10.5) can be rewritten as

$$E(\alpha, \beta) = K_6 \int D(x_1, y_1) f^*\left(\frac{d}{d_1}X_1 + \frac{d}{d_2}\alpha + 2p_0\sin\theta, \frac{d}{d_1}y_1 + \frac{d}{d_2}\beta + 2q_0\right)$$

$$\times \exp[i\phi(x_1, y_1, \alpha, \beta)]\, dx_1\, dy_1. \quad (10.7)$$

Diffuse transmittance can be represented as

$$D(x_1, y_1) = D_0 \exp[i\,\Phi(x_1 - vt, y_1)], \qquad (10.8)$$

where v is its velocity. In consideration of the statistical property of the phase factor, we get the intensity distribution (time averaging):

$$I(\alpha, \beta) = K_7 \int T'\left(\frac{d}{d_1}x_1 + \frac{d}{d_2}\alpha + 2p_0\sin\theta, \frac{d}{d_1}y_1 + \frac{d}{d_2}\beta + 2q_0\right) dx_1\,dy_1. \qquad (10.9)$$

If the filter is replaced exactly at its original position, that is, $d_1 = d_2 = d$, $p_0 = q_0 = 0$, Eq. (10.9) would become

$$I(\alpha, \beta) = K_7 \int T'(x, y)T(x + \alpha, y + \beta)\,dx\,dy, \qquad (10.10)$$

which is the expected cross correlation of the intensity transmittance of the input-image and the intensity pulse response of the system.

10.2.1.3 Characteristics of a Fresnel holographic filter

Because of the intensity processing, there is no influence of the image phase distortion and the phase response of the system.

The FHF has low requirements for in-plane and out-of-plane adjustments of the holographic filter. If only an in-plane shift of the filter is caused, that is, $d_1 = d_2 = d$, but $p_0 \neq 0$ and $q_0 \neq 0$, Eq. (10.10) would become

$$I(\alpha, \beta) = \int T'(x, y)T(x + \alpha + 2p_0, y + \beta + 2q_0)\,dx\,dy, \qquad (10.11)$$

which is the exact cross correlation but is also a whole in-plane shift relative to the coordinate system in the output plane. Additionally, by taking only an out-of-plane shift of the filter, that is, $p_0 = q_0 = 0$, but $d_1 \neq d$, we would obtain

$$I(\alpha, \beta) = \int T'(x, y)T\left(\frac{d}{d_1}x + \frac{d}{d_2}\alpha, \frac{d}{d_1}y + \frac{d}{d_2}\beta\right) dx\,dy, \qquad (10.12)$$

which is something similar to a scale-change correlation. For a 3-dB loss of performance, the allowable scale change is $\sim$0.5%; equivalently, $1 - d/d_1$ could be 0.5%. So, a few millimeters of an out-of-plane shift of the filter is permitted to obtain a harmless result.

The FHF has low requirements for the dynamic range of the hologram. Because of the Fresnel diffraction, no bright spot appears on the recording plane, so a low dynamic range of the holographic recording is needed.

The FHF has an increased signal-to-noise ratio that is due to spatial incoherent light.

10.2.2 Experimental demonstrations of a lensless intensity correlator

In an experimental demonstration, a 60-mW He–Ne laser is used to recording the FHF. The wavelength of the laser beam is 633 nm. In the lensless intensity correlator, the angle between the object beam and the reference beam is $\sim$30° and the parameter d is $\sim$50 cm. The letter C, as shown in Fig. 10.3, is chosen to be the detected signal. After holographic recording, the filter is inserted into the lensless correlation system. An object transparency,

Fig. 10.3. Detected signal.

Fig. 10.4. Input object signal.

as shown in Fig. 10.4, is placed in the input plane and illuminated by a spatial incoherent light source. The correlation results are shown in Fig. 10.5. From Fig. 10.5, we can see that the correlation result is satisfactory.

10.2.3 White-light intensity correlator with a volume Fresnel holographic filter

With the technique of the FHF, optical pattern recognition can be performed with a natural white-light source. It consists of two steps: First, a volume FHF (VFHF) must be made with a converging reference wave and a plane object wave coming from the opposite side of the recording plate in a holographic recording system; second, the pattern recognition result can be obtained at the output plane in a reflective optical correlator when the VFHF is placed in a relevant spatial filtering plane and the signal to be detected is placed in the input plane under the natural white-light illumination. The VFHF should play three roles in the optical correlation operation: phase transformation, reconstructed wavelength selection, and spatial filtering for correlation. Therefore the input signal does not need to be illuminated with monochromatic light. It could be either one with natural white-light illumination or a

Fig. 10.5. Experimental correlation result.

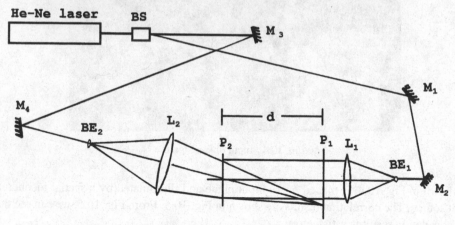

Fig. 10.6. Optical configuration for recording the reflective VFHF. BS, beam splitter; M_1, M_2, M_3, and M_4, mirrors; BE_1 and BE_2, beam enlargers; L_1 and L_2, lenses.

self-luminous object. With a TV camera and a TV screen to produce the input signal, this system can perform real-time processing. Thus a potential application is provided by this technique.

10.2.3.1 Making a volume Fresnel holographic filter

The scheme of the VFHF recording system is shown in Fig. 10.6. The detecting reference signal is placed at plane P_1 and illuminated by a collimated laser beam. The photosensitive plate with an emulsion thickness of z is placed at P_2. The converging reference beam that is focused onto point B in P_1 comes from the opposite side. According to Kirchoff's theorem, the complex light field inside the photographic emulsion that is due to the object light can be written as

$$O(\alpha, \beta, z) = A \exp\left[\frac{i2\pi}{\lambda_1}(d+z)\right] \int\int f_1(x, y) \exp\left\{\frac{i\pi}{\lambda_1(d+z)}[(\alpha-x)^2 + (\beta-y)^2]\right\} dx\, dy$$

$$\text{for } 0 \leq z \leq \Delta z, \quad (10.13)$$

and the complex light field that is due to the converging reference wave can be written as

$$R(\alpha, \beta, z) = B \exp\left\{\frac{-i\pi}{\lambda_1(d+z)}[(\alpha-h_0)^2 + \beta^2]\right\} \exp\left[\frac{-i2\pi}{\lambda_1}(d+z)\right]$$

$$\text{for } 0 \leq z \leq \Delta z, \quad (10.14)$$

where (x, y) and (α, β) are the coordinate systems of planes P_1 and P_2, respectively, $f_1(x, y)$ is the detecting signal, λ_1 is the wavelength of the laser beam, and A and B are complex constants. The above equations imply that the corresponding irradiance varies sinusoidally along the z direction within the emulsion. If the holographic recording is assumed to be linear in the developed photographic grain density, we will get a sequence of very thin holograms arranged in parallel in the photographic emulsion that act as reflecting planes. The term of the reflective function of the overall hologram relative to the holographic reconstruction can be written as

$$r(\alpha, \beta, z) = R(\alpha, \beta, z)O^*(\alpha, \beta, z) = C \exp\left\{\frac{-i\pi}{\lambda_1(d+z)}[(\alpha-h_0)^2 + \beta^2]\right\} \int\int f_1^*(x, y)$$

$$\times \exp\left\{\frac{-i\pi}{\lambda_1(d+z)}[(\alpha-x)^2 + (\beta-y)^2]\right\} dx\, dy \quad \text{for } 0 \leq z \leq \Delta z, \quad (10.15)$$

where C is a complex constant.

10.2.3.2 White-light optical correlation

The lensless white-light correlation system is sketched in Fig. 10.7. The object function $f_2(x, y)$ is placed at input plane P_1 and illuminated by an extended white-light source. The VFHF is appropriately placed at a proper position in plane P_2. Mirror M reflects the

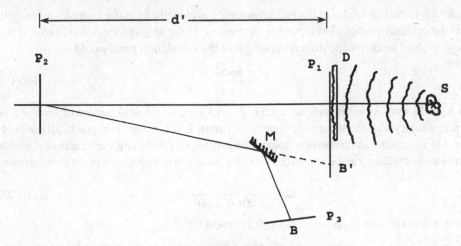

Fig. 10.7. Scheme of the lensless correlation system.

reconstructed wave front onto output plane P_3. This optical system is linear in irradiance. For the sake of decreasing and increasing interferences of different reflected wavelengths from the sequence of every thin reflecting plane of the volume hologram, only a single reconstructing wavelength that satisfies Bragg's law will be strongly reflected, and it is theoretically equal to the recording beam.

10.2.3.3 Discussion

The effects of the emulsion shrinkage: Although the single reconstructing wavelength is theoretically equal to λ_1, experimental observation is totally different. The chemical processing of photographic emulsion usually results in shrinkage of the emulsion thickness, which is ~15% and this will change the single reconstructing wavelength, that is, the variable z should be modified to kz, where k is the shrinkage factor, $k < 1$. Equivalently,

$$\lambda(d + kz) = k\lambda_1 \left(\frac{d}{k} + z \right) = \lambda_2(d' + z). \tag{10.16}$$

This expression implies that the shrunken filter's behavior seems to be that of a filter recorded with wavelength λ_2 at a distance of d' without shrinkage. As an example, suppose that the VFHF is recorded with a He–Ne laser (632.8 nm); the single reconstructing wavelength observed, λ_2, would be green (533 nm). Therefore, to get the exact optical correlation result in the system of Fig. 10.6, the distance d' should be modified to

$$d' = \frac{\lambda_1}{\lambda_2} d = \frac{d}{k}. \tag{10.17}$$

The effect of the wavelength and angular sensitivities of the VFHF: For Bragg incidence, according to coupled-wave theory, the diffraction efficiency of a phase reflection hologram can be 100% and the diffraction efficiency will decrease as the wavelength changes. Thus the reconstructing wavelength has a limited bandwidth, which will lead to chromatic dispersion and extend the correlation peak to a certain length. The wavelength sensitivity of the reflection hologram is theoretically

$$\Delta\lambda = \frac{3.5\lambda_2^2}{\pi n_0 T \sin\theta}, \tag{10.18}$$

where $\Delta\lambda$ is the bandwidth of the reconstructing wavelength, λ_2 is the central wavelength, n_0 is the refractive index of holographic medium, T is the medium thickness, and θ is the Bragg angle. The chromatic dispersion length of the correlation peak would be

$$L = \frac{h_0 \Delta\lambda}{\lambda_2}. \tag{10.19}$$

Let $h_0 = 55$ mm, $\lambda_2 = 530$ nm, $n_0 = 1.52$, $T = 15$ μm, and $\theta = 80°$; we will have $\Delta\lambda = 14$ nm, which yields the chromatic dispersion length $L = 1.4$ mm. It is practically acceptable. On the other hand, reflection holograms exhibit very little angular sensitivity, which decreases as the Bragg angle approaches 90°. Because of a cosine factor in the denominator,

$$2\delta_0 = \frac{3.5\lambda_2}{\pi n_0 T \cos\theta}, \tag{10.20}$$

with the same chosen parameters, Eq. (10.20) yields

$$2\delta_0 = 8°.$$

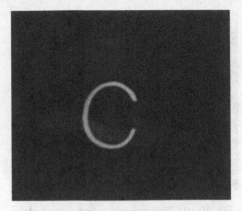

Fig. 10.8. Detected signal.

Fig. 10.9. Input object signal.

So the high degree of wavelength sensitivity permits illumination of the reflection VFHF with a white-light source. A low degree of angular sensitivity creates a proper space invariance for the system.

10.2.3.4 Experimental results

In experimental demonstrations, a He–Ne laser is used in the VFHF, making the system shown in Fig. 10.6. A transparency, as shown in Fig. 10.8, is chosen to be the detecting signal and is placed at plane P_1. The distance between P_1 and P_2 is set to be 360 mm, and the converging point of the reference beam is 55 mm away from the origin of the coordinates in plane P_3. The angle between the reference wave and the object wave within the emulsion is larger than 160°. The photographic plate used is a type D3 holographic plate made in China, which is photosensitive to the wavelength of 633 nm and is suitable for making a phase-volume hologram with high diffraction efficiency by means of a special plate process. After holographic recording and photochemical treatment we get the desired VFHF with higher diffraction. Then the VFHF is placed at plane P_2, as shown in Fig. 10.7, and the object signal, as shown in Fig. 10.9, is placed at P_1, which is 428 mm away from P_2, and illuminated by a 150-W high-pressure arc lamp. The distance d' is 1 1/5 times the distance

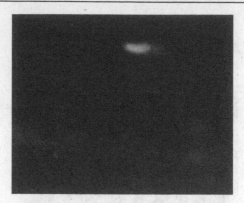

Fig. 10.10. Experimental correlation result.

because of a change in the reconstructed wavelength due to the emulsion shrinkage. The experimental optical correlation result observed at plane P_3 is shown in Fig. 10.10. It is obviously true that although the output correlation peak extends to a certain length because of the chromatic dispersion, it causes little harm to the practical pattern recognition. The experimental results are satisfactory.

10.2.4 Lensless intensity correlator with high discrimination

Because of the incoherent nature of the intensity correlation with a FHF, the input signals should be only real nonnegative functions, which causes a certain extension in the autocorrelation peak of such an object. Thus the discrimination of an optical intensity correlator is, in general, lower than that of an optical amplitude correlator. This problem becomes more serious for the detection of special kinds of objects, of which the Fourier spectra contain large zero-order components, such as an airplane or a tank. To improve the discrimination of the intensity correlator, a technique of optical differentiation preprocessing of the input object can be used. The differentiation preprocessing will eliminate or relatively reduce the zero-order and some lower-frequency components in the Fourier spectrum of the object. Thus the combination of the FHF intensity correlation system with the optical differentiation preprocessing system may not only keep the advantage of flexibility in the operation, but may also offer high discrimination of the object.

In order to realize an optical intensity correlation of the differential images, a spatial-differentiation filter must be first prepared for optical differentiation preprocessing of the input and the reference images. The scheme for fabricating the differentiation filter is shown in Fig. 10.11. A double-exposure holographic technique with a plane wave and a converging spherical wave is used. The plane wave is perpendicular to the photosensitive plate at the first exposure and changes by a small angle before the second exposure. The two exposure times are the same. After a suitable photochemical process, the 2-D differentiation filter can also be fabricated. With the differentiation filter, we can build up the optical intensity correlator of a differential image with the FHF to improve the discrimination of the optical pattern recognition system.

For an experimental demonstration, we chose the image shown in Fig. 10.12a as the original input-image. After optical differentiation preprocessing, the differential image is obtained, as shown in Fig. 10.12b. This differential image is used as the input to the FHF

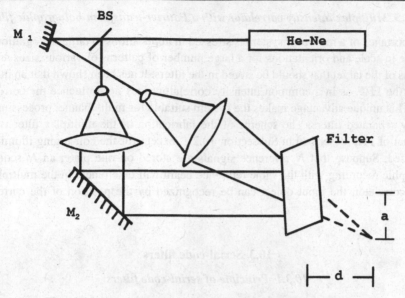

Fig. 10.11. Scheme for recording the spatial-differentiation filter.

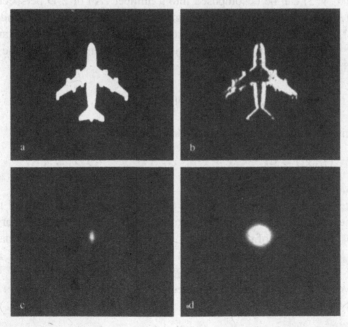

Fig. 10.12. (a) The original image; (b) the differential image; (c) the correlation result of the differential image; (d) the correlation result of the original input image.

optical intensity correlation system, and the autocorrelation result of Fig. 10.12b is shown in Fig. 10.12c. For comparison, Fig. 10.12d gives the intensity autocorrelation result of the original input image. We can see that the detectability of the correlation peak of Fig. 10.12c is evidently improved.

10.2.5 Multiplex intensity correlator with a Fourier-transform holographic filter

The importance of a multiplex system is stressed in applications to target recognition with a change in scale and orientations for a large number of patterns of various sizes and orientations of the target that should be stored in the filters. It has been shown that an in-plane shift of the FHF, as in a common intensity correlator, does not influence the correlation result. This unique advantage makes the system suitable for multichannel processing with spatially separated filters. The scheme of the fabrication of the multiplex filter is similar to that of FHF described in Subsection 10.2.1, except for the converging illuminating beam used. Suppose that N reference signals are stored on one plate; an N sequential holographic recording with the same reference beam will take place. In the multiplex intensity correlator, the input object can be recognized by the position of the correlation peak.

10.3 Serial-code filters

10.3.1 Principle of serial-code filters

An image can be regarded as a vector in hyperspace. For example, 128×128 picture may be considered as a vector in $M = 128 \times 128 = 16{,}384$-dimensional space. A certain 3-D object $O(x, y, z)$ can be decomposed into a number (N) of 2-D perspective images $f_i(x, y), i = 1, 2, \ldots, N$. The aim is to design such a bank of (K) independent vectors $g_k(x, y), k = 1, 2, \ldots, K$, that the inner products of any $f_i(x, y)$ and $g_k(x, y)$ are constant, i.e., the inner product of $f_i(x, y)$ and $g_k(x, y)$ is independent of i. Here the term inner product is defined as

$$\langle f, g \rangle = \int f(x, y) g^*(x, y) \, dx \, dy. \tag{10.21}$$

So the required g's should satisfy the following equations:

$$\langle f_i, g_k \rangle = C_k, \qquad i = 1, 2, \ldots, N; \qquad k = 1, 2, \ldots, K. \tag{10.22}$$

These vectors, g's, are the so-called SCF's.

For an image obtained from an unknown object, we correlate it with all these filters g's one by one, each time we have a datum of correlation. If all the values are the same, (i.e., they constitute a series of number 3, 3, ..., 3), then we say, Yes – the image is of the certain target; if the series of numbers is not the same but is in an irregular manner, we say No.

The filter vector $\{g_k(x, y)\}$ can be obtained by SDF method. Let $g_k(x, y)$ be a linear combination of f's:

$$g_1(x, y) = \sum_{i=1}^{N} a_i f_i(x, y). \tag{10.23}$$

According to Eqs. (10.22) the N unknown coefficients a_i $(i = 1, 2, \ldots, N)$ can be solved by the following N equations:

$$\langle f_i(x, y), g_1(x, y) \rangle = C, \qquad i = 1, 2, \ldots, N. \tag{10.24}$$

Then $g_1(x, y)$ is obtained.

The way in which we obtain the other independent vectors, $g_2(x, y)$, $g_3(x, y)$, etc., is quite different from that of the LAT. $g_2(x, y)$ should satisfy the same Eq. (10.4) and another equation, which is required by the independence between g_2 and g_1, i.e.,

$$\langle g_2(x, y), f_i(x, y) \rangle = C, \qquad i = 1, 2, \ldots, N,$$

$$\langle g_2(x, y), g_1(x, y) \rangle = 0. \tag{10.25}$$

The number of equations is now $N + 1$ (N is usually not more than 10^3), but $g_2(x, y)$, as a picture comprising millions of pixels, has millions of variables to be chosen freely; thus $g_2(x, y)$ can be solved from Eq. (10.5) by the selection of a proper set of $N + 1$ parameters.

Similarly, $g_3(x, y)$ is solved by

$$\langle g_3(x, y), f_i(x, y) \rangle = C, \qquad i = 1, 2, \ldots, N,$$

$$\langle g_3(x, y), g_1(x, y) \rangle = 0,$$

$$\langle g_3(x, y), g_2(x, y) \rangle = 0. \tag{10.26}$$

In such a way we obtain the complete set of SCF's $g_1(x, y)$, $g_2(x, y)$, ..., $g_K(x, y)$.

In the relevant optical system for target recognition, $g_k(k = 1, 2, \ldots, K)$ as correlators are placed in the filtering plane. With any input image it produces a light intensity value of correlation at the corresponding point in the output (correlation) plane. In the case in which there are a lot of pictures in the input plane, it is only at these points in the correlation plane that we can get a series of intensity values I_k as the filters are taken from 1 to K; these points correspond to the positions in the input plane where pictures of the specified target are located.

The intensity in the output plane is digitized and summed up into two buffers. The first buffer sums up the intensity values I_k and the second sums up I_k^2. Then the mean $\langle I \rangle$ and the standard deviation σ_I are determined and we have the ratio:

$$r = \frac{\langle I \rangle}{\sigma_I}. \tag{10.27}$$

The value of r represents the degree of how nearly the series of intensity data is lying in a horizontal line. Thus there will be a high peak of r ($r \gg 1$) at the position that corresponds to the image of a specified target. At other positions in the output plane that correspond to the input images of objects other than the specified targets, the light intensity obeys near-speckle-like statistics, so that $\langle I \rangle \approx \sigma_I$ and $r \approx 1$, much less than the peak. It was shown that searching for values of r above a certain threshold is equivalent to optimal linear detection of the constant-amplitude condition [16].

Generally speaking, it is not necessary that the series of intensity values, which specify the given target, be constants. We may choose an arbitrary series of values for a certain target, e.g., 1, 4, 1, 4, 1, 4 instead of 3, 3, 3, 3, 3, 3. This series of numbers plays a coding role in specifying an object, and this is what the term SCF means.

The error detection probability of this method can be considered in a rather simple way. Let L be the number of distinguishable gray levels of the output intensity I. If only one filter is used and only one point in the output plane is considered, it is obvious that, from a statistical point of view, of all possible objects there would be $1/L$ that fits the code number of this filter. Thus, when all K filters are used in sequence and all M_1 output points are

Fig. 10.13. Input scene containing both target and nontarget images.

considered, there are M_1/L^K of all possible objects that can fit the complete serial codings of the filters. If the total number of all possible input images is N_1 and from them N images are targets, the error detection probability E can be expressed.

For example, when $M_1 = 512 \times 512$, $N_1 = 10^4$, $N = 10^3$, $L = 8$, and $K = 15$, with

$$E = \frac{M_1(N_1 - N)}{L^K}, \qquad (10.28)$$

we will have $E \approx 1.0 \times 10^{-4}$. From Eq. (10.28) we know that E decreases dramatically with increasing K. Hence the SCF's exhibit excellent capability.

10.3.2 Digital simulations

The image of a 3-D object viewed from different directions depends on two parameters of the viewing angle, i.e., the azimuth angle θ and the zenith angle ϕ. We choose a toy tank as the target and deal with pattern recognition over images of 600 different views, which are selected according to the proper method [17]. These training images are digitized by an 1010-M PDS microdensitometer, and each image has 128×128 pixels. A set of $K = 15$ filters are designed to recognize these 600 views. To demonstrate 3-D recognition capability, we use the input scene of Fig. 10.13, which contains three images of different views of a toy tank, as well as two images of a toy automobile and an image of a toy airplane that represent false targets. The tank images in the top line are members taken from the original training set of 600 images. The optical correlation between this scene and each one of the 15 filters is completed by digital simulation. The values of r for different points in the correlation plane are plotted as shown in Fig. 10.14. It is clearly evident that there are three peaks and that their positions correspond to the locations of the three target images of Fig. 10.13. The false targets can be easily eliminated if a suitable threshold is chosen.

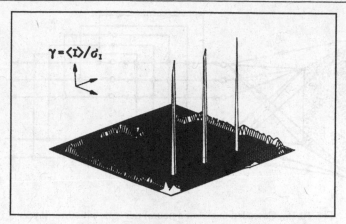

Fig. 10.14. Plot of the ratio $r = \langle I \rangle / \sigma_I$ for the input scene of Fig. 10.13 as the 15 filters are used.

10.4 Cascaded model of neural networks suitable for optical implementation

It is shown in Subsection 10.3.1 that the SCF's are easier to synthesize and that the number of required filters can be reduced greatly. But SCF's cannot guarantee that images that are outside the training set or that are partially hidden will be recognized.

In recent years, attention has been particularly paid to research on pattern recognition with the ability of error tolerance by means of artificial neural networks. Because of the massive parallelism and high information throughput of optics, research on artificial neural networks implemented optically has been attractive since the very beginning. However, the association ability of the Hopfield model based on the minimum Hamming distance is not provided with spatial rotation invariance. Therefore it is difficult to simply apply the Hopfield model to the classification and recognition of multiple 3-D targets with arbitrary spatial orientations. Although the layered feedforward neural networks based on a back propagation (BP) learning algorithm may suffice, the learning process of BP algorithms [18] is very time consuming for complicated 3-D targets and is often trapped into a local minimum so that a satisfactory result cannot be obtained. One of the reasons for this is that projections of 3-D targets with complex shapes viewed from arbitrary directions are generally very different.

The cascaded model of neural networks can be used for error-tolerant pattern recognition of multiple 3-D targets with arbitrary spatial orientations. The error-tolerance abilities of this model include not only the associative recognition of partially shaded targets or targets blurred by random noise in terms of the minimum Hamming distance, but also the associative recognition of projections of targets viewed from directions other than those of the training sets in terms of spatial rotation invariance. The learning process of this model is simpler, and this model is suitable for optical implementation. Computer simulations have shown that this model has good abilities of associative pattern recognition for patterns both inside and outside the training sets, as well as for patterns of partially shaded targets.

10.4.1 Structure and features of the cascaded model

The cascaded model presented in this chapter consists of two independent neural networks that are cascaded. The output layer of the first network is also the input layer of the second

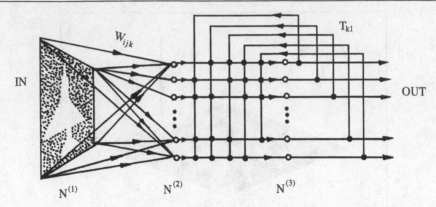

Fig. 10.15. Cascaded model of a neural network with three layers.

one, that is, the middle layer of the cascaded model. Hence it may also be called a three-layer cascaded neural network, as shown in Fig. 10.15. In Fig. 10.15, $N^{(1)}$ neurons of the input layer are usually arranged in the form of a 2-D matrix. The neuron states f_{ij} $[i = 1, 2, \ldots, I; j = 1, 2, \ldots, J; I \times J = N^{(1)}]$ may be $+1$ (represented by an open circle) or -1 (represented by a filled circle). Each set of $\{f_{ij}\}$ is considered to be an $N^{(1)}$-dimensional binary vector, and its 2-D matrix arrangement forms a binary projective image. The number of neurons in the middle layer is $N^{(2)}$. Generally speaking, $N^{(2)} \ll N^{(1)}$. The state of its neurons $\{C_k\}$ $[k = 1, 2, \ldots, N^{(2)}]$ may also be considered an $N^{(2)}$-dimensional vector, and C_k may adapt $+1$ or -1, as determined by summing the states of neurons in the input layer with weights W_{ijk} and by thresholding. W_{ijk} describes the interconnection weights from the neuron located in the ith row and the jth column of the input layer to the kth neuron in the middle layer. For an arbitrary projective image $\{f'_{ij}\}$ of a given target, $\{W_{ij}\}_k$ should make $\{C'_k\}$ the same binary vector $\{C_{mk}\}$, which is called the spatial-rotation-invariance code of the given target.

The number of neurons in the output layer is $N^{(3)}$. The network from the middle layer to the output layer is a Hopfield-type neural network. In general, $N^{(3)} = N^{(2)}$. For given spatial-rotation-invariance codes $\{C_{mk}\}$ $(m = 1, 2, \ldots, M)$ of M targets, the interconnection weights T_{kl} $[k, l = 1, 2, \ldots, N^{(2)}]$ may be found from these codes according to the Hebb rule.

From the above analysis it can be seen that, compared with a BP layered neural network, the cascaded model of neural networks has the following features:

(1) The learning process of each stage of the cascaded neural network can be accomplished separately according to the performance required by the whole system. It is evident that the learning process is simpler and has definite solutions. On the other hand, the BP algorithm treats layered neural networks as a whole system. Its learning process is more complex, time consuming, and has more indeterminate parameters (for example the number of neurons in the middle layers, layer number, gain parameters, and so on). In addition, the learning process is often trapped into a local minimum so that a convergent result cannot be reached.

(2) The first stage of the cascaded model is a feedforward-type network.

(3) The possibility of an optical implementation has been taken into account in the interconnection and learning algorithms of this cascaded model. For example, when

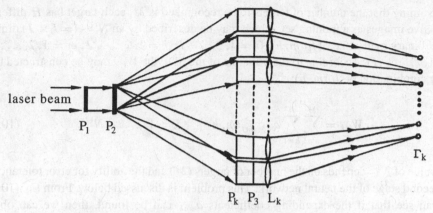

Fig. 10.16. Schematic diagram for the optical implementation of the interconnection weights W_{ijk}.

the configuration of the neurons is arranged as a matrix $\{f_{ij}\}$, the interconnection weights from the input layer to the middle layer W_{ijk} $[k = 1, 2, \ldots, N^{(2)}]$ can also be arranged as $N^{(2)}$ interconnection matrices $\{W_{ij}\}_k$.

The states of neurons in the middle layer may therefore be determined by the following summing and thresholding operation:

$$C_k = \text{sgn}\left(\sum_{i=1}^{I} \sum_{j=1}^{J} W_{ijk} f_{ij}\right) = \text{sgn}(\Gamma_k), \tag{10.29}$$

where $\text{sgn}(\Gamma_k)$ is the signum function.

The sum operation in Eq. (10.29) may be realized by an optical method, as shown in Fig. 10.16. In Fig. 10.16, P_1 is a transparency mask that is used to represent $\{f_{ij}\}$ through transparent pixels $(+1)$ and opaque pixels (-1) (in a real system, one may use an array of liquid-crystal switches or an optomagnetic spatial light modulator). P_2 is a beam splitter that can split a beam through P_1 into $N^{(2)}$ beams, P_k, that equal the light intensity. P_3 represents the $\{W_{ij}\}_k$ masks. L_k represents an array of convergence lenses. The light intensities in the focus will be proportional to

$$\Gamma_k = \sum_{i,j} W_{ijk} f_{ij}.$$

After thresholding the results are fed onto an optical neural network of a Hopfield model. Some experimental systems [19] similar to this have been reported.

10.4.2 Learning algorithm of the cascaded model of neural networks

10.4.2.1 Learning algorithm of an interconnection matrix $\{W_{ij}\}_k$ for the first stage of the model

As mentioned above, the aim of the first stage of the network is to implement spatial-rotation-invariance encoding for different projection images of a given set of targets. Their interconnection weights may be represented as $N^{(2)}$ interconnection matrices $\{W_{ij}\}_k$ $[k = 1, 2, \ldots, N^{(2)}]$. The function of $\{W_{ij}\}_k$ obviously is similar to that of the SDF that is usually employed in optical pattern recognition [20].

Assuming that the number of targets to be recognized is M, each target has H different projective images as a training set, which may be described by an $N^{(1)}$- $(= I \times J)$ dimensional binary matrix, i.e., $\{f_{ij}(m, h)\}$ $(i = 1, 2, \ldots, I; j = 1, 2, \ldots, J; m = 1, 2, \ldots, M; h = 1, 2, \ldots, H)$. According to the principle of the SDF the W_{ijk} may be constructed by a linear combination of $f_{ij}(m, h)$:

$$W_{ijk} = \sum_{m=1}^{M} \sum_{h=1}^{H} a_{mhk} f_{ij}(m, h) \qquad k = 1, 2, \ldots, N^{(2)}. \tag{10.30}$$

The value of $N^{(2)}$ depends on the number of targets (M) and the ability for error tolerance in the second stage of the neural network. This problem is discussed below. From Eq. (10.30) we can see that if the expanding coefficients a_{mhk} can be found, then we can obtain the W_{ijk}. To satisfy the requirement of spatial-rotation-invariance encoding, we can make

$$\sum_{i=1}^{I} \sum_{j=1}^{J} f_{ij}(m, h) W_{ijk} = C_{mk}, \qquad m = 1, 2, \ldots, M, \qquad k = 1, 2, \ldots, N^{(2)}, \tag{10.31}$$

where C_{mk} depends on only the given mth target and the kth neuron in the middle layer, no matter which projective image of the target would be. Its value may be $+1$ or -1. Therefore $\{C_{mk}\} = \{C_{m1}, C_{m2}, \ldots, C_{mN}\}$ is a set of spatial-rotation-invariance codes for the mth target. All the binary vectors $\{C_{mk}\}$ $(m = 1, 2, \ldots, M)$ form the storage vectors of the second stage of the neural network. The way to choose the value of C_{mk} is discussed in Subsection 10.4.2.2.

To find the expanding coefficients a_{mhk}, we take the inner product of one projection in the training set $\{f_{ij}(m', h')\}$ to the two sides of Eq. (10.30) and obtain

$$\sum_{m=1}^{M} \sum_{h=1}^{H} a_{mhk} \left[\sum_{i=1}^{I} \sum_{j=1}^{J} f_{ij}(m, h) f_{ij}(m', h') \right] = C_{m'k},$$

$$k = 1, 2, \ldots, N^{(2)}, \qquad m' = 1, 2, \ldots, M, \tag{10.32}$$

where the term in the brackets is an inner product (or overlap) between the hth projective vector of the mth target and the hth projective vector of the mth target. When the values of C_{mk} are chosen, Eq. (10.32) becomes a set of constant coefficient linear equations from which the a_{mhk} may be found. By substituting a_{mhk} into Eq. (10.30), we can obtain the corresponding W_{ijk} and make Eq. (10.31) hold.

When some noise is present or an input-image is not part of the training set, the output of the first stage of the neural network Γ_k [see Eq. (10.29)], in general, is not exactly ± 1. Although the first stage of the network may have a small capacity for error tolerance through thresholding, the target is not able to be recognized when the output codes Γ_k from the first stage have mistakes in their sign. To improve the ability for error tolerance of the whole cascaded network we employ a Hopfield-type network as the second stage with $C'_k = \text{sgn}(\Gamma_k)$ as its input. The state of the second stage of the network will converge to the feature codes $\{C_{mk}\}$ of the targets that are the nearest to $\{C'_k\}$, i.e., the correct result for error-tolerance pattern recognition.

Table 10.1. *Feature codes selected for each object*

Object	C_1	C_2	C_3	C_4	C_5	C_6	C_7	C_8	C_9	C_{10}	C_{11}	C_{12}	C_{13}	C_{14}	C_{15}	C_{16}
Liner	−1	−1	+1	+1	+1	−1	−1	−1	−1	+1	+1	+1	+1	−1	+1	−1
Fighter	+1	+1	−1	−1	+1	−1	−1	+1	−1	−1	+1	−1	+1	+1	+1	−1
Bomber	+1	−1	+1	−1	−1	+1	−1	−1	+1	−1	+1	+1	−1	+1	+1	−1
Rocket	−1	+1	−1	+1	−1	+1	+1	−1	−1	−1	−1	+1	+1	+1	+1	−1

10.4.2.2 Choice of feature codes

The interconnection matrix of the Hopfield model is given by

$$
T_{kl} = \begin{cases} \dfrac{1}{N^{(2)}} \displaystyle\sum_{m=1}^{M} C_{mk} C_{ml} & k \neq 1 \\ 0 & k = 1 \end{cases},
\tag{10.33}
$$

where $N^{(2)}$ is the number of neurons in the second stage of the network and M is the number of targets to be recognized and is also the number of the stored patterns (sets of codes) in the Hopfield network. The Hopfield model [21, 22] has been studied in more detail. To obtain a greater ability for error-tolerance, the choice of the stored patterns $\{C_{mk}\}$ should follow two rules; the first one is that the C_{mk} of every set of codes must adapt $+1$ or -1 with an equal probability, and the second is that the two sets of codes corresponding to different m must be orthogonalized, that is,

$$
\frac{1}{N^{(2)}} \sum_{k=1}^{N^{(2)}} C_{mk} C_{m'k} = \delta_{mm'}.
\tag{10.34}
$$

The codes that satisfy Eq. (10.34) are referred to as orthogonalized codes. Table 10.1 shows a choice of the feature codes for $M = 4$. The code length $N^{(2)}$ or the number of neurons in the middle layer depends on the requirement for error tolerance of the network. For a given number of targets M, the greater the $N^{(2)}$, the stronger the capacity for error tolerance. Computer simulations [23] have shown that when $N^{(2)} = M/0.25$ and orthogonalized codes are used, a Hopfield network can correctly recognize a set of input codes with 50% noise. Therefore the minimum value of $N^{(2)}$ is 16 when $M = 4$. A greater ability for error tolerance can be obtained if more neurons are used in the second stage of the neural network.

10.4.3 Gray-level compression of the mask

According to the above method, a bank of interconnection weights $\{W_{ijk}\}$ can be obtained; but the number of gray levels of $\{W_{ijk}\}$ calculated as such is up to 7000, which is too difficult to realize with an optical mask. According to the numerical distribution, the number of gray levels can be compressed down to 49 by a statistical method. In order to realize the bipolarity, positive and negative unipolar masks are made separately instead of a bipolar one. Then the computer-generated holograms of $\{W_{ijk}\}$ can be fabricated on a glass substrate with microelectronic techniques.

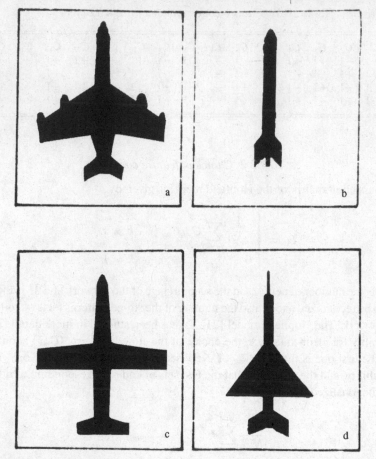

Fig. 10.17. Projective images of four aircraft at 0 orientation: (a), bomber; (b), rocket; (c), airliner; (d), fighter.

10.4.4 Property analysis of the model

As an example, taking four classes of aircraft (bomber, rocket, airliner, and fighter, as shown in Fig. 10.17) as the targets to be recognized, we have performed computer simulations for the cascaded neural network.

In the right-hand system (x_1, x_2, x_3), $(\theta_1, \theta_2, \theta_3)$ describe angles at which a target moves around the x_1, x_2, and x_3 axes, respectively. Figure 10.17 shows the projective images of the classes of aircraft at $\theta_1 = \theta_2 = \theta_3 = 0°$. $N^{(1)} = 10,000$ neurons are arranged as a square matrix, that is, $I = J = 100$. Figure 10.18 shows, as an example, one set of projective images after the targets were rotated. Each target produced 252 projective samples as its training set, according to the following method:

(1) Starting from the zero orientation, as shown in Fig. 10.17, keeping $\theta_1 = \theta_2 = 0°$, and changing θ_3 in steps of $10°$, we can obtain 36 projective samples for each target.
(2) Let $\theta_2 = 30°$ and repeat the operation of step (1) above for $\theta_1 = 30°$, $45°$, and $60°$, respectively; we can obtain 108 projective samples for each target.
(3) Let $\theta_2 = 45°$ and repeat the operation of step (2) above; we can obtain another 108 projective samples for each target.

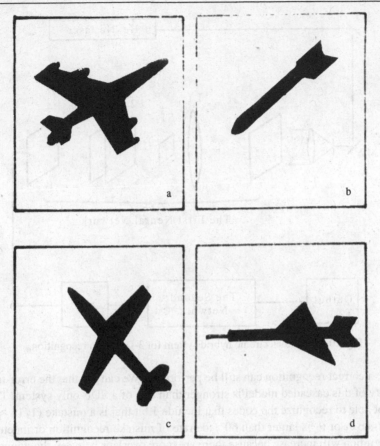

Fig. 10.18. Example of the projective images of four aircraft after spatial-rotation:
(a), bomber; (b), rocket; (c), airliner; (d), fighter.

By substituting the above binary projective samples $\{f_{ij}(m, h)\}$ ($m = 1, 2, 3, 4; h = 1, 2, \ldots, 252; i, j = 1, 2, \ldots, 100$) into Eqs. (10.32), (10.30), and (10.33), we can find the interconnection matrices $\{W_{ij}\}_k$ and $\{T_{kl}\}$ of the cascaded neural network. According to Eq. (10.29) and the iteration operation of the Hopfield model, computer simulations have been performed. When an input pattern is an arbitrary projective image of a given target, the recognized result is determined by the output code that corresponds to the stable state of the output layer. If the output code is one of four sets of codes listed in Table 10.1, the input pattern is that aircraft that corresponds to this set of codes, and vice versa.

The results of the computer simulations are as follows:

(1) It is expected that every one of the 1008 projective samples above is correctly recognized by this cascaded network.

(2) The simulation can also be performed correctly to recognize every projective view of the four targets rotated at an arbitrary angle θ_3 (for $\theta_1, \theta_2 \leq 60°$) that is outside the training sets. Therefore it shows that this cascaded model of a neural network has a good out-of-plane rotation invariance.

(3) In the case of partially shaded targets, when a shaded part of a projection is less than 30%, the rate of mistake-code outputs from the middle layer is up to 12.5%. In this

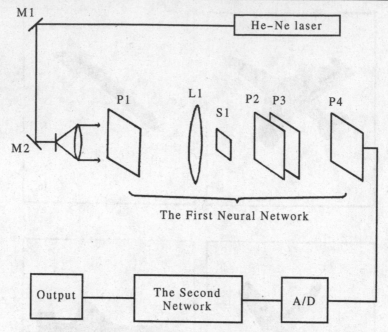

Fig. 10.19. Optoelectronic hybrid system for 3-D pattern recognition.

case, a correct recognition can still be performed. We can see that the error-tolerance ability of this cascaded model is stronger than that of a SDF-only system. The latter is not able to recognize the codes that include 1 bit that is a mistake ($1/16 \approx 6\%$).

When θ_1 or θ_2 is larger than $60°$, the rate of mistake recognition or indeterminate recognition will increase because there are more overlaps between the wings and the fuselage of the aircraft so that the differences among the four aircraft are small.

From the above analysis, the cascaded model of neural network proposed can be used to recognize given targets with arbitrary orientation according to their projective images. This model has the abilities of pattern recognition with spatial rotation invariance and error tolerance in the cases in which noise is present and there is a partially shaded input.

10.4.5 Optoelectronic hybrid system for three-dimensional pattern recognition

Figure 10.19 is an optoelectronic hybrid system [24]. The first neural network is implemented optically, and the second is simulated in a microcomputer, so it has the advantages of both the massive parallelism of the optical system and the flexibility of the computer. In Fig. 10.19 P1 is the input image, i.e., the input layer of the first network. L1 is the imaging lens, and S1 is an equal-light-intensity splitter. When a beam of light is incident upon input object P1, 4×4 images with equal intensity are obtained on P2. The computer-generated holograms of $\{W_{ijk}\}$ are used as the interconnection masks of the first network and are placed on P2. Just after the interconnection mask, we obtain the products of the 16 input images and the 16 interconnection weight functions. P3 is a lens array, and P4 is a detector array. The 16 products are converged by P3. The focus intensities represent 16 inner products, which are then received by a detector array and transformed by an analog-to-digital converter into a set of codes, which are the invariant feature codes of the object.

Thus the first stage of the network performs a heteroassociation that encodes arbitrary projection images into a set of feature codes $\{C'_j\}$.

In a microcomputer, $\{C'_j\}$ is used as the input pattern of the second network, which then iterates with interconnection matrix T_{ij} and converges to one of the M patterns of feature codes $\{C_{mn}\}$.

When a projection image (including images outside the training set) is input, the system can give a recognition result rapidly. The experimental results show that the system can correctly recognize more than 92% of the projection images of the four kinds of aircraft. At some perspective views, the differences among the four kinds of aircraft become small. In these cases, false alarms may be caused that are hard to avoid, even for human eyes.

10.5 Conclusion

Conventional pattern recognition is suitable for only in-plane rotation and scale-invariance recognition. In this chapter, we emphasize that the new approach of the FHF for pattern recognition, which we have studied in detail, offers convenience in certain applications for its real-time operation character and little effect of a displacement error in its alignment. The FHF has also been used in some intensity correlators to perform pattern recognition. An optoelectronic hybrid three-layer neural network is presented for pattern recognition of multiplex 3-D objects that rotate arbitrarily in space. The three-layer neural network is composed of two stages of different neural networks and has strong error and noise tolerances. The first layer is a heteroassociative network that encodes the input image into a set of feature codes; the SCF's are used as its interconnection weights. It is implemented optically. The second layer is an autoassociative network that provides error-tolerance recognition results by iteration; it is simulated by a microcomputer. The problems of optical implementation, such as the gray-level compression and the bipolarity, are also discussed.

This system combines optical recognition with neural networks. It has the advantages of massive parallelism, high speed, a large information capacity of optics, and the flexibility of the computer. So it is a practical for 3-D pattern recognition.

References

[1] A. B. VanderLugt, "Signal detection by complex spatial filtering," IEEE Trans. Inf. Theory **IT-10**, 139 (1964).

[2] J. L. Horner and P. D. Gianino, "Phase-only matched-filtering," Appl. Opt. **26**, 812–816 (1984).

[3] G. Mu, X. Wang, and Z. Wang, "Amplitude-compensated matched-filtering," Appl. Opt. **27**, 3461–3466 (1988).

[4] A. W. Lohman, "Matched-filtering with self-luminous objects," Appl. Opt. **7**, 561–566 (1968).

[5] G. Mu, X. Wang, and Z. Wang, "A new type of holographic encoding filter for correlation – a lensless intensity correlator," in *Holography Applications*, J. Ke and R. J. Pryputniewicz, eds., Proc. SPIE **673**, 546–549 (1987).

[6] Y.-N. Hsu and H. H. Arsenault, "Optical pattern recognition using circular harmonic expansion," Appl. Opt. **21**, 4016–4019 (1982).

[7] N. George, J. T. Thomason, and A. Spindel, "Photodetector light pattern detector," U.S. patent 3,689,772 (1972).

[8] D. Casasent and D. Psaltis, "Scale invariant optical correlation using Mellin transforms," Opt. Commun. **17**, 59–62 (1976).

[9] H. J. Caulfield, "Linear combination of filters for character recognition: a unified treatment," Appl. Opt. **19**, 3877–3879 (1980).

[10] G. F. Schils and D. W. Sweeney, "An optical processor for recognition of 3-D targets viewed from any direction," J. Opt. Soc. Am. A **5**, 1308–1310 (1988).

[11] S. Yin, C. Lu, and G. Mu, "3-D target recognition by using serial-code-filters," Optik **82**, 129–131 (1989).

[12] G. Mu, Z. Wang, D. Chen, and F. Wu, "A new technique for pattern recognition using Fresnel hologram and extended source," Optik **75**, 97–98 (1987).

[13] G. Mu, Z. Wang, K. Wang, and X. Wang, "Lensless optical pattern recognition using VFHF and natural source," Optik **76**, 139–141 (1987).

[14] G. Mu., Z. Wang, and W. Chen, "Optical intensity correlator with high discrimination," Optik **84**, 23–27 (1990).

[15] G. Mu, Z. Wang, and Z. Zhang, "Multiplex intensity correlator with Fourier-transform holographic filters," Chinese Lasers **81**, 76 (1989).

[16] D. W. Sweeney, E. Ochoa, and G. F. Schils, "Experimental use of iteratively designed rotation-invariant correlation filters," Appl. Opt. **26**, 3458–3465 (1987).

[17] D. Casasent, S. A. Liebowiz, and A. Mahalanobis, "Parameter selection for iconic and symbolic pattern recognition filters" in *Optical and Digital Pattern Recognition*, H. Liu and P. S. Schenker, eds., Proc. SPIE **754**, 284–303 (1987).

[18] D. E. Rumelhart, G. E. Hinton, and R. J. Williams, in *Parallel Distributed Processing*, D. E. Rumelhart and J. L. McCleland, eds. (MIT, Cambridge, MA, 1986).

[19] Y. X. Zhang, "Optical neural network and its application to pattern recognition," *Optonics and Lasers* **1**, 130 (1990).

[20] D. Casasent, "Unified synthetic discrimination function computational formulation," Appl. Opt. **23**, 1620–1627 (1984).

[21] D. J. Amit, H. Gutfreund, and H. Sampolinsky, "Statistical mechanics of neural networks near saturation," Ann. Phys. **173**, 30–67 (1987).

[22] W. A. Little and G. L. Siiaw, "Analytic study of the memory storage capacity of a neural network." Math. Biosci. **39**, 281 (1978).

[23] G. G. Mu and Y. X. Zhang, "Pattern recognition with multiple layer optical neural network," Pattern Recognition Artif. Intell. **3**, 53–57 (1989).

[24] Y. Sun, Y. X. Zhang, X. P. Yang, and G. G. Mu, "Optoelectronic network for pattern recognization with 100×100 units," Opt. Memory Neural Networks **2**, 1011–1015 (1993).

11

Applications of photorefractive devices in optical pattern recognition

Xiangyang Yang

11.1 Introduction

The possibility of writing dynamic holograms in photorefractive materials was proposed more than 20 years ago [1, 2]. With recent advances in the growth and preparation of various photorefractive materials, along with the maturation of associated device technologies such as spatial light modulators (SLM's) and detector arrays, the realization of working optical systems is becoming feasible. Recently, the interest in photorefractive-material-based optical systems has reemerged and many applications have been proposed and demonstrated [3–6].

The capability of recording efficient holograms with relatively low intensity levels without the need for chemical or thermal processing is a unique property of photorefractive materials that holds promise for a variety of optical pattern recognition tasks. A typical optical pattern recognition system performs operations of data (image) acquisition, feature extraction, and classification or identification. Photorefractive materials can be utilized in each of these operations. In the image acquisition, photorefractive materials can be used as the medium for spatial light modulation [7]. In the feature extraction phase, photorefractive nonlinear phenomena can be used to amplify certain image features selectively [8, 9]. The most common applications are in the classification and identification phase in which real-time optical correlators are constructed with photorefractive media [10, 11]. Volume holographic storage in these systems offers unique capabilities such as parallel page readout, instant access, and multichannel operations.

In this chapter, various applications of photorefractive materials in optical pattern recognition are reviewed. We begin our discussion in Section 11.2 with a brief introduction to the photorefractive phenomena and operations of two-wave and four-wave mixing. Then various photorefractive-material-based optical correlators are introduced and their performance potential as well as limitations are discussed. The application of photorefractive materials in novelty filters, in rf signal processing, and in the implementation of neural-network-based optical pattern recognition systems are described in the following sections.

11.2 Fundamentals of the photorefractive effect

11.2.1 Photorefractive effect

The photorefractive effect is a phenomenon in which the local refractive index of a medium is changed by a spatial variation of the light intensity. In this section, the fundamental of such an effect is briefly explained in simple terms. A thorough analysis on the photorefractive

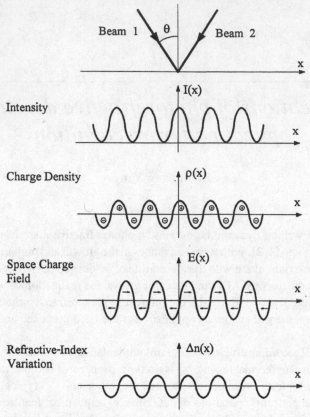

Fig. 11.1. Refractive-index grating produced by interference fringes in a photorefractive medium.

effect is beyond the scope and intention of this chapter, and interested readers are referred to Refs. 12 and 13.

We consider a simple model that is useful for explaining most of the observed photorefractive phenomena. Photorefractive materials generally contain donors and acceptors that arise from certain types of impurities of imperfections. The acceptors usually do not directly participate in the photorefractive effect, whereas the donor impurities can be ionized by absorbing photons. On light illumination, donors are ionized, by which the electrons are generated in the conduction band. By drift and diffusion transport mechanisms, these charge carriers are swept into the dark regions in the medium to be trapped. As a consequence, a space charge field is built up that induces the change in the refractive index.

We now consider the incidence of two coherent plane waves with equal amplitude interfering within the photorefractive medium, as illustrated in Fig. 11.1. The intensity of the interference fringes is given by

$$I(x) = I_0 \left[1 + \cos\left(\frac{2\pi x}{\Lambda} \right) \right], \tag{11.1}$$

where I_0 is the average intensity, $\Lambda = \lambda / \sin\theta$ is the period of the fringes, and λ is the wavelength of the light beams. In the bright regions close to the maximum intensity, photoionized charges are generated by the absorption of photons. These charge carriers diffuse away from the bright region and leave behind positively charged donor impurities. If these charge carriers are trapped in the dark region, they will remain there because there is almost no light to reexcite them. This leads to the charge separation, as shown in Fig. 11.1. It is seen that because of the periodic intensity distribution within the photorefractive medium, a space charge is created with the dark regions negatively charged and bright regions positively charged. The space charge continues building up until the diffusion current is counterbalanced by the drift current. The space charges produce a space charge field that, as depicted in Fig. 11.1, is shifted in space by a quarter of period (or $\pi/2$ in phase) relative to the intensity pattern. Owing to the Pockel effect, the space charge field induces a change in the refractive index, as given by [14]

$$\Delta n(t) = \Delta n_{\mathrm{sat}} \int_0^t \frac{A_1 A_2^* \exp[(t' - t)/\tau]\,\mathrm{d}t'}{I_0}, \tag{11.2}$$

where A_1 and A_2 are the amplitudes of the two incident waves, the asterisk denotes a phase conjugate, Δn_{sat} is the saturation index amplitude, and τ is the time constant. Both Δn_{sat} and τ are material-dependent parameters.

In general, the two beams can carry spatial information (e.g., images) with them. The interaction of the two beams then generates a volume index hologram in the refractive medium. While an optical beam propagates through the medium, it undergoes Bragg scattering by the volume hologram. If the Bragg scatterings are perfectly phase matched, a strong diffraction beam will reconstruct the original spatial information [13]. The formation and the diffraction of the dynamic holograms within the photorefractive medium can be explained by nonlinear optical wave mixing.

11.2.2 Two-wave mixing and four-wave mixing

There are two generic configurations for optical pattern recognition systems that use photorefractive materials, namely, two-wave mixing and four-wave mixing. Depending on the coherence between the read and the write beams, the four-wave-mixing configurations can be further classified into degenerate four-wave mixing and nondegenerate four-wave mixing. A brief discussion on these configurations is given next.

11.2.2.1 Two-wave mixing

In the two-wave-mixing configuration, as shown in Fig. 11.2, two coherent beams intersect in a photorefractive medium and create an index grating. The Bragg scattering involved in two-wave mixing is very similar to the readout process in holography. If we designate beam A_1 as the object beam and beam A_2 as the reference beam, the variation of photorefractive

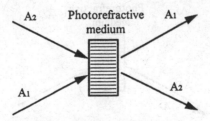

Fig. 11.2. Two-wave mixing in a photorefractive medium.

index of the hologram can be written as

$$\Delta n(x) \propto A_1^* A_2 \exp(-i\mathbf{K} \cdot \mathbf{r}) + A_1 A_2^* \exp(i\mathbf{K} \cdot \mathbf{r}), \tag{11.3}$$

where

$$\mathbf{K} = \mathbf{k}_2 - \mathbf{k}_1, \tag{11.4}$$

$\mathbf{k}_1$ and $\mathbf{k}_2$ are wave vectors of the two beams, respectively, and $\Delta n(x)$ is the refractive-index variation. When such a hologram is illuminated by the reference beam $A_2 \exp(-i\mathbf{k}_2 \cdot \mathbf{r})$, the diffracted beam is given by

$$o(x) = \eta A_1 A_2^* A_2 \exp(-i\mathbf{k}_1 \cdot \mathbf{r}), \tag{11.5}$$

where η is the diffraction efficiency. Note that the phase of A_2 cancels out and the diffracted beam is the reconstruction of the object beam $A_1 \exp(-i\mathbf{k}_1 \cdot \mathbf{r})$. Similarly, the reference beam A_2 can be reconstructed by the illumination of the hologram with object beam A_1.

In addition to a holographic analogy, two-wave mixing in most photorefractive crystals exhibits amplification, which is a unique feature not available in conventional holography. This occurs most efficiently in crystals in which the dynamic photorefractive index grating is 90° out of phase with respect to the intensity interference grating that produces it (see Fig. 11.1). The energy exchange is unidirectional, with the direction of the energy flow determined by the crystal parameters such as the crystal orientation and the sign of the photoionized charge carriers. Customarily the beam that loses energy is labeled as pump beam and the beam that becomes amplified as probe beam. Because the light energy is coupled from one beam to another, two-wave mixing is also known as two-beam coupling.

11.2.2.2 Four-wave mixing

In the four-wave-mixing configuration, two coherent beams write an index hologram and a third beam reads the hologram, creating the fourth (i.e., output) beam by diffraction, as illustrated in Fig. 11.3. To satisfy the Bragg condition, the third (read) beam must be counterpropagating relative to one of the two writing beams. If the read beam has the same wavelengths as the writing beams, the configuration is called degenerate four-wave mixing, whereas if the wavelengths of the read beam and the write beams are different, it is called nondegenerate four-wave mixing. Although degenerate four-wave mixing has been used in most of the applications demonstrated so far, nondegenerate four-wave mixing may be used in some cases in which the nondestructive reading of the hologram is required. This can be achieved by choosing a reading wavelength beyond the spectral response range of the photorefractive medium.

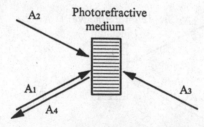

Fig. 11.3. Four-wave mixing in a photorefractive medium.

The index hologram generated by the two writing beams A_1 and A_2 are given by expression (11.3). When this hologram is illuminated by a counterpropagating reading beam $A_3 \exp(i\mathbf{k}_2 \cdot \mathbf{r})$, the diffraction is

$$o(x) = \eta A_1^* A_2 A_3 \exp(i\mathbf{k}_1 \cdot \mathbf{r}). \qquad (11.6)$$

It is seen that the diffracted beam counterpropagates with respect to the writing beam A_1. If both writing beam A_2 and reading beam A_3 are plane waves, the diffraction is a time-reversed replica of the beam A_1. Therefore four-wave mixing provides a convenient way for generating phase-conjugate waves. In general, however, all three waves can carry spatially modulated signals, and the amplitude of the diffracted beam represents the multiplication of these three images. If the diffracted beam is suitably manipulated, many image processing operations can be accomplished.

Beams A_1 and A_2 write a transmission hologram in the photorefractive medium. If beams A_1 and A_3 are used in the writing process, a reflection hologram is formed, which can be read with beam A_2, and it generates the same diffraction beam, as given by Eq. (11.6). The transmission and the reflection modes have different response times and diffraction efficiencies and therefore are suitable for different applications [15].

11.2.3 Multiplexing schemes

Multiple images or filters can be stored on a single piece of photorefractive medium by means of multiplexing. Once these images or filters are stored, they can be retrieved and serve as a library of reference images for pattern recognition. Because of the parallel readout and fast access capabilities of the holographic storage, an input can be compared with all the stored reference images at a very high speed. The three most commonly used multiplexing schemes in volume holographic storage are angular multiplexing [3, 16], wavelength multiplexing [10, 17], and phase-code multiplexing [18, 19]. All three multiplexing options are based on the Bragg-selective readout of thick holograms [20].

In the angular multiplexing scheme, the address of each image is represented by the incidence angle of the reference beam. To change the angle of the reference beam, a mirror mounted on a rotating step motor can be utilized [16]. For the rapid access of all the stored images, acousto-optic cells can be used to deflect the reference beam [3]. It should be noted that two acousto-optic cells must be used to compensate for the Doppler shift in frequency.

For wavelength multiplexing, both the object and the reference beams are fixed and only their wavelength is changed. The first demonstration of wavelength multiplexing with photorefractive materials was made to record three holograms corresponding to the three colors of an image [21]. The simultaneous replay of the three holograms reconstructs the colored image. The research on wavelength multiplexing is stimulated by the development of solid-state tunable laser diodes and specially doped photorefractive crystals sensitive to the laser diode wavelength range [22, 23].

In phase-code multiplexing, the reference beam consists of multiple-plane wave fronts. The relative phases among all these wave fronts are adjustable and represent the addresses of the stored images. Each image can be retrieved by illumination of the holograms with the exact same phase code as that used for the recording of the original image. The merits of phase-code multiplexing include the elimination of any mechanical movement or beam-steering requirement, fast access, high light efficiency, and fixed wavelength. When the generalized Hadamard codes are used [24, 25], the number of images to be multiplexed is limited only by the number of pixels of the spatial phase modulator. The limited number of pixels of currently available SLM's can be partially compensated for by combining rotation multiplexing [26].

11.2.4 Commonly used photorefractive materials

A photorefractive effect has been found in a large variety of materials. The most commonly used photorefractive materials in optical pattern recognition applications fall into three categories: electro-optic crystals, semi-insulating compound semiconductors, and photo-polymers.

Lithium niobate ($LiNbO_3$), barium titanate ($BaTiO_3$), and strontium barium niobate [SBN, $Sr_{(1-x)}Ba_xNb_2O_6$] are by far the three most efficient electro-optic crystals that exhibit a photorefractive effect at low intensity levels. Fe-doped $LiNbO_3$ has a large index modulation because of the photovoltaic effect. It is also available in relatively large dimensions. For example, a 3-cm^3 sample was used to record 5000 holograms [3]. Because of its strong mechanical qualities, $LiNbO_3$ can also be used to construct holographic disks [27]. Very recently, the sensitivity of $LiNbO_3$ was extended to near-infrared (670-nm) wavelengths by the addition of Ce dopants [22].

High diffraction efficiency can be achieved with $BaTiO_3$ crystals without the involvement of the photovoltaic effect [6]. This eliminates the phase distortion of the image beam and is preferred in some applications. However, $BaTiO_3$ crystals are not available in large dimensions. A 5 mm × 5 mm × 5 mm sample is considered to be big. Moreover, in order to reach the maximum index modulation, the sample must be cut at a certain angle of the crystallographic c axis [28]. This cut is difficult and it reduces the size of the sample. A phase transition exists near 13 °C; therefore the crystal must always be kept above this temperature.

SBN has a large electro-optic coefficient (r_{333}) that can be reached without a special crystal cut. The phase transition (which can be tuned by doping) is far from the room temperature. It is also subject to electric or temperature fixing [29, 30]. But the optical quality of SBN obtained up to now has been poorer than that of $LiNbO_3$ and $BaTiO_3$.

Several semi-insulating semiconductors, such as gallium arsenide (GaAs), gallium phosphite (GaP), and indium phosphite (InP), have demonstrated photorefractive effect and have been used in optical pattern recognition systems [31]. A prominent feature of these semiconductor crystals is their fast response to optical fields, i.e., the small value of time constant τ at low light intensity. Typically, a submillisecond response time can be achieved with GaAs at a modest laser power intensity of 100 mW/cm^2, which is 1 or 2 orders of magnitude faster than the more conventional materials such as $BaTiO_3$ under the same conditions [32]. The spectral response range of these materials is in the near-infrared wavelength. This can be an advantage or disadvantage, depending on the applications. One of the problems of these semiconductor photorefractive materials is the disparity between different samples. The photorefractive effect of the same material varies considerably among suppliers as well as from one ingot to another [33].

The photopolymer is a new type of photorefractive material. The photorefractive effect was first demonstrated in photopolymers in 1991 [34] with a very small diffraction efficiency. Recently, a photopolymer based on photoconductor poly(N-vinylcarbazalo) doped with a nonlinear optical chromophore has been developed and exhibits better performance than most inorganic photorefractive crystals [35]. Because of the relatively small thickness (typically tens of micrometers), only a limited number of images can be multiplexed in a single location. However, the photopolymer is suitable for the implementation of three-dimensional (3-D) optical disks [27] that can store a huge amount of data. The characterization of photopolymers can be found in Refs. 36 and 37.

11.3 Photorefractive correlators for two-dimensional pattern recognition

Since the introduction of the holographic matched filter [38], optical correlators have been developed as feasible systems for optical pattern recognition. The main advantage of optical correlators, compared with that of their digital counterparts, is that a high-resolution two-dimensional (2-D) Fourier-transform operation can be performed in parallel at a high speed. However, the overall speed of optical correlators has been limited by the speed and space–bandwidth product of the SLM's. To overcome such an limitation, photorefractive materials have been used as real-time dynamic holographic media for filter synthesis. Because of the large space–bandwidth product and multiplexing capability of photorefractive materials, these photorefractive correlators hold promise for many pattern recognition tasks.

11.3.1 VanderLugt-type correlator

There are two basic configurations of optical correlators: the VanderLugt correlator [38] and the joint transform correlator [39]. In the VanderLugt configuration, a matched filter is holographically written by the Fourier transform of the reference pattern and a plane wave. The Fourier transform of the input pattern is multiplied by the filter and then inversely Fourier transformed to produce the correlation between the reference and the input images. In 1978, Pepper *et al.* proposed that the matched filter can be synthesized by using a photorefractive crystal [40]. Since then a number of VanderLugt correlators based on photorefractive media have been demonstrated [10, 41–46].

11.3.1.1 VanderLugt correlator based on four-wave mixing

The nonlinear optical process involved in a photorefractive correlator is four-wave mixing. Figure 11.4 schematically illustrates both transmission-type and reflection-type photorefractive correlators. Let us first consider the transmission type. As shown in Fig. 11.4(a), the reference pattern $r(x, y)$ is displayed in a SLM that is placed in the front focal plane of the Fourier-transform lens L1. Its Fourier spectrum interferes with the reference plane wave and writes a holographic filter within the photorefractive medium. The input pattern $s(x, y)$ enters the correlator by being displayed in another SLM. The input beam propagates in the exact opposite direction of the reference plane wave. The input Fourier spectrum interacts with the dynamic holographic filter. As a result of four-wave mixing, the diffracted phase-conjugate wave is given by

$$O(\mu, v) = \eta R^*(\mu, v)S(\mu, v), \tag{11.7}$$

where η denotes the diffraction efficiency and S and R are Fourier transforms of the signal and the reference patterns, respectively. This phase-conjugate wave passes through lens L1 and is reflected by a beam splitter. The output at the back focal plane of lens L1 can be

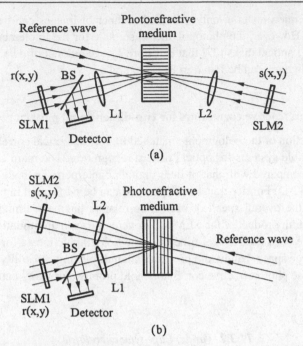

Fig. 11.4. VanderLugt correlators by means of four-wave mixing in a photorefractive medium: (a) transmission type, (b) reflection type.

written as

$$o(x, y) = \eta r(x, y) \star s(x, y) = \eta \iint_{-\infty}^{\infty} r^*(\alpha - x, \beta - y)s(\alpha, \beta) \, d\alpha \, d\beta. \qquad (11.8)$$

It is seen that the correlation between the input and the reference patterns is obtained.

If the reference plane wave is projected to the photorefractive medium from the opposite side of the reference pattern, a reflection-type holographic filter is induced. As shown in Fig. 11.4(b), when such a dynamic hologram is illuminated by the input beam, the diffracted beam passes through lens L1 and the correlation is again obtained at the output plane. The grating spacing of reflection holograms is typically of the order of the wavelength, which is smaller than that of the transmission holograms. The diffraction efficiency of a photorefractive medium is a bell-shaped function of the grating spacing, and the peak efficiency occurs usually at a grating spacing that corresponds to the reflection hologram. Therefore reflection holograms have a higher diffraction efficiency than transmission holograms. On the other hand, because the response time of a photorefractive medium decreases with the square of the grating spacing, the transmission holograms generally have a shorter response time in building up the dynamic hologram than that of the reflection hologram. Hence the selection of transmission-type or reflection-type correlators is application dependent.

The configurations shown in Fig. 11.4 can be modified to adapt to specific application requirements. For example, when the input and the output arms are exchanged in the transmission-type correlator shown in Fig. 11.4(a), a new correlator architecture can be obtained, which is schematically illustrated in Fig. 11.5. This architecture is helpful in understanding the reduction of shift invariance of photorefractive correlators.

Optical correlation by means of four-wave mixing in photorefractive media has been experimentally demonstrated [31, 45]. A schematic diagram of the GaAs-based optical

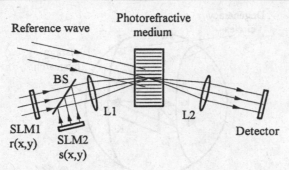

Fig. 11.5. Schematic diagram of a VanderLugt-type photorefractive correlator with the reference and the input patterns entering from the same path.

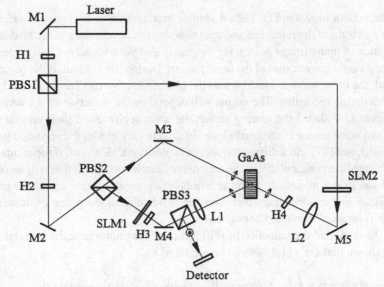

Fig. 11.6. Schematic of an optical correlator with semiconductor GaAs as the dynamic holographic medium. Beam expansion and collimation elements are omitted. M's, mirrors; H's, half-wave plates; PBS's, polarization beam splitters; L's, lenses.

correlator developed by a research group at the Jet Propulsion Lab is illustrated in Fig. 11.6 [45]. To overcome the small electro-optic efficiency of the GaAs and to enhance the output signal-to-noise ratio, polarization switching is used [31]. As shown in Fig. 11.6, the writing and the reading beams are in p polarization, but the diffracted beam is in s polarization. A polarization beam splitter reflects only the diffracted phase-conjugate beam and lets all stray light pass through. With a total laser intensity of 1.5 W/cm^2, approximately 1000 correlations per second can be achieved. This is far beyond the capability of current digital systems.

11.3.1.2 Shift invariance of photorefractive correlators

An inherent problem associated with the photorefractive correlator is the significant reduction in shift invariance. This is a result of the Bragg selectivity of volume holograms.

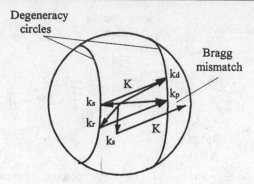

Fig. 11.7. Shift invariance along the degeneracy direction and Bragg mismatch along the nondegeneracy direction.

Consider the system shown in Fig. 11.5. A volume grating is recorded by the interference of the Fourier spectrum of the reference and a plane wave. Because an image can be considered as a summation of many image points, for simplicity and yet without the loss of generality, we consider a single image point of the input pattern. During the readout, if the input pattern is translated, the input wave is incident into the photorefractive medium at an angle different from that during recording. The output will depend on the direction of the wave-vector change. Figure 11.7 shows the grating vector, the wave vectors, and their normal surface. If the incident wave vector is changed along the degeneracy circle, the recorded grating is still Bragg matched [47] and a diffraction output is produced. However, if the incident wave vector is changed perpendicularly to the degeneracy circle, no diffraction will be observed because of the Bragg mismatch. In other words, the photorefractive correlator preserves shift invariance in the degeneracy direction (perpendicular to the plane of drawing), but loses the shift invariance in the other direction.

Gu *et al.* have studied the reduction of shift invariance in photorefractive correlators [48]. They have shown that Eq. (11.8) should be modified as

$$o(x, y) = \eta \iint_{-\infty}^{\infty} r^*(\alpha - x, \beta - y) s(\alpha, \beta) T \operatorname{sinc}\left(\frac{T}{2\pi}\xi\right) d\alpha \, d\beta, \qquad (11.9)$$

where T is the thickness of the photorefractive medium and ξ is a parameter determined by the system configuration. To examine the effect of finite thickness, we consider two extreme cases. First, if the thickness approaches zero (i.e., planar hologram), Eq. (11.9) becomes:

$$o(x, y) = \eta \iint_{-\infty}^{\infty} r^*(\alpha - x, \beta - y) s(\alpha, \beta) \, d\alpha \, d\beta. \qquad (11.10)$$

This is the same as Eq. (11.8), and full shift invariance is preserved. Next, if the thickness of the holographic medium approaches infinity, the sinc function in Eq. (11.9) becomes a dirac function. The integrand is nonzero only when the argument of the dirac function is zero. A detailed mathematical analysis showed that the integration is along a line in the degeneracy direction (perpendicular to the incident plane) [48]. As a result, the output of such system is simply one line of the 2-D correlation. In general, the correlation output is suppressed in the nondegeneracy direction. The degree of suppression is determined by the thickness of the recording medium.

A photorefractive medium with small thickness (e.g., ~50 μm) can be used to obtain 2-D shift invariance [49, 50]. However, a short interaction distance leads to a very low diffraction

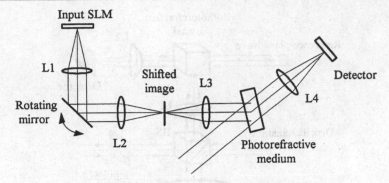

Fig. 11.8. Shift-invariant photorefractive correlator.

efficiency that may not be practical for many applications. Another method to gain shift invariance is to translate the input image along the nondegeneracy direction [51–53]. A possible configuration is shown in Fig. 11.8. A rotating mirror is placed at the back focal plane of the first Fourier-transform lens. When the mirror is rotated, the image at the focal plane of lens L2 is translated. It then interacts with the matched filter recorded in the photorefractive medium. At any specific position of the input image, the output is a one-dimensional (1-D) slice of the 2-D correlation between the input and the reference pattern. As the mirror rotates, the 1-D slice scans over the 2-D correlation plane, and the whole 2-D correlation is obtained after one sweep of the rotating mirror. To avoid the slow mechanical movement, other methods, such as an acousto-optic deflector, can be used to implement the image shift.

11.3.1.3 Optical-disk-based photorefractive correlators

Optical disk technology has matured and has found numerous applications in the past decade. To combine the large storage capacity of an optical disk and the parallel processing capability of an optical correlator, optical-disk-based photorefractive correlators have been developed at the California Institute of Technology [52–56]. Conventional optical disks can be used for the storage of a large number of reference images. If the optical disk is made of photorefractive materials, a large number of matched filters can be stored by means of multiplexing methods.

A schematic of the optical-disk-based photorefractive correlator is shown in Fig. 11.9. Reference images are stored on the disk in a 2-D format. A photorefractive crystal is used for dynamic matched-filter synthesis. During the filter recording phase, the disk illumination is blocked and the input SLM is illuminated. A hologram filter is formed in the photorefractive crystal by the Fourier spectrum of the input pattern and the reference plane wave. Once the hologram is recorded, the input beam is blocked and the disk is illuminated. Lens L1 takes the Fourier transform of the reference image that is within the field of view of the illuminating beam. Lens L2 transforms the light diffracted by the hologram to produce the correlation at the output plane. As discussed in the previous section, because of the finite thickness of the photorefractive crystal, the shift invariance is lost in one direction. Selecting this to be the along-track direction allows the disk rotation to restore 2-D shift invariance. All 1-D slices of the 2-D correlation are obtained sequentially by disk rotation. Therefore, instead of requiring a full 2-D detector at the output plane, a 1-D array is sufficient to detect each slice of the correlation output sequentially. The rotation of the disk also allows the system to search through a library of reference images in the along-track direction.

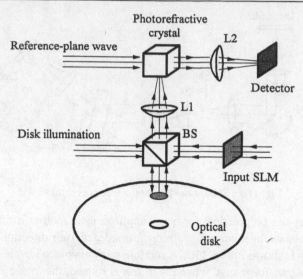

Fig. 11.9. Schematic of an optical-disk-based photorefractive correlator.

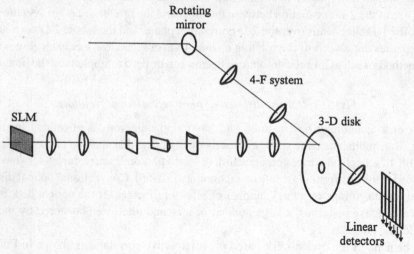

Fig. 11.10. Schematic of a 3-D photorefractive-disk-based optical correlator.

The system has been experimentally demonstrated and has been used to build a programmable associate memory. The correlation rate is limited primarily by disk speed and correlation peak detectability. If a 10-mW laser is used and more than 10^5 photons are required for correlation peak detection, the system has the potential to perform 400,000 correlations per second [53].

The storage capacity of the optical disk can be significantly increased if the information is stored in 3-D volume by means of angular multiplexing [55]. A schematic of the 3-D photorefractive-disk-based correlator is illustrated in Fig. 11.10. The 3-D disk can be implemented with any photorefractive materials that can be fabricated into a thick slab with sufficient area, such as the $LiNbO_3$ crystal and photopolymers. During the filter synthesis process, a reference image is displayed on the SLM. A pair of spherical lenses forms a telecentric system to remove the multiple diffraction orders and to reduce the size of the image.

Three cylindrical lenses are used to image the reference pattern in the along-track direction and Fourier transform it in the radial direction. The two spherical lenses, following the three cylindrical lenses, further demagnify the 1-D Fourier transform and 1-D image to match the spot size on the photorefractive disk. This pattern then interferes with a plane wave to form a hologram in the illuminated area of the 3-D disk. When the next reference pattern is displayed on the SLM, the incidence angle of the plane wave is changed such that another hologram can be stored in the same location by means of angular multiplexing. When the 3-D disk is rotated to another angular position, another set of holographic filters can be recorded.

During the pattern recognition phase, the input pattern is loaded into the SLM. Its 1-D Fourier transform interacts with the holographic filters stored in the 3-D disk. For each holographic filter, a 1-D slice of the 2-D correlation between the input and the original reference patterns is produced at the output plane. The angular multiplexing of these filters ensures that the 1-D slices of correlation between the input and all reference patterns do not overlap, but spread in the along-track direction. An array of 1-D line detectors detects these correlation slices. The 2-D correlations can be obtained by rotating the disk.

The system shown in Fig. 11.10 has been experimentally demonstrated [27]. It can achieve a high correlation rate without the need of a fast SLM. The correlation rate is limited primarily by the speed of the linear CCD detector arrays. One linear detector array is needed for each angular-multiplexed holographic filter. For example, if 50 holographic filters are multiplexed at each spot, then 50 linear detector arrays must be used. Assume that each linear CCD array has 500 pixels and operates at 30 MHz, the correlation rate is $\sim$6000 correlations per second. Less than 10 mW of light incident upon the disk is sufficient to realize this correlation rate. The correlation rate can be significantly increased by use of custom-designed detector arrays that perform the winner-take-all operation on the pixels and report only the location and the magnitude of the largest correlation peak off the chip.

11.3.2 Joint transform correlator

The joint transform correlator was first demonstrated by Weaver and Goodman [39] and independently by Rau [57] in 1966. Although real-time programmable joint transform correlators have been developed [58], their performance has been limited primarily by the small space–bandwidth product and narrow dynamic range of the currently available SLM's. To overcome these limitations, photorefractive materials can be used to record the joint transform power spectrum [11, 59–65].

Figure 11.11 illustrates the schematic of a photorefractive joint transform correlator based on four-wave mixing. The input and the reference patterns are loaded into two SLM's that are separated by a distance of $2h$ (center-to-center distance). The hologram generated within

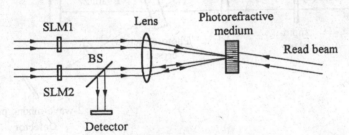

Fig. 11.11. Schematic of a photorefractive joint transform correlator.

the photorefractive medium at the focal plane can be expressed as

$$\Delta n(\mu, \nu) \propto |F[s(x-h, y) + r(x+h, y)]|^2$$
$$= |S(\mu, \nu)|^2 + |R(\mu, \nu)|^2 + S^*(\mu, \nu)R(\mu, \nu)\exp(-i4\pi\mu h)$$
$$+ S(\mu, \nu)R^*(\mu, \nu)\exp(i4\pi\mu h). \tag{11.11}$$

A plane wave propagating in the direction opposite to the input spectrum beam illuminates the hologram. The diffraction beam propagates against the reference spectrum beam and can be written as

$$o(\mu, \nu) = \eta S(\mu, \nu)R^*(\mu, \nu)\exp(i2\pi\mu h). \tag{11.12}$$

This diffracted light is inversely Fourier transformed by the lens and reflected to the output detector by the beam splitter. The output is given by

$$o(x, y) = \eta r(x+h, y) \star s(x+h, y). \tag{11.13}$$

The correlation peak appears at $x = -h$.

A prototypical photorefractive joint transform correlator has been developed by a research group at Thomson-CSF [60]. All optical components, including a double-frequency Nd:YAG laser, a He–Ne laser, a CCD detector, a photorefractive crystal, an electrically addressed SLM, and other optical elements, are packaged within a housing measuring 600 mm × 300 mm × 300 mm. It has been tested for applications in robotic vision, fingerprint identification, and human face recognition. The performance of the prototype can be found in Refs. 60 and 61.

A photorefractive joint transform correlator that combines four-wave mixing and two-beam coupling was recently developed by researchers at Tufts University and the U.S. Air Force Rome Laboratory [65]. As shown in Fig. 11.12, this correlator consists of two output correlation planes, called ports, that operate simultaneously and may have different discrimination characteristics. The input and the reference patterns are displayed side by side at the

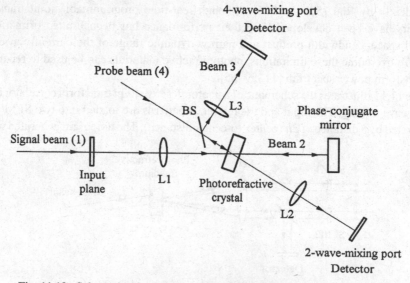

Fig. 11.12. Schematic of a two-port photorefractive joint transform correlator.

input SLM. The signal beam passes through lens L1 and projects the joint spectrum onto the photorefractive crystal. The probe beam, a plane wave, interferes with the signal beam and induces index gratings within the crystal. While passing through the photorefractive crystal, the probe beam interacts with the dynamic hologram and is selectively amplified or deamplified by the joint power spectrum. A subsequent Fourier transform of this beam produces cross correlation at port 1, the two-wave-mixing port.

A self-pumped phase-conjugate mirror is placed in the propagation path of the signal beam, which makes a replica of the outgoing signal beam reentering the crystal as the beam 2. Because the self-pumped phase-conjugate mirror provides self-alignment and aberration corrections, reentering beam 2 is Bragg matched with the index grating. Diffraction from the existing grating inside the crystal generates beam 3, which lies along the propagation path of the probe beam and is reflected by a beam splitter toward output port 2, the four-wave-mixing port. Lens L3 performs a Fourier transform and generates cross correlation at port 2.

A unique feature of the two-port joint transform correlator is its dual discrimination capability. With approximately equal intensity of the signal and the probe beams and suitable photorefractive gain, the two-wave-mixing port is capable of high discrimination whereas the four-wave-mixing port offers low discrimination. It has been shown that under suitable conditions the two-wave-mixing port can achieve peak-to-noise and signal-to-noise ratios that are better than those of the phase-only correlator, and the four-wave-mixing port performs similarly to the classical joint transform correlator. This leads to a potential application in which the correlator could be set up so that in one port a general class is detected (interclass recognition) and in the other port the specific item within a class is detected (intraclass recognition).

11.3.3 Optical wavelet transform correlator

The wavelet transform is a feasible alternative to the Fourier transform for pattern recognition [66]. When a wavelet transform is used, an image is decomposed into a summation of a series of base functions called wavelets. The wavelet transform contains the information, especially features of the input image that correspond to each of the wavelet functions. If the wavelet transforms, instead of the original images, are used to compute the correlation, higher noise immunity and discrimination capability can be expected [67, 68].

The wavelet transform of an image $s(x, y)$ is defined as

$$w_s(a_x, a_y, b_x, b_y) = \mathrm{WT}[s(x, y)] = \iint_{-\infty}^{\infty} s(x, y) h_{ab}(x, y) \, dx \, dy, \qquad (11.14)$$

where the wavelets $h_{ab}(x, y)$ are generated by the translation and dilation of the analyzing wavelet function $h(x, y)$, also called the mother wavelet by some researchers [69]; this function can be expressed as

$$h_{ab}(x, y) = \frac{1}{(a_x a_y)^{1/2}} h\left(\frac{x - b_x}{a_x}, \frac{y - b_y}{a_y}\right). \qquad (11.15)$$

For the convenience of optical implementations, Eq. (11.14) can be equivalently written as

$$w_s(a_x, a_y, b_x, b_y) = \frac{1}{(a_x a_y)^{1/2}} s(x, y) \star h\left(\frac{x}{a_x}, \frac{y}{a_y}\right). \qquad (11.16)$$

The frequency-domain representations of the wavelet transform can be obtained by taking the Fourier transform of Eq. (11.16) as

$$W_s = F[w_s(a_x, a_y, b_x, b_y)] = S(\mu, v)H_{ab}(\mu, v). \tag{11.17}$$

Similarly, the wavelet transform of the reference pattern $r(x, y)$ can be expressed as

$$W_r = R(\mu, v)H_{ab}(\mu, v). \tag{11.18}$$

If the wavelet transform of the reference pattern is used to synthesize the VanderLugt matched filter, the filter function is given by

$$\text{CMF}_W = R^*(\mu, v)H_{ab}^*(\mu, v)\exp(-i2\pi\alpha\mu). \tag{11.19}$$

Once the wavelet transform of the input image enters the VanderLugt correlator, the correlation between the wavelet transforms of the reference and the input images is obtained at the output plane as

$$o_W(x, y) = F[S(\mu, v)H(\mu, v)\text{CMF}_W] = w_s \star w_r(x - \alpha, y). \tag{11.20}$$

In order to simplify the operation of such a wavelet correlator, it is desirable to use the original images, instead of their Fourier transforms, as the inputs of the correlator. For this purpose, the terms in Eq. (11.20) can be reorganized and the optical field behind the matched filter can be written as

$$O(\mu, v) = S(\mu, v)H(\mu, v)\text{CMF}_W = S(\mu, v)\text{WMF}, \tag{11.21}$$

where

$$\text{WMF} = R^*(\mu, v)|H(\mu, v)|^2 \exp(-i2\pi\alpha\mu) \tag{11.22}$$

is called the wavelet matched filter [68, 70]. It is synthesized with the conventional matched filter and the square module of the wavelet spectrum.

The implementation of a wavelet correlator with photorefractive mediums has been investigated [68, 71, 72]. The first experimental demonstration was performed at Pennsylvania State University in 1992 [71]. The experimental setup is schematically shown in Fig. 11.13. During the filter synthesis process, the reference pattern is displayed on SLM1 at the input

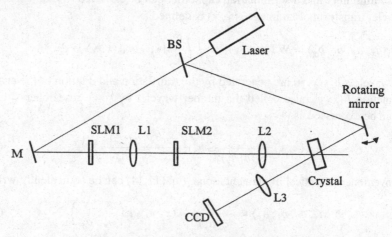

Fig. 11.13. Experimental setup of a photorefractive wavelet transform correlator.

plane. The power spectra of various wavelets with different dilation factors are sequentially displayed on SLM2, which is located at the back focal plane of lens L1. The incidence angle of the reference plane wave can be adjusted by the rotating mirror to angularly multiplex multiple wavelet functions. The Fourier transform of the reference pattern overlaps with the power spectrum of the wavelet on SLM2. Lens L2 images the SLM2 to the photorefractive crystal (Fe-doped $LiNbO_3$). The demagnification of lens L2 is adjusted to match the size of the wavelet spectrum with the entrance aperture of the crystal. The thickness of the crystal is suitably chosen such that it enjoys a reasonably large phase-matching tolerance as well as enough storage capacity.

In the process of pattern recognition, SLM2 is removed and the input scene is loaded into SLM1. The diffracted light is proportional to the product of the input spectrum and the wavelet matched filter. The correlation between the transforms of the input and the reference patterns is obtained at the output plane. Experimental results showed that the wavelet correlator performs better than the conventional correlator for cluttered images.

11.4 Photorefractive processors for radio frequency signal processing

Pattern recognition, in general, applies not only to the identification of 2-D imagery patterns but also to patterns in a generalized sense, such as speech and rf signals. Recently, photorefractive rf signal processors have been developed by making use of the dynamic beam-coupling property of the photorefractive media. In this section, the operations of photorefractive time-integrating correlators [73–75] and photorefractive adaptive notch filters [76–78] are briefly reviewed.

11.4.1 Photorefractive time-integrating correlator

The optical processing of rf signals can be implemented with the use of acousto-optical devices (AOD's) to impress the rf signal onto optical carriers [79]. In a time-integrating optical correlator, a linear detector array (usually a CCD array) is used to accumulate the photoinduced charges over a period of time. One of the drawbacks of these correlators is the undesirable background bias that builds up in each detector pixel, which limits both the integration time and dynamic range of the system. These problems can be overcome if the CCD detector array is replaced by a photorefractive medium.

The schematic of a photorefractive time-integrating correlator is illustrated in Fig. 11.14. The input and the reference rf signals are loaded into two AOD's. The diffracted light

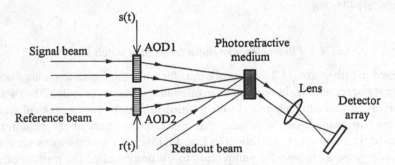

Fig. 11.14. Photorefractive time-integrating correlator.

beams is projected onto the photorefractive medium at an angle of 2θ. Imaging optics can be added to reduce the width of the light beams in order to match the entrance aperture of the photorefractive medium. The interference between the input and the reference beams form a dynamic index hologram within the medium. Recall Eq. (11.2); the refractive-index change is given by

$$\Delta n(t) = \Delta n_{\text{sat}} \int_0^t \frac{A_1 A_2^* \exp[(t' - t)/\tau] \, dt'}{I_0}, \tag{11.23}$$

where the time constant is inversely proportional to the total light intensity

$$\tau \propto (1/I_0). \tag{11.24}$$

The hologram is built up over a period of time. Refer to the configuration shown in Fig. 11.14; the amplitude of the index hologram can be written as [73]

$$\Delta n(t) = \Delta n_{\text{sat}} \exp\left(i\frac{2\pi x \sin\theta}{\lambda}\right) \int_0^t \frac{s(t' + x/v)r^*(t' - x/v)}{|s(t' + x/v|^2 + |r(t' - x/v)|^2} \exp[(t' - t)/\tau] \, dt', \tag{11.25}$$

where $s(t)$ and $r(t)$ are the input and the reference rf signals driving the two AOD's, respectively, v is the velocity of the acoustic wave, and λ is the wavelength of the light illuminating the AOD's. A weak reading beam at a different wavelength is Bragg tuned to read out the hologram. The diffracted output carries the correlation information and is imaged onto a linear array detector. Because the readout beam is coherent, further optical processing of the output can be performed if another optical processing unit is cascaded to the time-integration correlator.

The system shown in Fig. 11.14 may suffer from a possible resolution problem because of its inherent 1-D nature. Because the optical signals that write the index grating vary only in the x direction and the grating vector itself is also in the x direction, a cross-talk problem exists [74]. The solution to overcoming the cross talk is to place the two AOD's in the direction orthogonal to the index grating vector, i.e., in the y direction. In this geometry, the grating wave vector is in the x direction and the acoustic waves travel in the y direction. This eliminates the undesirable beam interaction and thereby eliminates the cross talk.

The photorefractive time-integrating correlator has been experimentally demonstrated. At the intensity level (incident to the crystal) of ~ 1.5 W/cm^2 and an integration time of ~ 1 s, correlations of both amplitude- and frequency-modulated rf signals up to 70 MHz were obtained with a BaTiO$_3$ crystal. Such a time-integrating correlator can be used to develop adaptive rf filters to track the interference sources and to suppress the interference noise adaptively [75, 80].

11.4.2 Photorefractive radio frequency notch filter

As discussed in Subsection 11.2.2.1, energy transfer from one beam to another occurs in nonlinear two-beam mixing. In order to have efficient two-beam coupling, the two beams involved must be mutually coherent. In other words, the difference in optical frequency between the two beams must be smaller than the response time of the photorefractive medium used, i.e., $\Delta\omega < 1/\tau$. As shown in Fig. 11.15(a), if the two beams are mutually coherent, energy transfers from the pump beam to the probe beam. The transfer efficiency depends on the degree of coherence of the two beams. If the two beams are incoherent, they

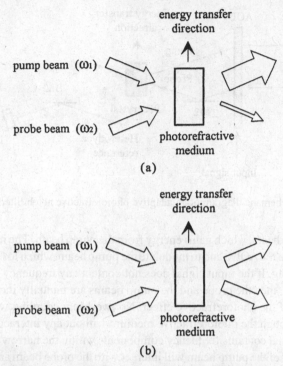

Fig. 11.15. Mutual coherence between the two beams is required for two-wave coupling: (a) coherent case, $\Delta\omega < 1/\tau$; (b) incoherent case, $\Delta\omega > 1/\tau$.

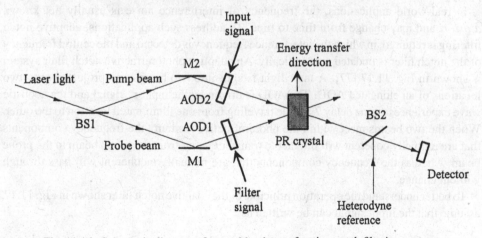

Fig. 11.16. Schematic diagram of a tunable photorefractive notch filtering system.

do not interfere with each other; hence the energy transfer does not occur, as illustrated in Fig. 11.15(b).

This effect can be utilized to construct an rf notch filter to eliminate the narrow-band noise from broadband signals. The operation of the photorefractive notch filter is schematically illustrated in Fig. 11.16. The central frequency of the notch filter, ω_r, is continuously tunable by means of a local reference waveform generator. The filter signal drives the AOD1 and

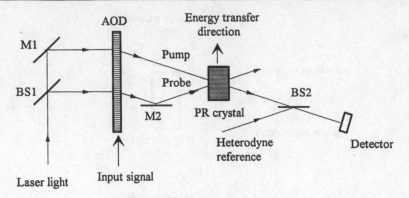

Fig. 11.17. Schematic diagram of an adaptive photorefractive notch filtering system.

modulates the probe beam, which gains energy from the pump beam. The input broadband rf signal is used to drive the AOD2 and to modulate the pump beam, which loses energy because of two-wave coupling. If the input signal does not contain any frequency components near the notch frequency, then the probe and the pump beams are mutually incoherent to within the response time of the photorefractive effect. No two-beam coupling will occur, and the two beams pass through the photorefractive medium without any interaction.

If the input rf signal contains frequency components within the narrow bandwidth of the notch filter, that part of the pump beam will interact with the probe beam and transfer optical energy from the signal-carrying pump beam into the probe beam. Therefore the noise in the output beam is strongly depleted. The rf output of the filter is obtained when the depleted pump beam is heterodyned with a local reference beam that is derived from the laser light.

In real-world applications, the frequency of interference noise is usually not known *a priori* and may change from time to time. To address such applications, adaptive notch filtering is required, in which the interference frequency is detected and the central frequency of the notch filter is updated automatically. An adaptive photorefractive notch filter system is shown in Fig. 11.17 [77]. A laser light is split into two beams and projected onto two locations of an elongated AOD. The AOD is driven by the input rf signal and the acoustic wave experiences a time delay T_a while traveling from one illuminated location to the other. When the two beams meet within the photorefractive medium, the frequency components that are mutually coherent will interact to transfer energy from the pump beam to the probe beam, whereas the frequency components that are mutually incoherent will pass through without change.

To better understand the operation principle of the adaptive notch filter shown in Fig. 11.17, assume that the input signal can be written as:

$$s_{\text{in}}(t) = n(t) + b(t), \tag{11.26}$$

where $n(t)$ denotes the narrow-band noise components and $b(t)$ represents the broadband signal components. If we choose the acoustic time delay T_a and the photorefractive time constant τ such that $1/\Delta f_n > T_a$ and $1/\Delta f_n > \tau$, where Δf_n is the bandwidth of the noise components, then the narrow-band components will couple with each other and transfer energy from the pump beam to the probe beam. If $1/\Delta f_b < T_a$ and $1/\Delta f_b < \tau$, where Δf_b is the bandwidth of the signal $b(t)$, the broadband components carried by the two beams do not interact and pass through without change.

The bandwidth of the notch filter is determined by the degree of the mutual coherence requirement of the two beams: $\Delta\omega < 1/\tau$. At a nominal intensity of 1 W/cm^2, the time constants of photorefractive media can vary widely from 10 μs to 1 s, depending on the particular material that is used. This leads to the corresponding bandwidths of the notch filter in the range from 1 Hz to 100 KHz. In a recent experimental demonstration in which a SBN crystal was used [78], an interference cancellation of 40 dB was obtained and an adaptivity to changing the frequency at a fast response time of 5.5 ms was achieved.

11.5 Photorefractive novelty filters

Novelty filters extract and display the parts of a scene that change with time. They are essential ingredients for the front-end visual systems of many animals [81]. An optical pattern recognition system loaded with novelty filters is useful in many applications such as motion detection and target tracking. Because of the 2-D parallel nature, optics is considered a good candidate for the implementation of novelty filters. As a matter of fact, a number of photorefractive novelty filters have been proposed and demonstrated over the past decade [32, 82–94].

Phase-conjugate interferometers have been used to implement novelty filters. Phase-conjugate mirrors, instead of plain optical mirrors, are used in these interferometers. One of the most important and interesting phenomena associated with phase-conjugate mirrors is time reversal. Figure 11.18 shows the schematic of a Michelson interferometer with two phase-conjugate mirrors. An incident light is split by the beam splitter into two arms and then reflected back by the phase-conjugate mirrors. According to the Stokes principle of time reversibility [95], when two phase-conjugated beams recombine at the beam splitter, a time-reversed replica of the incident beam is generated. This beam propagates backward, retracing the original incident-beam path. Thus there is no light at the output port. Total darkness is observed at the output end as a result of time reversal.

A novelty filter based on the phase-conjugate mirror interferometer was first demonstrated by Anderson and Erie in 1987 [82]. The optical system is schematically illustrated

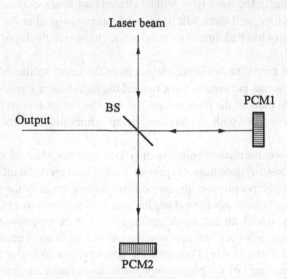

Fig. 11.18. Time reversal in a phase-conjugate Michelson (PCM) interferometer.

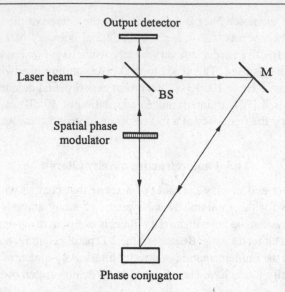

Fig. 11.19. Novelty filter based on a phase-conjugate interferometer.

in Fig. 11.19. The incident light is divided by a beam splitter. One resulting beam passes through an optical phase modulator that is loaded with input scenes. Both beams are subsequently incident upon the same self-phase-conjugate mirror. The generated phase-conjugate beams propagate back toward the beam splitter. Under steady-state conditions, the phase of the information carrying the beam is exactly canceled by its return trip through the spatial phase modulator, so that it emerges as a plane wave. Because of the time reversal, the two phase-conjugated beams coherently recombine at the beam splitter so as to travel back to the light source, with no light going to the output end of the novelty filter. However, if the image on the spatial phase modulator changes in time faster than the response time of the phase-conjugate mirror, the beam emerging from the spatial phase modulator will not be a plane wave. In particular, the wave front will be altered for those locations in the modulator that are changing in time, and there will be an instantaneous signal at the output. After the phase-conjugate mirror has had time to respond to the changes in the input scene, the output fades again.

It is interesting to note that, while the output from the beam splitter shows the novelty of the scene, the light that propagates back toward the light source carries the information corresponding to still parts of the input image, the so-called monotonous information. This information can be observed with an additional beam splitter placed directly in front of the plane-wave source.

The output contrast of the phase-conjugate-interferometer-based novelty filter is generally very high. If the phase-conjugate mirror is perfect, the contrast could be infinity. Because the speed of the photorefractive phase-conjugate mirror determines only the time required for an unchanging image to fade, not for a changing image to appear, any change in the image will appear instantly in such an novelty detection system. If the response time of the phase conjugator is too slow, however, the new and the old output images could become mixed with each other and unresolvable. Therefore the temporal resolution of the novelty filter is determined by the speed of the phase conjugator. To evaluate the overall performance of a novelty detection system, Liu and Cheng have defined an effective space–bandwidth

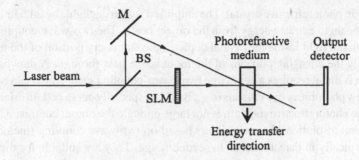

Fig. 11.20. Novelty filter based on two-wave coupling in photorefractive medium.

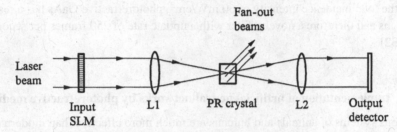

Fig. 11.21. Novelty filter based on beam fanning.

product of the novelty filter that is determined jointly by the response time of the phase conjugator, the changing speed of the input scene, and the numbers of pixels of the spatial phase modulator [32].

Two-wave coupling in photorefractive media can also be used to implement novelty filters. A novelty filter based on two-wave mixing is shown in Fig. 11.20 [84]. The information-carrying beam and a plane reference beam interact within the photorefractive medium and produce an index grating. Because of the 90° phase shift between the intensity interference fringes and the refractive-index grating, in steady state the object beam transfers energy to the reference beam and emerges from the crystal severely depleted. If the input scene suddenly changes, the two-wave coupling is momentarily defeated and the transmitted output image is intense at locations corresponding to image changes. The output becomes dark again after new index grating is established within the photorefractive medium.

A physical explanation for the two-wave-mixing depletion of the object beam is destructive interference: the transmitted output and diffracted beam by the reference beam destructively interfere at the output end. At the output plane, there are always two superimposed images: the real image passes through the photorefractive medium and the holographically reconstructed image read out by the reference beam. If the beam intensity ratio is suitably controlled, these two images can be of nearly equal amplitude but 180° out of phase. Thus they tend to cancel and result in dark output. Whenever there are any changes in the input scene, the transmitted image changes, but the holographically reconstructed image needs some time to respond. As a consequence, the destructive interference is temporally destroyed and the changing part of the input scene is observed at the output plane. In contrast to the phase-conjugate-interferometer-type novelty filters, both amplitude and phase modulators can be used in the two-wave-coupling novelty filter.

The simplest novelty filter is probably the one based on beam fanning, which was demonstrated by Ford et al. [87]. As illustrated in Fig. 11.21, it uses only one input beam projected into a high-gain photorefractive crystal. The amplified scattering light (beam fanning) serves as the probe beam to extract energy from the object beam. The two-wave-coupling process with this scattered light inside the crystal depletes the stationary portion of the transmitted beam and only the changing portions of the input scene pass through. A drawback of this configuration is that it requires a very large two-beam-coupling gain, which can be obtained with only a few photorefractive crystals (e.g., $BaTiO_3$) specially cut in certain orientations. If the gain of the photorefractive medium is not large enough, the output contrast will be poor.

The temporal resolution of novelty filters based on two-wave coupling (including beam fanning) is typically in the subsecond to second range. They are suitable for the detection of relatively slow changes. In some applications, such as high-speed target tracking, a fast update rate is critical and novelty filters based on phase-conjugate interferometers can be used. At the total incidence intensity of 60 mW/cm^2, photorefractive GaAs has a response time of 4 ms and therefore a novelty filter with a update rate of 250 frames per second can be built [32].

11.6 Implementation of artificial neural networks by photorefractive media

The nervous systems of animals and humans are much more effective than modern digital computers in general pattern recognition applications. For example, identifying characters and distinguishing voices are easily accomplished by human beings, but present significant difficulties for digital computers. Over the past two decades, much effort has been expended in the field of artificial neural networks to understand and exploit the computational principles utilized by the brain for high-speed pattern recognition [96]. The artificial neural networks developed so far are generally massive parallel computing systems in which a large number of simple processing elements are densely interconnected. Because of their massive parallel nature, neural networks are not adaptable to conventional digital computers. VLSI technology has been utilized to synthesize neural networks. The problems with VLSI implementation arise when the size of the network exceeds what can be accommodated on a single chip. For larger networks requiring a large number of chips, the problems associated with interconnecting the chips and sequencing the operations properly make the VLSI implementation difficult. In contrast, the massive interconnectivity and parallel processing capability are the main strengths of optics. The low accuracy of an analog–optical processing system can be compensated for by the robustness of neural networks. Therefore neural networks and optical systems are mutually complementary and the optical implementation is a promising approach for the construction of large-scale neural networks.

Although many approaches have been demonstrated over the past 10 years, our discussion in this chapter is focused on optical implementations with photorefractive media. The use of photorefractive media in optical neural network implementations potentially enables the high storage capacities that are required for densely interconnected networks of a large number of neurons. It also allows for the parallel updating of weighted interconnections by use of certain learning rules. The promise of these features has spurred considerable interest in the photorefractive-media-based neural network implementations [97–112].

The iterative equation for a single-layer neural network can be expressed as

$$y_j = g \left[\sum_{i=1}^{N} w_{ji} x_i \right],$$

(11.27)

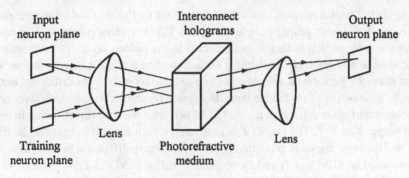

Fig. 11.22. Single-layer photorefractive neural network.

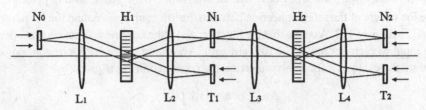

Fig. 11.23. Two-layer neural network trained by the back-error-propagation algorithm.

where x and y represent the input and the output pattern vectors, respectively, $g[\cdot]$ is a nonlinear operator, and w_{ji} is the interconnect weight between the ith input neuron and the jth output neuron. The basic learning rule can be characterized by the update equation:

$$w_{ji}(p+1) = w_{ji}(p) + \alpha_{ji}(p)x_i(p), \tag{11.28}$$

where $w_{ji}(p)$ is the interconnection weight at time p and $\alpha(p)$ is a multiplier that depends on a particular learning algorithm. Multiple-layer neural networks can be built by cascading several such single-layer structures.

One of the approaches for the implementation of a single-layer neural network is shown in Fig. 11.22. The neurons are arranged in planes, and neurons in the input plane are connected to the neurons in the output plane by means of holographic gratings recorded in a photorefractive medium. To establish the interconnect weights, the input patterns in the training set are sequentially displayed in the input plane and their corresponding target output patterns are displayed in the training neural plane by two SLM's located at the front focal plane of the Fourier lens. The interference of the input signal and the training signal redistributes photoexcited charges and thereby creates refractive-index gratings. The strength of the interconnection is determined by the amplitude of the refractive-index modulation of the holographic gratings. During the training process in general, the modification of a certain grating that connects two neurons can affect the connection between other neurons. This is undesirable in many neural network applications. It has been shown, however, that by placing the neurons in the input and the output planes on appropriate fractal grids, the connection between the ith input neuron and the jth output neuron can be modified without directly affecting the connections between other neurons [102]. After the neural network is trained, the SLM for the training pattern is turned off and the inputs are loaded into the input SLM. The diffracted light is focused by another Fourier lens onto the output plane.

One of the most widely used learning algorithms for training a fully adaptive multi-layer neural network is the back-error-propagation algorithm [113]. Figure 11.23 shows a two-layer optical neural network, which was proposed by Psaltis *et al.*, that is capable of implementing back-error-propagation learning [98, 99]. A training pattern $X^{(0)}$ is placed at the input plane N_0, which is then interconnected to the hidden layer N_1 by means of the dynamic volume hologram H_1. Simulating the action of an array of hidden neurons, a SLM placed at plane N_1 performs a soft thresholding operation on the light diffracted onto it to produce $X^{(1)}$, the output of the hidden layer. Hologram H_2 connects the hidden layer neurons at N_1 to the output plane N_2, where another SLM performs the final thresholding to produce the final output $Y = X^{(2)}$. The output Y is compared with the target output for the training pattern, and the error signal is generated at plane N_2. The undiffracted beams from N_0 and N_1 are recorded on SLM's at T_1 and T_2, respectively. The SLM's at T_1, T_2, and N_2 are then illuminated from the right-hand side to read out the stored signals, and the unmodulated light propagates back toward the left. Let $s_j^{(q)}$ be the total input to the jth neuron in plane N_q and $w_{ij}^{(q)}$ be the weight of the interconnection between the jth neuron at N_q and the ith neuron at plane N_{q-1} for $q = 1, 2$. Assume that the function $g[\cdot]$ is the nonlinear function that operates on the input to each neuron in the forward path. According to the back-error-propagation algorithm, the change of the interconnection matrix stored in H_2 is given by

$$\Delta w_{kj}^{(2)} \propto \delta_k g'\big[s_k^{(2)}\big] x_j^{(1)}, \tag{11.29}$$

where $g'[\cdot]$ is the derivative of $g[\cdot]$. Each neuron in N_2 is illuminated from the right by the error signal δ_k, and the backward transmittance of each neuron is proportional to the derivative of the forward output. It should be noted that the hologram recorded in H_2 is the outer product of the active patterns on planes N_2 and T_2. Therefore the change made in H_2 is what is described by expression (11.29). It can be proven that the changes in the interconnection weights stored in H_1 is

$$\Delta w_{kj}^{(1)} \propto \left\{ \sum_k \delta_k g'\big[s_k^{(2)}\big] w_{kj}^{(2)} \right\} g'\big[s_j^{(1)}\big] x_i^{(0)}. \tag{11.30}$$

In this way, interconnection weights in both layers are suitably modified based on the back-error-propagation rule.

Although it is possible to train the system shown in Fig. 11.23 with the back-error-propagation algorithm, several difficulties must be overcome to obtain a satisfactory performance. The hardware implementation is complicated by the need for bidirectional optical devices with different forward and backward characteristics. To overcome this problem, a local learning algorithm has been developed [106], in which a weight update for a certain layer depends on only the input and the output of that layer and a global scalar error signal. This learning procedure still guarantees that the network is trained by error descent, but is much easier to implement. Based on the local learning algorithm, a neural network for human face recognition has been demonstrated [107].

During the learning process, both positive and negative error signals occur and therefore both additive and subtractive modifications are required. To represent the positive and the negative error signals accurately, the Stokes effect can be used in the implementation of neural network learning [95]. Figure 11.24 shows a schematic of a single-layer neural net classifier implemented with a photorefractive crystal [104]. Initially the crystal contains no holograms, and the interconnections are built up by simply exposing the hologram with light source 1 with the training pattern displayed in the 2-D input SLM and the associated output pattern displayed in the 1-D reference SLM. The process is repeated for each pattern in the

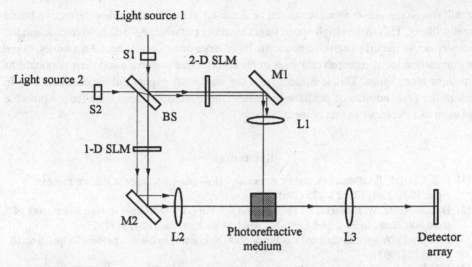

Fig. 11.24. Photorefractive neural net classifier trained by a back-error-propagation algorithm.

training set. After the initialization, the first pattern is loaded into the 2-D input SLM, and the reference-beam shutter is closed to interrogate the system. The reconstructed output pattern is then compared with the target output pattern to obtain the error signal. The weight modification is then performed in two steps. First, only those portions of the 1-D reference SLM corresponding to the positive portions of the error vector are opened and light source 1 is turned on to strengthen certain interconnections following the back-error-propagation rule. Then only those portions corresponding to negative elements of the error vector are loaded into the 1-D reference SLM and light source 2 is turned on. Because of the Stokes effect, the gratings generated by light source 1 and light source 2 are exactly 180° out of phase. Therefore the exposure of the photorefractive medium to light source 2 will reduce the diffraction of the dynamic hologram and hence weaken the appropriate interconnection weights. The training process repeats until the output for each pattern in the training set generates its corresponding target output. The system is then ready for pattern recognition applications.

Because of the large variety of neural network models, optical devices, and implementation approaches, many optical neural network implementations that use photorefractive materials have been proposed. It is impossible to describe all these implementations in a short section, and readers interested in the implementations uncovered by this chapter are referred to Refs. 5, 97, 100, 101, 103, 105, and 109–112.

11.7 Summary

This chapter surveys the research in optical pattern recognition with photorefractive media. The main advantages offered by the photorefractive media are dynamic holographic property, real-time processing, multiplexing capability, parallel access of multiple pages, and large storage capacity. A large number of matched filters can be stored in the form of dynamic holograms, and the correlation between the input and the reference data bank can be performed at a very high speed. By virtue of two-beam coupling, an adaptive notch filter can be developed to adaptively eliminate the interference noise from a broadband radio signal. The change in an input scene can be detected in real time by photorefractive-based novelty filters. This makes high-speed target tracking possible. As 3-D holographic media, photorefractive materials can also be used to build large-scale optical neural networks. There are currently a lot of research activities in photorefractive materials and their applications in pattern recognition. This is evident from the numerous papers published each year. In view of the great number of published papers in these areas, we apologize for the possible omission of references in this context.

References

[1] F. S. Chen, J. T. LaMacchia, and D. B. Fraser, "Holographic storage in lithium niobate," Appl. Phys. Lett. **13**, 223–225 (1968).

[2] D. L. Staebler, W. L. Burke, W. Phillips, and J. J. Amodei, "Multiple storage and erasure of fixed holograms in Fe-doped $LiNbO_3$," Appl. Phys. Lett. **26**, 182–184 (1975).

[3] F. H. Mok, "Angle-multiplexed storage of 5000 holograms in lithium niobate," Opt. Lett. **18**, 915–917 (1993).

[4] P. Yeh, A. E. Chiou, J. Hong, P. Beckwith, T. Chang, and M. Khoshnevisan, "Photorefractive nonlinear optics and optical computing," Opt. Eng. **28**, 328–343 (1989).

[5] Y. Owechko and B. H. Soffer, "Optical neural networks based on liquid crystal light valves and photorefractive crystals," in *Liquid-Crystal Devices and Materials*, P. S. Drzaic and U. Efron, eds., Proc. SPIE **1455**, 136–144 (1991).

[6] C. Alves, P. Aing, G. Pauliat, and G. Roosen, "Prospects of photorefractive memories for optical processing," Opt. Memory Neural Networks **3**, 167–190 (1994).

[7] J. Hong, S. Campbell, and P. Yeh, "Photorefractive spatial light modulator," in *Optical Information Processing Systems and Architectures*, B. Javidi, ed., Proc. SPIE **1151**, 476–484 (1989).

[8] A. E. Chiou and P. Yeh, "Symmetry filters using optical correlation and convolution," Opt. Eng. **29**, 1065–1072 (1990).

[9] T. Y. Chang, J. H. Hong, S. Campbell, and P. Yeh, "Optical image processing by matched amplification," Opt. Lett. **17**, 1694–1696 (1992).

[10] F. T. S. Yu, S. Wu, A. W. Mayers, and S. Rajan, "Wavelength multiplexed reflection matched spatial filters using $LiNbO_3$," Opt. Commun. **81**, 343–347 (1991).

[11] H. Rajbenbach, S. Bann, P. Réfrégier, P. Joffre, J.-P. Huignard, H.-S. Buchkremer, A. K. Jensen, E. Rasmussen, K.-H. Brenner, and G. Lohman, "Compact photorefractive correlator for robotic applications," Appl. Opt. **31**, 5666–5674 (1992).

[12] P. Günter and J. P. Huignard, eds., *Photorefractive Materials and their Applications*, (Springer-Verlag, Berlin, 1988), Vols. I and II.

[13] P. Yeh, *Introduction to Photorefractive Nonlinear Optics* (Wiley, New York, 1993).

[14] N. V. Kukhtarev, V. B. Markov, S. G. Odulov, M. S. Soskin, and V. L. Vinetskii, "Holographic storage in electrooptic crystals," Ferroelectrics **22**, 949–964 (1979).

[15] S. Wu, Q. Song, A. Mayers, D. A. Gregory, and F. T. S. Yu, "Reconfigurable interconnections using photorefractive holograms," Appl. Opt. **29**, 1118–1125 (1990).

[16] F. H. Mok, M. C. Tackitt, and H. M. Stoll, "Storage of 500 high resolution holograms in a $LiNbO_3$ crystal," Opt. Lett. **16**, 605–607 (1991).

[17] G. A. Rakuljic, V. Leyva, and A. Yariv, "Optical data storage by using orthogonal wavelength-multiplexed volume holograms," Opt. Lett. **17**, 1471–1473 (1992).

[18] C. Denz, G. Pauliat, G. Roosen, and T. Tschudi, "Potentialities and limitations of hologram multiplexing by using the phase-encoding technique," Appl. Opt. **31**, 5700–5706 (1992).

[19] X. Yang and Z.-H. Gu, "Three-dimensional optical data storage and retrieval system based on phase-code and space multiplexing," Opt. Eng. **35**, 452–456 (1996).

[20] H. Kogelnik, "Coupled wave theory for thick hologram gratings," Bell Syst. Tech. J. **48**, 2909–2947 (1969).

[21] F. T. S. Yu, S. Wu, A. Mayers, S. Rajan, and D. A. Gregory, "Color holographic storage in LiNbO$_3$," Opt. Commun. **81**, 348–352 (1991).

[22] S. Yin and F. T. S. Yu, "Specially doped LiNbO$_3$ crystal holography using a visible-light low-power laser diode," IEEE Photon. Technol. Lett. **5**, 581–582 (1993).

[23] S. Yin, H. Zhou, F. Zhou, M. Wen, Z. Yang, J. Zhang, and F. T. S. Yu, "Wavelength-multiplexed holographic storage in a sensitive photorefractive crystal using a visible-light tunable diode laser," Opt. Commun. **101**, 317–321 (1993).

[24] X. Yang, Y. Xu, and Z. Wen, "Generation of Hadamard matrices for phase-code multiplexed holographic memories," Opt. Lett. **21**, 1067–1069 (1996).

[25] X. Yang, Z. Wen, and Y. Xu, "Construction of Hadamard phase-codes for holographic memories," in *Photorefractive Fiber and Crystal Devices: Materials, Optical Properties, and Applications II*, F. T. Yu and S. Yin, eds., Proc. SPIE **2849**, 217–228 (1996).

[26] X. Yang, N. Li, Z. Wen, and Y. Xu, "Holographic storage using phase-code and rotation multiplexing," in *Photorefractive Fiber and Crystal Devices: Materials, Optical Properties, and Applications II*, F. T. S. Yu and S. Yin, eds., Proc. SPIE **2849**, 125–132 (1996).

[27] D. Psaltis, F. Mok, S. Li, and K. Curtis, "3-D holographic storage in image recognition," in *Euro-American Workshop on Optical Pattern Recognition*, B. Javidi and P. Refregier, eds., Vol. PM12 of SPIE Press Monographs Series (SPIE, Bellingham, WA, 1994), pp. 419–428.

[28] J. Hong, P. Yeh, D. Psaltis, and D. Brady, "Diffraction efficiency of strong volume holograms," Opt. Lett. **15**, 344–346 (1990).

[29] S. Redfield and L. Hesselink, "Enhanced nondestructive holographic readout in strontium barium niobate," Opt. Lett. **13**, 880–882 (1988).

[30] A. Kewitsch, M. Segev, A. Yariv, and R. Neurgaonkar, "Selective page-addressable fixing of volume holograms in Sr$_{0.75}$Ba$_{0.25}$Nb$_2$O$_6$ crystal," Opt. Lett. **18**, 1262–1264 (1993).

[31] L.-J. Cheng and K. L. Luke, "Photorefractive semiconductors and applications," in *Photorefractive Materials, Effects, and Applications*, P. Yeh and C. Gu, eds., Vol. CR48 of SPIE Critical Reviews Series (SPIE, Bellingham, WA, 1994), pp. 251–275.

[32] D. T. Liu and L.-J. Cheng, "Resolution of a target-tracking optical novelty filter," Opt. Eng. **30**, 571–576 (1991).

[33] L.-J. Cheng, J. Lagowski, M. F. Rau, and F. C. Wang, "Defects and photorefractive effects in GaAs," Rad. Eff. Defects Solids **111**, 37–43 (1989).

[34] S. Ducharme, J. C. Scott, R. J. Twieg, and W. E. Moerner, "Observation of the photorefractive effect in a polymer," Phys. Rev. Lett. **66**, 1846–1849 (1991).

[35] K. Meerholz, B. L. Volodin, Sandalphon, B. Kippelen, and N. Peyghambarian, "A photorefractive polymer with high optical gain and diffraction efficiency near 100%," Nature (London) **371**, 497–500 (1994).

[36] B. L. Volodin, Sandalphon, K. Meerholz, B. Kippelen, N. V. Kukhtarev, and N. Peyghambarian, "Highly efficient photorefractive polymers for dynamic holography," Opt. Eng. **34**, 2213–2223 (1995).

[37] U.-S. Rhee, H. J. Caulfield, J. Shamir, C. S. Vikram, and M. M. Mirsalehi, "Characteristics of the DuPont photopolymer for angular multiplexed page-oriented holographic memories," Opt. Eng. **32**, 1839–1847 (1993).

[38] A. B. VanderLugt, "Signal detection by complexed spatial filtering," IEEE Trans. Inf. Theory **IT-10**, 139–145 (1964).

[39] C. S. Weaver and J. W. Goodman, "A technique for optically convolving two functions," Appl. Opt. **5**, 1248–1249 (1966).

[40] D. M. Pepper, J. Auyeung, D. Fekete, and A. Yariv, "Spatial convolution and correlation of optical fields via degenerated four wave mixing," Opt. Lett. **3**, 7–9 (1978).

[41] J. O. White and A. Yariv, "Real-time image processing via four-wave mixing in photorefractive materials," Appl. Phys. Lett. **37**, 5–10 (1980).

[42] P. D. Foote, T. J. Hall, and L. M. Connors, "High speed two input real-time optical correlation using photorefractive BSO," Opt. Laser Technol. **18**, 39–42 (1986).

[43] M. G. Nicholson, I. R. Cooper, M. W. McCall, and C. R. Petts, "Simple computational model of image correlation by four-wave mixing in photorefractive media," Appl. Opt. **26**, 278–286 (1987).

[44] G. Gheen and L.-J. Cheng, "Optical correlators with fast updating speed using photorefractive semiconductor materials," Appl. Opt. **27**, 2756–2758 (1988).

[45] D. T. H. Liu and L. J. Cheng, "Real-time VanderLugt optical correlator that uses photorefractive GaAs," Appl. Opt. **31**, 5675–5680 (1992).

[46] F. T. S. Yu and S. Yin, "Bragg diffraction-limited photorefractive crystal-based correlators," Opt. Eng. **34**, 2224–2231 (1995).

[47] H. Lee, X.-G. Gu, and D. Psaltis, "Volume holographic interconnection with maximal capacity and mimimal cross talk," J. Appl. Phys. **65**, 2191–2193 (1989).

[48] C. Gu, H. Fu, and J.-R. Lien, "Correlation patterns and cross-talk noise in volume holographic optical correlators," J. Opt. Soc. Am. A **12**, 861–868 (1995).

[49] Q. B. He, P. Yeh, L. H. Hu, S. P. Lin, T. S. Yeh, S. L. Tu, S. J. Yang, and K. Hsu, "Shift-invariant photorefractive joint-transform correlator using $Fe:LiNbO_3$ crystal plates," Appl. Opt. **32**, 3113–3115 (1993).

[50] X. Yang, H. Szu, Y. Sheng, and H. J. Caulfield, "Optical Haar wavelet transforms of binary images," Opt. Eng. **31**, 1846–1851 (1992).

[51] C. Gu, J. Hong, and S. Campbell, "2-D shift-invariant volume holographic correlator," Opt. Commun. **88**, 309–314 (1992).

[52] D. Psaltis, M. A. Neifeld, and A. Yamamura, "Image correlators using optical memory disks," Opt. Lett. **14**, 429–431 (1989).

[53] D. Psaltis, M. A. Neifeld, A. Yamamura, and S. Kobayashi, "Optical memory disks in optical information processing," Appl. Opt. **29**, 2038–2056 (1990).

[54] M. A. Neifeld and D. Psaltis, "Programmable image associative memory using an optical disk and a photorefractive crystal," Appl. Opt. **23**, 4398–4409 (1993).

[55] K. Curtis and D. Psaltis, "Three-dimensional disk-based optical correlator," Opt. Eng. **33**, 4051–4054 (1994).

[56] A. A. Yamamura, M. A. Neifeld, S. Kobayashi, and D. Psaltis, "Optical disk-based artificial neural systems," Opt. Comput. Process. **1**, 3–12 (1991).

[57] J. E. Rau, "Detection of differences in real distribution," J. Opt. Soc. Am. **56**, 1490–1494 (1966).

[58] F. T. S. Yu and X. J. Lu, "A real-time programmable joint transform correlator," Opt. Commun. **52**, 10–16 (1984).

[59] F. T. S. Yu, S. Wu, S. Rajan, and D. A. Gregory, "Compact joint transform correlator with a thick photorefractive crystal," Appl. Opt. **31**, 2416–2418 (1992).

[60] H. Rajbenbach, "Dynamic holography in optical pattern recognition," in *Optical Pattern Recognition V*, P. P. Casasent and T.-H. Chao, eds., Proc. SPIE **2237**, 329–346 (1994).

[61] J. Rodolfo, H. Rajbenbach, and J.-P. Huignard, "Performance of a photorefractive joint transform correlator for fingerprint identification," Opt. Eng. **34**, 1166–1171 (1995).

[62] J. Khoury, M. Cronin-Golomb, P. Gianino, and C. Woods, "Photorefractive two-beam coupling nonlinear joint transform correlator," J. Opt. Soc. Am. B **11**, 2167–2174 (1994).

[63] J. Khoury, J. Kane, G. Asimellis, M. Cronin-Golomb, and C. Woods, "All-optical nonlinear joint transform correlator," Appl. Opt. **33**, 8216–8225 (1994).

[64] G. Asimellis, J. Khoury, J. Kane, and C. Woods, "Two-port photorefractive joint transform correlator," Opt. Lett. **20**, 2517–2519 (1995).

[65] G. Asimellis, M. Cronin-Golomb, J. Khoury, J. Kane, and C. Woods, "Analysis of the dual discrimination ability of the two-port photorefractive joint transform correlator," Appl. Opt. **34**, 8154–8166 (1995).

[66] C. K. Chui, *An Introduction to Wavelets* (Academic, Boston, 1992).

[67] X. Yang, N. P. Caviris, and M. Wen, "Optical wavelet correlators for cluttered target recognition," in *Optical Pattern Recognition V*, D. P. Casasent and T.-H. Chao, eds., Proc. SPIE **2237**, 402–418 (1994).

[68] X. Yang, "Optical wavelet joint transform correlator for automatic target recognition," J. Opt. **27**, 3–11 (1996).

[69] H. Szu, Y. Sheng, and J. Chen, "Wavelet transform as a bank of matched filters," Appl. Opt. **31**, 3267–3277 (1992).

[70] Y. Sheng, D. Roberge, H. Szu, and T. Lu, "Optical wavelet matched filters for shift-invariant pattern recognition," Opt. Lett. **18**, 299–301 (1993).

[71] M. Wen, S. Yin, P. Purwardi, and F. T. S. Yu, "Wavelet matched filtering using a photorefractive crystal," Opt. Commun. **99**, 325–330 (1993).

[72] J. Widjaja and Y. Tomita, "Optical wavelet-matched filtering by four-wave mixing in photorefractive media," Opt. Commun. **117**, 123–126 (1995).

[73] D. Psaltis, J. Yu, and J. Hong, "Bias-free time-integrating optical correlator using a photorefractive crystal," Appl. Opt. **24**, 3860–3865 (1985).

[74] J. Hong and T. Y. Chang, "Photorefractive time-integrating correlator," Opt. Lett. **16**, 333–335 (1991).

[75] R. M. Montgomery and M. R. Lange, "Photorefractive adaptive filter structure with 40-dB interference rejection," Appl. Opt. **30**, 2844–2849 (1991).

[76] J. Hong and T. Y. Chang, "Frequency-agile rf notch filter that uses photorefractive two-beam coupling," Opt. Lett. **18**, 164–166 (1993).

[77] J. Hong, "Adaptive notch filter using acousto-optics and photorefraction," U.S. patent 5,050,967 (1991).

[78] J. Hong and T. Y. Chang, "Adaptive notch filter using photorefractive two-beam coupling," IEEE J. Quantum Electron. **30**, 314–317 (1994).

[79] K. T. Staker, F. M. Dickey, M. L. Yee, and B. A. Kast, "Acousto-optic correlator for optical pattern recognition," in *Real-Time Optical Information Processing*, B. Javidi and J. J. Horner, eds. (Academic, Boston, 1994), Chap. 11.

[80] D. Psaltis and J. Hong, "Adaptive acoustooptic filters," Appl. Opt. **23**, 3475–3481 (1984).

[81] D. Falk, D. Bril, and D. Stork, *See the Light* (Harper & Row, New York, 1986), pp. 192–193.

[82] D. Z. Anderson and M. C. Erie, "Resonator memories and optical nolvelty filters," Opt. Eng. **26**, 434–444 (1987).

[83] D. Z. Anderson, D. M. Lininger, and J. Feinberg, "An optical tracking novelty filter," Opt. Lett. **12**, 123–125 (1987).

[84] M. Cronin-Golomb, A. M. Biernacki, C. Lin, and H. Kong, "Photorefractive time differentiation of coherent optical images," Opt. Lett. **12**, 1029–1031 (1987).

[85] R. Cudney, R. M. Pierce, and J. Feinberg, "The transient detection microscope," Nature (London) **332**, 424–426 (1988).

[86] N. S.-K. Kwong, Y. Tamita, and A. Yariv, "Optical tracking filter using transient energy coupling," J. Opt. Soc. Am. B **5**, 1788–1791 (1988).

[87] J. E. Ford, Y. Fainmain, and S. H. Lee, "Time-integrating interferometry using photorefractive fanout," Opt. Lett. **13**, 856–858 (1988).

[88] F. Vachess and L. Hesselink, "Synthesis of a holographic image velocity filter using the nonlinear photorefractive effect," Appl. Opt. **27**, 2887–2894 (1988).

[89] D. Z. Anderson and J. Feinberg, "Optical novelty filters," IEEE J. Quantum Electron. **25**, 635–647 (1989).

[90] J. A. Khoury, G. Hussain, and R. W. Eason, "Optical tracking and motion detection using photorefractive $Bi_{12}SiO_{20}$," Opt. Commun. **71**, 138–143 (1989).

[91] G. Hussain and R. W. Eason, "Velocity filtering using complementary gratings in photorefractive BSO," Opt. Commun. **86**, 106–112 (1991).

[92] F. T. S. Yu, S. Wu, S. Rajan, A. Mayers, and D. A. Gregory, "Optical novelty filter with phase carrier," Opt. Commun. **92**, 205–208 (1992).

[93] H. Rehn, R. Kowarschik, and K. H. Ringhofer, "Transient phase detection with barium titanate novelty filters," in *16th Congress of the International Commission for Optics: Optics as a Key to High Technology*, G. Akos, T. Lippenyi, G. Lupkovics, and A. Podmaniczky, eds., Proc. SPIE **1983**, 843–844 (1993).

[94] H. Rehn, R. Kowarschik, and K. H. Ringhofer, "Beam-fanning novelty filter with enhanced dynamic phase resolution," Appl. Opt. **34**, 4907–4911 (1995).

[95] G. G. Stokes, "On the perfect blackness of the central spot in Newtow's rings, and on the verification of Fresnel's formulae for the intensities of reflected and refracted rays," Cambridge Dubl. Math. J. **4**, 1 (1849).

[96] S. Haykin, *Neural Networks: A Comprehensive Foundation* (Macmillan, New York, 1994).

[97] Y. Owechke, G. J. Dunning, E. Marom, and B. H. Soffer, "Holographic associative memory with nonlinearities in the correlation domain," Appl. Opt. **26**, 1900–1910 (1987).

[98] K. Wagner and D. Psaltis, "Multilayer optical learning networks," Appl. Opt. **26**, 5061–5076 (1987).

[99] D. Psaltis, D. Brady, and K. Wagner, "Adaptive optical networks using photorefractive crystals," Appl. Opt. **27**, 1752–1759 (1988).

[100] L.-S. Lee, H. M. Stoll, and M. C. Tackitt, "Continuous-time optical neural network associative memory," Opt. Lett. **14**, 162–164 (1989).

[101] K. H. Fielding, S. K. Rogers, M. Kabrisky, and J. P. Mills, "Position, scale and rotation invariant holographic associative memory," Opt. Eng. **28**, 849–853 (1989).

[102] D. Psaltis, D. Brady, X.-G. Gu, and S. Lin, "Holography in artificial neural network," Nature (London) **343**, 325–330 (1990).

[103] C. Peterson, S. Redfield, J. D. Keeler, and E. Hartman, "Optoelectronic implementation of multilayer neural networks in a single photorefractive crystal," Opt. Eng. **29**, 359–368 (1990).

[104] J. Hong, S. Campbell, and P. Yeh, "Optical pattern classifier with perceptron learning," Appl. Opt. **29**, 3019–3025 (1990).

[105] G. J. Dunning, Y. Owechko, and B. H. Soffer, "Hybrid optoelectronic neural networks using a mutually pumped phase-conjugate mirror," Opt. Lett. **16**, 928–930 (1991).

[106] Y. Qiao and D. Psaltis, "Local learning algorithm for optical neural networks," Appl. Opt. **31**, 3285–3288 (1992).

[107] H.-Y. S. Li, Y. Qiao, and D. Psaltis, "Optical network for real-time face recognition," Appl. Opt. **32**, 5026–5035 (1993).

[108] J. Hong, "Applications of photorefractive crystals for optical neural networks," Opt. Quantum Electron. **25**, 551–568 (1993).

[109] F. T. S. Yu, S. Yin, and C.-M. Uang, "A content-addressable polychromatic neural net using a (Ce:Fe)-doped $LiNbO_3$ photorefractive crystal," Opt. Commun. **107**, 300–308 (1994).

[110] C. J. Cheng, P. Yeh, and K. Y. Hsu, "Generalized perceptron learning and its implications for photorefractive neural networks," J. Opt. Soc. Am. B **11**, 1619–1624 (1994).

[111] G. Lu, M. Lu, and F. T. S. Yu, "Multilayer associative memory and its hybrid optical implementation," Appl. Opt. **34**, 5109–5117 (1995).

[112] G. C. Petrisor, A. A. Goldstein, B. K. Jenkins, E. J. Herbulock, and A. T. Tanguay, "Convergence of backward-error-propagation learning in photorefractive crystals," Appl. Opt. **35**, 1328–1343 (1996).

[113] D. E. Rumelhart, G. E. Hinton, and R. J. Williams, "Learning internal representations by error propagation," in *Parallel Distributed Processing: Explorations in the Microstructures of Cognition*, D. E. Rumelhart and J. L. McClelland, eds. (MIT, Cambridge, 1986), Vol. 1, pp. 318–362.

12

Optical pattern recognition with microlasers

Eung Gi Paek

12.1 Introduction: microlasers, surface-emitting laser diode arrays, or vertical-cavity surface-emitting lasers

Optical pattern recognition has existed for approximately 45 years. The computational power of optics to perform 2-D (two-dimensional) Fourier transform, convolution and correlation at the speed of light is certainly unique over other technologies. Many different system architectures have been demonstrated to utilize the merits of optics.

However, the evolution of optical pattern recognition has been rather slow and its performance often has been challenged by its counterpart, digital computers. This was mainly due to the lack of suitable input/output devices such as SLMs (spatial light modulators) and detectors available in the market. Most of the input/output devices were developed for other consumer applications and their speeds were usually limited to conventional video rates (30 frames per second). With the rapid advances in digital computers that can perform near-video-rate 2-D Fourier transformation, higher speed devices are disperately needed in optical information processing.

New device possibilities exist in optical pattern recognition and information processing. Recent advances in optoelectronic device technology include surface emitting microlaser arrays, among others, which have a great potential for ultra-high speed (more than two orders of magnitude faster than video rate) optical pattern recognition and ATM header recognition due to their parallel computing capability.

The current status of the microlaser technology will be briefly reviewed and possible applications of new devices for optical pattern recognition and information processing will be described. Advances in these technologies would contribute to the underlying device infrastructure needed for advanced communications services, such as those sketched by the National Information Initiative or "information superhighway".

12.1.1 What is a surface-emitting laser?

Figure 12.1 compares the surface-emitting laser (SEL) diode [Fig. 12.1(a)] with the conventional edge-emitting semiconductor diode laser in a compact-disc player [Fig. 12.1(b)]. In a conventional edge-emitting laser diode, the laser beam emerges parallel to the active layer. Therefore a two-dimensional (2-D) arrangement of lasers is difficult. In contrast, the light from a SEL diode emerges perpendicular to the active layer, allowing a 2-D stack of many lasers on a planar wafer substrate. Iga and co-worked first proposed and demonstrated the feasibility of such SEL diodes in the late 1970's [1, 2].

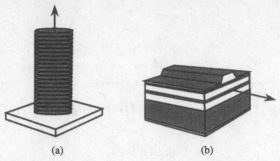

Fig. 12.1. (a) Vertical-cavity surface-emitting microlaser, (b) edge-emitting diode laser. (Reprinted from Ref. [4] by Jewell et al.)

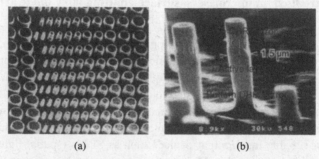

Fig. 12.2. Scanning electron micrographs obtained from the vertical-cavity surface-emitting microlasers. (Vertical-cavity SEL's, SEL diode arrays, or leave microlasers). Original acronym VCSEL's, SELDA's. (Reprinted from Ref. [4] by Jewell et al.)

12.1.2 What is a microlaser?

The original SEL's had a relatively large active volume (typically 10 μm^3), requiring a large driving current. In May 1989, a miniaturized surface-emitting microlaser with a very small active volume (typically 0.05 μm^3) compared to that of previous SEL's was developed [3–7]. As can be seen in Fig. 12.2, a small hairlike structure with a diameter of only 1.5 μm is an independent laser. The hairlike structure consists of an active layer surrounded by a pair of high-reflectivity mirrors on both sides. The laser cavity is formed along the direction normal to the wafer surface and laser light is emitted in either the top or the bottom direction. Therefore such lasers are often called vertical-cavity surface-emitting lasers (VCSEL's) to differentiate them from other SEL's that have the cavity along the wafer surface.

The small size of the active layer of the device, which consists of very thin (~100-Å) quantum-well layers, requires a very low driving current to turn on the laser. To compensate for the low gain through the thin active region, the resonators at both ends of the active medium should have extremely high reflectivity to minimize losses, typically 99.9% or a finesse of 3000. This high-reflectivity mirror and the efficient confinement of both electrons and photons within a waveguide define the two major difficulties in fabricating microlasers. We call such an array of microlasers a surface-emitting laser diode array (SELDA) or microlasers, in addition to the more commonly used term, VCSEL.

12.1.3 Why microlasers?

The microlaser has many good features that make it an excellent candidate for optical information processing. Some of the important features include:

- Individual microlasers can be as small as a few micrometers. In addition, the surface-emitting features allow 2-D integration of many lasers. Therefore $\sim 10^6$ lasers can be arranged in a 1-cm^2 chip area.
- Because of the small volume of the active medium, the threshold current can be very low (up to a few microamperes), greatly reducing the power requirements of a system (compared with edge emitters, whose threshold currents are 5–20 mA).
- The spectral linewidth of a microlaser is very narrow, typically less than 0.01 nm, for reasons that are explained below. Therefore a microlaser has a longer coherence length than a conventional Fabry–Perot laser diode.
- The beam profile of a microlaser is naturally circular. Therefore the beam can be directly used to illuminate optical components or spatial light modulators (SLM's) without requiring anamorphic prisms or lenses as in edge-emitting lasers. In this way, a conventional gas laser and a pinhole spatial filter can be replaced by a simple microlaser, making a compact and robust system.
- The manufacturing and test process of a microlaser is simple, once the crystal growth step is completed. Millions of microlasers can be obtained simultaneously. Moreover, microlasers can be easily tested in wafer form, whereas edge emitters must be cleaved before testing. Therefore microlasers have a potential for low-cost mass production.

However, current microlasers have the following limitations:

- The wavelengths of the microlasers are mostly limited to 0.75–1.0 μm. Therefore a holographic recording with a microlaser is currently difficult because most of the holographic recording materials are sensitive to visible wavelengths.
- Polarization of a microlaser is random or uncontrollable. This problem can be solved in a controllable manner by use of various methods, as explained in Subsection 12.2.5.
- The output power of a microlaser is not as high as that of an edge-emitting laser diode.
- The microlaser is still in its infancy; therefore extensive studies of packaging, addressing, and reliability are few compared with those of edge emitters.

Nonetheless, the microlaser is certainly one of the most important photonic devices developed in this decade and will play an increasingly important role in the future of optical information processing. Currently, active research is being pursued over the world to overcome these problems.

12.2 Status of microlasers

Below, the current status of microlasers related to optical information processing is summarized.

12.2.1 Low threshold current

An important parameter used to characterize the power requirements of any diode laser is its threshold current, the minimum current needed for lasing. A low threshold current of an 8.7-μA single quantum well [8] and a low threshold voltage 1.33-V VCSEL grown by

Table 12.1. *Some of the differences between edge emitters and surface emitters*

Parameter	Edge emitter	Surface emitter
Size (μm)	$300 \times 3 \times 3$	$10 \times 10 \times 5$
Active layer (μm)	$300 \times 3 \times 0.1$	$10 \times 10 \times 0.01$
Threshold current	5–20 mA	A few microamperes to several milliamperes
Reflector	Cleaved, 30% reflectivity	High reflectivity (99.9%)
Fabry–Perot spacing	Long cavity, small spacing	Short cavity, large spacing
Single mode	Transverse easier	Single longitudinal mode
Beam profile	Elliptical	Circular
Output power	up to 4 W (cw)	up to 100 mW (cw), 1 W (pulsed)
Polarization	Easy control	Requires special control
2-D integration	Difficult	Easy
Fabrication and testing	Requires cleave before test	Easy on-chip testing

metal–organic chemical vapor deposition were demonstrated [9]. An extremely large power conversion efficiency of over 50% was also achieved [10]. Also, a microlaser with a threshold current density of 80 A/cm^2 was demonstrated [11]. An ultralow threshold operation can be expected with a reduction in the active region volume until we meet the limitation that is due to nonradiative recombination and optical and electrical confinement [11, 12].

A thresholdless microlaser that does not have a threshold current like a light-emitting diode (LED) but still possesses narrow linewidths by photon recycling has also been reported [13].

12.2.2 Coherence

Coherence is one of the most important attributes of a light source for optical information processing and pattern recognition. The spectral linewidth of a microlaser is very narrow [14–16]. This narrow linewidth is attributed to the short cavity length of the microlaser that can support only a single longitudinal mode, an effect similar to that of a distributed-feedback laser or a gas laser with a Fabry–Perot etalon. In addition, the high-reflectivity of the reflector functions as a spectral filter. Therefore a microlaser has a longer coherence length than a conventional Fabry–Perot laser diode (see Table 12.1). A short linewidth of 65 MHz has been measured with heterodyne techniques for a microlaser (850–865-nm wavelengths) operated near threshold [15]. This corresponds to a coherence length of 5 m (0.00014-nm linewidth) and is comparable with that of an Ar-ion laser with an etalon.

Although microlasers operate inherently at a single longitudinal mode, the coherence length still decreases at high current levels. This is due to the multiple transverse modes at high currents. Figure 12.3 illustrates output-light power versus current (L–I) and voltage versus current (V–I) curves for a typical top-surface-emitting gain-guiding microlaser [17, 18]. The apparent kinks in both curves are associated with changes in the optical mode and corresponding carrier distributions. The first threshold corresponds to the TEM$_{00}$ lasing mode. The first plateau in the L–I characteristic is associated with the saturation of the TEM$_{00}$ mode. The subsequent threshold corresponds to the onset of the TEM$_{01*}$ mode. Lasing begins in the TEM$_{00}$ mode, but further increases in the current result in the occurrence of additional higher-order transverse modes. These higher-order transverse modes generate additional spectral modes that are typically separated by $\sim$0.7 nm, which greatly reduces the coherence length and the beam profile of an expanded beam.

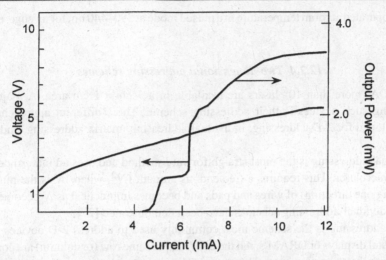

Fig. 12.3. Light output power versus current and voltage versus current curves of a typical microlaser (after Refs. 17 and 18).

Such a problem was solved with a spatial filtering technique, which was achieved by a simple reduction in the contact aperture size with respect to the width of the confinement (surrounded by the implantation) [18]. In this way, the kinks were eliminated and a single transverse mode operation leading to a high coherence length was achieved even at high current levels without sacrificing the output power (~63%) too much.

Also, each laser of the array operates independently; they are not phase locked with each other [19].

12.2.3 Visible microlasers

Current microlasers operate mainly in the wavelength range of 0.75–1.0 μm, which can be achieved with GaAs-based alloys. However, researchers are developing shorter-wavelength microlasers. This is motivated by the fact that the shorter wavelength would allow a larger capacity of an optical storage medium as well as easy alignment. The visible operation is crucial, particularly for holographic information processing, because most holographic recording materials are sensitive to only visible wavelengths.

The main difficulty in the fabrication of visible microlasers lies in the creation of mirrors with high reflectance and low loss in the visible region. Room-temperature lasing to wavelengths as short as 0.63 μm was demonstrated with InGaAlP alloys [20–23].

Recently an electrically pumped blue microlaser array was also demonstrated [24, 25]. Laser action was achieved at a wavelength of 484 nm under a pulsed-current injection at 77 K. The microlaser structure was composed of a CdZnSe/ZnSe multiple-quantum-well active layer, ZnSe cladding layers, and $SiO_2 TiO_2$ distributed Bragg reflectors (DBR's). This result shows promise of the microlaser for holographic information processing based on photorefractive crystals, many of which are sensitive at blue–green wavelengths. Likewise, additional work on blue–green lasers with homoepitaxial deposition of ZnSe is currently being pursued.

The recent demonstration of a blue laser diode by Nakamura at Nichia Chemical Ind. Ltd. (Tokushima, Japan) [26] shows great promise in the blue–green operation of the microlaser. The laser is II-VI based (GaN material with InGaN multiple quantum wells) and is currently operated at room temperature in a pulsed mode at 390–440 nm for an edge-emitting structure.

12.2.4 Two-dimensional addressing schemes

In a SELDA, more than 10^6 lasers are available in a 1 cm $\times$ 1 cm area. A major issue in operating such devices is their addressing scheme. Three different approaches have been developed for 2-D addressing: individual addressing, matrix addressing, and optical addressing.

Individual addressing is the most straightforward method and uses an independent wire for each microlaser. This requires a tremendous amount (N^2, where N is the number of lasers along one direction) of wires and pads and becomes impractical as N becomes large. Several individual-addressing schemes have been demonstrated [27].

Matrix addressing is the scheme most commonly used to address 2-D devices such as liquid-crystal displays or DRAM's. All the SEL's in the same row (or column) are connected together with their common external wire pad. These rows and columns are on the opposite side of the SEL's. To turn on a particular SEL at (i, j), a voltage is applied across the ith row and the jth column pad. This point-by-point addressing can be easily extended to line-by-line addressing by the simultaneous application of voltages across all the columns and a row. Such a matrix-addressing scheme requires only $2N$ electrodes and is easy to fabricate. A monolithic 32 $\times$ 32 matrix-addressable SELDA has been demonstrated by Orenstein et al., as shown in Fig. 12.4 [28]. Recently a high-density array of microlasers with a spacing of 30 μm between neighboring lasers was demonstrated [29].

Finally, an optical-addressing scheme of a SELDA is shown in Fig. 12.5. In this device, a 2-D image illuminating one side of a SELDA is detected by an array of heterojunction phototransistors (HPT's) and the current generated by each HPT turns on the corresponding laser. This method allows a complete parallel load of an image, without the need for electrical connections. A monolithic array of such an optically addressable SELDA has been demonstrated [30]. A similar concept had been developed for LED's by various groups and was used for optical information processing [31–33]. The output-light versus input-light relationship is described below in Subsection 12.5.2.

12.2.5 Polarization control

The output light from a microlaser is not linearly polarized. Therefore polarization control is one of the most important subjects to be resolved, especially for polarization-sensitive applications such as magneto-optic disks and coherent detection systems. Recently it was theoretically and experimentally predicted that the polarization of a microlaser grown on an $(n11)$-oriented substrate could be simply controlled by use of its intrinsic in-plane anisotropic gain distribution characteristics [34]. Stable polarization characteristics based on the prediction were realized for a conventional microlaser structure [35].

12.2.6 Multiple wavelengths

Multiple wavelengths and tuning are highly desirable in optical information processing and communications. A 2-D multicolor SELDA (MC-SELDA) is a 2-D array of microlasers, each with its own wavelength. Figure 12.6(a) shows a monolithic 2-D MC-SELDA originally demonstrated by Chang-Hasnain et al. [36]. The 7 $\times$ 20 array has a total of 140 microlasers, and each laser has a unique wavelength that is uniformly separated from its neighbors by

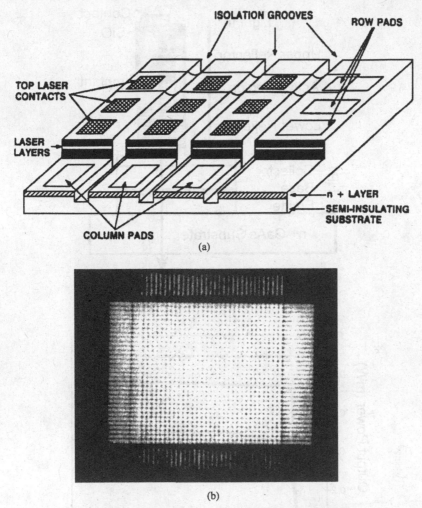

Fig. 12.4. Matrix addressing of a microlaser array (after Ref. 28).

0.3 nm, spanning a total wavelength of 43 nm. Such a wavelength variation was obtained by varying the laser cavity lengths when growing the wafers as shown in Fig. 12.6(b).

More recently, the same group demonstrated a record wavelength span of 62.7 nm by using a modified patterned-substrate growth technique in a molecular beam epitaxy system [37]. The authors claim highly uniform threshold currents with an average of ~2 mA with a high repeatability of wavelength spacing and a sharp wavelength-shift rate of 117.14 nm/mm.

12.2.7 Wavelength tuning

Continuously tunable lasers have a huge number of potential applications. These range from free-space optical interconnects and wavelength-division multiplexing for communications to holographic data storage and spectroscopy. The microlasers can be tuned in the same way as for edge-emitting lasers by varying the drive current applied to them. Recently a new method of wavelength tuning based on a deformable membrane mirror has been demonstrated [38]. As shown in Fig. 12.7, the mirror is fabricated at the end of a light-emitting cavity, with a small air gap separating the two. As the mirror moves back and

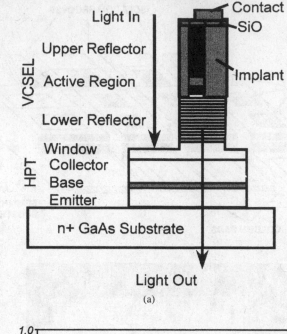

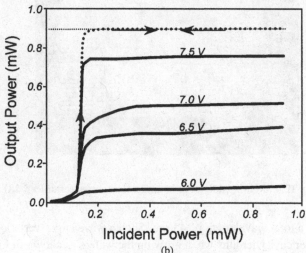

Fig. 12.5. Optically addressable integrated SELDA (after Ref. 30): (a) structure of the device, (b) light-output versus light-input relationship.

forth, the length of the cavity changes, resulting in a change in the resonant wavelength. A tuning range of 15 nm has been demonstrated using the method. Currently, the device requires extremely high threshold current, 37–64 mA, and thus had to be operated in pulsed mode, rather than continuous wave.

12.2.8 Efficiency

High electrical-to-optical power conversion (wall-plug) efficiency is one of the most important parameters for estimating power consumption. Microlasers suffer from poor wall-plug efficiencies because of excessive voltage drops across the DBR's. However, recent overall cw power efficiencies have approached 20% [39]. Also, dramatic improvements in

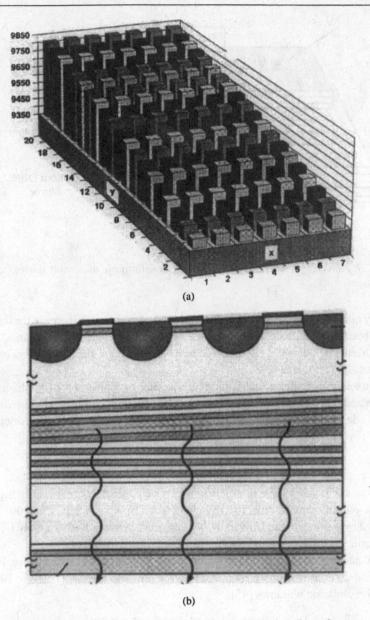

Fig. 12.6. 2-D MC-SELDA (after Ref. 36): (a) device, (b) wafer.

microlaser power conversion efficiency of up to 50% at low currents have been realized [40] with an index-guided top-emitting structure based on selective oxidation. The efficiencies demonstrated are at least comparable with those of edge-emitting lasers.

12.2.9 Modulation speed

The injection current modulation bandwidth of small microlasers has been predicted to be very high (>100 GHz) for the following reasons [41]:

- For a give injection current, the small volume of a microlaser leads to a large photon density and hence a short stimulated lifetime.

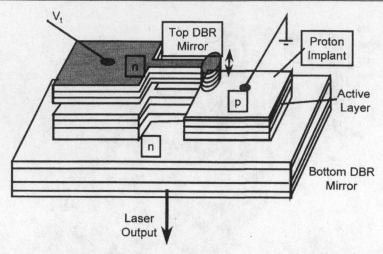

Fig. 12.7. Wavelength tuning of a microlaser by a deformable membrane mirror (after Ref. 38).

- Cavity quantum electrodynamic effects have been expected to increase the differential gain, because of an enhancement of the emission rate into the lasing mode. Recent measurements indicate a fast intrinsic response, with a 3-dB modulation bandwidth of more than 50 GHz [42].
- Recent analyses indicate that a microlaser, under the constraint of nonlinear gain or current density limitations, has the same intrinsic modulation bandwidth as conventional edge-emitting lasers with the same cavity losses and photon density [43].

12.2.10 High-power output

In spite of the small size of the microlaser, it can be operated at reasonably high-power levels. A cw output of more than 100 mW with a wall-plug efficiency of 20% has been demonstrated by a group at the University of California at Santa Barbara; they tailored the laser's operation to increased temperatures [44].

Also, 1-W pulsed operation has been demonstrated with a top-surface-emitting 100 μm × 100 μm broad-area laser with a grid contact segregating the laser into a 10 × 10 array of 8 μm × 8 μm emission windows [45].

12.3 Optical correlators with microlasers

12.3.1 Introduction

Pattern recognition has long been one of the major applications of optics. The capability of computing a complicated operation, such as a 2-D Fourier-transform, convolution, and correlation based on the wave nature of optics, has provided a strong tool for this application. Correlation is an operation that calculates the similarity of two patterns and is widely used for pattern recognition. It is represented by an equation

$$g(x, y) = \int\int f(\xi, \eta) h(\xi - x, \eta - y) \, d\xi \, d\eta,$$

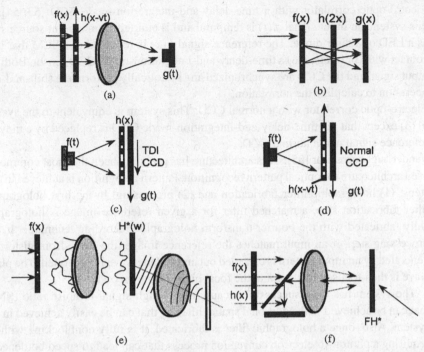

Fig. 12.8. Classification of optical correlators. TDI, time delay and integration.

where f, h, and g refer to an input, a filter, and a correlation output, respectively. Therefore summation and shift are two basic operations in calculating correlations.

12.3.2 Classification of optical pattern recognition systems

Optical correlators have been around since the early 1950's. Although most of the current optical correlators are based on the coherent optical processors (either the VanderLugt or the joint transform correlator), there are many other types of correlators, as is shown in Fig. 12.8. Below a brief description of each of the systems is provided. More detailed analyses can be found in various references [46–48].

(a) Kretzmer-type correlator [49]: This is probably the oldest and most straightforward correlator developed in the early 1950's. In the system, summation is achieved spatially by an optical lens and is detected by a single detector one by one. Also, a shift operation is achieved when one (or both) of the inputs is mechanically moved along both the x and the y directions. Such a movement can be achieved by an acousto-optic beam deflector in which the signal flows through the one-dimensional medium. Unfortunately, there are no fast 2-D image shifters, currently.

(b) Shadow casting [50–52]: In this system architecture, both shift-and-add operations are done in an all-optical fashion simultaneously. The problems with such a system are diffraction effects and no room for the spatial filtering operation. Microlasers, even with their natural 2-D nature, can find limited application for this type of correlator because of the diffraction effect that is due to the high coherence of the light from microlasers.

(c)   Electro-optic correlator with a time-delay-and-integration-mode CCD [53, 54]: In
      this system, an input signal $f(t)$ is temporal and is emitted from a point source such
      as a LED or a laser diode. The reference signal $h(x)$ is recorded in a SLM that is in
      contact with (or imaged to) a time-delay-and-integration-mode CCD sensor. Both the
      input signal and the CCD are synchronized and systolically perform the shift-and-add
      operation to calculate the correlation.

(d)   Electro-optic correlator with a normal CCD: This system is equivalent to the system
      in (3) except that the time-delay-and-integration-mode CCD is replaced by a moving
      reference signal and a normal CCD.

(e)   VanderLugt correlator [55]: This architecture has probably been the most commonly
      used architecture for optical pattern recognition. Pattern recognition is achieved in two
      steps: (1) holographic filter fabrication and (2) processing. In the first holographic
      filter fabrication step, a matched filter for a given reference image is holographi-
      cally fabricated with the Fourier-transform holographic recording technique. In the
      processing step, if an input matches the reference image, the wave-front distortion
      generated by an input pattern is canceled out and a plane wave is generated. The plane
      wave is then focused to a point at the focal plane of the second lens.

      The system has been successful because of the high signal-to-noise ratio (SNR)
      that can be achieved with additional spatial filtering that can be easily achieved in the
      system. Also, once a holographic filter is fabricated, it is fully nonblocking without
      requiring a photon-to-electron conversion process that can lead to speed bottleneck.

      However, the system requires an input from high-optical-quality recording materi-
      als and a holographic filter to preserve shift-invariant recognition. The filter has to be
      positioned exactly in the original position in which it was recorded. Also, the holo-
      graphic filter must be positioned exactly in the focal plane of the Fourier-transform
      lens. Recent advances in high-quality SLM's and *in situ* holographic recording ma-
      terials allow a convenient implementation of such an optical correlator with high
      performance.

(f)   Joint transform correlator [56, 57]: The joint transform correlator has become popular
      these days because it is simple to implement and real-time operation is possible with
      currently available devices (SLM's and 2-D CCD's) [58].

      The operation of the joint transform correlator can be considered a two-step Fourier-
      spectrum (modulus square of a Fourier-transform) generation. At first, both an input
      and a reference image are located in the input plane (spatial domain) side by side.
      The Fourier spectrum of the two inputs is recorded in a high-resolution recording
      material. The Fourier spectrum of the two inputs is Fourier transformed again by
      the second Fourier-spectrum generator. The first-order diffraction outputs from the
      second Fourier-spectrum generator are the correlations of the input and the reference
      images.

      Optics can perform a real-time Fourier-spectrum generation quite well in a simple
      and low-cost setup consisting of a laser, a SLM, a 2-D image sensor, such as a CCD,
      and a spherical lens.

      The advantages of the system are that it allows easy real-time implementation of a
      correlator with currently available devices and it does not require accurate filter posi-
      tioning. Also, as in a VanderLugt correlator, various spatial filtering operations such
      as bipolar phase-only filtering can be incorporated [59–61]. Moreover, one can use
      different wavelengths for hologram recording and readout without requiring careful
      alignment.

The disadvantages are that it requires a fast and high-resolution holographic recording material and the correlator is not all-optical, requiring intermediate photon-to-electron conversion processes.

The recent photorefractive semi-insulating multiple-quantum-well SLM developed by Partovi *et al.* [62] has many desirable features suitable for such correlators. It has high speed (several microseconds for 280 mW/cm^2 at 600–850 nm), high sensitivity (0.8 μJ/cm^2 to allow the use of a low-power laser diode), high efficiency (3%), low operational voltage (20 V), a large index change (0.06), and a reasonable resolution (50 lines/mm).

A high-speed joint transform correlator that uses this device has been successfully demonstrated [63]. The system uses a low-power laser diode (3 mW) and is capable of 3×10^5 correlations/s. When the system is incorporated with a fast ferroelectric liquid-crystal SLM and a fast detector, several thousand correlations can be achieved within a second. This speed is $\sim$2 orders of magnitude faster than that of digital computers.

Although the above correlators are described in one-dimensional terms, many combinations of the above architectures are possible [64, 65].

The classical holographic pattern recognition system in (e) has had limited application because it is bulky and is too sensitive to misalignment, input recording materials, and filters. Such problems can be greatly alleviated with a SELDA. Two recently developed correlators based on microlasers are introduced below.

12.3.3 Multichannel optical correlator based on a mutually incoherent microlaser array

Yang and Gregory [66] demonstrated a multichannel optical correlator by using the mutually incoherent property of a microlaser array to improve the performance of an optical correlator with respect to immunity to coherent noise and a high SNR as well as high light throughput. As shown in Fig. 12.9, an optical correlator is incoherently duplicated by an array of light sources to increase the SNR.

In the system, the Fourier spectra due to different microlasers can overlap one another if the separations between neighboring microlasers are adjusted to match the period of a SLM in the filter plane. Although a matched filter was synthesized with a single reference pattern, it is duplicated at many periodic locations in the Fourier plane and matches the multiple-input spectra. In this way, an efficient space–bandwidth product in the filter plane is increased and sharper correlation peaks are obtained.

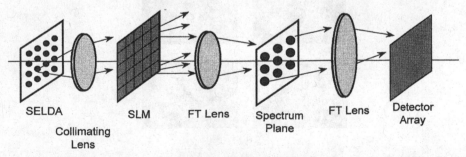

SELDA   SLM   FT Lens   Spectrum Plane   FT Lens   Detector Array

Collimating Lens

Fig. 12.9. Multichannel optical correlator with a mutually incoherent microlaser array (after Ref. 66). FT, Fourier-transform.

Such an approach would be extremely useful for increasing the reliability of a conventional optical correlator. The only practical issues here are generation of reference beams that are coherent with object beams to record holograms and the removal of undesired fringes in an autocorrelation peak. The authors suggested a time-division-multiplexing (TDM) scheme as a way to eliminate interference fringes.

12.3.4 Compact and robust incoherent correlator

As explained above, the light from a microlaser is highly monochromatic. However, the phases of the light from the microlasers are not locked with each other. These two unique coherence properties (temporally highly coherent and spatially incoherent) make a SELDA an ideal light source for implementing a compact and robust incoherent correlator.

Figure 12.10(a) shows an example of the microlaser-based incoherent correlator [67]. The light from each SELDA is collimated by lens L_1 and illuminates the hologram to reconstruct holographic images on the output plane, which is located at the focal plane of lens L_2. The image generated by each SEL is shifted by the amount that corresponds to the position of the SEL. The reconstructed images generated by the light from different SEL's are added up incoherently because each laser operates independently, averaging out the phase-sensitive interference terms. The eventual summation of all the reconstructed images generated by all the SEL's in the input plane gives the correlation between the input and the reference image stored on the hologram. Because the system does not involve any moving parts (e.g., rotating diffuser) or bulky optical components, the whole system can be miniaturized and integrated with semiconductor technologies.

Figure 12.10(b) shows the correlation output obtained from the SELDA correlator. A holographic filter was fabricated for the input pattern (Bell logo) and was tested for the two input patterns (the Bell logo and a Chinese character meaning light). For the correct

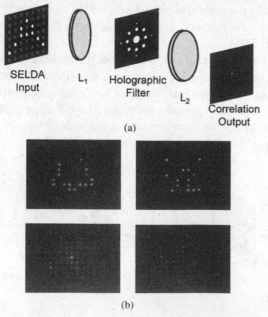

Fig. 12.10. Compact and robust incoherent correlator: (a) system, (b) experimental results. (Reprinted from Ref. [67] by Paek et al.)

input [Bell logo], a bright autocorrelation peak appears at the center of the correlation output (bottom left figure). On the other hand, for the incorrect input [Chinese character (middle right figure)], a cross-correlation is obtained (bottom right figure). As shown in the figure, the cross-correlation signal is much weaker than the autocorrelation peak, allowing a satisfactory discrimination between the two input patterns.

12.4 Holographic memory readout with microlasers

12.4.1 Introduction

Volume holographic memory has been extensively investigated in the past as a way of massively storing media to allow fast random-access page-organized memory [68–72]. This volume holographic memory has reawakened interest recently because of the immense demand for storage media with fast access and a large storage capacity for applications such as multimedia. However, the holographic system normally requires bulky and complicated beam deflectors to steer the beam direction from one to another, corresponding to the desired page. Moreover, speed and resolution of a beam deflector are quite limited.

12.4.2 Compact and ultrafast holographic memory with a SELDA

Such traditional problems can be solved with a SELDA [73]. As is explicitly shown in Fig. 12.11, a SELDA combined with a simple collimating lens can function as an efficient multiple-beam steerer to change the beam direction from one to the other. Such beam steering can be achieved within one nanosecond which is the switching speed of the microlasers. This ultrafast beam steering is used to reconstruct angular-multiplexed holographic memories in which each image is recorded using the second beam (reference beam) propagating along a different angle. By using this simple and compact optical setup, any frame can be randomly accessed within one billionth of a second.

In the experimental results shown in Fig. 12.11, memories are recorded in a volume hologram ($LiNbO_3$ crystal, 0.01% Fe doped) that can provide a large storage capacity of 10^{11}–10^{12} bits because of the third dimension available in a volume storage medium. The figures on the right-hand side show the images retrieved from the volume hologram by the light from a SELDA. Each laser is separated from the adjacent one by 70 μm, corresponding to an

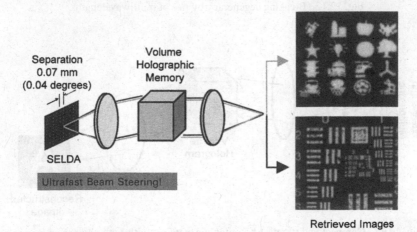

Retrieved Images

Fig. 12.11. Compact and ultrafast holographic memory with a microlaser array.

angular separation of ~0.040°. As can be seen in the figure, the two independent lasers separated by only 70 μm reconstruct totally different images. In this way, each tiny microlaser can be matched to a separate page, allowing the array to form a selective address generator.

12.4.3 Combined angular and wavelength multiplexing with a two-dimensional MC-SELDA

In general, angular multiplexing can be achieved along one direction in a plane formed by both reference and object beam directions because of the degeneracy effect. As can be seen in Fig. 12.12(a), the Bragg condition is satisfied by waves that have any of the directions that form the surface of a cone (the shaded area in the figure) whose axis is normal to the grating planes [74]. This degeneracy causes cross-talk along the direction perpendicular to the Bragg direction. However, this cross-talk can be greatly reduced by use of different wavelengths along the degenerate directions, as shown in Fig. 12.12(b). Such a concept has been recently demonstrated with five different wavelengths to record a total of 2000 holograms: five different wavelengths with 400 angularly multiplexed images for each wavelength [75].

The concept can also be implemented in a compact and ultrafast fashion by a 2-D MC-SELDA, as shown in Fig. 12.13. The SELDA has the same wavelength along the angular-

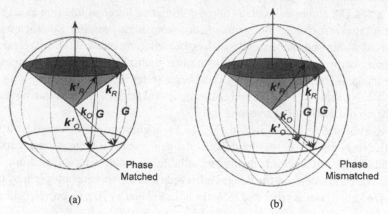

Fig. 12.12. Breaking degeneracy by use of multiwavelengths.

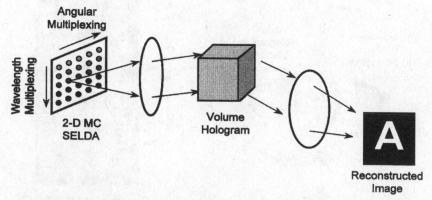

Fig. 12.13. Compact and ultrafast holographic memory with both angular and wavelength multiplexing by a 2-D MC-SELDA.

multiplexing direction and different wavelengths along the degeneracy direction. In addition, such a combinational multiplexing with a MC-SELDA has an additional angular separation effect along the degenerate direction because each microlaser lies in a different position. In this way, the storage capacity of a volume hologram can be greatly increased in a compact and ultrafast readout system within the total limit imposed by the available number of charge carriers in a medium and the theoretical limit of V/λ^3, as was proved by Lee et al. [76] and Psaltis et al. [77].

12.4.4 Incoherent–coherent multiplexing with a SELDA

An incoherent–coherent multiplexing approach to reduce the noise due to multiplexing by use of an array of microlasers has been developed by Jenkins and Tanguay [78]. The concept of the method can be understood as follows:

When multiple gratings are recorded simultaneously by a coherent light source, all the cross terms are added coherently. In other words, the amplitudes of all the gratings are summed first and squared later. As a result, the number of noise gratings (or the number of cross-talk terms) becomes N^2, where N is the number of gratings. On the other hand, as in this proposed method, if each grating is recorded sequentially one by one, the cross terms are incoherently added (each grating amplitude is squared first and summed over all the gratings later), and thus the number of noise gratings becomes N; therefore cross-talk can be greatly reduced. Such an incoherent recording can be achieved simultaneously with an array of independent (not phase locked with each other) coherent light sources, such as a SELDA. Jenkins and Tanguay applied this noise reduction method to photonic interconnection and holographic optical elements for 2-D wavelength-division-multiplexing applications.

This technology of using microlasers for holographic storage will be extremely useful when suitable holographic recording materials become available in the future. Recent developments in biological holographic recording materials, such as bacteriorhodopsin, show a great new promise for future computers [79].

12.5 Microlasers for holographic associative memory

12.5.1 Introduction

Recent advances in neural networks opened many new possibilities for optical information processing for broad application areas [80–84]. Optics, especially coherent optics, has found an excellent match in implementing the neural networks that require parallel and analog computing.

Figure 12.14 illustrates an associative memory for word-break recognition to generate a readable text from a continuous string of words [85]. In the system, an input word stream is presented at the input plane of the system. Autocorrelation peaks that appear at places where there is a match between the input and the memory words prerecorded in a hologram are detected. All the spurious sidelobes are removed by a threshold operation in the correlation plane. Next, the separation between the peaks is magnified along the word direction. This stretched correlation output is reflected back to illuminate the hologram and reconstruct the whole memory at the output plane. The output through a window that is situated at the origin of the output plane is the desired readable text, with all the errors corrected and spaces inserted between words.

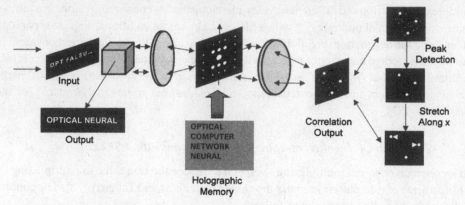

Fig. 12.14. Holographic associative memory for word-break recognition. (Reprinted from Ref. [85] by Paek et al.)

12.5.2 Holographic neurons

Such a neural network system normally consists of three parts: (1) a recognition part to compare an input with all the memories, (2) a reconstruction part to retrieve the corresponding memory, and (3) nonlinear thresholding elements to make decisions. Recognition (pattern recognition in Subsection 12.3) and reconstruction (holographic memory in Subsection 12.4) have already been described. Below we focus on the third part, nonlinear thresholding elements, or so-called optical neurons.

There has been much research done on optical neurons. Recently an integrated vertical-cavity surface-emitting microlaser array with HPT's to yield optically controlled lasers has been demonstrated by Chan *et al.* [30]. The structure of the device is shown in Fig. 12.5, as explained above. Figure 12.5(b) shows the output versus input optical power relationship for various bias voltages. As shown in the figure, the device displays a suitable nonlinear threshold function as an optical neuron with the threshold input optical power of 0.1 mW (at the wavelength of 855 nm). It also provides an optical gain of approximately 5, making it an excellent candidate for an optical neuron array.

Besides its nonlinear threshold function, the output light from the optical neuron is coherent. It is a laser; therefore it can reconstruct any type of hologram. In Fig. 12.15(a), the neuron with an input power above threshold is fired to reconstruct the corresponding memory. Figure 12.15(b) shows an experimental result. If the output-light signal is below threshold, nothing appears in the output plane (left). On the other hand, as soon as the input light is increased above the threshold level, the neuron is fired and the corresponding holographic memory is retrieved (right).

12.5.3 Microlaser-based holographic associative memory

When the optical neurons are combined with both recognition and reconstruction parts, which were demonstrated above, in a simple two-lens system, a compact, ultrafast, and highly efficient neural network system may be implemented in the future. Furthermore, the bulk lenses can be replaced by planar zone plates, allowing the integration of the whole system into the smallest scale possible.

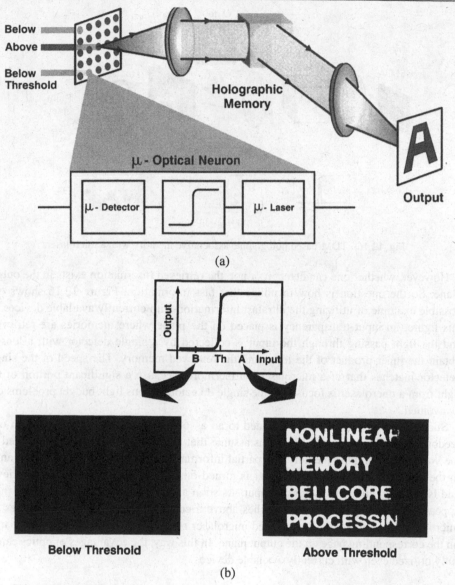

Fig. 12.15. Holographic memory readout with an array of holographic neurons: (a) system, (b) experimental results.

12.5.4 Holographic associative memory with time-division multiplexing

In previous work on holographic memory readout with a SELDA, it has been shown that the memory is capable of retrieving a 2-D information with approximately 10^5 bits within 1 ns, with a total storage capacity of 10^{11}–10^{12} bits in 1 cm^3. However, the question is how to detect such ultrafast information, 10^{14} bits/s or 100 Tbits/s. This is certainly difficult, first because the speed of the current 2-D image sensor is much slower than 1 Gframe/s, and second because the light emitted from each microlaser for only 1 ns has to be distributed over a 2-D image, creating a light budget problem.

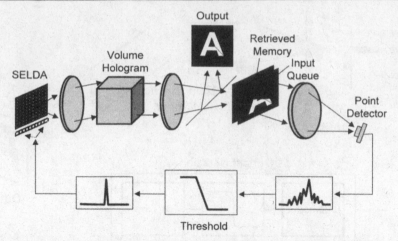

Fig. 12.16. TDM-based holographic associative memory with a microlaser.

However, whether one can detect it or not, the retrieved information exists in the output plane. So the question is how to utilize that fast information. Figure 12.16 shows one possible example of utilizing the ultrafast information with currently available devices. In this figure, an input transparency is placed on the plane where memories are retrieved, and the light passing through the input is collected by a single detector with a lens to obtain the inner product of the input and the retrieved memory. The speed of the single detector matches that of a microlaser. Furthermore, because a significant portion of the light from a microlaser is focused to the single detector, serious light budget problems can be avoided.

Such a TDM system can be extended to an associative memory by the addition of a feedback loop, as in Fig. 12.16. Let us assume that a whole encyclopedia is recorded on the volume holographic memory and partial information on one of the pages is presented in the input plane. As each microlaser is turned on, the corresponding page is retrieved and is instantly compared with the input. As soon as the memory that matches the input appears, the inner product value reaches above a certain threshold level and latches the microlaser, The light from the matched microlaser reconstructs the complete information on the corresponding page on the output plane. In this way, the advantages of optics can be fully utilized even with currently available devices.

12.6 Integration and packaging

To answer the issue of how to align and package a complex optical system in free-space optics, a planar integration technology has been developed [86–90]. As can be seen in Fig. 12.17, the free-space coherent optical processor in Fig. 12.17(a) can be integrated to a planar module, as shown in Fig. 12.17(b).

The main motivation for this approach is to culminate the well-established high-precision lithographic technology that has been extremely successful in leading to the mass production of semiconductor components. In this way, optical systems may be designed with a computer-aided design tool and mass produced by a foundry with the same or a similar technology as that of conventional semiconductors. The main benefits of using such an integrated module are

- Compatibility with semiconductor fabrication technology.

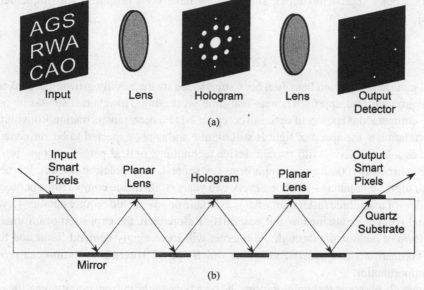

Fig. 12.17. Planar integration of a coherent optical processor (after Ref. 86): (a) coherent optical processor, (b) planar integration.

- A compact and robust system that does not require any alignment.
- Because all components are on a common substrate, the whole system is stable with respect to changes in temperature, dust, humidity, etc.
- High-precision (close to 1-μm accuracy) prealignment that can avoid time-consuming alignments each time.
- Possibility of placing all alignment-critical components on one side of the substrate.
- Easy integration with other semiconductor components that share the same technology.
- Easy mass production.

However, the approach also raises several issues to be resolved, i.e.,

- Aberrations that are due to the oblique angles of light propagation or off-axis holographic optical elements.
- Limited optical path lengths are available on a thin substrate.
- A high-resolution lithographic technique is required for fabricating high-quality (large space–bandwidth product) diffractive optical elements to replace conventional bulk lenses in this approach. This is even more serious in this case than in conventional diffractive optical components in free-space optics because the light path length is short and all the other components need to be scaled down accordingly.

A hybrid integration of a microlaser chip and a planar optics substrate has been implemented [89]. In the system, a top-surface-emitting microlaser array (9-μm diameter, 85-μm spacing, and 850-nm wavelength) was bonded to a single quartz glass substrate by the flip-chip bonding technique with In solder bumps. The lenses were lithographically fabricated by the diffractive optical element technique with four phase levels. They have a square shape with a size of 500 μm × 500 μm. Their focal lengths match the thickness of the glass substrate, which is 3 mm. Surrounding areas are coated with a Ge film to absorb any undesired stray light. The deflection angle relative to the normal is limited by the resolution of the lithographic process; it was 6.2° in this experiment.

Various other approaches along similar lines have been proposed and implemented [91–93].

12.7 Conclusion

Optical pattern recognition has often been superseded by the rapidly growing digital computer technology that approaches near-real-time processing. However, it should be noted that the computational power of optics to compute 2-D Fourier transformation, convolution, and correlation at the speed of light is still unique and is not expected to be surpassed by other technologies. Even with current device technology, optical pattern recognition can achieve more than 1000 image comparisons per second. Extrapolating the developmental speed of digital computers – doubling every two years, 1000 image comparisons per second (~2 orders of magnitude faster than what is available now) – would require another 15 years or so and will not be attained by the year 2010. Before then, the explosion of information in the form of multimedia through the Internet will increasingly demand faster and faster image comparisons and understanding, for which optical pattern recognition can provide many opportunities.

Fortunately photonic technology forms the backbone of the current communication technology. The current transmission speed of an optical signal is moving beyond 1 THz, much faster than the clock speed of a digital computer. Total available optical bandwidths can be as high as 25 THz near the communication wavelengths. These bandwidths give a clear edge to optics-based information technology for storage, displays, and processing. In this regard, optical pattern recognition has a bright future if it is closely tied to other photonic technologies and takes advantage of new photonic devices. For example, the developmental speed of SLM's has been much slower than that of photonic devices for communications.

Undoubtedly, surface-emitting microlasers will have rich new possibilities for pattern recognition as the device technology becomes mature enough to allow low (of the order of microwatts) power consumption per laser, high reliability, integration with other devices, and suitable addressing schemes for high-density arrays, etc.

Current optical pattern recognition systems need to be greatly improved in terms of not just speed, but of accuracy. This accuracy requires the development of elaborate measurement methods to calibrate and test performances. Such test and measurement issues need to be seriously addressed in the future before any optical systems are to be used for practical purposes.

References

[1] K. Iga, S. Ishikawa, S. Ohkouchi, and T. Nishimura, "Room-temperature pulsed oscillation of GaAlAs/GaAs surface-emitting injection laser," Appl. Phys. Lett. **45**, 348–350 (1984).

[2] H. Soda, K. Iga, C. Kitahara, and Y. Suematsu, "GaInAsP/InP surface emitting injection lasers," Jpn. J. Appl. Phys. **18**, 2329–2330 (1979).

[3] J. L. Jewell, A. Scherer, S. L. McCall, Y. H. Lee, S. Walker, J. P. Harbison, and L. T. Florez, "Low threshold electrically pumped vertical-cavity surface-emitting microlasers," Electron. Lett. **25**, 1123–1124 (1989).

[4] J. L. Jewell, J .P. Harbison, and A. Scherer, "Microlasers," Sci. Am. Nov. 86–94 (1991).

[5] J. L. Jewell, J. P. Harbison, A. Scherer, Y. H. Lee, and L. T. Florez, "Vertical-cavity surface-emitting lasers: design, growth, fabrication, characterization," IEEE J. Quantum Electron. **27**, 1332–1346 (1991).

[6] Y. H. Lee, B. Tell, K. F. Brown-Goebeler, R. E. Leibenguth, and V. D. Mattera, "Deep-red continuous wave top-surface-emitting vertical-cavity AlGaAs superlattice lasers," IEEE Photon. Technol. Lett. **3**, 108–109 (1991).

[7] Y. H. Lee, B. Tell, K. F. Brown-Goebeler, J. L. Jewell, and J. M. Hove, "Top-surface-emitting GaAs four-quantum-well lasers emitting at 0.85 μm," Electron. Lett. **26**, 710–712 (1990).

[8] G. M. Yang, M. H. McDougal, and P. D. Dapkus, "Ultralow threshold VCSEL's fabricated by selective oxidation from all epitaxial structure," in *Conference on Lasers and Electro-Optics*, Vol. 15 of 1995 OSA Technical Digest Series (Optical Society of America, Washington, DC, 1995), paper CPD4.

[9] K. D. Choquette, R. P. Schneider Jr., and K. M. Geib, "Low threshold voltage vertical-cavity lasers fabricated by selective oxidation," Electron. Lett. **30**, 2043–2044 (1994).

[10] K. L. Lear, K. D. Choquette, R. P. Schneider Jr., S. P. Kilcoyne, and K. M. Geib, "Selectively oxidized vertical-cavity surface-emitting lasers with 50% power conversion efficacy," Electron. Lett. **31**, 208–209 (1995).

[11] A. Hayashi, T. Mukaihara, N. Hatori, A. Matsutani, F. Koyama, and K. Iga, "Lasing characteristics of low-threshold oxide confinement InGaAs-GaAlAs vertical-cavity surface-emitting lasers," IEEE Photon. Technol. Lett. **7**, 1234–1236 (1995).

[12] H.-J. Yoo, J. R. Hayes, N. Andreadakis, E. G. Paek, G. K. Chang, J. P. Harbison, L. T. Florez, and Y. S. Kwon, "Low series resistance vertical-cavity front-surface-emitting laser diode," Appl. Phys. Lett. **56**, 1942–1944 (1990).

[13] T. Numai, M. Sugimoto, I. Ogura, H. Kosaka, and K. Kasahara, "Current versus light output characteristics with no definite threshold in *pnpn* vertical to surface transmission electro-photonic devices with a vertical-cavity," Jpn. J. Appl. Phys. **30**, L602–L604 (1991).

[14] G. Bjork, A. Karlsson, and Y. Yamamoto, "On the linewidth of microcavity lasers," Appl. Phys. Lett. **60**, 304–306 (1992).

[15] U. Mohideen, R. E. Slusher, F. Janke, and S. W. Koch, "Semiconductor-microlaser linewidths," in *International Quantum Electronics Conference* (Optical Society of America, Washington, DC, 1994), p. 153.

[16] G. R. Olbright, R. P. Bryan, W. S. Fu, R. Apte, D. M. Bloom, and Y. H. Lee, "Linewidth, tunability, and VHF-millimeter wave frequency synthesis of vertical-cavity GaAs quantum-well surface-emitting laser diode arrays," IEEE Photon. Technol. Lett. **3**, 779–781 (1991).

[17] T. E. Sale, *Vertical-Cavity Surface-Emitting Laser* (Wiley, New York, 1995).

[18] R. A. Morgan, G. D. Guth, M. W. Focht, M. T. Asom, K. Kojima, L. E. Rogers, and S. E. Callis, "Transverse mode control of vertical-cavity top-surface-emitting lasers," IEEE Photon. Technol. Lett. **4**, 374–377 (1993).

[19] H. J. Yoo, J. R. Hayes, E. G. Paek, A. Scherer, and Y. S. Kwon, "Array mode analysis of two-dimensional phased arrays of vertical-cavity surface-emitting lasers," IEEE J. Quantum Electron. **26**, 1039–1051 (1990).

[20] J. A. Lott and R. P. Schneider Jr., "Electrically injected visible (639–661 nm) vertical-cavity surface-emitting lasers," Electron. Lett. **29**, 830–832 (1993).

[21] R. P. Schneider Jr., K. D. Choquette, J. A. Lott, K. L. Lear, J. J. Figiel, and K. J. Malloy, "Efficient room-temperature continuous-wave AlGaInP/AlGaAs visible (670 nm) vertical-cavity surface-emitting laser diodes," IEEE Photon. Technol. Lett. **6**, 313–316 (1994).

[22] R. Thornton, Y. Zou, J. Tramontana, M. Hagerott Crawford, R. P. Schneider, and K. D. Choquette, "Visible (670 nm) vertical-cavity surface-emitting lasers with indium tin oxide transparent conducting top contacts," in *IEEE Lasers and Electro-Optics Society 1995 Annual Meeting* (IEEE, New York, 1995), Vol. 2, pp. 108–109.

[23] M. H. Crawford and R. P. Schneider Jr., "Performance of high-efficiency AlGaInP-based red VCSELs," in *Conference on Lasers and Electro-Optics*, Vol. 15 of 1995 OSA Technical Digest Series (Optical Society of America, Washington, DC, 1995), pp. 168–169.

[24] S. Yoshii, T. Yokogawa, A. Tsujimura, Y. Sasai, and J. L. Merz, "Demonstration of blue vertical-cavity surface-emitting laser diode," in *1995 53rd Annual Device Research Conference Digest* (IEEE Electron Devices Society, 1995), pp. 138–139.

[25] H. Jeon, V. Kozlov, P. Kelkar, A. V. Nurmikko, D. C. Grillo, J. Han, M. Ringle, and R. L. Gunshor, "Optically pumped blue-green vertical-cavity surface-emitting laser," in *Conference on Lasers and Electro-Optics*, Vol. 15 of 1995 OSA Technical Digest Series (Optical Society of America, Washington, DC, 1995), pp. 33–34.

[26] See for example F. Su, "Blue laser will triple optical data storage capability," OE Rep. No. 152 (SPIE, Bellingham, WA, 1996).

[27] A. Von Lehmen, C. Chang-Hasnain, J. Wullert, L. Carrion, N. Stoffel, L. Florez, and J. Harbison, "Independently addressable InGaAs/GaAs vertical-cavity surface-emitting laser arrays," Electron. Lett. **27**, 583–585 (1991).

[28] M. Orenstein, A. C. Von Lehmen, C. Chang-Hasnain, N. G. Stoffel, L. Florez, J. P. Harbison, and L. T. Florez, "Matrix-addressable vertical-cavity surface-emitting laser array," Electron. Lett. **27**, 437–438 (1991).

[29] Y. Zou, R. Thornton, J. Tramontana, M. Hibbs-Brenner, and R. Morgan, "High density, high power arrays of vertical-cavity surface-emitting lasers operating at 850 nm," in *IEEE Lasers and Electro-Optics Society 1995 Annual Meeting* (IEEE, New York, 1995), Vol. 2, pp. 443–444.

[30] W. K. Chan, J. P. Harbison, A. C. Von Lehmen, L. T. Florez, C. K. Nguyen, and S. A. Schwarz, "Optically controlled surface-emitting lasers," Appl. Phys. Lett. **58**, 2342–2344 (1991).

[31] K. Matsuda, H. H. Adachi, T. Chino, and J. Shibata, "Integration of InGaAsP/InP optoelectronic bistable switches with a function of optical erasing," IEEE Electron. Devices Lett. **11**, 442–444 (1990).

[32] D. S. Chemla, D. A. B. Miller, and P. W. Smith, "Nonlinear optical properties of GaAs/GaAlAs multiple quantum well material: phenomenon and applications," Opt. Eng. **24**, 556–564 (1985).

[33] K. Kasahara, Y. Tashiro, N. Hamao, M. Sugimoto, and T. Yanase, "Double heterostructure optoelectronic switch as a dynamic memory with low-power consumption," Appl. Phys. Lett. **52**, 679–681 (1988).

[34] T. Ohtoshi, T. Kuroda, A. Aiwa, and S. Tsuji, "Dependence of optical gain on crystal orientation in surface-emitting lasers with strained quantum wells," Appl. Phys. Lett. **65**, 1886–1887 (1994).

[35] M. Takahashi, P. Vaccaro, K. Fujita, T. Watanabe, T. Mukaihara, F. Koyama, and K. Iga, "An InGaAs-GaAs vertical-cavity surface-emitting laser grown on GaAs(311) substrate having low threshold and stable polarization," IEEE Photon. Technol. Lett. **8**, 737–739 (1996).

[36] C. Chang-Hasnain, J. P. Harbison, C. E. Zah, M. W. Maeda, L. T. Florez, N. G. Stoffel, and T. P. Lee, "Multiple wavelength tunable surface-emitting laser arrays," IEEE J. Quantum Electron. **27**, 1368–1376 (1991).

[37] Y. Wupen, G. S. Li, and C. J. Chang-Hasnain, "Multiple-wavelength vertical-cavity surface-emitting laser arrays with a record wavelength span," IEEE Photon. Technol. Lett. **8**, 4–6 (1996).

[38] M. S. Wu, E. C. Vail, G. S. Li, W. Yuen, and C. J. Chang-Hasnain, "Tunable micromachined vertical-cavity surface-emitting laser," Electron. Lett. **31**, 1671–1672 (1995).

[39] M. G. Peters, D. B. Young, F. H. Peters, J. W. Scott, B. J. Thibeault, and L. A. Coldren, "17.3% peak wall plug efficiency vertical-cavity surface-emitting lasers using lower barrier mirrors," IEEE Photon. Technol. Lett. **6**, 31–33 (1994).

[40] K. L. Lear, K. D. Choquette, R. P. Schneider Jr., S. P. Kilcoyne, and K. M. Geib, "Vertical-cavity surface-emitting lasers with 50% power conversion efficiency," in *Conference on Lasers and Electro-Optics*, Vol. 15 of 1995 OSA Technical Digest Series (Optical Society of America, Washington, DC, 1995), p. 55.

[41] H. Yokoyama and S. D. Brorson, "Rate equation analysis of microcavity lasers," J. Appl. Phys. **26**, 1492–1499 (1990).

[42] G. Shtengel, H. Temkin, P. Brusenback, T. Uchida, M. Kim, C. Parsons, W. E. Quinn, and S. E. Swirhun, "High-speed vertical-cavity surface-emitting lasers," IEEE Photon. Technol. Lett. **5**, 1359–1363 (1993).

[43] A. Karlsson, R. Schatz, and G. Bjork, "On the modulation bandwidth of semiconductor microcavity lasers," IEEE Photon. Technol. Lett. **6**, 1312–1314 (1994).

[44] C. T. Troy, "Vertical-cavity laser hits new power levels high-temperature operation broadens applications," Photon. Spectra, p. 24 (1993).

[45] R. A. Morgan, K. Kojima, M. T. Asom, G. D. Guth, and M. W. Focht, "1W (pulsed) vertical-cavity surface-emitting laser," Electron. Lett. **29**, 206–207 (1993).

[46] See, for example, S. Jutamulia, ed., Special issue on optical information processing, Proc. IEEE **84**(5) (1996).

[47] F. T. S. Yu and D. A. Gregory, "Optical pattern recognition," Proc. IEEE **84**, 733–752 (1996).

[48] Y. Ichioka, T. Iwaki, and K. Matsuoka, "Optical information process in and beyond," Proc. IEEE **84**, 694–719 (1996).

[49] E. R. Kretzmer, Bell Teleph. J. **31** (1951).

[50] L. S. E. Kovasznay and A. Arman, "Optical autocorrelation measurement of two-dimensional random patterns," Rev. Sci. Instrum. **28**, 793–797 (1957).

[51] D. McLachlan Jr., "The role of optics in applying correlation functions to pattern recognition," J. Opt. Soc. Am. **52**, 454–459 (1962).

[52] J. Tanida and Y. Ichioka, "Optical logic array processor using shadowgrams," J. Opt. Soc. Am. **73**, 800–809 (1983).

[53] M. A. Monahan, K. Bromley, and R. P. Bocker, "Incoherent optical correlators," Proc. IEEE **65**, 121–129 (1977).

[54] K. Bromley, "An optical incoherent correlator," Opt. Acta **21**, 35–41 (1974).

[55] A. VanderLugt, "Signal detection by complex spatial filtering," IEEE Trans. Inf. Theory **IT-10**, 139–145 (1964).

[56] C. S. Weaver and J. W. Goodman, "A technique for optically convolving two functions," Appl. Opt. **5**, 1248–1249 (1966).

[57] J. E. Rau, "Detection of differences in real distributions," J. Opt. Soc. Am. **56**, 1490–1494 (1966).

[58] F. T. S. Yu and X. J. Lu, "A real-time programmable joint transform correlator," Opt. Commun. **52**, 10–16 (1984).

[59] J. L. Horner and P. D. Gianino, "Phase-only matched filtering," Appl. Opt. **23**, 812–816 (1984).

[60] D. Psaltis, E. G. Paek, and S. S. Venkatosh, "Optical image correlation with binary spatial light modulator," Opt. Eng. **23**, 698–704 (1984).

[61] B. Javidi and J. L. Horner, *Real-time optical information processing* (Academic, New York, 1994).

[62] A. Partovi, A. M. Glass, D. H. Olson, G. J. Zydzik, H. M. O'Bryan, T. H. Liu, and W. H. Knox, "Cr-doped GaAs/AlGaAs semi-insulating multiple quantum well photorefractive devices," Appl. Phys. Lett. **62**, 3088–3090 (1993).

[63] A. Partovi, A. M. Glass, T. H. Chiu, and D. T. H. Liu, "High-speed joint-transform optical image correlator using GaAs/AlGaAs semi-insulating multiple quantum wells and diode lasers," Opt. Lett. **18**, 906–908 (1993).

[64] D. Psaltis, "Two-dimensional optical processing using one-dimensional input devices," Proc. IEEE **72**, 962–974 (1984).

[65] E. G. Paek, C. H. Park, F. Mok, and D. Psaltis, "Acousto-optic image correlators," in *Hybrid Image Processing*, D. P. Cosasent and A. G. Tescher, eds., Proc. SPIE **638**, 25–31 (1986).

[66] X. Yang and D. A. Gregory, "Multichannel optical correlator based on a mutually incoherent microlaser array," Opt. Lett. **20**, 2405–2407 (1995).

[67] E. G. Paek, J. R. Wullert II, M. Jain, A. Von Lehmen, A. Scherer, J. Harbison, L. T. Florez, H. J. Yoo, and R. Martin, "Compact and robust incoherent holographic correlator using a surface-emitting laser-diode array," Opt. Lett. **16**, 937–939 (1991).

[68] P. J. van Heerden, "Theory of optical information storage in solids," Appl. Opt. **2**, 393–400 (1963).

[69] D. Psaltis and F. Mok, "Holographic memories," Sci. Am. 70–76 (Nov. 1995).

[70] For recent trends of holographic storage, see the special issue on optical memory, Appl. Opt. **35**(14) (1996).

[71] F. H. Mok, M. G. Tackitt, and H. M. Stoll, "Storage of 500 high-resolution holograms in a LiNbO$_3$ crystal," Opt. Lett. **16**, 605–607 (1991).

[72] J. H. Hong, I. McMichael, T. Y. Chang, W. Christian, and E. G. Paek, "Volume holographic memory systems: techniques and architectures," Opt. Eng. **34**, 2193–2203 (1995).

[73] E. G. Paek, J. R. Wullert II, M. Jain, A. Von Lehmen, A. Scherer, J. Harbison, L. T. Florez, H. J. Yoo, and R. Martin, "Compact and ultrafast holographic memory using a surface-emitting microlaser diode array," Opt. Lett. **15**, 341–343 (1990).

[74] R. J. Collier, C. B. Burckhardt, and L. H. Lin, *Optical Holography* (Academic, New York, 1971), pp. 466–467.

[75] S. Campbell and P. Yeh, "Sparse-wavelength angle-multiplexed volume holographic memory system: analysis and advances," Appl. Opt. **35**, 2380–2388 (1996).

[76] H. Lee, X. Gu, and D. Psaltis, "Volume holographic interconnections with maximal capacity and minimal crosstalk," J. Appl. Phys. **65**, 2191 (1989).

[77] D. Psaltis, X. Gu, and D. Brady, "Fractal sampling grids for holographic interconnections," in *Optical Computing'88*, P. Chavel, J. W. Goodman, and G. Robin, eds., Proc. SPIE **963**, 70 (1988).

[78] B. K. Jenkins and A. R. Tanguay Jr., "Incoherent/coherent multiplexed holographic recording for photonic interconnections and holographic optical elements," U.S. patent 5,121,231 (June 9, 1992).

[79] R. R. Birge, "Protein-based computers," Sci. Am. 90–95, March (1995).

[80] J. J. Hopfield, "Neural networks and physical systems with emergent collective computational abilities," Proc. Natl. Acad. Sci. USA **79**, 2554 (1982).

[81] D. Psaltis and N. Farhat, "A new approach to optical information processing based on the Hopfield model," in *Technical Digest of the ICO-13 Conference*, International Commission for Optics, Sapporo, Japan, 1984), p. 24.

[82] N. Farhat, D. Psaltis, A. Prata, and E. G. Paek, "Optical implementations of the Hopfield model," Appl. Opt. **24**, 1469–1475 (1985).

[83] K. Wagner and D. Psaltis, "Optical neural networks: an introduction to a special issue by the feature editors," Appl. Opt. **32**, 1261 (1993).

[84] F. T. S. Yu, T. Lu, X. Yang, and D. A. Gregory, "Optical neural network with pocket-sized liquid-crystal televisions," Opt. Lett. **15**, 863–865 (1990).

[85] E. G. Paek and A. Von Lehmen, "Real-time holographic associative memory for identifying words in a continuous letter string," Opt. Eng. **28**, 519–525 (1989).

[86] J. Jahns and A. Huang, "Planar integration of free-space optical components," Appl. Opt. **28**, 1602–1606 (1989).

[87] K. Brenner and F. Sauer, "Diffractive-reflective optical interconnects," Appl. Opt. **27**, 4251–4254 (1988).

[88] J. Jahns, Y. H. Lee, C. A. Burrus Jr., and J. L. Jewell, "Optical interconnects using top-surface-emitting microlasers and planar optics," Appl. Opt. **31**, 592–597 (1992).

[89] J. Jahns, R. A. Morgan, H. N. Nguyen, J. A. Walker, S. J. Walker, and Y. M. Wong, "Photonic integration of surface-emitting microlaser chip and planar optics substrate for interconnection applications," IEEE Photon. Technol. Lett. **4**, 1369–1372 (1992).

[90] J. Jahns and B. Acklin, "Integrated planar optical imaging system with high interconnection density," Opt. Lett. **18**, 1594–1596 (1993).

[91] W. B. Veldkamp, "Microoptics and microelectronics for image processing," in *1993 IEEE International Solid-State Circuits Conference Digest of Technical Papers* (IEEE, New York, 1994), pp. 130–131.

[92] W. B. Veldkamp, "Wireless focal planes 'On the road to amacronic sensors'," IEEE J. Quantum Electron. **29**, 801–813 (1993).

[93] K. Rastani, M. Orenstein, E. Kapon, and A. C. Von Lehmen, "Integration of planar Fresnel microlenses with vertical-cavity surface-emitting laser arrays," Opt. Lett. **16**, 919–921 (1991).

13

Optical properties and applications of bacteriorhodopsin

Q. Wang Song and Yu-He Zhang

13.1 Introduction

Light has its meaning because of human eyes. Although most optical-related technologies are originated from, inspired by, or do favors for the human visual system, none of them has ever surpassed or even successfully mimicked the sophistication of it. On the imaging side, lenses are widely used in most optical devices; on the receiving side, where the photon and molecule interface is located, however, things become much more complicated. This is a field in which optics, electronics, and biochemistry inevitably interact. The ultimate goal of surpassing the eye system is some distance away, but fundamental investigations have been under way. Finding and characterizing appropriate materials is a fundamental research. In this respect, nothing has come nearer to the material found in the human visual receptor, both in function and form, than bacteriorhodopsin (BR).

BR is the light-harvesting protein in the purple membrane of a micro-organism called *Halobacterium salinarium*. On the absorption of light, BR converts from a dark-adapted state to a light-adapted state. The subsequent absorption of light by the latter generates a photocycle that makes it a useful material in optical applications. The absorption spectra of the key intermediates of the BR photocycle are shown in Fig. 13.1 [1]. The photochemical reactions that have drawn the attention of optics researchers are the **bR** (572 nm) ⇔ **M** (412 nm) and **bR** (576 nm) ⇔ **K** (620 nm) transitions. The high cyclicity, high photosensitivity, long-term stability, and the ease of modification by chemical, physical, or genetic methods make it a promising material in optical data storage and data processing applications [2, 3].

A variety of research has been accomplished in the characterization and the applications of BR. A comprehensive review in the biochemical level has been given in Ref. 1. This chapter concentrates on the review of optical applications. In the characterization part, only optical properties are discussed.

13.2 Optical characterization

The wavelength dependence of the refractive index, the absorption, and the diffraction efficiency of BR film was investigated and reported by Birge *et al.* [4]. The investigation provided a quantitative analysis of the photorefractive and photodiffractive properties associated with the **bR** ⇔ **M** photoreaction. The absorption spectra of **bR** and **M** are shown in Fig. 13.1. Because the deprotonated chromophore has a strongly blue-shifted absorption band, the **M** state plays an important role in many applications that require a large

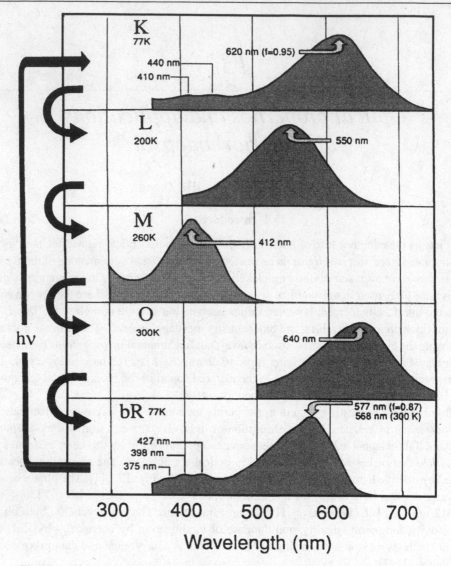

Fig. 13.1. Electronic (one-photon) absorption spectra of light-adapted BR and selected intermediates in the photocycle. Band maxima are indicated in nanometers [1].

shift in the absorption maximum or the refractive index for optimal function. To trap the **M** state, a reduced temperature of $-40\,^{\circ}\mathrm{C}$ is used in the investigation. When a Kramers–Kronig transformation is applied, the dispersion in the refractive index of **bR** can be derived with information on the absorption spectra, which can be easily measured. The diffraction efficiency of a holographic grating stored in a BR film depends on the difference of the refractive index (as well as the absorption) changes in the **bR** and the **M** states. Therefore the dispersion in the refractive change from pure **M** to pure **bR**, i.e., $\Delta n(\mathbf{bR}) - \Delta n(\mathbf{M})$, was calculated and is shown in Fig. 13.2.

The diffraction efficiency of the holographic grating was analyzed by use of the coupled-wave theory equations of Kogelnick [5]. The total diffraction efficiency was considered as the summation of the diffraction efficiencies contributed from the refraction (η_{phase}) and the

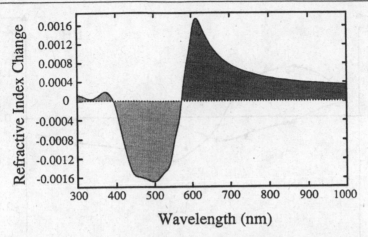

Fig. 13.2. Change in the refractive index associated with the photochemical generation of pure **bR** from a solution containing pure **M** for a solution concentration of 2×10^{-3} M in 65% glycerol:water (v/v) at $-40\,°C$. The refractive-index change in this region of the spectrum is linearly proportional to concentration, and thus a more concentrated solution will exhibit a proportionally larger change in refractive index [4].

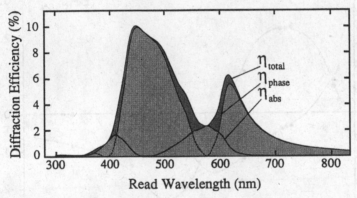

Fig. 13.3. Effect of the read wavelength on the diffraction efficiency of a 30-μm film of BR assuming a protein concentration of 2×10^{-3} M [4].

absorption (η_{abs}). The consideration holds as far as the weak diffraction is concerned. The effect of the read wavelength on the diffraction efficiency of a 30-μm BR film is shown in Fig. 13.3. The integrated diffraction efficiency from 300 to 800 nm is dominated by refractive-index contributions (η_{phase}) that are maximum in regions of minimal **bR** and **M** absorption. The maximum in the refractive component occurs at 451 nm (η_{phase}) whereas the maximum in the absorption component occurs at 575 nm (η_{abs}). The maximum integrated diffraction efficiency is at $\sim$440 nm (η_{totals}). An adequate diffractive performance for most applications was predicted for the write wavelength in the 380–420- and 500–650-nm regions and for read wavelengths from 380 to 740 nm.

Experiments were set up by Zeisel and Hampp [6], who used a modified Michelson interferometer to measure the spectral dependence of the light-induced changes of the refractive index Δn and the absorption Δa in the wildtype BR and a BR variant, **BR**$_{D96N}$. The **BR**$_{D96N}$ differs from the wildtype by a single amino acid exchange, aspartic acid

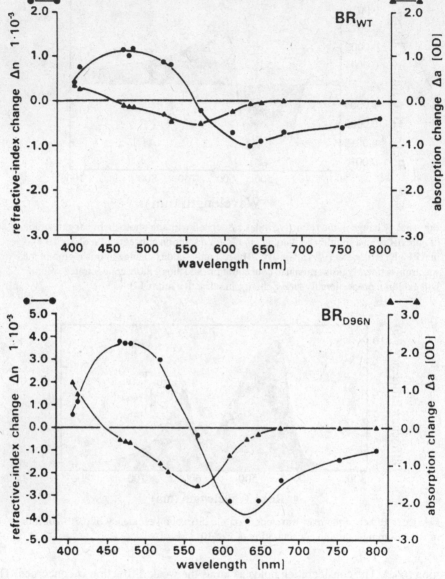

Fig. 13.4. Wavelength-dependent changes of the refractive index Δn (●) and the absorption Δa (▲) for BR_{WT} and $\boldsymbol{BR_{D96N}}$. The thickness of the films is 25 μm, and the iOD for the samples is 3.2 and 3.6, respectively [6].

96 → asparagine. The light from a Kr laser with a 568-nm wavelength and an intensity of 20 mW/cm² was used to pump the BR from the **bR** state to the **M** state. The probe light beam was from either a Kr laser for the wavelengths of 406, 413, 468, 476, 520, 539, 568, 647, 676, 752, and 799 nm or from a He–Ne laser for 633 nm. The refractive-index change was measured in terms of the shift of the interference pattern caused by the interferometer. The thickness of the films used in the experiment was 25 μm.

The measured absorption changes in optical density (OD) units and the changes of the refractive index are plotted in Fig. 13.4 for a wildtype BR_{WT} and a $\boldsymbol{BR_{D96N}}$ film with

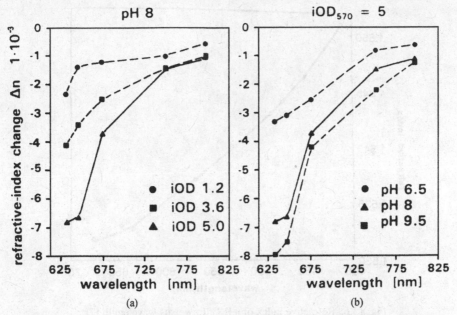

Fig. 13.5. Refractive-index changes in the red and the near-IR (600–800 nm) that are dependent on (a) the OD at a constant pH value of 8, (b) an increasing pH at a constant iOD value of $iOD_{570} = 5$ and a thickness of 25 μm [6].

initial OD's (iOD's) of $iOD_{570} = 3.2$ and $iOD_{570} = 3.6$, respectively. The curve for BR_{WT} in Fig. 13.4 reverses the shape of the predicted curve in Fig. 13.2. The reason is that the change obtained in Fig. 13.4 is for the transition from the **bR** state to the **M** state, whereas the calculation in Fig. 13.2 is based on the opposite transition. The spectral dependences of Δn and Δa were similar for both types of BR films. But the amplitudes obtained with BR_{D96N} films were ~4 times higher than those observed in the BR_{WT} films. Figure 13.5 shows the more detailed analysis of the influence of iOD_{570} and the pH value on the refractive-index changes for BR_{D96N} films in the wavelength range from 600 to 800 nm. The measurements showed that the refractive-index modulation at red and near-IR wavelengths depends more on the pH value of the BR_{D96N} film than on its OD and that it increases with increasing pH.

The calculated diffraction efficiency for BR_{D96N} showed that the maxima for the diffraction efficiency were found at the 647- (~6%) and 468-nm wavelengths. The experimental measurements were ~20% lower than the theoretical prediction by the Kramers–Kronig transformation and the Kogelnick diffraction theory, and a maximal diffraction efficiency of 4.7% was found at 647 nm.

The determination of the refractive index (instead of the refractive-index change) of a BR film was reported by use of a modified critical-angle method [7]. Figure 13.6 shows the measured refractive index of a pure BR film. The dispersive behavior is similar to that of glass in the visible region. The variation of refractive index predicted by the Kramers–Kronig transformation was not observed in this region and was possibly due to the fact that the refractive-index change induced by the absorption change was smaller than 0.003 [4–6], which is comparable with the experimental error of approximately ±0.002 throughout the measurements.

The intensity-dependent refractive index of chemically enhanced BR was investigated by Song *et al.* [8]. The Z-scan method was used in the experiment with two wavelengths

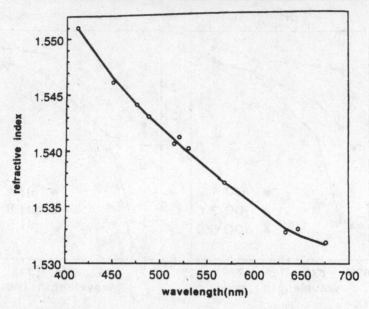

Fig. 13.6. Refractive index of a BR film versus wavelength [7].

of 476 and 633 nm. The results of the measurements showed that there was a saturation intensity I_s of ~3 mW/cm^2 at 476 nm and ~4 mW/cm^2 at 633 nm, and also that (1) at low illumination intensities near I_s, the refractive-index change is negative at 633 nm and positive at 476 nm; (2) at intermediate intensities ($I_s < I < I_s$), the refractive index changes a little or has no changes at all within the measurement accuracy; and (3) at high intensities ($> 10^3\, I_s$), the refractive-index changes were negative at both wavelengths. This observation led to a conclusion that, at intensities lower than the saturation intensity, the linear effect predicted by the Kramers–Kronig transformation was the major cause of the refractive-index changes; at intensities much higher than the saturation intensity, the thermal effect was the dominant effect. The experimental results are shown in Fig. 13.7.

To confirm the influence of the thermal effect futher, an experiment was carried out with a Michelson interferometer [9]. From the data of the measurements, the thermo-optic coefficient with regard to the refractive-index change was calculated to be negative, $6.89 \times 10^{-4}\,^\circ\mathrm{C}^{-1}$. The investigation provided solid evidence to support the previous suggestions that the large optical nonlinearity of the BR film at high illumination was of a thermal origin. The study also indicated that if the thermal expansion of the film was controlled by a rigid holding scheme, the usable thermal nonlinearity could be increased by ~4 times.

13.3 Wave mixing and phase conjugation

Wave mixing and phase conjugation have been intensively investigated with a variety of materials. Unlike photorefractive crystals, BR films are isotropic when no special orientation procedures or light illumination are applied. The reported phase-conjugate reflectivities are at most ~30%, and the diffractive efficiencies are even lower. The anisotropic properties of BR can be developed by illumination of linearly polarized light or by special procedures during the formation of the BR film, but they have not been fully utilized to generate any encouraging result.

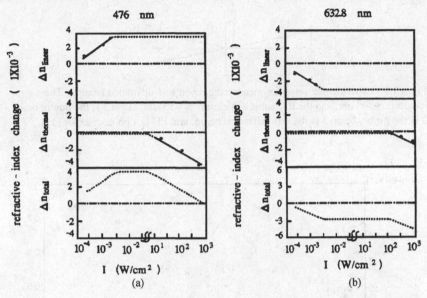

Fig. 13.7. Measured contributions to total change in the refractive-index change in BR as a function of incident intensity: (a) 476 nm, (b) 633 nm [8].

Werner *et al.* presented a series of investigations in wave mixing and phase conjugation with BR [10–12]. With a solid BR-doped polymer film ∼300 μm thick and liquid samples ∼100 μm thick, they measured the reflectivity of the phase-conjugate signal of a four-wave-mixing scheme as a function of the angle between the signal and the pumps. The results were dependent on not only the forms (liquid or solid) of the material but on the wavelength of the pumping beams as well. With a wavelength of 514.5 nm and a solid slide, there was no angle dependence on which the phase-conjugate reflectivity could change. When the wavelength is changed to ∼700 nm, the reflectivities vary from ∼26% at $\theta \approx 2°$ to zero near $\theta \approx 32°$, where θ is the angle between the probe and the pump beams inside the sample. The reflectivity by use of a liquid sample with a 514.5-nm wavelength assumes a similar angle dependence but with less magnitude (reflectivity was ∼0.1%). The resultant conclusions from the above experiments include the following: (1) The resolutions of the solid slide are at least $\lambda/2n = 0.17\,\mu$m, whereas the resolution of the liquid sample is limited to ∼4.6 μm; (2) the effect at 524 nm is resonant and arises from the intensity dependence of the population of the **bR** and the **M** states, whereas in the near IR the mechanism of the effects involves a diffusion process that washes out the grating as the grating period becomes short.

A two-wave-mixing experiment with a setup as shown in Fig. 13.8 was also carried out by this group. Two ∼700-nm beams with an angular separation of ∼2° were intersected to write a refractive-index grating with spacing $\Lambda = 120\,\mu$m in a 450-μm-thick BR film that was in a polyvinyl alcohol polymer matrix. A self-diffracted first-order beam (beam 3 in Fig. 13.8) was created. By using extra illumination with 442-nm light that enhances the **M** → **bR** transition and another beam with a 632.8-nm wavelength that causes the opposite **bR** → **M** transition, it was found that the induced grating was affected by an interplay between the population of the **bR** and the **M** states of the BR. The probe-beam and pump-beam wavelength was 770 nm. The results in Fig. 13.9 show that the uniform extra illumination of the 442- or the 632.8-nm light decreases the visibility of the grating because of the drive

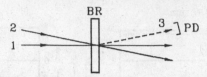

Fig. 13.8. Experimental setup for measurements on self-diffracted beam 3. The wave-length is ~700 nm, and the BR sample thickness is 450 μm; beam 1 is the pump, beam 2 is the probe, beam 3 is the self-diffracted beam, and PD is a photodetector [11].

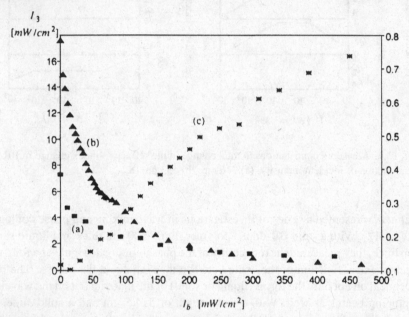

Fig. 13.9. Intensity of first-order diffracted beam 3 (in the two-wave-mixing experiment of Fig. 13.8, with a wavelength of 770 nm) as a function of the intensity I_b of an additional illumination with wavelengths of (a) 442 nm, (b) 632.8 nm, (c) 632.8 and 442 nm simultaneously, where I_b was the varied 442-nm intensity and the 632.8-nm intensity was held constant (354 mW/cm^2). The vertical scale (intensity of I_3) for (a) and (b) is given on the left-hand side and for (c) on the right-hand side. All beam cross sections were approximately 1 mm^2, and the angle between the two writing beams was 1.3° in air. The input intensities I_1 and I_2 were (a) 11.5 and 2.8 W/cm^2, (b) 23 and 5.6 W/cm^2, (c) 17.2 and 4.2 W/cm^2 [11].

of the sample toward saturation (by the 632.8-nm light) or the conversion back to the initial **bR** state (with the 442-nm light). The illumination of both wavelengths simultaneously causes a partial cancellation of this erasure, as shown in curve (c) of Fig. 13.9. These results show that the mechanism of the nonlinearity involves the light-induced population of the **BR** states in addition to the dependence on the grating period of the diffusive effects.

13.4 Real-time holography

With broadband and high cyclicity, BR qualifies as an efficient dynamic holographic media. This material can easily be made into large-area or volumetric media, and it exhibits a high

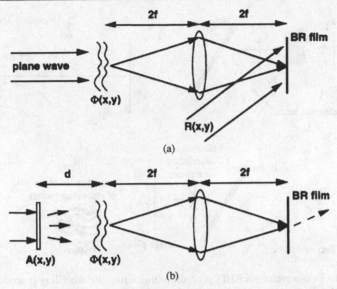

Fig. 13.10. (a) Recording of the hologram with phase aberrator $\Phi(x, y)$, (b) readout of
the hologram with object $A(x, y)$ through the phase aberrator [14].

intrinsic resolution (>4000 line pairs/mm) and a good photocyclicity (>10^6) and can be
designed with variable storage times from milliseconds to hundreds of seconds.

Using a technique proposed and originally demonstrated with conventional holographic
film by Kogelnick and Pennington [13], Downie proposed an application of BR to the
problem of one-way coherent imaging through a thin time-varying phase aberrator [14].

The schematic setup of the recording and the reading of the hologram with a phase
aberrator is shown in Fig. 13.10. In the recording process, the hologram interfered by the
reference beam $R(x, y)$ and the image of the phase aberrator $\Phi(x, y)$ was stored in the thin
BR film. The recorded reference beam $R(x, y)$ was then read out by the image of the object
$A(x, y)$ through the same phase aberrator used in the recording process. The reconstructed
image after the BR film was simply the first-order diffraction of the aberrated object, but
with the phase aberrations on the wave canceled out by the phase-conjugate function stored
in the hologram. Therefore the diffracted wave front that emerges from the hologram is a
replica of the original, undistorted object wave front,

To suppress the scattering noise presented by the BR film because of graininess, Downie
used a polarization hologram in which the interfering object and the reference beams had
orthogonal linear polarizations [15]. Unlike the ordinary holographic materials, BR films
exhibit photoinduced anisotropy as birefringence and dichroism [16]. This allows two orthogonally polarized beams to interact and to form a polarization hologram in the BR film.
The wavelength used was 514 nm in the recording and the reconstruction of the hologram.
The diffraction efficiency was of the order of 0.15% with an exposure of $\sim$10 mJ/cm^2,
and the reconstructed image resolution was greater than 20 line pairs/mm. A time-varying
aberration was also simulated when the fixed aberration was moved manually. The cycle
time was reported to be a few seconds.

An adaptation of resonant holographic interferometric spectroscopy (RHIS) was presented that permits real-time, two-dimensional, background-free species concentration measurements in turbulent flow fields [17]. As shown in Fig. 13.11, two simultaneous holograms
were recorded in the BR film at different laser wavelengths, one tuned near an absorption

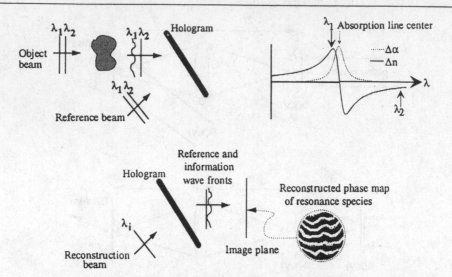

Fig. 13.11. Two-wavelength RHIS procedure illustrating the recording procedure and tuning of lasers 1 and 2 (top) and the reconstruction process (bottom). Note that the reconstruction can be accomplished with any available laser λ_i [17].

line (λ_1) and the other tuned off this feature (λ_2). The phase contributions to the interferogram due to the background are subtracted out during the reconstruction because the two object beams with different wavelengths pass through exactly the same path, and therefore the resulting interference fringes correspond uniquely to the density of the species under interrogation. Figure 13.12 demonstrates the resulting field of view by choosing the strong Na D_2 line near 589 nm as a resonant feature for the real-time RHIS. The BR film used in the experiment was 46 μm thick by 20 mm in diameter and had an OD of 1.1 at 633 nm. The YAG-pumped dye lasers were used with laser 1 tuned near the resonance and laser 2 tuned ~0.03 nm away. The results demonstrated that phase variations due to temporal fluctuations in bulk thermal, pressure, and density gradients were completely subtracted out.

Real-time holographic lensless imaging was demonstrated with BR film as the holographic medium [18]. The reconstructed image had a resolution of ~80 lines/mm and a space–bandwidth product of ~2×10^6. Figure 13.13 shows the details of the reconstructed image with a He–Ne laser wave as the image carrier. As a dynamic holographic material, the thermal decay and the readout-with-erasure effects in BR inevitably decrease the diffraction efficiency and the phase-conjugate reflectivity of the hologram. When the initial and the steady states were separated during the readout process, a linear dependence between the initial peak reflectivity and the reading-light intensity was revealed, as shown in Fig. 13.14. Experimental results also demonstrated that the evolution of the diffraction efficiency during the recording process was not a monoexponential procedure. Instead, a peak value of the diffraction efficiency was reached before decreasing to the steady state. These investigations suggested that an optimal performance might be reached by manipulation of the duration of the reading and recording laser pulses.

13.5 Spatial light modulators

Optically addressed spatial light modulators (SLM's) were investigated and demonstrated by use of BR films [19]. The modulation of the local absorption of the BR film was controlled by

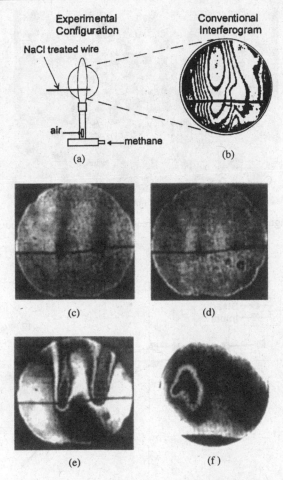

Fig. 13.12. (a) Field of view and Na-seeding arrangement, (b) conventional double-pulse interferogram of holographic images taken before and after lighting the flame, (c) reconstructed holographic images of laser 1 only (tuned near resonance), (d) laser 2 only (tuned away from resonance), (e) RHIS interferogram with both lasers on, (f) RHIS interferogram of salt droplet in flame, demonstrating mixed flame imaging [17].

the combined use of yellow (**bR → M**) and blue (**M → bR**) light. It was observed that the **M** lifetime of BR_{D96N}, which was a BR variant in which the amino acid Asp96 was replaced by Asn, was strongly dependent on both the pH value of the medium and its ionic strength [20]. Its increased light sensitivity resulted in high **M** populations, even at low light intensities.

A numerical simulation was carried out by Thoma *et al.* [19], who used a rate equation and the Beer–Lambert law to show that the transmission of the BR_{D96N} film was influenced by a number of parameters that included the iOD, the **M** lifetime τ_M, and the wavelength and the intensity of the control light (Fig. 13.15). It is obvious that when the iOD decreases the transmission curves are shifted to lower intensities that imply a more transparent material [Fig. 13.15(a)]. Figure 13.15(b) shows that increasing the **M** lifetime increases the sensitivity of the film to the light intensity. A direct implication from Fig. 13.15(b) was that a BR_{D96N} film with an unlimited **M** lifetime would be purely controlled by light illumination. The numbers on the curves in Figs. 13.15(c) and 13.15(d) are intensities of the red and the

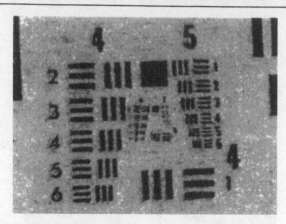

Fig. 13.13. Magnified central part of the reconstructed image observed from a CCD imager [18].

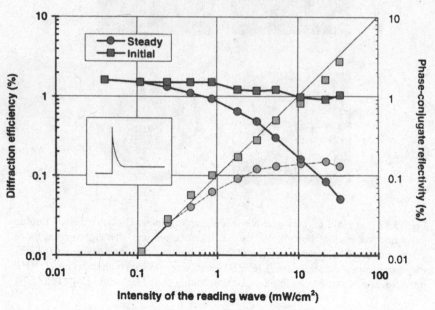

Fig. 13.14. Initial peak and steady-state diffraction efficiencies and phase-conjugate reflectivities with respect to the reading-light intensity. Shown in the inset is the time response of the readout process. The writing intensities are 9.4 and 11.2 mW/cm^2 [18].

blue light fields, respectively. The transmittance for a yellow information-carrying wave of 568 nm was increased with red light of 633 nm (**bR → M**) and decreased with blue light of 413 nm (**M → bR**), as shown in Fig. 13.15(c). The transmittance for a blue information-carrying wave of 413 nm could be controlled with 568-nm yellow light whose intensities are indicated on the horizontal axis of Fig. 13.15(d).

The photocontrol property simulated in Fig. 13.15 was used in an experiment to demonstrate the feasibility of nonlinear filtering with the BR film. A Fourier-transform system was used in the experimental setup. The pattern of the to-be-processed object was carried by a blue light wave of 413 nm, whereas the control pattern was carried by a yellow light

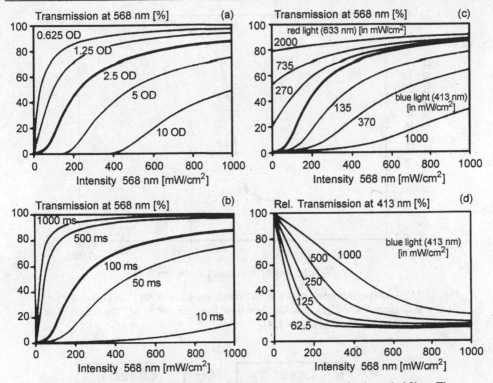

Fig. 13.15. Simulation of the intensity-dependent transmission of polymethyl films. The influences of the sample parameters, i.e., (a) iOD, (b) τ_M, and light-controlled absorption changes for (c) yellow, (d) blue light are shown [19].

wave of 568 nm. The spectra of both wavelengths overlapped on the Fourier plane where the BR film was placed. Depending on the pattern of the control beam, wherever the two wavelengths overlapped, the transmittance decreased and the output of the corresponding Fourier components of the to-be-processed object weakened.

Chemically enhanced BR film was used in another SLM system to transform the incoherent image to a coherent one [21]. The film used had an OD of ~5.4 at 570 nm, an **M** lifetime of 7 s, a thickness of ~200 μm, and an aperture of 20 mm. Figure 13.16 is the experimental measurement of the relative transmission intensities of the 514-nm readout light as a function of the intensity of the incoherent white light that illuminates to control the transmittance of the BR film. The dynamic range shown in the figure was ~200. The SLM system was capable of transforming an image with a resolution of 100 line pairs/mm. This number is limited mainly by the optical system used in the experiment. The intrinsic resolution of the BR film is of an order of magnitude higher.

13.6 Optical correlation and pattern recognition

The high resolution and cyclicity make BR a competitive medium for optical correlation and pattern recognition. In a series research carried out by Hampp *et al.*, a so-called **M**-type hologram recording process was used to implement a dual-axis joint Fourier-transform correlator [22–24]. In **M**-type holography, the BR film in the **bR** state is pumped to the **M** intermediate by a green light, and the hologram is written with blue light that causes the

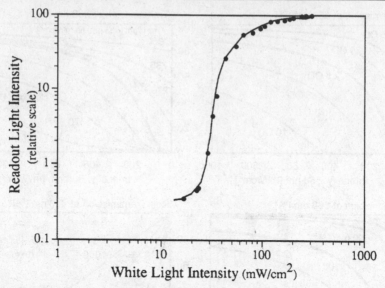

Fig. 13.16. Relative transmission of a weak coherent readout beam (524 nm) as a function of incoherent white-light intensity measured from the BR SLM. The intensity of the readout beam was 0.16 mW/cm² [21].

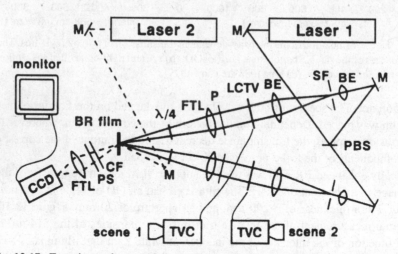

Fig. 13.17. Experimental setup of the dual-axis joint Fourier-transform correlator with video inputs [23].

M → bR transition. Because the BR variant *BR*$_{D96N}$, which had a prolonged **M** lifetime, was used, the photochemical transitions were virtually purely photocontrolled. The diffraction efficiency of *BR*$_{D96N}$ was twice as much as that of the wildtype BR, although there was no recording sensitivity difference between them for an **M**-type recording. One advantage of the **M**-type hologram was that the hologram was confined to an area where the pump beam was illuminating, and spatial filtering was easily realized in this way.

In an experimental setup as shown in Fig. 13.17, the beam from laser 1 with a 413-nm wavelength was split into two beams that were oppositely circularly polarized before reaching the BR film. The pumping beam with a 530-nm wavelength from a second laser was linearly polarized and served as the readout beam with the Bragg incidence angle. The polarization of the correlation output from this polarization hologram was orthogonal to that of the scattering noise; thus a linear polarizer behind the BR film filtered out the noise and improved the signal-to-noise ratio 21-fold [22]. The pumping beam also functioned as a low-pass filter that was confined to a required diameter. This low-pass filtering eliminated the autocorrelation signal resulting from the pixel structure of the two liquid-crystal televisions (LCTV's).

This BR-film-based joint transform correlator was capable of doing dynamic pattern recognition. In an experimental demonstration, two identical pictures were placed in front of both television cameras (TVC's in Fig. 13.17). When one of the pictures was moved, the correlation peak dynamically followed the position of the object. The SNR was at least 45 dB, which was limited by the resolution of the framegrabber. Under the intensities of 25 mW/cm^2 (413 nm) on the LCTV's and of 60 mW/cm^2 in the pump–readout beam (530 nm), the hologram rise and decay rates were higher than the LCTV cross-talk-free frame rate of 20 pictures/s. This demonstrated the real-time capability of the described correlator. A PC-based system was set up to process the correlation peaks, and the speed of the moving object was represented by an arrow vector indicating the direction of the movement and the actual speed through its length [24].

13.7 Holographic switching and optical interconnection

Holographic spatial switching by BR films was presented and demonstrated [25, 26]. The independent and arbitrary switching was realized by a programmable microhologram array (MA) for which BR film served as the dynamic recording and control medium. In the experimental demonstration a BR film modified by hydroxilamine with a thickness of 57 μm and a size of 40 mm × 40 mm was used. Each microhologram in the MA was recorded with a 633-nm He–Ne laser. The incidence angle of the two writing beams was controlled by a combination of an acousto-optical deflector, a beam splitter, and a mirror, and this angle determined the diffraction direction of the corresponding input light channel. A sequence of control signals provided successive recordings of the microholograms in the array. This resulted in parallel connections between channels corresponding to a preset law. To reset the connection, old microholograms in the MA were erased and replaced with new ones. The signal carrier was a He–Cd laser beam of 440-nm wavelength modulated by rectangular pulses. The light switched by the microhologram was reflected by a mirror to a photodetector and was registered. The diffraction efficiency was $\sim$2.5% with a power of 0.7 mW for each writing beam and 1 mW for the reading beam.

The experimental results showed that the switched beam can be deflected by the microhologram in the angular range of more than 10° to any position with a nonuniformity of less than 3 dB and a diffraction efficiency of more than 1.5%. The results suggested a possibility of the deflection of more than 200 independent directions along each orthogonal coordinate, provided that the beam section was $\sim$1 mm. The switch provided a capability of routing more than 40×10^4 channels. The experiment also showed that a rectangular pulse of 20-ms duration in the recording beam provided enough power to reset the switch.

13.8 Photoreceptor/artificial retina

On photoexcitation, BR generates a narrow electric pulse while starting the photocycle. This phenomenon is somewhat similar to what happens in the human visual system, and

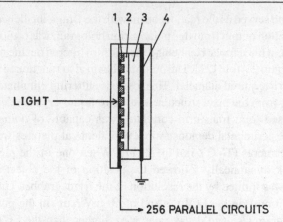

Fig. 13.18. Junction structure of the **bR**-based pixelized photoreceptor; 1, 256-pixel ITO electrode (100-μm thickness, 10-Ω/cm^2 area resistance); 2, BR-containing PM LB film; 3, polymer electrolyte gel comprising 4% carboxymethylchitin and 2-M aqueous KCl (300-μm thickness); 4, gold counterelectrode coated on a glass. The light-receiving area of the photoreceptor is 31 nm × 31 mm [29].

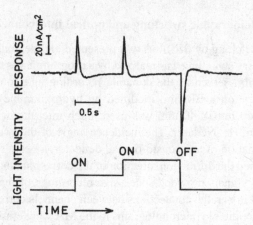

Fig. 13.19. Response profile of photoelectric signals from a **bR**-based photoreceptor [29].

researchers have made use of this photovoltage effect to build artificial retinas. The BR-based photoreceptors being studied were categorized into two types [27]: One used the temporal differential responsivity of oriented BR to create a differential enhancement [28–31], and the other implemented a pseudo-X-cell architecture by using optical lithography or an etching technique to generate excitatory and inhibitory regions [32, 33]. Miyasaka *et al.* reported one of the first photoreceptors based on the temporal differential responsivity of BR [28, 29]. The device shown in Fig. 13.18 consists a thin multilayered film of BR (less than 70 nm thick with 6 to 14 monolayers) that was coated by the Langmuir–Blodgett technique on a two-dimensionally arrayed indium tin oxide electrode. An electrolyte gel was then sandwiched between the protein film and a gold-coated glass plate. The most significant function of this device was its differential responsivity to the light. Figure 13.19 is the typical profile of the time response of a **bR**-induced photocurrent. The photoinduced current was believed to have been rectified in the cathodic direction, which resulted from the electrochemical interface regardless of the orientation of the purple membrane. The action spectrum of the transient photocurrent coincides with the broad absorption spectrum of **bR** and indicates that the origin of photoresponse is **bR**. The response time of the device was reported to be no faster than 1 ms but has a potential of the order of 10^{-8} s. Excellent linearity of the photocurrent with respect to the incident light intensity, which covers a range from 10^{-4} to 10^{-1} W/cm^{-2}, was confirmed and was attributed to the quantum conversion mode of the monolayer of BR. But the relation between the peak transient photocurrent and the light intensity differential was found to be nonlinear and was probably due to the circuit electric constants.

For the purpose of demonstration, the light-induced current from each pixel was converted to a dc voltage and amplified and finally output to a light-emitting device. The experiments showed that the intensity and the edge profile of the detected image displayed by the light-emitting-device pixels were dependent on the direction and the rapidity of the motion. Because of the differential responsivity of the BR film, the photoreceptor is capable of detecting a change in illumination resulting from a mobile object and ignoring the constant illumination from the still portion of the objects. The stability of the device was also investigated. The results indicated that the photoreceptor maintains its photoelectric activity without a detectable output decay for a period of 12 months under ambient conditions. The device was considered relatively thermally stable with no significant suppression in the photoelectric output at temperatures up to 70 °C.

The photoelectric properties of oriented BR can be used to implement photoreceptors that mimic the ganglion receptive field. There are two methods used in the preparation of the photoreceptor cells, one based on etching and the other based on photochemical lithography. In the etching method, two identically oriented films were etched and combined in an opposite direction, as shown in Figs. 13.20a–13.20c [27]. The illumination of the central region will

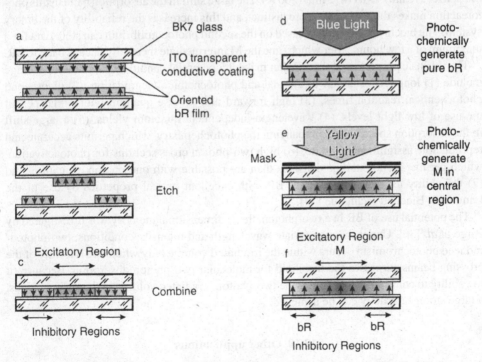

Fig. 13.20. Two methods of simulating an X-cell ganglion receptive field by using the photovoltaic properties of oriented BR films [27].

induce a photovoltage that is opposite to that of the circular outer region, and thus the photo-electric response of the cell mimics the spatial profile of the ganglion X-cell receptive field.

In the method based on photochemical lithography [33] as shown in Figs. 13.20d–13.20f, the excitatory and inhibitory regions analogous to those created by etching were obtained by the masked illumination of blue (400–420-nm) and yellow (550–600-nm) lights in a flexible manner.

13.9 Optical memory

Under proper conditions the photocycle of BR can be manipulated into binary photochemi-cal reactions, and this bistable property can be utilized for optical switching and binary data storage applications [34]. One binary system was formed at liquid-nitrogen temperatures, at which the photoequilibrium was between the primary photoproduct **K** and the light-adapted form of the native protein, **bR**: **bR** (576 nm) $\Leftrightarrow$ **K** (620 nm) (77 K). The formation times asso-ciated with both the forward and the reverse photoreactions were less than 5 ps. The storage duration by operating Langmuir–Blodgett-based BR film at 77 K was infinite. The absolute absorptivity measurements would result in bit errors due to uncertainties under lower conver-sion ratios without sufficient illumination power. A differential read data enhancement tech-nique was developed that provided high reliability without requiring high laser powers. This approach uses two lasers to monitor the state of the device and uses differential absorption and a pair of reference bits to assign the state of the bits that make up an individual word [34].

Another photoequilibrium coupling **bR** and **M** at higher temperatures ($\sim$200 K) can also be used for moderate speed (microsecond) photonic switching and binary storage applica-tions: **bR** (572 nm) $\Leftrightarrow$ **M** (412 nm) (200 K). The large shift in the absorption band in this pho-toreaction makes the two states more distinct, and this increases the reliability of the binary system. The biochrom [35] that is based on the second photoequilibrium can store data for a maximum of a few hours, after which time the **M** intermediate thermally decays back to **bR**.

The intrinsic properties of the BR that make it a potential candidate for photonic storage include (1) long-term stability to thermal and photochemical degradation, (2) picosecond photochemical reaction times, (3) high forward and reverse quantum yields that permit the use of low light levels, (4) wavelength-independent quantum yields, (5) a large shift in the absorption spectrum accompanying the photochemistry, which permits accurate and reproducible assignments of state, (6) high two-photon cross sections for photoactivation, which permits higher storage densities than are possible with one-photon excitation, and (7) the ability to form thin films of BR with excellent optical properties by use of the Langmuir–Blodgett technique [34].

The potential use of BR in a two-photon, three-dimensional memory was investigated by Birge *et al.* [36]. Under the appropriate wavelength and intensity conditions, two-photon-induced photochromism occurs within the irradiated volume only where the intensity of the crossing beams surpasses the threshold for molecular two-photon absorption. Because of the ability to control the location of the two-photon-irradiated volume in three dimensions, a large storage capacity can be achieved.

13.10 Other applications

As reported in Ref. 8, the thermal effect will result in a negative refractive-index change in BR film when the illumination light intensity exceeds a threshold under which the saturation

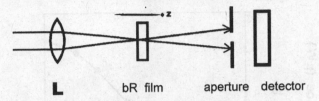

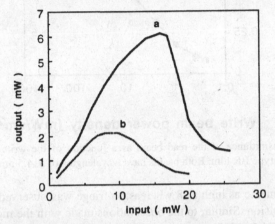

Fig. 13.21. Experimental setup for Z-scan measurement [37].

Fig. 13.22. Optical-limiting behavior when the focal length of the lens is 10 cm: curve a, $\lambda = 632.8$ nm; curve b, $\lambda = 514.5$ nm [37].

effect takes place. Under Gaussian beam illumination this thermal effect induces a refractive-index change profile that assumes a negative lens. This negative lensing effect was used to demonstrate optical-limiting behavior [37]. The experimental setup was very simple, as shown in Fig. 13.21. The BR film was placed in the focal plane. No device in the system (including the BR film) had to be moved. The output from the BR film increases with an increase in the input intensity at lower intensity. However, when the incident intensity exceeds a thresholding point, the output decreases rapidly and is eventually shut down. Figure 13.22 gives the experimental results. Although the unit shown in the horizontal axis in Fig. 13.22 is in milliwatts, the actual focused beam intensities on the BR film exceed 200 W/cm^2 when the thresholding happens.

A logarithmic dependence of the transmittance of the BR film on the incident light intensity below the saturation region was reported by Werner et al. [10] and was confirmed and used for optical logarithmic transformation of speckle images by Downie [38]. The transmission behavior with respect to the light intensity of a 514.5-nm wavelength is given in Fig. 13.23. A wildtype BR with chemical additives was used in the experiments. A two-tone image was stored in the BR film and was read out for statistic evaluation. The experiments confirmed that because of the logarithmic transformation nature of the BR film, the multiplicative, signal-dependent noise of a speckle image has been transformed into additive, signal-independent noise.

A solid dry BR film, fabricated by Takei and Shimizu, was reported to have significantly less scattering and less OD than BR in suspension in the 700–800-nm region [39]. This BR film was then inserted into a Fabry–Perot cavity to generate interference fringes under the irradiation of a diode laser emitting at 792 nm. The experiment showed that the fringes

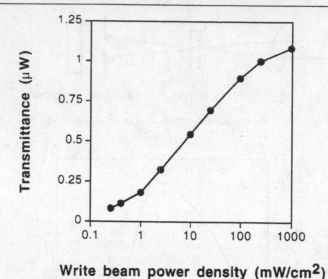

Fig. 13.23. Transmittance of the read beam as a function of the write-beam power density for wild-type BR film. Both beams have wavelengths of 514.5 nm [38].

so generated had a finesse as high as 8 whereas no fringe was observed if the cavity was filled with BR suspension. Similar to the observations made with the modified Michelson interferometer [6], a fringe shift was observed when the corresponding area was irradiated by a pumping beam of 543.5-nm wavelength. The advantage of the Fabry–Perot cavity was that it was structurally more robust and immune to vibrations and thus more suitable for the device configuration.

13.11 Conclusions

We have reviewed the optical properties and applications of the bio-optical material bacteriorhodopsin. The light-induced photocycle with its accompanying changes in absorption and refraction offers many interesting properties. The excellent holographic performance of the protein derives from the large change in refractive index that occurs following the appropriate light excitation. The good resolution, high cyclicity and sensitivity, potential low cost, and the ability to be modified by genetic, physical, and chemical methods to suit different requirements make bacteriorhodopsin very attractive for many optical applications. The modifications and new applications of this bio-optical material comprise ongoing and active research. This review is by no means complete, and we apologize for any negligence.

Acknowledgment

This work is supported in part by the U.S. Air Force under contract F30602-96-C-0066 and NASA grant NAG 2-1084.

References

[1] R. R. Birge, "Photophysics and molecular electronic applications of the rhodopsins," Ann. Rev. Phys. Chem. **41**, 683–733 (1990).

[2] D. Oesterhelt, C. Brauchle, and N. Hampp, "Bacteriorhodopsin: a biological material for information processing," Q. Rev. Biophys. **24**, 425–478 (1991).

[3] C. Brauchle, N. Hampp, and D. Oesterhelt, "Optical applications of bacteriorhodopsin and its mutated variants," Adv. Mater. **3**, 420–428 (1991).

[4] R. R. Birge, K. C. Izgi, J. A. Stuart, and J. R. Tallent, "Wavelength dependence of the photorefractive and photodiffractive properties of holographic thin films based on bacteriorhodopsin," Proc. Mater. Res. Soc. **218**, 131–141 (1991).

[5] H. Kogelnick, "Coupled wave theory for thick hologram gratings," Bell Syst. Tech. J. **48**, 2909–2947 (1969).

[6] D. Zeisel and N. Hampp, "Spectral relationship of light-induced refractive index and absorption changes in bacteriorhodopsin films containing wildtype BR_{WT} and the variant BR_{D96N}," J. Phys. Chem. **96**, 7788–7792 (1992).

[7] C. P. Zhang, Q. W. Song, C. Y. Ku, R. B. Gross, and R. R. Birge, "Determination of the refractive index of a bacteriorhodopsin film," Opt. Lett. **19**, 1409–1411 (1994).

[8] Q. W. Song, C. P. Zhang, R. B. Gross, and R. R. Birge, "The intensity-dependent refractive index of chemically enhanced bacteriorhodopsin," Opt. Commun. **112**, 296–301 (1994).

[9] Q. W. Song, C. P. Zhang, C. Y. Ku, M. C. Huang, R. B. Gross, and R. R. Birge, "Determination of the thermal expansion and thermo-optic coefficients of a bacteriorhodopsin film," Opt. Commun. **115**, 471–474 (1995).

[10] O. Werner, B. Fischer, A. Lewis, and I. Nebenzahl, "Saturable absorption, wave mixing, and phase conjugation with bacteriorhodopsin," Opt. Lett. **15**, 1117–1119 (1990).

[11] O. Werner, B. Fischer, and A. Lewis, "Strong self-defocusing effect and four-wave mixing in bacteriorhodopsin films," Opt. Lett. **17**, 241–243 (1992).

[12] O. Werner, R. Daisy, B. Fischer, and A. Lewis, "Forward phase conjugation by three-wave mixing with bacteriorhodopsin," Opt. Commun. **92**, 108–110 (1992).

[13] H. Kogelnick and K. S. Pennington, "Holographic imaging through a random medium," J. Opt. Soc. Am. **58**, 273–274 (1968).

[14] J. D. Downie, "Real-time holographic image correction using bacteriorhodopsin," Appl. Opt. **33**, 4353–4357 (1994).

[15] Sh. D. Kakichashvili, "Polarization recording of holograms," Opt. Spectrosk. (USSR) **33**, 171–173 (1972).

[16] N. M. Burykin, E. Ya. Korchemskaya, M. S. Soskin, V. B. Taranenko, T. V. Dukova, and N. N. Vsevolodov, "Photoinduced anisotropy in bio-chrom films," Opt. Commun. **54**, 86–70 (1985).

[17] J. E. Millerd, N. J. Brock, M. S. Brown, and P. A. DeBarber, "Real-time resonant holography using bacteriorhodopsin thin films," Opt. Lett. **20**, 626–628 (1995).

[18] Y. H. Zhang, Q. W. Song, C. Tseronis, and R. R. Birge, "Real-time holographic imaging with a bacteriorhodopsin film," Opt. Lett. **20**, 2429–2431 (1995).

[19] R. Thoma, N. Hampp, C. Brauchle, and D. Oesterhelt, "Bacteriorhodopsin films as spatial light modulators for nonlinear-optical filtering," Opt. Lett. **16**, 651–653 (1991).

[20] A. Miller and D. Oesterhelt, "Kinetic optimization of bacteriorhodopsin by aspartic acid 96 as an internal proton donor," Biochim. Biophys. Acta **1020**, 57–64 (1990).

[21] Q. W. Song, C. P. Zhang, R. Blumer, R. B. Gross, Z. P. Chen, and R. R. Birge, "Chemically enhanced bacteriorhodopsin thin-film spatial light modulator," Opt. Lett. **18**, 1373–1375 (1993).

[22] N. Hampp, R. Thoma, D. Oesterhelt, and C. Brauchle, "Biological photochrome bacteriorhodopsin and its genetic variant Asp96 $\rightarrow$ Asn as media for optical pattern recognition," Appl. Opt. **31**, 1834–1841 (1992).

[23] R. Thoma and N. Hampp, "Real-time holographic correlation of two video signals by using bacteriorhodopsin," Opt. Lett. **17**, 1158–1160 (1992).

[24] R. Thoma and N. Hampp, "Adaptive bacteriorhodopsin-based holographic correlator for speed measurement of randomly moving three-dimensional objects," Opt. Lett. **19**, 1364–1366 (1994).

[25] A. L. Mikaelian and V. K. Salakhutdinov, "High-capacity optical spatial switch based on reversible holograms," Opt. Eng. **31**, 758–763 (1992).

[26] V. K. Salakhutdinov and A. L. Mikaelian, in *Photonic Switching*, A. M. Goncharenko *et al.*, eds., Proc. SPIE **1807**, 442–451 (1992).

[27] Z. P. Chen and R. R. Birge, "Protein based artificial retinas," Trends Biotechnol. **11**, 292–300 (1993).

[28] T. Miyasaka, K. Koyama, and I. Itoh, "Quantum conversion and image detection by a bacteriorhodopsin-based artificial photoreceptor," Science **255**, 342–344 (1992).

[29] T. Miyasaka and K. Koyama, "Quantum conversion and image detection by a bacteriorhodopsin-based artificial photoreceptor," Appl. Opt. **32**, 6371–6379 (1993).

[30] R. R. Birge, "Protein-based optical computing and memories," IEEE Computer **25**, 56–67 (1992).

[31] H. Sasable, T. Furuno, and K. Takimono, "Photovoltaics of photoactive protein/polypeptide LB films," Synth. Metals **28**, C787–792 (1989).

[32] Z. Chen, H. Takei, and A. Lewis, "Optical implementation of neural networks with wavelength-encoded bipolar weight using bacteriorhodopsin," in *Proceedings of the International Joint Conference on Neural Networks* (San Diego, 1990), pp. 803–807.

[33] H. Takei, A. Lewis, Z. Chen, and I. Nebenzahl, "Implementing receptive fields with excitatory and inhibitory optoelectrical responses of bacteriorhodopsin films," Appl. Opt. **30**, 500–509 (1991).

[34] R. R. Birge, C. F. Zhang, and A. F. Lawrence, "Optical random access memory based on bacteriorhodopsin," in *Molecular Electronics*, F. T. Hong, ed. (Plenum, New York, 1989), pp. 369–379.

[35] N. N. Vsevolodov and G. R. Ivanisky, "Biological photosensory complexes as technical information carriers," Biofizika **30**, 883 (1985).

[36] R. R. Birge, R. B. Gross, M. B. Masthay, J. A. Stuart, J. R. Tallent, and C. F. Zhang, "Nonlinear optical properties of bacteriorhodopsin and protein-based two-photon three-dimensional memories," Mol. Cryst. Liq. Cryst. Sci. Technol. **3**, 133–147 (1992).

[37] Q. W. Song, C. P. Zhang, R. B. Gross, and R. R. Birge, "Optical limiting by chemically enhanced bacteriorhodopsin films," Opt. Lett. **18**, 775–777 (1993).

[38] J. D. Downie, "Optical logarithmic transformation of speckle images with bacteriorhodopsin films," Opt. Lett. **20**, 201–203 (1995).

[39] H. Takei and N. Shimizu, "Nonlinear optical properties of a bacteriorhodopsin film in a Fabry-Perot cavity," Opt. Lett. **19**, 248–450 (1994).

14

Liquid-crystal spatial light modulators

Aris Tanone and Suganda Jutamulia

14.1 Introduction

Liquid-crystal (LC) spatial light modulators (SLM's) are the by-products of display technology, and they can be divided into two groups. The first are those that evolved from the inexpensive liquid-crystal televisions (LCTV's), and the second are those that were specially designed optical modulators such as liquid-crystal light-valve (LCLV-) and ferroelectric liquid-crystal- (FLC-) based modulators. The earlier applications of the LCTV-based SLM's [1–8] were aimed at finding a substitute for the already available yet expensive SLM's, such as the magneto-optic SLM (MOSLM), the microchannel SLM, or the LCLV.

In this chapter, we first discuss the building block of an optical processor and the functions of a SLM, followed by the explanation about the classification and the optical and the electro-optical properties of LC materials. After giving an explanation of the structure of a LCTV, a LCLV, and a FLC SLM, we then provide some applications of the LCTV in real-time optical signal processing, and finally we provide a list of the problems and the future trends of research and applications of the LC SLM.

14.1.1 Building blocks of an optical processor

No matter how complicated it is, the block diagram of a typical optical processor can always be represented by the block diagram depicted in Fig. 14.1. It consists of three main parts: the input system, the processor, and the output system [9].

The input system consists of two parts. The first is the light source, which provides the optical carrier, and the second is the signal transducer or transport medium, which imposes information on the system. In the early days this information simply meant a two-dimensional (2-D) image recorded on a piece of film. Photographic film poses the utmost resolution (down to 1 μm) and sensitivity, which are much better than those of any other type of transport mediums known to date. The principal drawbacks, however, are its need of wet processing, which is also a time-consuming process, and its nonreusable characteristics [9, 10]. Furthermore, when it is implemented in an optical system that requires high-precision processing, the film or the photographic plate has to be placed inside a liquid gate for index matching, which reduces the phase variation because of the irregular thickness of the film or the plate. In line with the progress in optical processing, the information now may come from another independent system, which may be an electronic system, an optical system, or a real-world scene [11]. A SLM then becomes an indispensable tool as it offers a high-speed capability of inputting a large volume of information into the optical system.

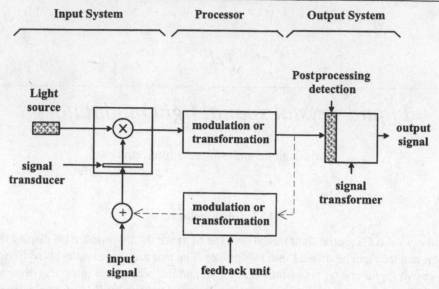

Fig. 14.1. Block diagram of an optical processor.

The processor is the most important part of an optical processor. It may either modulate or impose a transformation on the incident wave front. Basically it may simply consist of the SLM, which provides the input signal, and a combination of spatial filter, polarizer, phase plate, and other components.

The output system has the function of detecting the postprocessed signal and presenting it in an easy to grasp form, such as a displayed image on a screen or digitally stored 2-D images.

Among the three systems that comprise an optical processor, the output system is the most developed part, owing to the progress in the optoelectronics field and fiber-optic communications. This is in contrast to the least-developed signal transducer, which, for quite a long time, has relied mainly on film or a photographic plate. Aside from the introduction of the laser in the 1960's, little progress in transport mediums has been made during the past three decades compared with that of the output elements. Most of the SLM's known to date were developed during the past 30 years, with the greater part being in the past 10 years. It should be mentioned that attached to these systems are the optical components such as lens, prism, and others that are responsible for the transmission of optical signals between each block. This also includes free space, which is an important element in the optical transmission system. When the processed signal is fed back to the input system, as shown by the dashed line in Fig. 14.1, the optical processor is said to have a feedback unit. There are two possibilities. If the feedback unit is purely an optical system, then we have a pure optical processor. However, if the feedback unit consists of some electronic circuitry, then we have a hybrid processor.

14.1.2 Spatial light modulator

A SLM may have three basic functions. It can be a signal transducer that is a part of the input system that impresses the input to an optical processor, it can be a part of the processing unit that modulates or transforms the incident beam, and sometimes it can also be used as a temporary storage element. These basic functions may be represented by the box in the

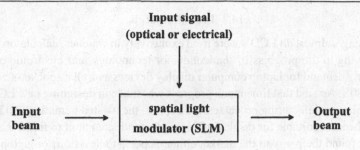

Fig. 14.2. Basic functions of a SLM.

block diagram shown in Fig. 14.2. An electronic or optical input signal is used to create a spatial distribution across the SLM. This spatial distribution will then modulate the input light beam, which can either be transmitted through or reflected by the SLM, depending on whether it is a transmission-type or a reflection-type modulator. The result is a signal-bearing modulated output, which may be an intermediate output that requires further processing or the expected final output of the processor. Recently, Ichioka et al. [11] introduced a new classification of SLM's according to this functional approach; thus we may have either an input SLM, a processor SLM, or an output SLM.

The modulation can be done through some of the following mechanisms. The amplitude or intensity may be modulated by the absorption inside the SLM, the phase by the refractive-index change, and the polarization by the rotation inside the SLM. When the change is controlled or inflicted by electrical means, we have an electrically addressed SLM (EA-SLM), and if the change is inflicted optically, then we have an optically addressed SLM (OA-SLM). The separation of SLM's into the EA-SLM's and the OA-SLM's is the general classification widely used to date.

An EA-SLM is usually constructed with a pixel structure. The incident optical wave front is thus individually modulated at each pixel by the electrical signal. The advantage of an EA-SLM is its capability of hybrid processing. However, the pixel structure then introduces the multiple diffraction spots in a coherent processing system, which reduces the processed energy of the information carrier. Furthermore, there is some inefficiency in space utilization because of the presence of the dead zone between electrodes.

In contrast to an EA-SLM, an OA-SLM is made up of a continuous structure. Generally, the addressing optical image produces an electric charge distribution over the material that generates a secondary effect for electro-optic modulation. In this context, an OA-SLM can be described as a combination of a detector and a modulator. The advantage of an OA-SLM is its capability to modulate a light beam by using another light beam. The common disadvantage among OA-SLM's is the expensive fabrication cost, which subsequently impedes the progress of their development. However, there are some existing SLM's, such as the photorefractive crystal, that are simply bulk crystals that do not require a complex fabrication process.

An introduction and extensive references to the types of SLM's, such as the acousto-optic modulator, MOSLM's, microchannel SLM's, or LCLV's, can be found in Refs. 2 and 9–12. Besides their extremely low cost when compared with that of the various above-mentioned SLM's, LCTV's attracted the interest of researchers because of their other advantages such as a large range of gray levels and sufficient contrast ratios that were not available in those binary, yet expensive, SLM's.

14.2 Liquid crystals

In the early stage, almost all LCD's were used exclusively in watches, calculators, and toys. However, owing to the progress in semiconductor technology and electronic computers, there was a big demand for laptop computer display devices as well as pocket-sized TV's in the early 1980's. Around that time much progress was made in designing new LC materials and new display manufacturing processes. As a result, the twisted nematic (TN) LCD technology also became suitable for displaying high information content to fulfill that demand, and LCTV's found their way to the market either as pocket televisions or laptop computer displays. In parallel with that, the increasing demand for speed and throughput in a digital information processing system also suggested the need for novel technologies in addition to microelectronics. This demand then triggered the new fields in optical information processing for which the SLM with real-time inputting capability was badly needed. Because of its commercial availability and low cost, some of the LCTV's that were then being modified ended up on an optical bench as SLM's instead of in their original function as TV's [2–8].

14.2.1 Liquid-crystal classifications

It is commonly known that matter exists in three phases: solids, liquids, and gases. Solids may either be crystalline or amorphous. A crystalline solid has a regular arrangement of the molecules over a large distance compared with molecular dimensions, which is called the long-range orderness. When a crystalline solid is heated, it is transformed into an isotropic liquid at its melting point. However, there are certain substances with elongated (typically cigar-shaped) molecules that do not directly pass from a crystalline solid to an isotropic liquid and vice versa, but rather they adopt an intermediate structure that flows like a liquid but still possesses the anisotropic physical properties similar to those of crystalline solids. Thus we have a structure that is not as ordered as a solid crystal but that also is not as disordered as the usual isotropic liquid state. This phase between solid and liquid is termed liquid-crystal, liquid-crystalline phase, mesophase, or mesomorphic phase, and the materials are called mesomorph, liquid crystalline, or mesomorphic substance [13]. The material has the optical properties of an ordered crystal in that temperature range.

LC materials are broadly classified into three classes according to the translational or orientational orderness in their molecular arrangement. These are the smectic, the nematic, and the cholesteric classes.

Smectic comes from the Greek word *smectea*, for soap, as in this phase the material tends to have mechanical properties akin to those of soap. There are various textures with rather mundane labels, smectic A, B, C, ..., I, not according to any microscopic property, but rather in the chronological order of their discovery. The smectic-class LC is depicted in Fig. 14.3(a), in which the molecules are parallel and their centers are stacked in parallel layers. In some smectic substances the molecules are further ordered by being arranged in a row within the individual layers, whereas in others the molecules are randomly distributed in the layers [14, 15]. The plane of the layer is one molecule thick.

Nematic comes from the Greek word *nema*, for thread. In the nematic phase, the molecules tend to align themselves with their long axes parallel, but their molecular centers of gravities are unordered and random, as in the case of the isotropic liquid, as shown in Fig. 14.3(b). Certain threadlike defects that have only one-dimensional ordering are commonly observed in this material.

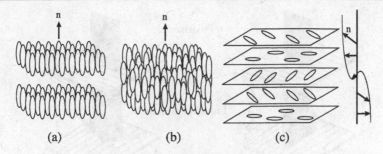

Fig. 14.3. LC classification: (a) smetic, (b) nematic, (c) cholesteric structures.

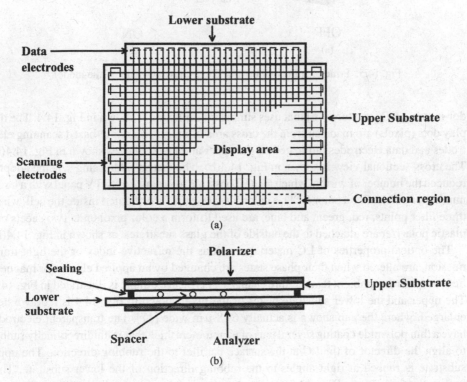

Fig. 14.4. Views of a LCD: (a) fundamental LCD structure, (b) its cross-sectional view.

The cholesteric description comes from the fact that the third class of LC's has a molecular structure that looks like many cholesterol esters, in which the molecules align themselves in a separate plane with the long axes parallel to each other within each plane. However, the directors, as shown by the arrows in Fig. 14.3(c), are twisted gradually at a well-defined angle. Here the vector $\bar{n}$ is used to show the macroscopic direction of the aligned LC molecules, and it is designated as a director.

14.2.2 Optical and electro-optical properties of twisted nematic liquid crystals

Most LCD's use nematic LC molecules that are intentionally twisted. Basically, the cell of a TN LCD is made of two glass substrates containing a thin TN LC layer inside. A typical

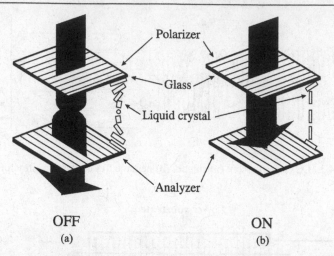

OFF

(a)

ON

(b)

Fig. 14.5. Principle of operation of a TN display (normally white mode).

dot-matrix LCD structure [16] that uses simple multiplexing is shown in Fig. 14.4. The display dots (pixels) are made up from the cross area of indium tin oxide-based scanning electrodes and data electrodes fabricated on the upper and lower glasses shown in Fig. 14.4(a). The cross-sectional view is shown in Fig. 14.4(b). The matrix-addressing scheme adopted reduced the number of wiring, which made it possible to produce a LCTV panel with a larger number of pixels. For a color LCD, a color filter layer is incorporated inside the cell, where three filter points, red, green, and blue, are used to form a color pixel unit. Two sheets of a plastic polarizer are attached to the outside of the glass substrates, as shown in Fig. 14.4(b).

The optical properties of LC materials, such as the refractive index or the light transmission, are altered when their phase states are changed by an applied electric or magnetic field or when the ambient temperature changes. A typical example is illustrated in Fig. 14.5. The upper and the lower substrate plates of the pixel shown in Fig. 14.4 is redrawn here enlarged, where the gap shown is actually $6-8$ μm wide [17]. The transparent electrodes have a thin polymide coating several tens of nanometers thick that is unidirectionally rubbed to align the director of the LC at the surface parallel to the rubbing direction. The upper substrate is rubbed at right angles to the rubbing direction of the lower substrate. Thus in the unactivated state [Fig. 14.5(a)], the director undergoes a continuous 90° twist in the region between the substrate, which is indicated by the orientation of the cylinders. Because the polarizer sheets on the top and the bottom have been aligned perpendicularly to each other, the linearly polarized light from the upper polarizer propagates through the layer and rotates its polarization plane in step with the twisted structure. At the bottom of the layer it is polarized parallel to the transmission axis of the lower polarizer.

Applying a certain amount of voltage across the upper and the lower electrodes orients the optic axis in the central portion of the LC layer predominantly parallel to the electric field, and the twisted structure disappears [Fig. 14.5(b)]. The polarization direction of the light is no longer rotating, and light passing through the cell intersects the second polarizer in the cross position where it is absorbed, causing the activated portion of the display to appear dark. With the addition of a reflector behind the lower polarizer, this mode is used in watches and calculators. The only difference with a LCTV is that a seven-segment-type display is used in watches and calculators, compared with the dot-matrix structure in a LCTV display.

14.3 Electrically addressed spatial light modulators

The LCTV was originally produced as a pocket television. With proper modification, a commercially available LCTV can be used as an EA-SLM. Along with the advancement in the fabrication technology of the display devices, the simple matrix structure was replaced by the active-matrix thin-film-transistor (TFT) structure [18]. The name active matrix comes from the fact that in order to increase the contrast ratio between V_{on} and V_{off} in the dot-matrix structure when the number of pixels increases, a TFT has to be placed as a switching element at each pixel. The voltage across each pixel could then be held independently for each frame instead of being scanned, and as a whole, the active-matrix display offers a higher contrast ratio [19].

14.3.1 Background

In order to find the effect of a TN LC on incident polarized light along the twist axis (z axis), the material is divided into N incremental slices orthogonal to the z axis. Each slice is assumed to be a wave plate with a phase retardation and an azimuth angle. Thus, locally, it can be treated as a uniaxial crystal whose optical axis is parallel to the direction of the molecules. The propagation of linearly polarized light along the twisted axis can then be described by use of Jones calculus [12], and the analytical model and result are given by Lu and Saleh [20, 21].

Under an applied electric field in the direction of the z axis, all molecules are assumed to tilt by an angle θ that tends to align with the applied field, as shown in Fig. 14.5(b). The tilt angle is a function of the root-mean-square (rms) value of the applied voltage V_{rms} [20]:

$$\theta = \begin{cases} 0 & V_{rms} \leq V_c \\ \dfrac{\pi}{2} - 2\tan^{-1}\left\{\exp\left[-\left(\dfrac{V_{rms} - V_c}{V_0}\right)\right]\right\} & V_{rms} > V_c \end{cases}, \tag{14.1}$$

where V_c is a threshold voltage below which no tilting of the molecules occur and V_0 is a constant. When $V_{rms} - V_c = V_0$, the tilt angle is 49.6°. It is clear that for $V_{rms} > V_c$, the angle θ increases with an increase of V_{rms}, reaching a saturation value of $\pi/2$ for large V_{rms}.

On the other hand, the extraordinary refractive index seen by a light beam propagating along the electric-field direction (z axis) is given by

$$\frac{1}{n_e^2(\theta)} = \frac{\cos^2\theta}{n_o^2} + \frac{\sin^2\theta}{n_e^2}, \tag{14.2}$$

where n_o and $n_e = n_e(0)$ are the ordinary and the extraordinary refractive indices of the molecules in the absence of the electric field. Therefore, in the presence of the applied field that gives $0 < \theta < \pi/2$, the value $n_e(\theta)$ can be found. The Jones matrix of the LCD that describes the transformation of the polarization states across the LC layer may be shown to be a function of one variable, β. This variable is defined as half of the phase retardation and is related to the applied voltage by means of the tilt angle θ through

$$\beta = \frac{\pi d}{\lambda}\left[n_e(\theta) - n_o\right]. \tag{14.3}$$

This analysis is derived with an assumption that the tilt angle is independent of the position z. At the boundaries, i.e., $z = 0$ and $z = d$, θ varies because of the boundary effect that complicates the analysis. The experimental results obtained with the simplified model seem to be adequate [20, 22, 23].

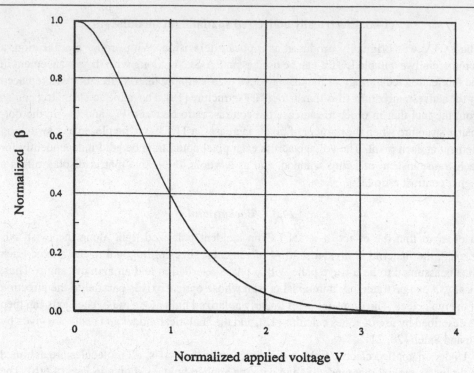

Fig. 14.6. Normalized β as a function of the normalized voltage of the LCTV, assuming $n_o = 1.5$ and $\Delta n = 0.2$.

The maximum value of β occurs in the absence of an electric field and is given by

$$\beta_{\max} = \frac{\pi d}{\lambda} [n_e - n_o]$$

$$= \pi r, \tag{14.4}$$

where $r = \Delta n d / \lambda$ and $\Delta n = n_e - n_o$. For a specific LCTV panel, the value of β and $\beta_{\max}$ can be determined if n_e, n_o, the cell thickness d, and the wavelength λ are known. The relation between the normalized parameter $\beta/\beta_{\max}$ and the normalized applied voltage $(V_{\mathrm{rms}} - V_c)/V_o$ is shown in Fig. 14.6 for $\Delta n = 0.2$. As shown in Ref. 20, when $\Delta n = 0.1$ and 0.3, the curves almost coincide, thus showing that the function is almost independent of n_e. Since then β has been considered a scaled version of V_{rms}, and the variation of intensity transmittance and the phase shift for light passing through a TN LCD are determined as functions of β.

The amplitude transmittance and the phase shift introduced by the device as functions of β can be derived with the help of Jones calculus. It is assumed that the LC material is sandwiched between a polarizer and an analyzer, making angles ψ_1 and ψ_2 with the x axis, as shown in Fig. 14.7. If we denote the incident-wave and the transmitted-wave complex amplitudes by $\bar{E}$ and $\bar{E}'$, with the following Jones vectors [12, 23]

$$\bar{E} = \begin{pmatrix} E_x \\ E_y \end{pmatrix} = \begin{pmatrix} \cos \psi_1 \\ \sin \psi_1 \end{pmatrix}, \tag{14.5}$$

$$\bar{E}' = \begin{pmatrix} E'_x \\ E'_y \end{pmatrix}, \tag{14.6}$$

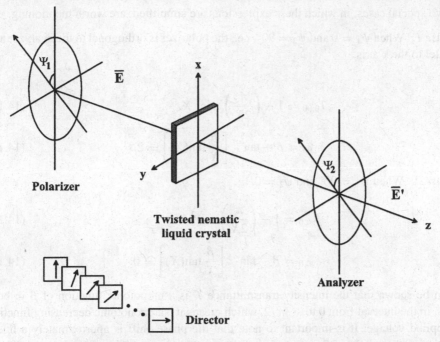

Fig. 14.7. General configuration of LCTV with an arbitrary polarizer–analyzer pair.

then Jones calculus gives

$$\begin{pmatrix} E'_x \\ E'_y \end{pmatrix} = \exp(-j\phi) \begin{pmatrix} \cos^2 \psi_2 & \sin \psi_2 \cos \psi_2 \\ \sin \psi_2 \cos \psi_2 & \sin^2 \psi_2 \end{pmatrix}$$

$$\times \begin{pmatrix} \dfrac{\pi}{2} \dfrac{\sin X}{X} & \cos X + j\beta \dfrac{\sin X}{X} \\ -\cos X + j\beta \dfrac{\sin X}{X} & \dfrac{\pi}{2} \dfrac{\sin X}{X} \end{pmatrix} \begin{pmatrix} \cos \psi_1 \\ \sin \psi_1 \end{pmatrix}. \quad (14.7)$$

The first matrix and the third vector shown above represent the effect of the polarizer and the analyzer, respectively. The Jones matrix in the middle was derived following the standard procedures as described in Ref. 12. The parameter X is given by

$$X = \left[\psi^2 + \left(\frac{\Gamma}{2} \right)^2 \right]^{1/2}, \quad (14.8)$$

where ψ is the twisted angle, $\Gamma = 2\pi/\lambda(n_e - n_o)l$ is the total phase retardation across the LC layer of thickness l, and $\Gamma/2 = \beta$.

The intensity transmittance T and the phase shift δ are given by

$$T = \left[\frac{\pi}{2X} \sin X \cos(\psi_1 - \psi_2) + \cos X \sin(\psi_1 - \psi_2) \right]^2$$

$$+ \left[\frac{\beta}{X} \sin X \sin(\psi_1 + \psi_2) \right]^2, \quad (14.9)$$

$$\delta = \beta - \tan^{-1} \frac{\dfrac{\beta}{X} \sin X \sin(\psi_1 + \psi_2)}{\dfrac{\pi}{2X} \sin X \cos(\psi_1 - \psi_2) + \cos X \sin(\psi_1 - \psi_2)}. \quad (14.10)$$

It is clear that for a given ψ_1 and ψ_2, both T and δ are functions of one variable, β, only.

Two special cases, in which these expressions are simplified, are worth mentioning.

Case 1: When $\psi_1 = 0$ and $\psi_2 = 90°$, i.e., the polarizer is orthogonal to the analyzer and parallel to the x axis,

$$T_{0,90} = 1 - \left(\frac{\pi}{2X}\right)^2 \sin^2 X, \tag{14.11}$$

$$\delta_{0,90} = \beta + \tan^{-1}\left[\frac{\beta}{X}\tan(X)\right] \cong 2\beta. \tag{14.12}$$

Case 2: When $\psi_1 = 90°$ and $\psi_2 = 0$,

$$T_{90,0} = 1 - \left(\frac{\pi}{2X}\right)^2 \sin^2 X, \tag{14.13}$$

$$\delta_{90,0} = \beta - \tan^{-1}\left[\frac{\beta}{X}\tan(X)\right] \cong 0. \tag{14.14}$$

It can be shown that the intensity transmittance T is a monotonic function of β in both cases, in the interval from 0 to $\sqrt{3}\pi/2$, which means it is a monotonic decreasing function of applied voltage. It is important to note that the phase shift is approximately a linear function with a slope of 2 in case 1, but there is no phase shift in case 2. Accordingly, we may conclude that for case 1 the device can be used as both an intensity and a phase modulator, but for case 2 it can be used as a pure intensity modulator. Further discussion on the optimal parameters for a phase modulator or intensity modulator can be found in Refs. 20 and 24.

In some cases, the twisted angle of the crystal is less than 90°. Ohkubo and Ohtsubo [24] introduced a general twisted angle α and evaluated the intensity transmittance T at $(\varphi1, \varphi2) = (\pm45°, \pm45°)$ and $(\varphi1, \varphi2) = (0, \pm45°)$. By using the $T_{0,45}$ and $T_{0,-45}$, we can determine the twist direction as either a clockwise or a counterclockwise direction. The twist direction of the molecular directors is an important factor for the application that utilizes the polarization-modulation property of the LC [25].

14.3.2 Liquid-crystal television spatial light modulators

The early version of the LCTV was simple and could be disassembled easily. However, as the fabrication technology progressed, the structure of the LCTV became complicated. Some LCTV's have their electronic driver tightly attached to the back of the panel, and there is no way to separate them. However, projection TV's produced later can be modified easily as SLM's as they came with three separate LC panels, as shown in Fig. 14.8. The color projection TV's come in various sizes and configurations [26–28].

An Epson projection display provided these three sets of LC panels that can be disassembled easily from the unit. A recent review shows that, aside from the FLC-type LCD's, most of the reports on the use of LCTV's for optical information processing [29–34] have depended largely on the Epson projection LCTV's.

The LCD size for the Epson VPJ-700 is 19.8 mm × 25.6 mm, and it comprises 220 × 320 pixels. The pixel size is 60 μm × 55 μm and the center-to-center spacing is 80 μm × 90 μm. One advantage of the LCD for a projection display is that every pixel in the LCD is used to

Table 14.1. *Recent LC SLM data*

Parameter	Kubota *et al.* [27]	Kopin Toolkit [36]
Display size (inch-diagonal)	1.9	0.75
Pixel number	1472 × 1024	640 × 480
Pixel pitch (μm^2)	29 × 24	24 × 24

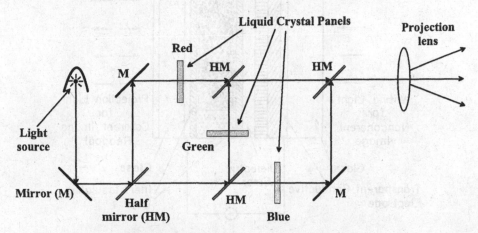

Fig. 14.8. Projection display with three LC panels.

modulate only one wavelength. In the old color LCTV's, it took three subpixels to form one image pixel. The result is essentially a 3:1 resolution advantage for the three monochromatic LCD's over a color LCTV with the same total number of pixels [29].

The basic concerns about LCTV SLM's include the contrast ratio, sizes and total number of pixels, and the processing speed. The contrast ratio is defined as the ratio between the transmitted light intensity of the On state versus the Off state. Most of the applications of LCTV's in optical information processing have simply adapted the video input capability of the LCTV, and there was only a limited effort to modify the electronic driver circuit of the LCTV [35] as compared with the work in the field of a FLC SLM, in which the driver was designed toward a high processing capability in a megahertz frame rate. As of 1994, pixel numbers from 1920 × 480 to 1840 × 1035 with pixel sizes that vary from $14\,\mu m \times 44\,\mu m$ to $40\,\mu m \times 40\,\mu m$ and contrast ratios of 150–400 for a TFT LCD have been reported [11]. These numbers given in Ref. 11 are mostly from the literature of Japanese LCD manufacturers. Recently, Kopin Corporation in Taunton, MA produced an active-matrix LCD evaluation kit [35]. It is a 0.75-in. (1.9-cm) diagonal display, with 640 × 480 pixels, 24-μm pitch, and a pixel aspect ratio of 1:1. The pixel pitch is comparable with that reported by Kubota *et al.* [27], i.e., $29\,\mu m \times 24\,\mu m$. Table 14.1 highlights their differences.

14.4 Optically addressed spatial light modulators

The research on LCD's, the forerunner of the LC-based SLM, started in the late 1960's, so it took another 30 years to evolve to its present state. In the early years, there were some attempts to use an electron beam to address the LC cell attached to a cathode-ray tube [37]

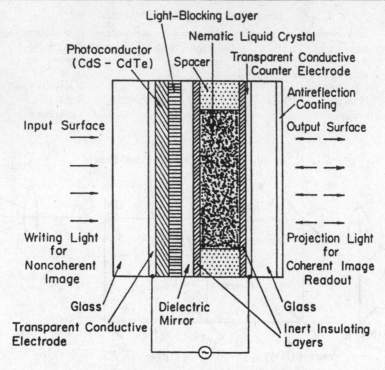

Fig. 14.9. Side view of a LCLV and its operation.

and to utilize the dynamic scattering mode [38] of the nematic LC. The first LC-based SLM, i.e., the Hughes LCLV [39], was introduced in 1970, and the FLC-based SLM [40] was introduced in the mid-1980's.

14.4.1 Liquid-crystal light valves

The LCLV is a reflective TN LC cell. It takes advantage of the pure birefringence of the LC material in order to modulate the output laser beam while in the On state. Operation of the LCLV is clarified in the side view shown in Fig. 14.9. There are two principal layers: the photoconductor layer on the right-hand side, and the nematic LC layer on the left-hand side. These two layers are separated by a light-blocking layer and a dielectric mirror in the center. An incoherent image is formed on the photoconductor (CdS–CdTe) layer to control the applied alternating voltage to the LC layer in response to the input intensity at every point in space. Laser light illuminating the back of the LCLV is reflected back, but is modulated by the birefringence of the LC layer at every point. In order to achieve this effect, the molecular alignment has a 45° twist between the two surfaces of the LC layer. Because the two beams are optically isolated, a readout at high optical intensity provides gain; hence the SLM may be called a light valve [41].

The analytical model of an LCLV is given by Lu and Saleh [21], who used the Jones calculus described above. The LCLV is considered as two cascaded transmissive cells, and they show that the LCLV could be operated either as a spatial intensity or a phase-only modulator. Recently Mukohkaza *et al.* [42] reported the use of a parallel-aligned FLC SLM. Instead of a twisted alignment, the LC molecules are aligned in parallel, as shown in

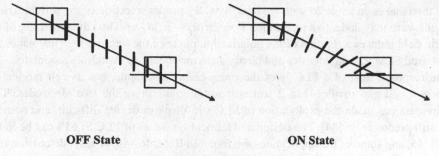

OFF State   **ON State**

Fig. 14.10. Parallel-aligned LC molecules.

Upper electrode and analyzer

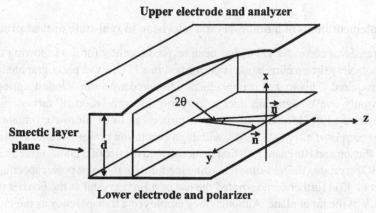

Lower electrode and polarizer

Fig. 14.11. FLC device structure.

Fig. 14.10. It eliminates the complication of polarization rotation introduced by the twisted structure.

In addition to incoherent-to-coherent conversion and other coherent processing applications, the LCLV is also used as a wavelength converter. An infrared image can be generated when an infrared beam is passed through the LCLV, which is modulated by a visible image. This application is required for the test of military and remote-sensing equipment, which generally operate in the 3–5-μm and 8–12-μm bands [43].

14.4.2 Ferroelectric liquid-crystal spatial light modulators

A FLC is usually referred to as the smectic-C LC. As shown in Fig. 14.3(a), the smectic LC's are arranged in layers. The material has a nonzero dipole or ferroelectricity. When placed between two close glass plates, the director n can be switched between two uniform orientations, both of which are in the plane of the LC film and separated by an angle 2θ, where θ is the tilt angle of the molecules within the smectic layers [40]. When an electric field is applied along the x axis shown in Fig. 14.11, a torque is produced that switches the molecular direction into the $+\theta$ direction. When the polarity is reversed, the molecules can be switched into the $-\theta$ direction.

When linearly polarized light entering at an angle θ coincides with the director of the crystal, the wave travels with the extraordinary refractive index n_e without retardation. When an electric field is applied to flip the molecules into the $-\theta$ direction, the polarization plane then makes an angle 2θ with the optic axis. By proper selection of the cell thickness, the light wave may undergo a retardation $\Gamma = 2\pi(n_e - n_o)d/\lambda_0$. Thus the switching of the electric field induces a flipping of the polarization plane of the light wave. This makes the FLC-based SLM a good intensity and binary state modulator or switching modulator.

The response time of a FLC is of the order of microseconds, but its cell fabrication requires a cell gap smaller than 2 μm without shortcircuiting the two electrodes. This requirement has made the exploitation of FLC's in displays device difficult, and nematic LC's still predominate [44]. The design and fabrication issues of FLC SLM's can be found in Ref. 45, and some of the applications are found in Refs. 46–48. A short discussion on a surface-stabilized FLC is given in Ref. 48. The tabulation data on an OA-SLM is also given in Ref. 11.

14.5 Implementations of a liquid-crystal television in real-time optical processing

The joint transform correlator (JTC) has been revived recently for the following reason. It is known that every time a correlation is performed in a JTC, a wet photographic recording process is required. This makes it impractical compared with the VanderLugt correlator, which uses only one holographic filter to perform the correlation of various inputs and a fixed reference [49]. However, the JTC has turned out to be more convenient than the VanderLugt correlator after OA-SLM's with high-resolution power became available.

In 1981, Pichon and Huignard [50] first demonstrated the use of a photorefractive bismuth silicate (BSO) crystal as the real-time recording medium for the joint power spectra. In 1985, Loiseaux *et al.* [51] further demonstrated the use of a BSO crystal at the Fourier plane and a BSO LCLV at the input plane. Although they employed a transparency as the input to the BSO LCLV, in principle the transparency could be replaced with a video monitor. The BSO crystal could resolve the joint power spectra with 1000 lp/mm spatial frequency (lp stands for line pairs). The resolution of the BSO LCLV was 12 lp/mm with a 25 mm × 25 mm usable area equivalent to 300 × 300 pixels. These efforts realized the real-time square-law conversion in a JTC.

In 1984, Yu and Lu [52] demonstrated the use of a LCLV as the square-law detector for the power spectrum and a computer-controlled MOSLM in the input plane in a real-time programmable JTC. The space–bandwidth product (SBP) or the pixel number of the MOSLM was limited to 48 × 48 pixels (currently 256 × 256 pixels). However, because the object shape was usually more regular than the joint transform power spectra, it required fewer numbers of pixels than its power spectra. The resolution of the LCLV was 15 lp/mm. Because the resolution of the LCLV was less than that of photographic film or a BSO crystal, a long focal transform lens was required. The scale of the Fourier transform was proportional to the focal length of the lens [53]. This work introduced the new concept of a real-time computer-programmable JTC.

14.5.1 Programmable joint transform correlators

In 1987, Yu *et al.* [4] demonstrated another new architecture by using a LCTV as an EA-SLM for displaying both the input and the power spectra. A microcomputer was used to generate object and reference patterns simultaneously on the LCTV. A collimated coherent beam then illuminated the LCTV, which had been disassembled and immersed in a liquid gate. The LCTV first displayed the functions $f(x, y)$ and $g(x, y)$ side by side. The joint transform of the object and the reference patterns was performed by a lens. If the object and the reference patterns were identical, the joint transform spectrum would consist of a fringe structure. A CCD camera functioned as a square-law detector for recording the joint power spectra of the input shown in the LCTV. The detected pattern was then displayed on the LCTV, and the CCD detected the correlation output. The detecting area of the CCD was usually approximately 10 mm × 10 mm, consisting of 512 × 512 pixels. Thus the resolution was ~51.2 lp/mm. However, because the video output was connected to and displayed by a LCTV that had ~120 × 240 pixels (currently a LC display with more than 1000 × 1000 pixels is available), the effective resolution of the CCD–LCTV combination was ~12 lp/mm. Although the effective resolution was of the same order as that of the LCLV, an additional magnifying lens could be employed in this new architecture. This work then opened two new opportunities:

(1)   To exclude the use of a long focal lens: the JTC could be very compact,
(2)   To detect the power spectra with a CCD; thus a variety of digital processing can be done before the second Fourier transform.

   It should be noted that the processing speed was ignored in favor of exploring the contrast and the resolution limit of the LCTV. A schematic diagram of the preliminary experimental setup is shown in Fig. 14.12. In the experiment, the fringe structure was recorded by a vidicon onto a video tape. The recorded tape of the fringe structure was then replayed with the same LCTV. The correlation signal, if it existed, was then displayed on the TV monitor. In this preliminary investigation, five fringes or more produced recognizable correlation spots.
   An illustration of these results is shown in Fig. 14.13. Figure 14.13a shows the images of two microcomputer-generated fighters as a simulation of the input scene and the reference object. Figures 14.13b and 14.13c are the left half of the correlation output and the extended

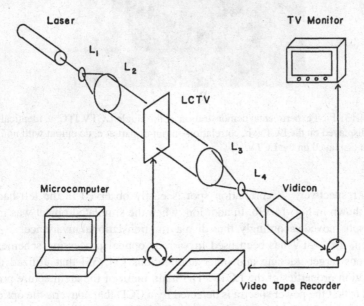

Fig. 14.12. Experimental setup.

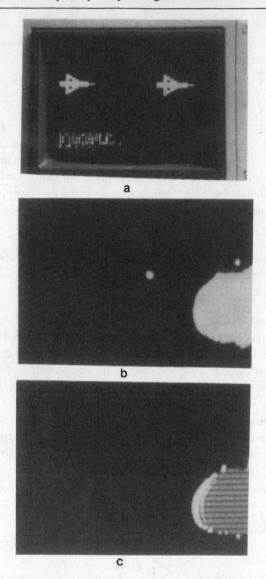

Fig. 14.13. First experimental demonstration of the single LCTV JTC: a, identical patterns displayed on the LCTV; b, correlation output of part a; c, dc output with no input pattern displayed on the LCTV.

dc patterns, respectively. A correlation spot is easily observed in the left-hand part of the picture shown in Fig. 14.13b. In addition, when the simulated object was moved, the correlation spot moved accordingly, thus displaying translational invariance.

Since then, the LCTV has been used in various optical processing schemes, such as in autonomous target tracking and other applications [54, 55] that utilized the amplitude modulation properties of the LCTV. The main merit of the architecture proposed by Yu *et al.* [4] is that the power spectra is detected by a CCD that converts the optical signals into electronic signals. Thus the electronic signals can be processed by a computer before they are sent to the LCTV for the second optical Fourier transform. Many algorithms have been proposed to process the power spectra electronically [56].

From the concept of the diffraction pattern of Young fringes, we see that the JTC is similar to the method of double-exposure speckle photography [57]. In this method, the speckle field scattered by an object is recorded. After the object is slightly translated, the speckle field will not change but only shifts a distance proportional to the shift of the object. The shifted speckle field is then recorded on the same film. The developed film contains two identical but shifted speckle fields $f(x, y) + f(x - b, y)$, where b is the displacement. The diffraction pattern of the film will be the superposition of Young fringes from every pair of the initial and the shifted speckles. The generated Young fringes are again recorded on a film. The diffraction pattern of the Young fringes will provide correlation spots at $(-b, 0)$ and $(b, 0)$. Thus the translation can be detected. This method has also been applied to astronomy to resolve binary stars [58]. If advantage is taken of the newly available architecture of Yu *et al.* [4], the double-exposure speckle photography can be performed in real time [59–62].

14.5.2 Real-time phase modulators

Most of the applications of the LCTV as a spatial modulator discussed above have utilized the amplitude and the polarization modulations of the LC panel. But as was shown in Subsection 14.5.1, when incident light is transversely polarized parallel to the front director of a TN LC cell, assuming the cell is operated below its optical threshold voltage, phase modulation occurs [5, 6, 20, 22].

There are two ways, however, to observe the phase modulation in a LCTV SLM. The first way is binary phase-only spatial light modulation, which can be achieved when the second analyzer is oriented to be perpendicular to the bisector of the two transmission states of the LCTV [63], as shown in Fig. 14.14. Thus it still utilizes the polarization-modulation properties of the TN LC. The application of this binary phase modulation in conjunction with a point diffraction interferometer to correct the phase nonuniformity of the LCTV was reported in Ref. 64. The second way is the use of the controlled birefringence, as discussed in Subsection 14.3.1, in which the first experimental demonstration of this property was reported by Yu *et al.* in 1987 [5]. In this subsection, we review some of the applications of the phase-modulation properties of the LCTV, starting with the use of π phase modulation to generate a real-time kinoform, which is easily achieved with the commercially available LCTV [65]. Some of the applications of phase-modulation properties of the LCTV SLM in real-time optical processing can be found in Refs. 66–71. The experimental verifications of the simultaneous amplitude and phase modulation were also discussed in the works of

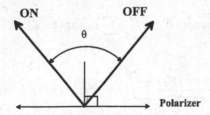

Fig. 14.14. Orientation of the polarizer to achieve bipolar phase-only modulation.

Zhang *et al.* [30] and Neto *et al.* [31]. Recently, by using the fractional Talbot effect, Serrano-Heradia *et al.* presented a novel technique for measuring the phase modulation in a LCTV panel that is not sensitive to environmental disturbance and eliminates the measurement ambiguity between complimentary phases φ and $2\pi - \varphi$ [72].

A kinoform is a computer-generated phase hologram [66, 70–73] in which the phase distribution is directly recorded on the hologram. Unlike most of the computer-generated holograms, the kinoform does not rely on the nonzero diffraction order to reconstruct a complex wave field. The kinoform acts as a Fresnel lens that changes the phase of the illuminating wave by its optical thickness variation. When the kinoform is displayed on a LCTV that operates in the phase-modulation mode, the thickness variation of a conventional kinoform is substituted by the birefringent effect of the LC molecules, which varies as a function of applied voltage.

Because the kinoform algorithm is well established [74, 75], a program was written to generate a 128×128 pixel phase distribution from a real object function. A pseudorandom phase array is used with the original object to reduce the effect of losing the amplitude information in the kinoform. The calculated phase values are then converted into gray-scale values, which are saved in a file for later use. In the simulation, the resulting gray scales are converted back to phase variations, for which a short Fast Fourier-transform routine is added to the program for the simulation of the reconstruction process. A mismatch factor α is then introduced by which if $\alpha = 1$ a perfect match results and if $0 < \alpha < 1$, a phase mismatch results. A perfect match represents the LC panel that has a modulation depth of 2π, whereas a phase mismatch represents a phase-modulation depth of less than 2π.

Three-dimensional plots of computer-simulated results of the reconstructed images when the modulation depth is 2π and π are shown in Figs. 14.15(a) and 14.15(b), respectively. Note that the output images have been thresholded to show the surface profile of the reconstructed letter B. However, for the unthresholded image, the reconstructed letter is displayed with some intensity variation, as shown in Fig. 14.15(d). The noise of the left-hand letter in Fig. 14.15(b) is clearly the nonthresholded mirror image, whereas for the perfect-match cases shown in Fig. 14.15(a) or 14.15(c), the noise structure is clearly suppressed.

We must note that the presence of a false image for phase mismatch has been discussed by Kermisch [76] and others [76–79], following the publication of a kinoform by Lesem *et al.* [73]. Because the modulation phase for the kinoform is a 2π modulo of the phase distribution $\varphi(x, y)$, it can be represented by a nonlinear limiter input–output model [74], as shown in Fig. 14.16. The dashed line has been added in the figure to facilitate the Fourier series expansion of the phase distribution. If the phase variation of the original wave front is Z, the reconstructed wave front from the kinoform would be equal to $\exp(jZ')$, where we assume that Z' is a periodic function for which $\exp(jZ')$ can be expanded into a Fourier series, such as

$$\exp(jZ') = \frac{1}{2\pi} \sum_m c_m \exp(jmZ), \tag{14.15}$$

where

$$c_m = \frac{\sin(m\pi/N_a)}{m\pi/N_a} \sum_k (1/N_a) \exp[j2\pi k(1 - m)/N_a] \tag{14.16}$$

and N_a is the total number of the discrete levels. If we introduce a parameter γ to take into account the phase mismatch for the kinoform reconstruction, then when the phase function

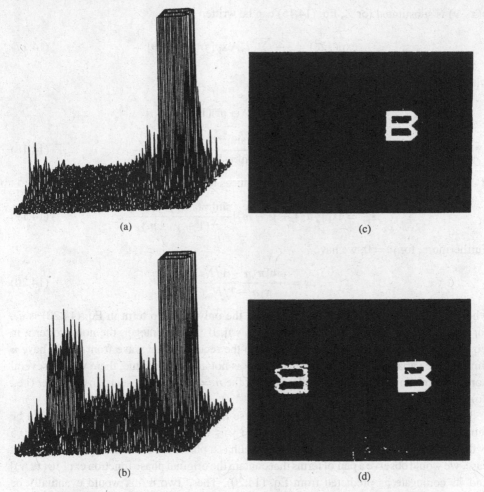

(a)

(c)

(b)

(d)

Fig. 14.15. Computer-simulated results of kinoform reconstruction with a LCTV that has (a) 2π, (b) π phase-modulation depth. (c) and (d) are the 2-D representations of (a) and (b), respectively.

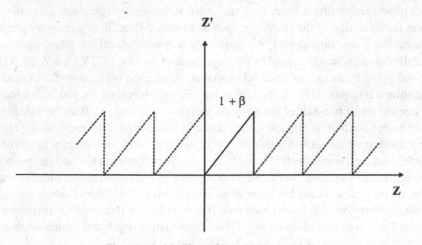

Fig. 14.16. Nonlinear input–output model.

$\varphi(x, y)$ is substituted for Z, Eq. (14.15) can be written as

$$\exp(jZ') = \frac{1}{2\pi} \sum_m c'_m \exp[jm\varphi(x, y)], \tag{14.17}$$

with

$$
\begin{aligned}
c'_m &= \frac{1}{N_a} \exp[j\pi(1 + \gamma - m)(1 - 1/N_a)] \\
&\times \frac{\sin(m\pi/N_a)}{m\pi/N_a} \frac{\sin[\pi(1 + \gamma - m)]}{\sin[\pi(1 + \gamma - m)/N_a]}.
\end{aligned} \tag{14.18}
$$

If N_a becomes infinitely large, Eq. (14.18) reduces to

$$c'_m = \exp[j\pi(1 + \gamma - m)]\frac{\sin[\pi(1 + \gamma - m)]}{\pi(1 + \gamma - m)}. \tag{14.19}$$

Furthermore, for $\gamma = 0$, we have

$$c_n = \frac{\sin[\pi(n - 1/N_a)]}{\pi(n - 1/N_a)}, \tag{14.20}$$

where $n = (1 - m/N_a)$. Note that if $\gamma = 0$, the only nonzero term in Eq. (14.20) is c_1, for which we see that $\exp(jZ') = \exp[j\varphi(x, y)]$. If γ is an integer, the nonzero term in Eq. (14.20) is shifted to $n = 1 + \gamma$, in which the reconstructed wave front would have n times phase variations. On the other hand, if γ is not an integer value, there will be several nonzero terms in Eq. (14.20). Needless to say, the $n = 0$ term represents a zero-order (i.e., dc) term.

The effect of the LC panel's having a phase-modulation depth lower than 2π can be represented by the mismatch factor γ for $-1 < \gamma \leq 0$. The physical meaning of $\gamma = -0.5$ is that the kinoform generated by the LC panel has a phase-modulation depth of π. In that case, we would observe a pair of terms that contain the original phase function $\exp[j\varphi(x, y)]$ and its conjugate as predicted from Eq. (14.20). These two terms would eventually be responsible for the real and virtual images reconstructed from the kinoform. Thus they predict the presence of a false (conjugate) image, and the intensity of the false image is inversely proportional to the number of quantization levels N_a. For a conventional process in which photographic film is used, the phase mismatch mainly depends on some physical factors in the formation of the kinoform, such as exposure time, illuminating wavelength, and others. For a real-time device, the mismatch is dependent on the phase-modulation depth of the device, which is limited by the physical nature of the LCTV [5, 6, 8, 20, 71]. To avoid overlapping in the case of on-axis holograms, shifting the object from the optical axis of the kinoform is required [77]. If the SBP of the original object is $a \times a$ pixels, this shifting means a reduction of one-half of the pixel numbers on each axis and thus the reduction of the SBP to one-fourth of its original value. This also means that if the reconstructed image occupies the first or the second quadrant, the false image will then occupy the third and the fourth quadrants, respectively. The only crossing point between the two images would be the origin of the Cartesian coordinate; thus it provides well-separated images. In reality, however, we may calculate the kinoform of an object with a SBP of $b \times a$, where $b < 1/2a$, which should also give a well-separated pair. If $b = 1/2a$, the image will superpose along the y axis. Thus, note that when we use a phase-modulation depth that is smaller than 2π, besides the smaller phase modulation, the reconstructed image suffers from low diffraction efficiency and a smaller SBP compared with those of the 2π modulation case.

The experimental setup consists of a standard optical Fourier-transform architecture. The calculated kinoform is sent as a gray-scale distribution (which controls the voltage applied to each pixel physically) by a Datacube AT-428 image board to display on the LCTV. The LCTV is used as a phase modulator. After Fourier transformation, the magnified image is picked up by a CCD camera and displayed on a TV monitor.

The LCTV used is a Seiko LVD-202 with the front and the back polarizers removed. The phase-modulation depth of this LCTV is smaller than 2π if the LC panel is modulated by a full swing of the gray-scale level from 0 to 255. The π phase modulation is determined from the reduction of the zero-order intensity level to its minimum value when a Ronchi grating pattern that has a maximum gray level of 180 is displayed on the LCTV operating in the phase-modulation mode [22, 80]. The reduction of zero-order intensity of the grating is obtained through a series of experiment in which various gray scales are used while the brightness control voltage of the LCTV panel is adjusted. The 180 gray scales were determined to be the optimum value for the π-rad modulation depth.

Figure 14.17 depicts a set of images, each with a different number of quantization levels, i.e., 2, 4, 8, and 16, reconstructed with the error-reduction method [75]. The laser source was a He–Ne laser, and the wavelength was 633 nm. To avoid the bright dc spot when recording these reconstructed images, we selected two adjacent pairs of images that had the same image qualities in terms of brightness and similarity to their original object for which the bright spots were blocked. From the results, we note that when the number of quantization levels decreases to 2, the real image and its conjugate images have the same intensity levels.

Because the kinoform image quality has been well established previously [74, 81, 82], there were no attempts to evaluate it other than from visual judgment. Figure 14.18 shows the typical LCTV kinoform used to reconstruct the images of Fig. 14.17.

Because the phase modulation is inversely proportional to the wavelength, the longer the wavelength, the smaller the phase-modulation depth [5, 20]. When we replaced the 633-nm He–Ne laser with a 488-nm Ar laser, a modulation depth of 2π could easily be achieved. The reconstructed images from a π kinoform and a 2π kinoform are shown at the top and the bottom of Fig. 14.19, respectively. When the modulation depth was 2π, the false image simply disappeared.

The LCTV panel can also be used for optical beam steering if we write a discrete blazed grating onto the LCTV panel. Figure 14.20 shows an example in which an input image (character G) in a $4f$ optical processor is steered across the output plane [80]. The LCTV panel was placed at the Fourier plane of the optical system, and three phase gratings with different periods were written to the LCTV panel. The image in the middle (row 4) was the original image without the grating structure. The images at the top and the bottom were the images when the grating spatial frequencies were the highest with opposite slopes. The presence of false images is due to the limited modulation depth ($<2\pi$) of the LCTV panel. Thus a better panel with a full modulation depth will overcome this problem. When the pixel pitch can be reduced further and the pixel mismatch between the image board and the LCTV panel can be overcome, we will expect a better deflection angle, as shown here, where the total angle is $\sim$0.1°. For practical applications such as reading or writing a multiplex hologram [83], additional efforts are needed to remove the residual diffraction

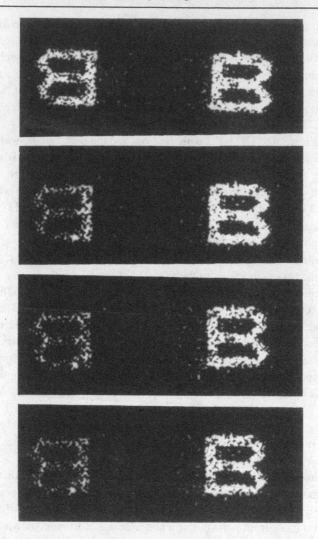

Fig. 14.17. Reconstructed images for quantization levels of 2, 4, 8 and 16.

orders, unless the modulation depth of the LCTV panel is 2π and the pixel mismatch is eliminated.

14.5.3 Application in camera

In another application, we have also demonstrated the use of a LC switch, which can be considered a single-pixel SLM, in a single-lens-reflex (SLR) camera to replace a flipping mirror [84]. In a SLR camera, a mirror is used to direct 100 percent of the incoming light to either the view finder or film. We used a LC switch that is a single-pixel SLM to change the polarization of light between 0° and 90°. A polarizing beamsplitter then directs all of the light, after it passes through the LC SLM either to the view finder or film. This system is good only for the photography using linearly polarized light. If the incoming light is not polarized, then only 50 percent of the light can be used. A regular beamsplitter will always split the incoming light 50-50, into two directions.

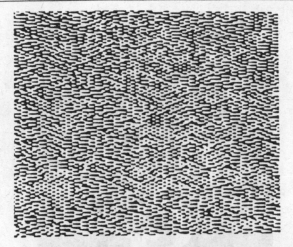

Fig. 14.18. Typical kinoform image.

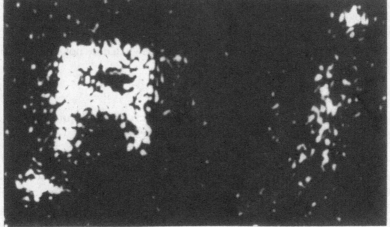

Fig. 14.19. Reconstructed images when the modulation depths are π (top) and 2π (bottom) with a 488-nm laser.

Fig. 14.20. Beam-steering example when a $4f$ optical processing system is used.

We applied this LC switch to an ophthalmoscope taking a picture of retina. The illuminating light on the retina was linearly polarized. The merits are:

- the flipping mirror generates sound but the LC switch gives off no sound – the sound may stimulate the patient to move the eye,
- the flipping mirror generates vibration but the LC switch does not – the exposure will be clear and crisp.

14.6 The applications, the problems, and the future of liquid-crystal spatial light modulators

We have shown that LC-based SLM's have become useful tools in optical information processing. In a simple application, we can use a uniform light beam to illuminate the SLM the way we illuminate a photographic transparency. Then we may generate an optical signal $A(x, y)$ by simply displaying a transmittance function $A(x, y)$ on the SLM. If we use an electrically addressed LC SLM, then the system is simply an LC projection display. Since the spectral transmission band of LC materials covers infrared region, an infrared scene can be generated using a proper infrared light source. The simulated infrared scene can be used for testing infrared equipment [85].

An optically addressed SLM can function as a wavelength converter where the write-in and readout wavelengths are different, an incoherent-to-coherent converter where the write-in light is incoherent but the readout light is coherent, an optical amplifier where the write-in intensity is low but the readout intensity is high, etc. The applications of optically addressed SLM's are discussed in Section 14.4.1.

In yet another application, recently we developed a technique to remove the pixel structure from an image projected by a LC projection display. A LC projection display is actually an electrically addressed LC SLM. It must have a pixel structure to form an image. However, this pixel structure then contributes to the formation of multiple Fourier spectra in the frequency plane. The question is: can we remove this pixel structure from the projected image to obtain a movie-like image?

It is well-known fact that a single hole in the frequency plane acts as a spatial filter for the optical signal. By passing only one spectrum order through the hole in the frequency plane, the pixel structure can be removed. The only problem with this treatment, which blocks out most of the spectrum orders, is we lose most of the energy. The projected image will be very dim. To correct this, we have covered each spectrum order with a different phase filter. Every spectrum order was delayed by phase filter with different thickness, and was no longer coherent to each other. As a result, the pixel structure was removed, but no intensity was lost. We first discussed this method in Ref. 86, and then presented the experimental result in Ref. 87. This technique can significantly improve the quality of LC projection display, which may improve the low-resolution LC display for overhead projector and make it looks like a high-resolution one.

Despite the fact that the applications discussed were simply proof-of-principle, they still demonstrate their promising aspects including the FLC SLM.

As we move toward the 21st century, we believe that parallel processing is going to replace most of the serial processing schemes adopted to date. Thus, instead of using a fiber optic to transfer the serial data, we will expect to use a fiber bundle to transfer the image in the information freeway [11]. This will all become possible because of parallelism and the interconnection of optics that can be done with the speed of light. For that purpose, the use of SLM's will be indispensable, and parallel processing and interconnections can then be done almost in real time.

At the moment, the LCTV-based SLM's are driven by the video frame rate of 30 frames/s. This is not a surprise as the LCTV's were primarily designed for display purposes, and the input devices, such as CCD camera, also work at the video rate. Thus one of the problems that has to be solved in order to boost the processing speed is to increase the frame rate of the LC SLM. All the research on the FLC SLM seems to be geared toward these goals, as the FLC offers a faster switching speed.

The second problem is to increase the number of pixels. Because of the need in high-definition TV, we have seen that the pixel numbers for the high-definition TV standard [88], i.e., 1920×1080 or 1280×720, have been reported [11]. With the progress in the electronic fabrication of high-definition TV displays, we expect to see a better-quality LC SLM, such as that reported in Ref. 42, which can be widely available for the optical processing research community. However, unless there is a mass-production market, the price of these specially designed LC SLM's will prevent their widespread use in the optical processing community for quite some time.

References

[1] F. T. S. Yu and S. Jutamulia, *Optical Signal Processing, Computing, and Neural Networks* (Wiley, New York, 1992), Chap. 5.

[2] J. A. McEwan, A. D. Fisher, P. B. Rolsma, and J. N. Lee, "Optical-processing characteristic of a low-cost liquid crystal display device," J. Opt. Soc. Am. A **2**, 8 (1985).

[3] H. K. Liu, J. A. Davis, and R. A. Lily, "Optical data-processing properties of liquid crystal television spatial light modulators," Opt. Lett. **10**, 635–637 (1985).

[4] F. T. S. Yu, S. Jutamulia, T. W. Lin, and D. A. Gregory, "Adaptive real-time pattern recognition using a liquid crystal TV based joint transform correlator," Appl. Opt. **26**, 1370–1372 (1987).

[5] F. T. S. Yu, S. Jutamulia, T. W. Lin, and X. L. Huang, "Real-time pseudocolor encoding using a low-cost liquid crystal TV," Opt. Laser Technol. **19**, 45–47 (1987).

[6] N. Konforti, E. Marom, and S.-T. Wu, "Phase-only modulation with twisted nematic liquid crystal spatial light modulators," Opt. Lett. **13**, 251–253 (1988).

[7] T. H. Barnes, T. Eiju, K. Matsuda, and N. Ooyama, "Phase-only modulation using a twisted nematic liquid crystal television," Appl. Opt. **28**, 4845–4852 (1989).

[8] H. K. Liu and T. S. Chao, "Liquid crystal television spatial light modulator," Appl. Opt. **28**, 4772–4780 (1989).

[9] J. Tsujiuchi, Y. Ichioka, and T. Minemoto, *Optical Information Processing* (Ohms, Tokyo, 1989).

[10] F. T. S. Yu, *Optical Information Processing* (Wiley, New York, 1983).

[11] Y. Ichioka, T. Iwaki, and K. Matsuoka, "Optical information processing and beyond," Proc. IEEE **84**, 694–714 (1996).

[12] A. Yariv and P. Yeh, *Optical Waves in Crystal* (Wiley, New York, 1984).

[13] E. Kaneko, *Liquid Crystal TV Displays: Principles and Applications of Liquid Crystal Displays* (KTK Science, Tokyo, 1987).

[14] D. Demus, "Types and classification of liquid crystals," in *Liquid Crystals Applications and Uses*, B. Bahadur, ed. (World Scientific, Singapore, 1990), Vol. 1, Chap. 1.

[15] J. L. Fergason, "Liquid Crystals," Sci. Am. **211**, 77–85 (1964).

[16] S. Morizumi, "Materials and assembling process of LCDs," in *Liquid Crystals Applications and Uses*, B. Bahadur, ed. (World Scientific, Singapore, 1990), Vol. 1, Chap. 7.

[17] T. Scheffer and J. Nehring, "Twisted nematic and supertwisted nematic mode LCDs," in *Liquid Crystals Applications and Uses*, B. Bahadur, ed. (World Scientific, Singapore, 1990), Vol. 1, Chap. 10.

[18] T. Yanagisawa, K. Kasahara, Y. Okada, K. Sakai, Y. Komatsubara, I. Fukui, N. Mukai, K. Ide, S. Matsumoto, and H. Hori, "A 3.1-in TFT-addressed color LCD," Proc. Soc. Inf. Disp. **26**, 213–216 (1985).

[19] W. den Boer, F. C. Luo, and Z. Yaniv, "Microelectronics in active-matrix LCDs and image sensors," in *Electro-Optical Displays*, M. A. Karim, ed. (Marcel Dekker, New York, 1992).

[20] K. Lu and B. E. A. Saleh, "Theory and design of the liquid crystal TV as an optical spatial phase modulator," Opt. Eng. **29**, 240–246 (1990).

[21] K. Lu and B. E. A. Saleh, "Complex amplitude reflectance of the liquid crystal light valve," Appl. Opt. **30**, 2354–2362 (1991).

[22] A. Tanone, "Study of phase modulation properties of twisted-nematic liquid crystal television as applied to computer generated hologram and signal processing," Ph.D. dissertation (Pennsylvania State University, University Park, PA, 1993).

[23] G. R. Fowles, *Introduction to Modern Optics* (Dover, New York, 1975), Chap. 2.

[24] K. Ohkubo and J. Ohtsubo, "Evaluation of LCTV as a spatial light modulator," Opt. Commun., **102**, 116–124 (1993).

[25] H. Sakai and J. Ohtsubo, "Image substraction using polarization modulation of liquid crystal television," Appl. Opt. **31**, 6852–6858 (1992).

[26] N. Kimura, "Reflective liquid crystal display," Kogaku **24**, 606–610 (1995).

[27] K. Kubota, M. Imai, and T. Matsumota, "Liquid crystal projection display," Kogaku **24**, 611–616 (1995).

[28] H. Hori, "Technology trends for direct-view type TFT-LCDs," Kogaku **24**, 617–623 (1995).

[29] J. C. Kirsch, D. A. Gregory, M. W. Thie, and B. K. Jones, "Modulation characteristics of the Epson liquid crystal television," Opt. Eng. **31**, 963–970 (1992).

[30] Z. Zhang, G. Lu, and F. T. S. Yu, "Simple method for measuring phase modulation in liquid crystal television," Opt. Eng. **33**, 3018–3022 (1994).

[31] L. G. Neto, D. Roberge, and Y. Sheng, "Programmable optical phase-mostly hologram with a coupled-mode modulation liquid-crystal television," Appl. Opt. **34**, 1944–1950 (1995).

[32] R. Dou and M. K. Giles, "Simple technique for measuring the phase property of a twisted nematic liquid crystal television," Opt. Eng. **35**, 808–812 (1996).

[33] J. L. McClain Jr., P. S. Erbach, D. A. Gregory, and F. T. S. Yu, "Spatial light modulator phase depth determination from optical diffraction information," Opt. Eng. **35**, 951–954 (1996).

[34] A. Serrano-Heradia, G. Lu, P. Purwosumarto, and F. T. S. Yu, "Measurement of the phase modulation in liquid crystal television based on the fractional-Talbot effect," Opt. Eng. **35**, 2680–2684 (1996).

[35] J. Aiken, B. Bates, M. G. Catney, and P. C. Miller, "Programmable liquid-crystal TV spatial light modulator: modified drive electronics to improve device performance for spatial-light modulation operation," Appl. Opt. **30**, 4605–4609 (1991).

[36] LVGA Evaluation Toolkit brochure, Kopin Corporation, Taunton, MA.

[37] J. A. Van Raalte, "Reflective liquid crystal television display," Proc. IEEE **56**, 2146–2149 (1968).

[38] See, for example, Ref. 13, Subsection 2.2.

[39] J. Grinberg, A. Jacobson, W. P. Bleha, L. Miller, L. Fraas, D. Boswell, and G. Myer, "A new real time non-coherent to coherent light image converter using the Hughes liquid crystal light valve," Opt. Eng. **14**, 217–225 (1975).

[40] N. Collings, W. A. Crossland, P. J. Ayliffe, D. G. Vass, and I. Underwood, "Evolutionary development of advanced liquid crystal spatial light modulators," Appl. Opt. **28**, 4740–4747 (1989).

[41] D. Armitage, J. I. Thackara, and W. D. Eades, "Photoaddressed liquid crystal spatial light modulators," Appl. Opt. **28**, 4763–4771 (1988).

[42] N. Mukohkaza, N. Yoshida, H. Toyoda, Y. Kobayashi, and T. Hara, "Diffraction efficiency analysis of a parallel-aligned nematic-liquid-crystal spatial light modulator," Appl. Opt. **33**, 2804–2811 (1994).

[43] M. S. Welkowsky, R. A. Forber, C. S. Wu, and M. E. Pedinoff, "Visible-to-infrared image converter using the Hughes liquid crystal light valve," in *Spatial Light Modulators and Applications II*, U. Efron, ed., Proc. SPIE **825**, 193–197 (1987).

[44] T. Ikeda and O. Tsutusmi, "Optical switching and image storage by means of azobenzene liquid-crystal films," Science **268**, 1873–1875 (1995).

[45] D. J. McKnight, K. M. Johnson, and R. A. Serati, "256 × 256 liquid-crystal-on-silicon spatial light modulator," Appl. Opt. **33**, 2775–2784 (1994).

[46] Y. Kobayashi, T. Takemori, N. Mukohzaka, N. Yoshida, and S. Fukushima, "Real-time velocity measurement by the use of a speckle-pattern correlation system that incorporates a ferroelectric liquid-crystal spatial light modulator," Appl. Opt. **33**, 2785–2794 (1994).

[47] D. C. O'Brien, R. J. Mears, T. D. Wilkinson, and W. A. Crossland, "Dynamic holographic interconnections that use ferroelectric liquid-crystal spatial light modulators," Appl. Opt. **33**, 2795–2803 (1994).

[48] K. M. Johnson and G. Moddel, "Motivations for using ferroelectric liquid crystal spatial light modulators in neurocomputing," Appl. Opt. **28**, 4888–4899 (1989).

[49] A. VanderLugt, "Signal detection by complex spatial filtering," IEEE Trans. Inf. Theory **IT-10**, 139–145 (1964).

[50] L. Pichon and J. P. Huignard, "Dynamic joint-Fourier transform correlator by Bragg diffraction in photorefractive $Bi_{12}SiO_{20}$ crystals," Opt. Commun. **36**, 277–280 (1981).

[51] B. Loiseaux, G. Illiaquer, and J. P. Huignard, "Dynamic optical cross-correlator using a liquid crystal light valve and a bismuth silicon oxide crystal in the Fourier plane," Opt. Eng. **24**, 144–149 (1985).

[52] F. T. S. Yu and X. J. Lu, "A programmable joint transform correlator," Opt. Commun. **52**, 10–20 (1984).

[53] J. W. Goodman, *Introduction to Fourier Optics* (McGraw-Hill, New York, 1968).

[54] E. C. Tam, F. T. S. Yu, D. A. Gregory, and R. D. Juday, "Autonomous real-time object tracking with an adaptive joint transform correlator," Opt. Eng. **29**, 314–320 (1990).

[55] F. T. S. Yu, X. Li, E. Tam, S. Jutamulia, and D. A. Gregory, "Detection of rotational and scale varying objects with a programmable joint transform correlator," Proc. SPIE **1053**, 167–176 (1989).

[56] S. K. Rogers, J. D. Kleine, M. Kabrinsky, and J. P. Mills, "New binarization techniques for joint transform correlation," Opt. Eng. **29**, 1088–1093 (1990).

[57] E. Archbold, J. M. Burch, and A. E. Ennos, "Recording of in-plane surface displacement by double exposure speckle photography," Opt. Acta **17**, 883–898 (1970).

[58] A. Labeyrie, "Attainment of diffraction limited resolution in large telescopes by Fourier analyzing speckle pattern in star images," Astron. Astrophys. **6**, 85–87 (1970).

[59] B. Bates and P. C. Miller, "Liquid crystal television in speckle metrology," Appl. Opt. **27**, 2816–2817 (1988).

[60] B. Bates, P. C. Miller, and L. Wang, "Liquid crystal TVs in speckle metrology: optimum conditions for bipolar phase modulation," Appl. Opt. **28**, 1969–1971 (1989).

[61] A. Ogiwara, H. Sakai, and J. Ohtsubo, "Real-time optical correlator for doubly exposed clipped speckle," Opt. Commun. **78**, 213–216 (1990).

[62] A. Ogiwara, H. Sakai, and J. Ohtsubo, "Real-time optical joint transform correlator for velocity measurment using clipped speckle intensity," Opt. Commun. **78**, 322–326 (1990).

[63] J. A. Davis, R. A. Lilly, K. D. Krenz, and H. K. Liu, "Applicability of the liquid crystal television for optical data processing," in *Nonlinear Optics and Applications*, P. A. Yeh, ed., Proc. SPIE **613**, 245–248 (1986).

[64] E. C. Tam, S. Wu, A. Tanone, F. T. S. Yu, and D. A. Gregory, "Closed-loop binary phase correction of an LCTV using a point diffraction interferometer," IEEE Photon. Technol. Lett. **2**, 143–146 (1990).

[65] A. Tanone, Z. Zhang, C.-M. Uang, F. T. S. Yu, and D. A. Gregory, "Phase modulation depth for a real-time kinoform using a liquid crystal television," Opt. Eng. **32**, 517–521 (1993).

[66] E. C. Tam, F. T. S. Yu, A. Tanone, D. A. Gregory, and R. D. Juday, "Data association multiple target tracking using a phase mostly liquid crystal television," Opt. Eng. **29**, 1114–1121 (1990).

[67] D. A. Gregory, J. A. Loudin, J. C. Kirsch, E. C. Tam, and F. T. S. Yu, "Using the hybrid modulating properties of liquid crystal television," Appl. Opt. **30**, 1274–1378 (1991).

[68] D. A. Gregory, J. C. Kirsch, A. Tanone, S. Yin, P. Andres, F. T. S. Yu, and E. C. Tam, "Analysis of phase modulation in an LCTV based joint transform correlator," Micro. Opt. Technol. Lett. **6**, 211–214 (1993).

[69] E. C. Tam, F. T. S. Yu, S. Wu, A. Tanone, S.-D. Wu, J. X. Li, and D. A. Gregory, "Implementation of kinoforms using a continuous-phase SLM," in *OSA Annual Meeting*, Vol. 15 of 1990 OSA Technical Digest Series (Optical Society of America, Washington, DC, 1990), p. 259.

[70] J. Amako and T. Sonehara, "Computer-generated hologram using TFT active matrix liquid crystal spatial light modulator (TFT-LCSLM)," Jpn. J. Appl. Phys. **29**, L1533–L1535 (1990).

[71] J. Amako and T. Sonehara, "Kinoform using an electrically controlled birefringent liquid crystal spatial light modulator," Appl. Opt. **32**, 4622–4628 (1991).

[72] A. Serrano-Heradia, G. Lu, P. Purwosumarto, and F. T. S. Yu, "Measurement of the phase modulation in liquid crystal television based on the fractional-Talbot effect," Opt. Eng. **35**, 2680–2684 (1996).

[73] L. B. Lesem, P. M. Hirsch, and J. A. Jordan Jr., "The kinoform: a new wavefront reconstruction device," IBM J. Res. Develop. **13**, 150–155 (1969).

[74] W.-H. Lee, "Computer-generated holograms: techniques and applications," in *Progress in Optics*, E. Wolf, ed. (North-Holland, Amsterdam, 1978), Vol. 16, 119–232.

[75] J. R. Fienup, "Iterative method applied to image reconstruction and to computer-generated holograms," Opt. Eng. **19**, 297–305 (1980).

[76] D. Kermisch, "Image reconstruction from phase information only," J. Opt. Soc. Am. **60**, 15–17 (1970).

[77] J. W. Goodman and A. M. Silvestri, "Some effects of Fourier-domain phase quantization," IBM J. Res. Develop. **14**, 478–484 (1970).

[78] W. J. Dallas, "Phase quantization – a compact derivation," Appl. Opt. **10**, 673–674 (1971).

[79] W. J. Dallas, "Phase quantization in holograms – a few illustrations," Appl. Opt. **10**, 674–676 (1971).

[80] A. Tanone, Z. Zhang, C. M.-Uang, and F. T. S. Yu, "Optical beam steering using a liquid-crystal television panel," Micro. Opt. Technol. Lett. **7**, 285–289 (1994).

[81] S. Jacobsson, S. Hard, and A. Bolle, "Partially illuminated kinoforms: a computer study," Appl. Opt. **26**, 2773–2781 (1987).

[82] O. Bryngdhal and F. Wyroski, "Digital holography – computer generated holograms," in *Progress in Optics*, E. Wolf, ed. (Elsevier, Amsterdam, 1990), Vol. 28, 1–86.

[83] S.-D. Wu, Q. Song, A. Meyers, D. A. Gregory, and F. T. S. Yu, "Reconfigurable interconnections using photorefractive holograms," Appl. Opt. **29**, 1118–1125 (1990).

[84] S. Jutamulia, H. Niwa, and S. Toyoda, "Use of liquid crystal switch in single-lens-reflex camera," Proc. SPIE **2263**, 421–429 (1994).

[85] S. Jutamulia, G. M. Storti, W. M. Seiderman, J. Lindmayer, and D. A. Gregory, "Infrared signal processing using a liquid crystal television," Opt. Eng. **30**, 178–182 (1991).

[86] S. Jutamulia, S. Toyoda, and Y. Ichihashi, "Removal of pixel structure in liquid crystal projection display," Proc. SPIE **2407**, 168–176 (1995).

[87] X. Yang, N. Li, and S. Jutamulia, "Liquid crystal projection image depixelization by spatial phase scrambling," Proc. SPIE **2650**, 149–159 (1996).

[88] I. Gorog, "Displays for HDTV: direct-view CRT's and projection systems," Proc. IEEE **82**, 520–536 (1994).

15

Representations of fully complex functions on real-time spatial light modulators

Robert W. Cohn and Laurence G. Hassebrook

"O tempora, o mores!"

Cicero, Orations

"The medium is the message."
Marshall McLuhan, *Understanding Media: The Extensions of Man*

15.1 Introduction

Today, a majority of the approaches to optical pattern recognition involve coherent optical correlators that use real-time, electrically addressed spatial light modulators (SLM's) in both the input and the filter planes (as illustrated in Fig. 15.1.) These devices differ dramatically from the medium, photographic film, that was used in the original VanderLugt correlator [1]. Today SLM's operate at from video frame rates [30 frames per second (fps)] to up to 10,000 fps, as opposed to film, which is cumbersome to develop in place. Electrically addressable SLM's have resolutions (i.e., spatial bandwidths) of 10^4 to 10^6 pixels whereas holographic film can easily have a resolution of 10^9. The modulation characteristics can be quite varied as well. Originally VanderLugt used film that produced only intensity variations. SLM's of interest are generally thought of as phase-only or amplitude-only [as illustrated in Fig. 15.2(a)]. In practice, devices are found to have coupling between amplitude and phase [as shown in Fig. 15.2(b)] [2–4]. This has led to the term coupled-amplitude-phase modulation, with special cases being amplitude-mostly and phase-mostly. SLM's may also be limited in that not all phase values between 0 and 2π or all amplitude values between zero and unity can be obtained. SLM's that can achieve only a finite number of levels are limited in the sense that the phase or amplitude is quantized [see Figs. 15.2(c) and 15.2(d)]. These differences between the earlier fixed-pattern SLM's and today's real-time programmable SLM's challenge our assumptions and lead us to consider new approaches to representing information in optical processors.

Thus a major consideration in the realization of optical processing systems is how to represent complex values with limited-range SLM's. The discussion in the literature of these issues is pervasive, but in only a few publications is complex-valued representation the central topic. Our goal is to review optoelectronic processor developments with a particular focus on complex representations. These considerations are essential in light of the rapid advances in SLM's, their applications, and in related innovations in the design and fabrication of diffractive optics and computer-generated holograms.

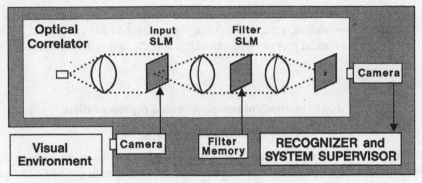

Fig. 15.1. Optoelectronic correlator. The optical correlator is a coprocessor in the complete mixed analog–digital processor. In addition to high-level inferences and decision making, the system supervisor coordinates, controls, and processes data into and out of the optical correlator.

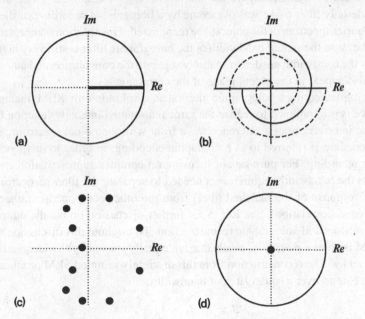

Fig. 15.2. Operating characteristics of SLM's. The modulation types shown are (a) phase-only and amplitude-only (thick line), (b) coupled-amplitude-phase that has continuous phase ranges of 2π (solid curve) and 4π (dashed curve), (c) quantized, (d) biamplitude-phase that specifically shows amplitudes of unity and zero.

A recurring theme throughout this chapter is the trade-off between the ability to represent desired complex functions as accurately as possible and what is possible in real-time computer-assisted processors. We show that there is currently a gap between the performance possible with real-time processors and what has been achieved with off-line optimization. The gap can be sizable for representations of complex-valued composite functions. Much of the current research reported here is directed at reducing this gap. We also consider how such improvements can enhance the capabilities of optoelectronic processors.

This chapter is organized into three sections: early methods of representing complex-valued modulation, methods of representation suited to current SLM's, and a discussion of the interplay between optical processor functionality and the ability to represent complex values.

15.2 Early methods of complex-valued representation

15.2.1 Holographic encoding

The intimate dependence of the representation on the target modulator can be appreciated by a comparison of the original VanderLugt optical correlator with current SLM-based optoelectronic correlators. Consider the problem solved by the VanderLugt correlator: An object can be identified in the input scene through linear optical filtering. The frequency-plane filter is generally complex-valued but the frequency-plane modulator varies in only its amplitude transmittance. (Throughout we define amplitude as meaning real positive values. For arbitrary modulation we use the term complex amplitude.) The limitation of the amplitude-only filter plane was overcome by a holographic recording of the complex-conjugate Fourier spectrum of the object to be recognized. This solution is acceptable for film correlators because the spatial bandwidth of the holographic filter is still very high after one accounts for the bandwidth needed to spatially separate the convolution and autocorrelation terms that also appear in the output plane of the correlator.

Consider implementing a holographic filter on an amplitude-only SLM, that is, the SLM transmittance is programmed to realize the same real-valued intensity function that is produced by the interference of a reference wave front with the object spectrum. This mathematical procedure is referred to as holographic encoding, in order to distinguish it from holographic recording. For purposes of focusing on complex representations, we specifically discuss the bandwidth requirements needed to separate the filter reconstruction (i.e., the impulse response of the matched filter) from the other components (rather than separating the cross correlation). See Ref. 5 for further discussion on bandwidth restrictions for both correlators and holographic reconstruction. Throughout this discussion we assume that the SLM is an $n \times n$ array of pixels that are equally spaced by the identical increment Δ in both x and y. The reconstruction from this discretely sampled SLM produces periodic replicas that extend over a square area of bandwidth,

$$B = B_x B_y = 1/\Delta^2, \tag{15.1}$$

where $B_x = B_y = 1/\Delta$ [see Fig. 15.3(a)]. We often refer to B as the nonredundant bandwidth of the modulator. If the desired reconstruction (i.e., the impulse response) is b units long by b units wide, then the on-axis term will be $2b \times 2b$ (because of autocorrelation of the impulse response with itself) and the conjugate impulse response will be $b \times b$. Figure 15.3(b) illustrates this condition for the case in which the usable filter bandwidth is maximized. In this case the usable bandwidth of the SLM is $B/16$, that is, the SLM is only as effective as an $n/4 \times n/4$ pixel SLM that does not require holographic encoding. Because of the already limited resolution of SLM's, this loss in effective resolution would seriously reduce the effective digital computation rate of optical Fourier transformers. The usable bandwidth can be increased (say, if more bandwidth is allowed in y than in x) but it may be less than desirable to work with nonsquare or variable-resolution SLM's (as well as imagers and framegrabbers). Therefore we conclude that holographic encoding in most applications uses only a small fraction of the bandwidth of the SLM.

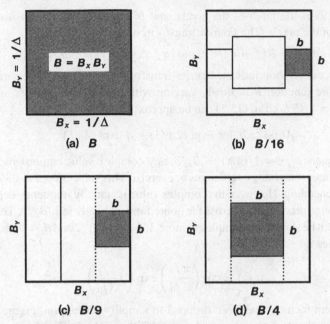

Fig. 15.3. Available spatial bandwidth of SLM's and usable spatial bandwidth (shaded regions) of various encoding algorithms. The plots are shown in the Fourier-transform or reconstruction plane of the SLM: (a) The available spatial bandwidth B of an $n \times n$ pixel array of pitch Δ in both x and y. (b) The usable spatial bandwidth $B/16$ of holographic encoding for the SLM in (a) under the typical design constraint that the usable portion of the reconstruction has a square aspect. (c) The usable bandwidth $B/9$ for Burckhardt's method and under the same design constraint as that of (b). (d) The usable bandwidth $B/4$ for Florence–Juday's method 1 and also method 2 when the superpixel is arranged as a 2×2 pixel array.

15.2.2 Detour-phase-encoding methods

Consider another class of encoding algorithms specifically referred to as detour-phase computer-generated holography [6]. The methods as originally described by Brown and Lohmann are designed for binary amplitude modulation. The complex amplitude of any given cell (i.e., a cluster of elemental pixels or a superpixel) is effectively achieved by use of the clear area of each pixel to represent amplitude and the position of the clear area to represent phase. Fine control over phase and amplitude requires a resolution much higher than the cell; thus this method has a usable bandwidth that is much less than that set by the resolution of the modulator. If the amplitude of the modulator can be continuously varied, then a complex-valued modulation can be constructed with four [7] or even three pixels [8]. Thus the usable bandwidth would be $B_x/3$ in one direction, which gives $B/9$ total usable bandwidth for a SLM that is a square array of pixels and under our assumption that the reconstruction is square [see Fig. 15.3(c)]. However, it can be argued that the usable bandwidth is even less.

Consider Burckhardt's method [8] in which three adjacent pixels are used to represent one complex value. The transmittance is written as

$$a(x) = br(x) + cr(x + \Delta) + dr(x + 2\Delta), \tag{15.2}$$

where $r(x)$ is a function (such as a rect) that describes the subaperture of a pixel (identical for each pixel), Δ is the pitch of the pixels, and $b, c,$ and d are real positive amplitude transmittances of the pixels. The Fourier transform of this superpixel is

$$A(f_x) = R(f_x)[b + c\ \exp(j2\pi f_x\Delta) + d\ \exp(j4\pi f_x\Delta)], \tag{15.3}$$

where the uppercase symbols indicate Fourier-transformed variables. The Fourier transform of the subaperture function R is slowly varying with spatial frequency f_x and is usually ignored. For $\Delta = 1/(3f_0)$, Eq. (15.3) can be approximated as

$$A(f_0) \simeq b + c\ \exp(j2\pi/3) + d\ \exp(j4\pi/3). \tag{15.4}$$

Thus at the frequency $f_0 = 1/(3\Delta) = B_x/3$, any complex value can be produced by the selection of the three weighting coefficients $b, c,$ and d. This value is used as the design value for purposes of encoding. However, the complex value is actually frequency dependent, and its value can change dramatically across the nonredundant bandwidth $B_x/3$. The magnitude of the problem can be shown by a simple example. If $b = c = 1/2$ and $d = 0$, then Eq. (15.3) for all frequencies is

$$A(f_x) \simeq \cos\left(\frac{\pi f_x}{3f_0}\right) \exp\left(j\frac{\pi f_x}{3f_0}\right), \tag{15.5}$$

where the element factor R has been dropped to simplify discussion. Figure 15.4(a) is a graphical construction of the results of relations (15.4) and (15.5). The designed value at f_0 is $A(f_0) = 0.5\angle60°$. This is illustrated in Fig. 15.4(a) as a result of adding the two phasors

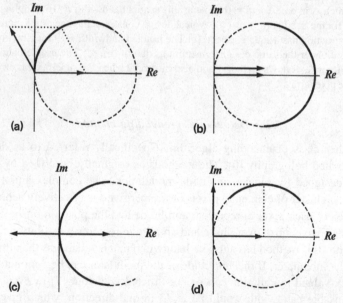

(a)

(b)

(c)

(d)

Fig. 15.4. Frequency dependence of various detour-phase-encoding algorithms for (a) Burckhardt's method, (b) Florence–Juday (method 1 only), (c) Florence–Juday (method 1 or 2), (d) Lee's method and also Florence–Juday method 2 when the superpixel is arranged as a 4×1 pixel array. The frequency-dependent complex amplitude is represented by the circular arc (solid curve). The desired complex value is the intersection of the two dotted lines [in (a) and (d) only]. The length of the arc corresponds to the values that would be found across the range of spatial frequencies b in the usable band (the shaded regions in Fig. 15.3).

together. Also illustrated is the solid circular arc that describes the locus of the complex values over the entire nonredundant bandwidth $B_x/3$. This shows that the complex values vary from $0.866\angle 30°$ to $0\angle 90°$ over the nonredundant band from $f_0/2$ to $3f_0/2$. Thus the approximation is very poor at the edges of the band. Depending on the accuracy required, it may be necessary to reduce the usable bandwidth further.

Two methods of encoding complex values that are similar to detour-phase methods were introduced by Florence and Juday [9]. Their approach is based on the use of either a phase-only modulator or the use of amplitude modulators that are overlayed with a four-phase-level diffractive optic. These methods have an advantage in that the reconstruction is centered on the optical axis. They are implemented with either two pixels or a 2×2 array of pixels, thus resulting in a usable (square) bandwidth of $B/4$ [see Fig. 15.3(d)]. The increased bandwidth is offset by the increased frequency sensitivity of the effective complex modulation of the superpixels. The first method defines the modulation of the superpixel as

$$a(x) = \text{r}(x)\exp(j\phi_b) + \text{r}(x + \Delta)\exp(j\phi_c), \tag{15.6}$$

where ϕ_b and ϕ_c can be programmed to any desired phase over a 2π range. The modulation for the second method is

$$a(x) = b\text{r}(x) + c\text{r}(x + \Delta)\exp(j\pi/2) + d\text{r}(x + 2\Delta)\exp(j\pi) + e\text{r}(x + 3\Delta)\exp(j3\pi/2), \tag{15.7}$$

where each pixel is programmed in amplitude and is offset by a fixed phase shift. Equation (15.7) is written in one dimension for convenience, but the effective modulation of the superpixel is identical if the pixels are arranged in a 2×2 array. The concept is also interesting in that it would also be possible to customize each diffractive optic to compensate for any deviations in flatness of the amplitude modulator.

The effective complex modulation of Florence and Juday's first method is determined from the Fourier transform of Eq. (15.6) at $f_0 = 0$ to be

$$A(0) = 2\cos\left(\frac{\phi_b - \phi_c}{2}\right)\exp\left(j\frac{\phi_b - \phi_c}{2}\right). \tag{15.8}$$

Thus the difference between the two phases determines the effective amplitude and the average of the phases determines the effective phase. The second method produces an effective complex modulation of

$$A(0) = b - d + j(c - e). \tag{15.9}$$

We encode the desired complex value by first determining in which quadrant of the complex plane the desired value lies. This information is used to select either b or d and c or e to be nonzero. The amplitudes of the two nonzero transmittances are then set to produce the desired complex value.

The usable spatial-frequency range is greater for Florence–Juday methods but, as a result, the frequency-dependent errors are increased. This can be shown by a comparison of the analysis of Burckhardt's method with the corresponding analysis of Florence–Juday methods. Recall that there is a phasor associated with each pixel of the superpixel. The phasing between the phasors determines the effective amplitude. The locus of all possible phasings (for two phasors of amplitude 1/2) describes a circle of radius 1/2 centered at $(1/2, 0)$ [see Fig. 15.4(b)]. The nonrepetitive bandwidth for a two-pixel superpixel is $B_x/2$. Over this frequency range, one phasor can vary by π with respect to the other phasor. The resulting complex values [shown by the solid curve in Fig. 15.4(b)] cover half of the circular locus. Figure 15.4(b) specifically shows the result for the two phasors that are in phase at $f_0 = 0$ (for either method 1 or 2). The encoded complex value is $A(0) = 1\angle 0°$, but the amplitude varies from 0.707 to 1 and the phase varies from $-45°$ to $45°$ over the usable frequency range. A second example is given in Fig. 15.4(c), in which the value 0 is encoded by method 1. Once again the locus of complex values is contained in a 90° range. However, the phase does not vary continuously. Instead it is contained between 45° and 90° and $-45°$ and $-90°$. Figure 15.4(d) shows the locus for method 2 if a 4×1 pixel array is used as a superpixel. This construction also describes Lee's delayed-sampling method [7].

It would be interesting to compare the effects of these frequency-dependent errors with other systematic errors (such as phase quantization). For example, with binary phase-only devices, it is common to use a single value of the phase to represent any complex value on half of the complex value. Thus localization to a quadrant by methods 1 and 2 may produce results at the band edges that are somewhat better than those obtained with binary phase-only modulators. We do not attempt these analyses here.

Summarizing to this point, note that although holographic- and group-oriented encoding methods can be used to approximate desired complex values, the usable bandwidth never exceeds $B/4$ for typical square systems. Thus, even under the best conditions, a loss by a factor of 4 is a serious penalty when we are considering using real-time SLM's that are composed of a relatively small number of pixels.

15.2.3 Multiple spatial light modulators

Are there other approaches that can be used to encode complex information that use the bandwidth more efficiently? The answer is yes. One method is to combine modulators, either in cascade or in parallel [10–12]. The Florence–Juday method 1 algorithm can be implemented by the placement of one phase-only SLM in each arm of a Michelson interferometer. In this implementation the complex coefficients are not frequency dependent. Juday and Florence note that any tilt between the two SLM's must be eliminated [10]. This system does use the bandwidth more efficiently; however, it does not use the full-bandwidth B. This comparison is based on considering that the two SLM's have a total of B pixels; this leads to a total usable bandwidth of $B/2$. The cost of additional optics and mechanical alignment further reduces the attractiveness of two-SLM approaches.

15.3 Methods of representing complex values on current spatial light modulators

15.3.1 Synthesis of Fourier transforms by use of time-integrated spectrum analyzers

The methods described to this point are all designed for coherent spatial filtering operations. It is worth briefly considering time-integrating spectral analysis [13, 14] as a method of synthesizing a desired complex function [15, 16]. The approach interferometrically synthesizes all spatial frequencies of the input signal in sequence and time integrates them on a detector. Speeds in excess of video rates appear possible where acousto-optic Bragg cells and high-frame-rate cameras are used. The method is a hybrid optoelectronic approach in that the complex modulation exists as only an electronic or digital signal. For example, in an optoelectronic correlator [17], the spectrum synthesized by the processor would be electronically multiplied by the reference spectrum, and then the product is used as the input signal to the processor.

The basic concept can be shown by considering the Fourier transform of

$$s(x;t) = \left[\frac{a(t)}{2}\right]^{1/2} \left\{ \delta\left(x + \frac{v}{2}t\right) \exp\left[\frac{j\phi(t)}{2} + \frac{jn\pi}{4}\right] \right.$$
$$\left. + \delta\left(x - \frac{v}{2}t\right) \exp\left[-\frac{j\phi(t)}{2} - \frac{jn\pi}{4}\right]\right\}. \tag{15.10}$$

The function corresponds to two point sources that are counterpropagating in the x direction with a velocity of $v/2$. The parameter n is an integer that is used to add quadrature phase shifts to $\phi(t)$. The need for this is explained below. The function $a(t)\exp[j\phi(t)]$ is the temporal signal for which the Fourier transform is desired. The two impulses are Fourier tranformed by a lens to produce two plane waves that interfere with amplitude:

$$S(f_x;t) = [2a(t)]^{1/2} \cos[\pi f_x vt + \phi(t)/2 + n\pi/4], \tag{15.11}$$

which is the Fourier transform of Eq. (15.10). The magnitude squared of Eq. (15.11) [followed by the standard trigonometric identity $\cos^2 A = 1/2(1 + \cos 2A)$] gives the time-varying interference fringe

$$I(f_x;t) = a(t) + a(t)\cos[2\pi f_x vt + \phi(t) + n\pi/2]. \tag{15.12}$$

We assume that $a(t)$ is a time-limited function of duration T. The time integration of Eq. (15.12) gives

$$I(f_x) = T\langle a(t)\rangle + \text{Re } \mathcal{F}(a(t)\exp\{j[\phi(t) + n\pi/2]\})|_{f=vf_x}, \tag{15.13}$$

where the first term is written in terms of the time average of the amplitude and the second term contains the Fourier transform of the complex signal. For $n = 0$ the second term in Eq. (15.13) gives the real part of the Fourier transform $A(vf_x)$, and for $n = 1$ the second term gives the imaginary part of the transform. In practical implementations it may be desirable to suppress the first term by also forming the intensity patterns for $n = 2$ and $n = -1$ and subtracting them from the respective $n = 0$ and $n = 1$ intensity patterns. The integration function is simply achieved by the accumulation of the intensity patterns on a CCD imager for a period of time T followed by a readout of the image.

The synthesizer can be implemented with the arrangement shown in Fig. 15.5. Crossed one-dimensional Bragg cells permit the synthesis of two-dimensional images. The amplitude and the phase shifts are produced by point modulators that precede the deflectors. The traveling point sources can be directly produced by a Bragg deflector. These spots would be observed at the focal point between the lens and the camera in Fig. 15.5. For a stationary spot, a cw signal is applied. The spot diameter depends on the length of the Bragg aperture that is illuminated by both the laser and the cw acoustic wave. The number of resolvable spots is determined by the time–bandwidth product (the delay of the cell multiplied by the bandwidth of the acoustic transducer). For example, when standard values for tellurium dioxide (TeO$_2$) Bragg cells are used, the slow shear wave velocity is $v/2 = 620$ m/s [18]. An 8-mm aperture gives a delay of 12.8 μs, a transducer bandwidth of 40 MHz, and a time–bandwidth product of 512. If the frequencies are input in random sequence, then each frequency must be allowed to fill the full aperture in sequence. This can result in a very slow frame rate. For example, if the synthesis of a 256 × 256 image is desired, then the aperture

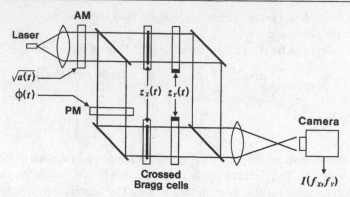

Fig. 15.5. Time-integrated spectrum analyzer. The Fourier transform of the temporal signal $a(t) \exp[j\phi(t)]$ appears as the time-integrated image recorded by the camera. The Fourier transform of the desired complex signal can be synthesized by either stepped or continuously chirped sinusoids for the functions $z_x(t)$ and $z_y(t)$. The camera is in the image plane of the crossed Bragg cells. AM, amplitude-only point modulator; PM, phase-only point modulator.

can be reduced to 6.4 μs. (A small amount of time, which we ignore, is also required for shuttering the laser when the cell is filled.) This corresponds to only ~2.4 fps to synthesize the real or the imaginary part by Eq. (15.13). Two to four frames are required for synthesizing the entire complex image. Electromechanical x–y scanners can scan a field of this size at a rate of roughly 30 fps [19].

In comparison, much faster scanning is possible when the Bragg cells are driven with counterpropagating chirp signals. With a chirp signal the Bragg cell also produces a spot at the focal plane in Fig. 15.5. However, the spot now sweeps a raster line at a constant velocity. Spatial resolution approaching the time–bandwidth product is possible for sweep times of the order of the aperture delay of the cell. As with the cw synthesizer, the chirp synthesizer also images the phase modulation [20] of the Bragg cells onto the camera. An analysis that is similar to the above analysis is presented to show this result. The modulation produced by interfering the two waves is

$$s(x;t) = \left[\frac{a(t)}{2}\right]^{1/2} \left\{ \exp\left[j\pi\alpha\left(x + \frac{v}{2}t\right)^2 \right] \exp\left[\frac{j\phi(t)}{2} + \frac{jn\pi}{4} \right] \right.$$
$$\left. + \exp\left[j\pi\alpha\left(x - \frac{v}{2}t\right)^2 \right] \exp\left[-\frac{j\phi(t)}{2} - \frac{jn\pi}{4} \right] \right\}. \qquad (15.14)$$

Repeating the algebraic steps used for analyzing the random-frequency addressing system gives

$$I(x;t) = a(t) + a(t)\cos[2\pi\alpha x v t + \phi(t) + n\pi/2], \qquad (15.15)$$

$$I(x) = T\langle a(t)\rangle + \mathrm{Re}\,\mathcal{F}(a(t)\exp\{j[\phi(t) + n\pi/2]\})_{|f=\alpha v x}. \qquad (15.16)$$

The intensity is written as a function of the x coordinate that is in the plane of the Bragg cell. With a $1\times$ imaging system, x can be replaced by f_x, which shows that Eqs. (15.13) and (15.16) are essentially identical and that chirped Bragg modulations can be used in place of the cw modulation.

The advantage of the chirped method in terms of speed is decided. This can be shown by a comparison with the above example. Using the same Bragg cell as described in the example above, we consider the speed required for synthesizing $A(\alpha v f_x)$. In the architecture shown in Fig. 15.5, the amplitude and the phase modulation simultaneously illuminate a continuum of positions across the aperture of the Bragg cell. For this reason the chirp must fill the aperture and must be at least twice the time delay of the aperture to form the desired fringe. For the first T_a seconds, the chirp fills the aperture. Then, after the signal $a(t)$ is gated on T_a more seconds, the function is gated off. For two-dimensional transforms the process is repeated for each line of the two-dimensional modulation. The bandwidth of the chirp is limited by the transducer of the Bragg cell to 40 MHz. Because the signal is applied for only half the sweep, the bandwidth used is 20 MHz and the full Bragg cell aperture of 8 mm, or a T_a of 12.8 μs, is illuminated to obtain a resolution of 256. The sweep time is then 25.6 μs, which gives a frame rate of approximately 152 fps for a 256×256 image. Full complex transforms could then be synthesized at rates of 38 to 76 spectra per second, depending on whether two or four frames are used. CCD cameras that have frame rates in excess of 200 fps are commercially available [21].

In terms of speed and resolution, Bragg cell techniques compare favorably with direct video-addressed SLM's. However, the cost of Bragg cells and associated rf electronics can be substantially greater than those of SLM's that are video addressed. The usable bandwidth is controlled primarily by the time–bandwidth product of the cell and the number of time integrations required for synthesizing the complex function. This processor is the first that we have described that uses time-sharing or time-sequential operations to perform an operation. This reduces the effective temporal bandwidth, which is compensated for by the high bandwidth of rf systems. Differences between baseband/video electronics and rf electronics and between temporal and spatial information processing complicate comparisons between these approaches. Nonetheless, the Bragg cell synthesizer is an important approach for representing complex modulation, especially given the extremely high dynamic range and signal-to-noise ratio (SNR) demonstrated by acousto-optic processors [13–17].

All schemes for encoding complex-valued functions that have been described so far use less than the full spatial bandwidth of the SLM. For holographic methods, this bandwidth is $B/9$, for group-oriented-encoding methods the most usable bandwidth is $B/4$, and for the two-SLM interferometric approach the usable bandwidth is $B/2$. The time-integrating synthesizer, as shown in Fig. 15.5, contains several light modulators, both point and spatial. If we focus attention on only the SLM's (for which we consider a crossed pair to represent one SLM) there are two SLM's used, which reduces the bandwidth to $B/2$. Additionally, the two to four frames required for synthesizing a complex spectrum reduces the temporal bandwidth over that possible with SLM's that produce complex modulation in a single frame. Next we describe methods of encoding complex-valued functions that use the entire spatial bandwidth B and that perform this processing by programming the SLM with only one frame of data.

15.3.2 Full-bandwidth methods of encoding: encoding by global optimization

One general approach to encoding uses global optimization to produce a desired diffraction pattern. The modulation function for each pixel is varied until a solution meets the design constraints. Many successful designs have been developed with iterative techniques such as simulated annealing [22, 23], genetic [23, 24], and several other global optimization algorithms. Excellent results in terms of diffraction efficiency, accuracy, and noise have been demonstrated for the design of fixed-pattern diffractive optics [25–29]. These are all time-consuming for applications that require encodings in real or even near real time. These approaches have also been adapted to encode composite pattern recognition filters to available modulators [30, 31]. Space does not permit a full discussion of these methods. However, an additional discussion of specific algorithms is presented as it relates to the possibility of achieving real-time encoding with global optimization. Also, we review some important general results on the properties of globally optimal encodings in Subsection 15.3.4.

15.3.3 Full-bandwidth methods of encoding: point-oriented encoding

The approach that we are primarily interested in is point-oriented encoding. In contrast to global optimization, algorithms can be devised that run at real-time rates on serial electronic processors. These approaches are in keeping with the requirements of real-time optical and optoelectronic processors that typically can process serial video signals in real time, given that only a few numerical operations are required per pixel. For these point-oriented systems, the optical performance is not necessarily optimal. However, acceptable optical performance coupled with real-time operation would greatly enhance the flexibility and adaptability needed to use optoelectronic processors for real-time autonomous tasks.

15.3.3.1 Carrier-based methods

The term point-oriented encoding originally referred to various methods of modulating a spatial carrier of a fixed frequency. The earliest methods were designed for amplitude transparencies drawn by pen plotter. Dallas shows a basic method of modifying the pulse width and the pulse position of a square wave to achieve any desired complex modulation [32]. Another variant for phase-only modulators was developed by Kirk and Jones [33]. The effective complex modulation is understood in terms of Fourier series analysis. For example, in the Kirk and Jones method, a periodic carrier of spatial frequency f_0 is modulated in amplitude α and phase ψ_α. This signal is encoded as the phase-only function,

$$a(x, y) = \exp[j\psi(x, y)] = \exp\{j[\alpha \cos(2\pi f_0 x) + \psi_\alpha(x, y)]\}. \tag{15.17}$$

The Fourier series expansion of Eq. (15.17) produces a component of complex amplitude

$$a_c \equiv a_c \exp(j\psi_c) = J_0(\alpha) \exp(j\psi_\alpha), \tag{15.18}$$

which will reconstruct on the zero diffraction order. The function $J_0(\alpha)$ is the zero-order Bessel function. Thus a_c is proportional to the complex amplitude of the dc or zero-order far-field diffraction pattern. Any desired value of amplitude a_c between 1 and 0 can be implemented when $J_0(\alpha)$ is inverted to find the appropriate value of α. However, the method requires the carrier to have a spatial frequency in excess of the bandwidth of the complex modulation to avoid interference from the other diffraction orders. More importantly, if this method is applied to a spatial modulator that is composed of an array of pixels, at least two (and preferrably more) pixels are required for synthesizing one period of the carrier. Thus the method can be viewed as group oriented, in which a group consists of the amplitude multiplied by a period of the carrier. The method is point oriented in that each encoding can be calculated in sequence at each point with reference to only the complex amplitude modulation at each point. We refer to these methods as carrier-based encoding to distinguish them from the more recent methods of point-oriented encoding that we discuss next.

15.3.3.2 Pixel-based methods

The first successful point-by-point-encoding method was probably the well-known phase-only matched spatial filter [34]. The desired fully complex function is the matched spatial filter $H^+(f_x) = S^*(f_x)$, where $S(f_x)$ is the spectrum of the object that we design the filter to recognize. The phase-only filter is produced by the trivial encoding operation $H^0(f_x) = \exp\{j \arg[S^*(f_x)]\}$. It is apparent that each pixel of the filter-plane SLM can be programmed independently of all others and that the full spatial bandwidth B is available for information processing. The encoding usually produces very good correlation peaks. In their initial paper, Horner and Gianino also generalized the meaning of the phase-only filter to include the phase-only encoding of composite recognition filters [34]. Shortly thereafter, Casasent and Rozzi showed that this nonlinear transformation of a complex-valued composite filter into a phase-only function severely degrades the performance of the filter for recognition tasks [35]. The nonlinear effects can be better appreciated by consideration of the effects on the Fourier-transform reconstruction of a phase-only composite function. A powerful demonstration is to phase-only encode the composite of two lens functions [36]. Davis and Cottrell showed that intermodulation can dramatically enhance stronger intensities and reduce weaker intensities in the reconstruction. Additionally, spurious peaks are formed at sum and difference frequencies. The observations of Casasent and Rozzi led to the development of optimization algorithms for compensating for the interactions between the individual functions in the composite filter [30, 31]. Although the phase-only filter has not been generally successful as a fast method of point-oriented encoding, it works quite well for noncomposite recognition. This is fortunate, in that several encoding algorithms have followed that extend the performance and the applicability of point-oriented encoding.

15.3.3.3 Maximum correlation intensity matched filter

The first major advancement after the introduction of the phase-only filter was the recognition that the actual SLM's available are in general neither phase-only nor amplitude-only but coupled [2–4]. Typically amplitude is a function of phase. At this time both Juday [4] and Farn and Goodman [37] developed solutions to the problem of designing noncomposite correlation filters. The objective of their designs is to maximize the intensity of the correlation peak when the training image is placed in the input of the correlator. The intensity of the correlation peak can be maximized by maximization of the magnitude of each frequency component $S(f_x)H(f_x)$. Thus at each frequency the phase of the spectrum $S(f_x)$ should be conjugate to the phase of the filter and the filter magnitude should be as large as possible. (If the SLM could produce any value, then the filter magnitudes would be infinite.)

When the modulation characteristic is coupled, as shown in Fig. 15.6, the analysis also reduces to a problem of maximizing the amplitude of each frequency component $S(f_x)H(f_x)$ individually. In this case, ϕ, the phase of $S^*(f_x)$, does not necessarily equal the phase of the optimal realizable filter $H^0(f_x)$. Instead, $G(\phi)$, the point on the complex-valued modulator characteristic that produces the largest amplitude in the direction of $\arg(S^*)$, is selected [37]. As shown in Fig. 15.6, the component of H^0 in the direction of S^* is $H^0 \cos(\delta\phi)$, where $\delta\phi$ is the phase angle between S^* and $H^0 = G(\phi)$. Thus the intensity of the correlation

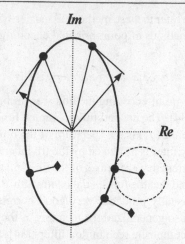

♦ Desired complex value
● Encoded complex value

Fig. 15.6. Maximum correlation peak intensity and minimum Euclidean distance optic filter (MEDOF) encoding methods. The maximum intensity method (upper half of the complex plane) selects the point on the modulation characteristic that produces the largest amplitude component in the direction of the desired complex value. The arrow represents the projection onto the desired direction. The desired amplitude is infinite. The MEDOF-encoding method (lower half of the complex plane) maps the desired values to the closest point on the modulation characteristic. The dashed circle demonstrates for one point that the mapping is the closest possible. The maximum correlation method is a special case of MEDOF encoding.

peak is not maximized by the selection the conjugate phase that is prescribed for a classical matched or inverse filter. The function $G(\phi)$ can be found for any modulator characteristic in advance of specifying the spectrum of the reference image. The desired complex filter H^+ is rapidly encoded when the single function call $H^0(f_x) = G[\arg(S^*)]$ is performed for each frequency f_x.

In most recognition problems the phase of the correlation peak is not of concern. With a fully complex SLM, scaling the reference spectrum by $\exp(j\alpha)$ has no effect on the correlation. This provides 1 degree of freedom that can be used to maximize the intensity of the correlation peak further. When this degree of freedom is used, the optimization problem can be generalized for a given spectrum $S(f_x)$ to find the value of α in $G(\phi + \alpha)$ such that the intensity of the correlation peak is maximum. For the maximum intensity filter, the magnitude of the correlation peak can be written as

$$|c(0)| = \int |S(f_x)| \cdot |G\{\arg[S^*(f_x)] + \alpha\}| \, df_x; \qquad \alpha \in [0, \, 2\pi], \qquad (15.19)$$

and the integral is performed for all values of α to find the value of α that maximizes the correlation. Farn and Goodman noted that Eq. (15.19) is a correlation integral [37]. He suggested that a fast way to perform the optimization over α is to sample $G(\phi)$ discretely at a small number of angles over the complex plane. This can lead to a small discrete (circular) convolution that can be rapidly evaluated with the fast Fourier transform (FFT). The values of $S(f_x)$ would also be discretized and sorted into appropriate phase bins. The correlation peak amplitude $|c(0)|$ would then be approximated as

$$|c_j(0)| \approx \sum_{i=1}^{n} |S_b(\phi_i)||G(\phi_i + \alpha_j)|; \qquad \alpha_j = \frac{2\pi j}{n}, \tag{15.20}$$

where $|S_b(\phi_i)|$ is the sum of the magnitudes of $S(f_x)$ in the ith range of phases. For n, a small number, most of the numerical computation would involve the sorting from spatial frequencies f_x to angles ϕ_i.

It is quite conceivable that these numerical operations can be performed at real-time serial video rates. Consider that there are N pixels to be encoded and n phase bins. The sorting would require $N \ln_2(n)$ comparisons if the search is done hierarchically starting from the most significant to the least significant phase bit. There would be N additions in the binning process and of the order of $4n \ln_2(n)$ multiplies (three FFT's and the multiplication of the FFT's of the magnitudes of S_b and G). Following the determination of α^0, the optimal value of α_j, the SLM is programmed with the encoding algorithm $H^0(f_x) = G[\arg(S^*) + \alpha^0]$. This can be performed with N look-up-table operations. Consider the number of operations needed to perform the encoding algorithm with values of $N = 256 \times 256$ and $n = 64$. At 30 fps, the generalized encoding algorithm requires approximately 12 million logical comparisons, 2 million additions and look-up-table operations, and 50,000 multiplies per second. This can be compared with commercially available, application-specific serial image processing chips from Sumitomo Metals that run at up to 50-MHz processing rates.

15.3.3.4 Minimum Euclidean distance noncomposite filters

Juday generalized the Farn and Goodman [37] and Juday [4] encoding method for non-composite correlation filters to filters that are optimal for a variety of metrics, including the SNR in the correlation plane, the peak-to-correlation energy, and a metric that contains both the SNR and the peak-to-correlation energy, called peak to total energy [38]. Juday showed that the earlier method of maximizing intensity is also a special case of his generalized method. The encoding procedure is referred to as the minimum Euclidean distance optimal filter (MEDOF). The MEDOF minimizes the sum of the distances between the optimal fully complex filter H^+ and the optimal realizable filter H^0. Juday demonstrated that for noncomposite filters the correlation metric is usually the global maximum. The method suggests a two-step search procedure similar to that of Farn and Goodman. The first step is to determine the point-by-point mapping function $G(r, \phi)$ that minimizes the distance $d(r, \phi)$ for any given complex value $H = (r, \phi)$, where the ordered pair represents magnitude and phase. The second step is to find the values of the 2 degrees of freedom of gain γ and rotation α of the optimal filter function H^+ that minimize the sum of the distances:

$$d_T(\gamma, \alpha) = \int d(\gamma|H^+|; \arg H^+ + \alpha) \, df_x, \tag{15.21}$$

where distance is a real positive quantity. Note that, instead of scaling H^+, it is convenient to incorporate these factors into the mapping function as $G(\gamma r, \phi + \alpha)$ and the distance function as $d(\gamma r, \phi + \alpha)$. Because the distance function d can be precomputed and stored as a look-up table, it is possible to develop fast global searches by using a binning approach similar to that of Farn and Goodman in approximation (15.20). The result is a two-dimensional convolution of the form

$$d_T(\gamma_k; \alpha_l) \approx \sum_{i=1}^{n} \sum_{j=1}^{m} b_{i,j} \, d\left(\gamma_k \left| H_{i,j}^+ \right|; \arg H_{i,j}^+ + \alpha_l\right), \tag{15.22}$$

where $b_{i,j}$ is the number of times a continuous value of H^+ is discretized into the value $H_{i,j}^+$ in the ith magnitude and jth phase bin. The analysis of the number of operations is identical to that for Farn and Goodman's method if the product nm replaces n. Obviously more operations are required for searching the larger two-dimensional space. Nonetheless, it is still conceivable that the encoding algorithm can be completed in real time. As a numerical example, consider that the complex plane is discretized into $n = 16$ gains and $m = 64$ phases. At 30 fps, the number of operations would increase to 20 million logical comparisons and 1.2 million multiples per second. This calculation rate still appears reasonable in terms of commonly available electronics. Certainly, a designer can trade-off the resolution of the bins and the precision of the arithmetic in order to find the optimum more accurately or to complete the calculations more quickly. Of course, for the greatest speed the optimization step can be eliminated. Another approach for reducing digital computation would be available if the scene is varying slowly compared with the system frame rate. The optimization procedure could then be spread out over several frames. Alternatively, an approach could be applied in which the parameters γ and α are adaptively adjusted based on observed changes in the correlation peak. This second approach is applicable when a correlator is being used for object tracking. If the adaptive approach is used in recognition and target acquistion, we question whether γ and α might be adapted away from optimal so as to sharpen a false target.

15.3.3.5 Minimum Euclidean distance composite filters

Recently MEDOF approaches have been adapted and modified for encoding complex-valued composite functions, including composite recognition filters [39–41] and diffractive optic spot-array generators [42, 43]. Although these algorithms do apply the closest mapping step, there appear to be further developments required for approaching the performance possible with global optimization. Two issues requiring further study involve the use of the degrees of freedom available in many composite function designs and also in recognizing that mappings other than the closest mapping may sometimes optimize certain metrics of interest.

One interesting development is the incorporation of the MEDOF into the iterative design of the synthetic discriminant function (SDF) filters by Montes-Usategui et al. [39]. This method appears to be a generalization to arbitrary modulation characteristics of Bahri and Kumar's successive forcing algorithm, which is designed for phase-only characteristics [44]. Both encoding methods are forms of projection-onto-constraints algorithms [45, 46] (the most widely known method is the Gerchberg–Saxton algorithm [47]).

The Montes-Usategui method consists of two complementary projection operations $H_k^0 = p_i H_k^+$ and $H_{k+1}^+ = p_2 H_k^0$, where p_i are the two projection operations and H_k^0 is the MEDOF encoding (without optimization) of H_k^+, a filter function that satisifies the SDF design equations [48]. These equations are

$$c_i(0) = \int S_i(f_x) H^+(f_x) \, \mathrm{d}f_x; \qquad i = 1, 2 \cdots M, \tag{15.23}$$

where S_i represents the M training images and $c_i(0)$ are the correlation peak heights desired for each training image. Equation (15.23) shows that the H^+ usually is not unique because the frequency space is infinite dimensional and the correlation space is finite dimensional. For application on a frequency-plane SLM that has N pixels, the design equations are written as

$$c_i(0) = \sum_{j=1}^{N} S_i(j)H^+(j); \qquad i = 1, 2 \cdots M, \tag{15.24}$$

where j represents frequency. Typically N is much greater than M. Thus, from the perspective of linear algebra, the filter H^+ and the spectra of each of the training images S_i can be viewed as N-dimensional vectors and the set of the M values of c_i can be viewed as an M-dimensional vector. The training vectors S_i span (at most) an M-dimensional space. Therefore there is an $N-M$-dimensional linear vector space that is orthogonal to the subspace of the training vectors [49]. The SDF filter can be adjoined with any arbitrary vector from the orthogonal subspace and satisfy Eq. (15.24). Montes-Usategui *et al.* use these $N-M$ degrees of freedoms to adapt the original SDF filter so that H^+ satisfies Eq. (15.24) and the errors in encoding H^+ are as small as possible.

Their encoding algorithm is specified by the two projection operations p_1 and p_2. The first operator performs the standard point-by-point MEDOF encoding $H_k^0 = G[|H_k^+|, \arg(H_k^+)]$ of the fully complex generalized SDF function H_k^+, where $G(\cdot, \cdot)$ is the minimum distance mapping function. The second operator maps the encoded function H_k^0 into a new generalized SDF function H_{k+1}^+ under the constraint that the total distance between H_k^0 and H_{k+1}^+ is as small as possible. The solution to this optimization problem is the SDF function added together with the vector component of H_k^0 that lies in the subspace that is orthogonal to the space of the training images. Alternately repeating the two projection operations usually leads to a generalized SDF function that approaches the operating curve more closely with each iteration. In some cases the convergence of the solution process may stop; however, the solution never diverges in succeeding iterations. As with the generalized SDF, the resulting solution is not unique and depends strongly on the starting vector. Kumar and Carlson have also proposed a similar approach that includes a minimum distance mapping in each iteration [40]. Their metric includes terms for maximizing the filter energy. In passing, we note that the numerical requirements for iteratively computing the generalized SDF are substantially greater than the MEDOF algorithm applied to noncomposite filters. This is due to the iterative nature of the algorithms and the large numer of multiplications associated with the training images [39].

15.3.3.6 Pseudorandom encoding

Another point-oriented-encoding method that uses the entire usable bandwidth of the SLM is the method of Cohn and Liang, which is referred to as pseudorandom encoding [50]. This method simply uses the average value of a random variable to represent the desired complex value. Thus, for a phase-only SLM, the random variable is the phase and the statistics of the phase are chosen so that the average value of the phase is the desired complex value. The method is extendable to many other types of modulation characteristics, including coupled-amplitude-phase modulators and biamplitude-phase modulators [42]. The ability to computer generate random variables that have the desired statistical properties provides the second degree of freedom needed to treat each pixel of the SLM individually and as if each is fully complex.

In methods involving randomness, noise is always of concern. However, in the pseudo-random-encoding method the noise level is (1) closely linked to the diffraction efficiency of the fully complex modulation, (2) reduced in the Fourier observation plane because of the natural averaging that occurs because of wave-front superposition, and, most importantly, (3) the noise that is generated is diffused to an average uniform level over the entire usable bandwidth B [50, 51]. The third observation is especially important compared with other systematic or deterministic methods (including the MEDOF) that tend to generate noise or spurious terms at harmonically related frequencies [36, 42, 52]. The harmonics usually do not appear at every frequency, i.e., they occupy a bandwidth less than B. As a result, it is often possible that pseudorandom encoding will produce peak noise levels that are substantially smaller than those for the deterministic methods.

We present a few examples of the encoding procedure, followed by a comparison of how well pseudorandom and MEDOF methods encode a composite function. Rather than draw a hard distinction with MEDOF, the comparisons show that a blend of the methods often produces a performance that is substantially better than either alone.

The theoretical background for pseudorandom encoding is presented in greatest detail in Ref. 50. Here we directly present, by way of example, specific encoding algorithms. The simplest example of encoding is for biamplitude-phase modulation (see Fig. 15.7). The desired complex modulation is $a_c = (a_c, \psi_c)$ and the resulting modulation by the SLM is $a = (a, \psi)$, where the ordered pairs are the polar representations of the complex quantities and the modulations are functions of position. The pseudorandom-encoding design statement is used in general to select $a_c = \langle a \rangle$, where

$$\langle a \rangle = \int a\, p(a)\, \mathrm{d}a \tag{15.25}$$

is the ensemble average of the complex-valued random variable a and $p(a)$ is the probability density function (pdf). The density function is selected by design and thus varies with position.

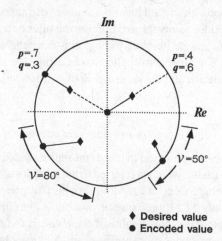

Fig. 15.7. Pseudorandom-encoding methods. The biamplitude-phase-encoding method (upper half of the complex plane) randomly selects unity with a probability of p and zero with a probability of $q = 1 - p$. The dashed lines indicate the statistically possible alternative mappings of the desired complex values. The pseudorandom phase-only encoding method (lower half of the complex plane) randomly selects a value of phase over limits of spread v centered around the phase of the desired complex value.

Example 1: Pseudorandom-encoding for biamplitude-phase modulation [32]. For the specific case of the biamplitude modulator, any desired phase ψ_c can be directly produced by the modulator. Therefore we directly select $\psi_c = \psi$. However, the amplitude of the modulator $a = |a|$ can be set to only either 1 or 0. Nonetheless, any amplitude between 1 and 0 can be realized on average by proper selection of the pdf. Specifically consider the set of density functions for the binomial distribution

$$p(a) = q\delta(a) + p\delta(a - 1); \qquad p = 1 - q \in [0, 1], \qquad (15.26)$$

where there is a probability p that the random variable takes on a value of 1 and a probability $q = 1 - p$ that the random variable takes on the value of 0. Evaluating Eq. (15.25) with the pdf from Eq. (15.26) gives the well-known weighted average

$$\langle a \rangle = 1p + 0q = p. \qquad (15.27)$$

This result provides a simple and direct formula for encoding any desired amplitude between 0 and 1 as $a_c = |a_c| = p$.

The encoding procedure uses the standard uniform random-number generator ran(*iseed*) that produces random numbers between 0 and 1. If the value of ran is less than p, then a is set to 1 and ψ is set to ψ_c; otherwise, if ran is between p and 1 then a is set to zero. The procedure is applied in sequence to the N pixels of the SLM. At the ith pixel the threshold value p_i is selected to equal the desired amplitude a_{ci}. The encoding algorithm is illustrated in Fig. 15.7. The modulator amplitude that is randomly selected is connected to the desired complex value by a solid line and the amplitude that is not selected is connected to the desired value by a dashed line.

For any type of pseudorandom encoding the expected intensity of the diffraction pattern takes the general form [42, 50]

$$\langle I(f_x) \rangle = \left| \sum_{i=1}^{N} A_{ci} \right|^2 + \sum_{i=1}^{N} (\langle |A_i|^2 \rangle - |A_{ci}|^2), \qquad (15.28)$$

where $A_{ci}(f_x)$ is the Fourier transform of the ith pixel located at position x_i in the modulator plane. Equation (15.28) shows that the observed intensity pattern consists of two components. The first term represents the desired diffraction pattern

$$I_c(f_x) = |\mathcal{F}[a_c(x)]|^2, \qquad (15.29)$$

where $\mathcal{F}[\cdot]$ is the Fourier transform operator. The second term represents the average level of background (i.e., speckle) noise that is produced as a result of the random encoding. The noise energy is spread over an area of broad extent that corresponds to the diffraction pattern of the aperture of a single pixel.

For the biamplitude-phase modulator (assuming that the pixels are point sources of infinitesimal extent) the expected intensity is [42]

$$\langle I(f_x) \rangle = \left| \sum_{i=1}^{N} A_{ci} \right|^2 + \sum_{i=1}^{N} p_i q_i. \qquad (15.30)$$

It is clear that the second term describing the noise will be smaller if the desired amplitudes a_{ci} are clustered near either 1 or 0, and this term will be larger if the amplitudes are clustered near 0.5. The level of the noise, as indicated by the second term of Eq. (15.30), indicates to what degree noise is affecting the accuracy of a design. Note that this term can be calculated with a small number of operations from the desired values of the modulation and without performing FFT's. This metric can be used to decide in advance of encoding whether the encoding of a particular function will give acceptable results. For designs in which there are design freedoms, the metric also can provide guidance in selecting the function that encodes with optimal performance from a set of acceptable fully complex functions.

Example 2: Pseudorandom encoding for phase-only modulation [50]. For phase-only SLM's the phase ψ is a random variable. The statistics of ψ are chosen so that the desired phase $\psi_c = \langle \psi \rangle$ and the amplitude $a_c = \langle \exp[j(\psi_c - \langle \psi \rangle)] \rangle$. Formulas are easily developed by specializing Eq. (15.25) to

$$\langle a \rangle = \int p(\psi) \exp(j\psi) \, d\psi \qquad (15.31)$$

for a phase-only modulator. The form of the expectation is essentially a Fourier transform or (from the field of probability and statistics) a characteristic function [53]. Average or effective amplitude control between 0 and 1 can be obtained by proper selection of a family of pdf's. A particularly useful family is the uniform family

$$p(\psi; \nu) = \frac{1}{\nu} \text{rect} \left(\frac{\psi - \langle \psi \rangle}{\nu} \right); \qquad \nu \in [0, 2\pi]. \qquad (15.32)$$

Evaluation of Eq. (15.31) by using the pdf's from Eq. (15.32) gives any desired complex value according to

$$\langle a \rangle = \text{sinc}(\nu/2\pi) \exp(j \langle \psi \rangle). \qquad (15.33)$$

This leads to the direct method of encoding a_c by performing the simple look-up-table operation,

$$\nu = 2\pi \, \text{sinc}^{-1}(a_c), \qquad (15.34)$$

and then, using a uniform random-number generator ran(*iseed*) of unity spread and zero mean, calculating the value of phase modulation:

$$\psi = \langle \psi \rangle + \nu \, \text{ran}(iseed). \qquad (15.35)$$

The expected value of ψ satisfies the pseudorandom-encoding design condition that $a_c = \langle a \rangle$. The lower half of Fig. 15.7 illustrates the phase-only encoding algorithm. Note that the random spread ν increases as the distance between the desired value and the phase-only curve increases.

Example 3: Pseudorandom encoding for amplitude-coupled-phase modulators. One way to extend phase-only encoding to amplitude-phase modulators is to compensate for the amplitude weighting $a(\psi)$ in the amplitude-coupled modulation $a = a(\psi) \exp(j\psi)$. This approach produces encoding formulas that are similar in form to the ones for phase-only encoding. This method is especially useful for phase-mostly modulators for which the amplitude varies by a small amount as a function of phase. The amplitude compensation method turns out to be a type of histogram equalization procedure [54]. For the coupled modulators Eq. (15.25) takes the form of an amplitude-weighted average

$$\langle a \rangle = \int a(\psi) p(\psi) \exp(j\psi) \, d\psi \equiv a_0 \exp(j\psi_0) \qquad (15.36)$$

of the complex exponential. As with Eq. (15.3), the randomness permits us to realize arbitrary values of amplitude and phase. However, because this new average is amplitude-weighted, ψ_0 is not equal to $\langle \psi \rangle$ (also a_0 is not equal to $\langle a \rangle$). There is also a coupling between the resulting amplitude and the phase in Eq. (15.36) that requires a two-dimensional search if we were to use pdf's of the form of Eq. (15.32). However, the simultaneous two-dimensional search can be reduced to a sequential one-dimensional search by the selection of a family of pdf's $p(\psi)$ that compensate for the amplitude coupling $a(\psi)$. The effective pdf is of the form $p_{\text{eff}}(\psi) \propto a(\psi)p(\psi)$, where the correct scale factor ensures that the cumulative distribution function $P(\psi)$ has a total probability of 1 for $\psi = \infty$. In order to encode the modulation, a random-number generator of density $p(\psi)$ is needed. It can be produced by transforming the uniform random variable $s \in [0, 1]$ according to

$$\psi = P^{-1}(s), \tag{15.37}$$

which is the inverse of the distribution function. This method is illustrated by the development of two encoding formulas.

Example 3a: Encoding for a prespecified amplitude coupling. An amplitude-coupling function of a modulator [see Fig. 15.3(b), dashed line] is linear with phase such that

$$a(\psi) = m\psi + b; \qquad \psi \in [-2\pi, 2\pi], \tag{15.38}$$

where m is the slope and $b = a(0)$. The effective density function desired is a rect function similar in form to Eq. (15.32), having a spread v, and it is centered on the desired phase $\psi_c = \psi_0$ such that

$$p_{\text{eff}}(\psi) \propto \text{rect}\left(\frac{\psi - \psi_0}{v}\right). \tag{15.39}$$

For this case the family of pdf's that compensates for $a(\psi)$ is specifically

$$p(\psi) = \frac{1}{\psi + b/m} \left[\ln\left(\frac{\psi_0 + v/2 + b/m}{\psi_0 - v/2 + b/m}\right)\right]^{-1} \text{rect}\left(\frac{\psi - \psi_0}{v}\right), \tag{15.40}$$

and the transformed random variable found with Eq. (15.37) is

$$\psi = \frac{(\psi_0 + v/2 + b/m)^s}{(\psi_0 - v/2 + b/m)^{s-1}} - b/m. \tag{15.41}$$

Evaluating Eq. (15.36) with Eqs. (15.38) and (15.40) gives a closed-form expression for effective modulation:

$$a_c = \langle a \rangle = mv \left[\ln\left(\frac{\psi_0 + v/2 + b/m}{\psi_0 - v/2 + b/m}\right)\right]^{-1} \text{sinc}\left(\frac{v}{2\pi}\right) \exp(j\psi_0). \tag{15.42}$$

Note that the compensation of the amplitude coupling also eliminated the bias drift between ψ_0 and ψ_c. The spread needed to achieve the desired amplitude can now be found by a one-dimensional search over the spread v (for a fixed value of phase ψ_0.) Also note that, by the selection of the pdf's to compensate for the amplitude coupling, Eq. (15.42) is similar in form to Eq. (15.32) for pseudorandom phase-only encoding.

Example 3b: Encoding when the amplitude-coupling function is not prespecified. For actual SLM's the coupling can be quite different, and can in fact (for liquid-crystal-type SLM's in particular) be continuously varied from phase modulating to amplitude modulating as a function of the polarization of the illumination [55]. Another current limitation is that most SLM's available today can barely produce a 2π phase range, whereas in example 3a, a 4π range was assumed. This does not constrain the encoding method if the amplitude coupling is viewed as a periodic function of phase, as illustrated in Fig. 15.3(b), solid line. The periodic assumption permits the development of encoding formulas in which modulator values can be randomly selected around the discontinuity at $-\pi$ on the complex plane. Using a discrete pdf as the random selection function leads to the especially simple result given below.

The identical design procedure as in example 3a is followed for discrete binary random variables. The effective density function $p_{\text{eff}}(\psi_0)$ has equal values at $\psi = \psi_0 \pm v/2$ and is zero otherwise. For these pdf's the effective complex amplitude is

$$a_c = \langle a \rangle = \frac{2a(\psi_0 + v/2)\, a(\psi_0 + v/2)}{a(\psi_0 - v/2) + a(\psi_0 - v/2)} \cos(v/2) \exp(j\psi_0), \qquad (15.43)$$

and the phase random variables are generated by a simple threshold test on the random number s. These results are especially useful in that this closed-form result applies to any function $a(\psi)$ for which ψ has a range of at least 2π. The result is quite similar to the deterministic Florence–Juday method 1 encoding in Eq. (15.8). The key difference is that the deterministic method uses two pixels to represent one complex number, thus sacrificing the usable bandwidth. The random-encoding method encodes the complex value with a single pixel.

15.3.3.7 Partial pseudorandom encoding

Partial encoding was first described by Hassebrook *et al.* in applications to encoding composite pattern recognition filters to phase-only SLM's [41]. In that study it was found that the error probabilities are minimized over encoding by either pseudorandom encoding or MEDOF encoding if some pixels are encoded by MEDOF and the rest are encoded by the pseudorandom method. (It should be noted that the MEDOF formula for phase-only SLM's is identically Horner's phase-only filter. Furthermore, a search over the parameters of gain γ and rotation angle α is unnecessary as the parameters have no effect on the distance metric.) A similar comparison was developed for the problem of encoding fully complex designs of spot-array generation functions of the form

$$a_c(x) \propto \sum_{k=1}^{M} \exp[j(2\pi f_k x + \phi_k)] \qquad (15.44)$$

to biamplitude-phase SLM's [42]. For this type of SLM, the MEDOF can be optimized by searches over the single parameter γ that scales the magnitude a_{ci} of the desired complex values (see Fig. 15.8). For MEDOF-only encoding, complex values that have amplitudes between 0 and 0.5 are encoded to 0, and complex values that have amplitudes between 0.5 and γ are encoded according to the phase-only filter. For this discussion the complex values are normalized so that γ is equivalent to the maximum amplitude of the complex values.

For partial pseudorandom encoding the gain parameter γ can also be varied to optimize performance. The parameter γ rescales the complex values from a maximum amplitude of unity to a value greater than unity. Those complex values a_{ci} for which $a_{ci} > 1$ are encoded by the MEDOF and the other values inside the unit circle are encoded by the pseudorandom method, as illustrated in Fig. 15.8.

Thus in partial pseudorandom encoding the value of gain γ is used to vary the number of complex values that are partially pseudorandom encoded and the number that are MEDOF

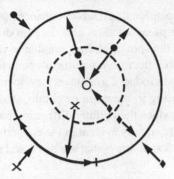

- ● Random biamplitude
- ◆ MEDOF-only
- ✕ Random phase-only

Fig. 15.8. Partial pseudorandom-encoding methods. The random phase-only and random biamplitude methods are augmented so that any desired amplitude that exceeds unity is mapped to unity by the MEDOF algorithm. The MEDOF algorithm for biamplitude modulation is also shown for comparison. The dashed circle of radius 1/2 represents the breakpoint for mapping to zero or unity.

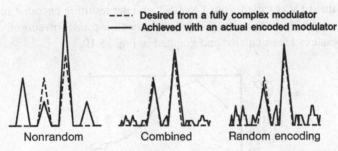

---- Desired from a fully complex modulator
—— Achieved with an actual encoded modulator

Nonrandom   Combined   Random encoding

Fig. 15.9. Accuracy improvement of partial pseudorandom encoding of composite functions over either pseudorandom only or nonrandom alone.

encoded. A specific value of γ in excess of unity usually optimizes the encoding of the desired complex function, and the optimal value depends on the function that is to be encoded. An explanation for the improvement of partial encoding over either MEDOF or pseudorandom encoding alone has been given by Hassebrook et al. in terms of finding a balance between systematic and random errors [41]. Consider increasing γ from unity to a value in excess of unity. This will reduce the random errors produced by pseudorandom encoding for two reasons. First, fewer values are now pseudorandom encoded, and second, those values that are randomly encoded are closer to the modulation characteristic, thus producing less random noise. However, the increase in γ also introduces systematic errors [56, 57] for the deterministic MEDOF encoding of the amplitudes in excess of unity. Thus there can be a value of γ that tends to balance the contributions due to random and systematic errors.

A conceptual drawing that represents the effect of the MEDOF, pseudorandom, and partial encoding on the diffraction pattern of a simple spot generator is shown in Fig. 15.9. The desired diffraction pattern would be two spots of different intensities. The MEDOF encoding, being a deterministic method, tends to produce the undesired harmonics. There is also a type of nonlinear competition that tends to enhance the stronger spot and to reduce the weaker spot [36, 52]. The pseudorandom encoding produces a broadly spread noise background, and the errors in the spot intensities are due to the influence of randomness in the encoding procedure, rather than from nonlinear competition. However, with partial encoding the harmonics can be introduced as long as they do not exceed the noise level. As the harmonics rise with increasing γ, the noise level falls until for one particular value of γ the two errors balance to produce the optimal diffraction pattern for the various partial encodings. These tendencies are equally evident in the encoding of complicated functions, including the encoding of 8×8 spot arrays that are discussed next.

15.3.3.8 Comparisons of various encodings of an 8×8 spot-array generator

The complex function to be encoded is a 300×300 pixel modulation that produces 64 spots of equal intensity in an 8×8 array [42]. Ideally, the background noise is zero, so one goal of the design is to make the noise as small as possible. The phases of the spots ϕ_k in expression (15.44) are not considered important for the application of the device, but these phase degrees of freedoms are used to specify the unique values of the a_{ci} to be encoded. The phases are selected randomly, rather than as identical, which increases the diffraction efficiency of the complex function from 5%, for all phases identical, to 22%. This function is encoded by both MEDOF-only and partial pseudorandom encoding. For each case the value of γ is varied from 1 to 1.82, and the resulting encoded function a_i is Fourier transformed with a 300×300 point discrete Fourier-transform subroutine. Various performance measures are calculated and graphed in Fig. 15.10.

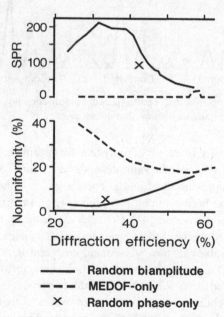

Fig. 15.10. Performance of encoding a composite function for an 8×8 spot-array by the partial pseudorandom and MEDOF methods. For the plots of the signal-to-peak-noise ratio (SPR) and nonuniformity, the diffraction efficiency is roughly a linear function of the gain factor γ. The two crosses represent the largest value of the SPR and the lowest value of nonuniformity for partial phase-only encoding as a function of γ.

A conservative measure of signal-to-background noise is the average intensity of the 64 spots divided by the maximum noise intensity in the entire discrete Fourier-transform file. This measure is referred to as the signal-to-peak-noise ratio (SPR). The average signal intensity to average noise intensity ratio is typically 1 to 2 orders of magnitude larger than the SPR. Nonuniformity is defined as a relative error; specifically, the standard deviation of the 64 spot intensities is divided by the average intensity of the desired spots. One measure of the diffraction efficiency is η_e, which measures the percentage of energy ending up in the desired diffraction pattern compared with the total energy in the diffraction pattern. It is calculated as the sum of the 64 spot intensities divided by the total energy in the Fourier-transform plane. The average intensity transmittance of the modulator is $\eta_t = N_{on}/N$, where N_{on} is the number of modulator pixels set to unity amplitude. The energy utilization efficiency is then $\eta = \eta_t \eta_e$. Although γ is not shown, note that the plotted diffraction efficiency η increases monotonically as a function of γ.

Figure 15.10 shows that the best pseudorandom-encoded design outperforms the best MEDOF-only design in SPR (212 versus 19.5) and nonuniformity (2.5% versus 17.8%). The efficiency η of the best MEDOF-only encoding is higher than the best pseudorandom encoding (59% versus 32%), as is η_e (88% versus 62%), but even this difference can be reduced by sacrificing some uniformity and some SPR. For example, even at an efficiency of 57% the pseudorandom-encoded design outperforms the MEDOF-only design in the SPR and nonuniformity.

A realistic illustration of the differences between MEDOF-only and partial pseudorandom encoding (as opposed to the qualitative comparison shown in Fig. 15.9) is shown in Fig. 15.11. The diffraction pattern shown is the one that maximized the SPR as a function of γ for each encoding. Note the differences in the background noise that have been

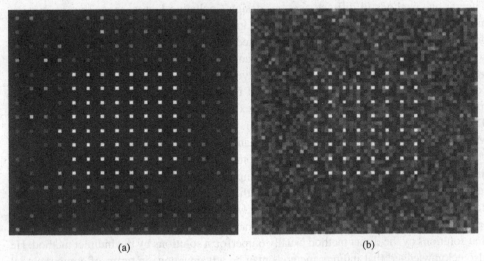

(a)   (b)

Fig. 15.11. Gray-scale image of the 8×8 spot-array resulting for (a) the MEDOF, (b) partial pseudorandom biamplitude encoding. The results are for the encoding that produces the largest value of the SPR, which is 19.5 for (a) and 212 for (b). In order to show the noise background, the image in (a) is saturated with respect to the peak intensity by a factor of 26 and the image in (b) by a factor of 212. By spreading the noise out over the entire usable bandwidth, the partial-encoding method demonstrates peak noise that is lower by a factor of more than 10:1.

brought out by saturation of the maximum gray scale in each intensity pattern by a factor that is roughly the same value as its respective SPR. The MEDOF design shows a series of harmonically related noise orders. Orders like this are quite common in many binary diffractive optics designs today. To summarize, pseudorandom encoding can be viewed as an attempt to maximize the entropy of the approximation errors in encoding. In selecting gain γ some intermodulation products are accepted as long as an overall better performance in SPR, uniformity, etc., is achieved.

Thus the minimum distance criteria, by themselves, do not optimally encode composite functions. Algorithms such as partial encoding can build on the MEDOF to improve performance, but they too do not necessarily produce globally optimal solutions. However, the major objective of encoding is the ability to handle complex numbers with ease and speed in real-time systems. Improvements in performance must be tempered with the computation rate of the optoelectronic system's digital processor. The remainder of this article presents the contrasting issues of optimal performance and scenarios for the use of complex-value encoding in real-time adaptive optoelectronic processors.

15.3.4 Optimality

The pixel-oriented methods (with the exception of noncomposite correlation filters) do not meet certain optimality requirements. With the goal of developing improved methods of point-oriented encoding, we briefly review important results for the design of phase-only diffractive optics.

15.3.4.1 Direct versus indirect optimization

The various methods of diffractive optics design (including lenses, spot-array generators, and pattern recognition filters) are surveyed and generalized by Mait [58]. There are two general approaches to design, which are referred to as direct and indirect. The direct method solves for the optimal design subject to the constraints imposed by the modulation characteristic. A single optimization problem solves the entire problem. The indirect method consists of two steps: (1) the specification of the desired complex function through optimization and (2) the encoding of the complex function to the modulation characteristic. The encoding usually also requires optimization, such as the searches over γ and α in the MEDOF, the maximum correlation intensity, and partial pseudorandom-encoding methods described above. The maximum correlation intensity method could also be classified as a direct method, because the optimal fully complex solution is not explicitly specified but rather it is embedded within the derivation of the algorithm. However, most practical direct designs do not provide explicit solutions as they require trade-offs between various metrics (e.g., SPR, uniformity and diffraction efficiency, or peak intensity and SNR). Mait noted that solutions by the direct method usually outperform solutions by the indirect method. He also acknowledged that indirect methods may be advantageous in terms of computational speed and ease of use.

Thus for the indirect methods that we have mainly been focusing on, there are two stages at which optimization can lead to improvements. First, an optimal complex-valued function can be designed. If possible, the design would include an optimization that reduces the total distance between the composite function and the modulation characteristic. For composite functions in particular, the available degrees of freedom (e.g., phase and amplitude weightings of the individual functions) permit such an optimization. Second, the encoding can be optimized to improve fidelity. This was demonstrated for the example of partial encoding on the biamplitude SLM in which a particular value of γ optimizes the SPR. It appears likely that nearly optimal values of γ can be estimated by modeling rather than by a time-consuming search. Such a result would be desirable for real-time applications. However, an even larger improvement would result if it were possible to develop a noniterative and real-time method of selecting the values of the degree of freedoms used in specifying the composite function.

15.3.4.2 Least upper bound on diffraction efficiency

The amount of the improvement possible has been related to diffraction efficiency by Wyrowski [59]. He shows that for the phase-only modulation characteristic there is a least (or tightest) upper bound on the diffraction efficiency of

$$\eta_{opt} = \bar{a}_c^2 / \overline{a_c^2}, \tag{15.45}$$

where the overbar represents the average over the N pixels of the modulation a_c. The numerator is the squared average amplitude transmittance, and the denominator is the average intensity transmittance of the modulation. The denominator can also be interpreted as the variance of the amplitude transmittance. Relationships between the variance and the average indicate that the maximum diffraction efficiency can never exceed unity for a phase-only modulator, or for that matter for a fully complex modulator. The efficiency is maximized by making the values of a_c as close to each other as possible. Thus minimizing the variance of the modulation amplitudes maximizes the diffraction efficiency [56, 57]. Unity efficiency can be achieved if the amplitude modulation can be made white. This objective is identical to the problem in holographic recording in which the diffusers are used to level the recorded spectrum [32, 60, 61]. The upper bound for phase-only spot-array generators (for which the phase of the spots provides design freedoms) is found through numerical optimization to be $\sim$95%–99% [52, 62].

In general the diffraction efficiency of a fully complex function is [51, 59]

$$\eta = \overline{a_c^2}. \tag{15.46}$$

This could possibly be larger than unity if the modulator provides gain. However, with the usual assumption that the device is passive, the maximum amplitude $a_{c\,max}$ does not exceed unity. In pseudorandom encoding for phase-only SLM's, the maximum amplitude is assumed normalized to unity so that

$$\eta_{pr} = \overline{a_c^2} / (a_{c\,max})^2, \tag{15.47}$$

where $a_{c\,max}$ normalizes the maximum value of effective modulation $|\langle a \rangle|$ to unity amplitude. This result follows from Eqs. (15.28) and (15.29) [51]. This situation corresponds to unity gain γ. For partial pseudorandom encoding the gain is larger than unity and the diffraction efficiency increases to a value larger than η_{pr}, as illustrated by Fig. 15.10 and for similar performance curves for phase-only partial pseudorandom encodings.

Wyrowski also found that the gain that maximizes the diffraction efficiency in Eq. (15.45) is [59]

$$\gamma_{opt} = \bar{a}_c / \overline{a_c^2}. \tag{15.48}$$

This gain is also greater than unity and it is obtained when the desired function is mapped according to Horner's phase-only encoding. Thus the nonlinear process of phase-only encoding effectively produces gain. Wyrowski also noted that, although this mapping does produce maximum efficiency, some reduction in efficiency is needed to meet the requirements of uniformity. However, example designs by Mait [58] and by Gale *et al.* [52] show that perfectly uniform spot arrays can be designed with an efficiency that is from a few tenths of a percent to a few percent less efficient than the upper bound. It is satisfying to note that as the diffraction efficiency in Eq. (15.45) approaches unity, the gain in Eq. (15.48) also approaches unity. The diffraction efficiency of pseudorandom encoding in Eq. (15.47) also approaches unity. This can be visualized as a compression of the horizontal axis in Fig. 15.10 that describes the performance of a continuum of partial pseudorandom-encoding algorithms.

It is illustrative to consider the following simple example. The amplitudes of a_{ci} are $a_{c1} = 1$ and $a_{c2} = 0.5$. The diffraction efficiency from Eq. (15.46) for the full complex function or from Eq. (15.47) for pseudorandom encoding is $\eta = \eta_{pr} = 0.625$. The diffraction efficiency from Eq. (15.45) for the phase-only encoding would be $\eta_{opt} = 0.90$. The gain that maximizes diffraction efficiency from Eq. (15.48) is $\gamma_{opt} = 1.20$. However, if $a_{c1} = 1$ and $a_{c2} = 0.8$, then $\eta_{pr} = 0.90$, $\eta_{opt} = 0.988$, and $\gamma_{opt} = 1.10$, which illustrates how the diffraction efficiency of pseudorandom-encoding approaches that of the phase-only encoding as the desired complex modulation becomes more spatially uniform.

These results on efficiency and its upper bound are important for guiding the optimization of an encoding algorithm that must be computed in real or near real time. Because the diffraction efficiency can be determined in terms of the modulation, there should be ways to approach optimal performance by performing operations only in the modulation plane. Also, the upper bound (if it can be precomputed off-line) provides useful information for developing stopping criteria during time-critical searches. Finally, knowledge of the optimal gain may provide guidance on selecting the correct gain that optimizes partial pseudorandom encoding and other algorithms. All the results in this section have been developed for phase-only SLM's. It would be helpful if these results could be generalized to coupled-amplitude-phase SLM's.

15.3.4.3 Fast methods of selecting phase degrees of freedom

Although optimality of design has been a major emphasis of late, in real-time applications good designs that meet the time constraints of the digital processing unit must be accepted. The question is then, how can the performance of encoding be brought closer to the theoretical limits? In the last section it was mentioned that when the phases ϕ_k are picked randomly in expression (15.44) it is possible to increase diffraction efficiency over that possible if all the phases are identical. This method had been reported earlier by Burckhardt and was given the name random-phase coding [63]. This method can be improved on by reselection of the phases, a number of times, each time with a different random seed. For each iteration expression (15.44) is calculated, followed by calculation

of the diffraction efficiency with Eq. (15.47). The average of several runs is ~16% efficiency and the best run gave the 22% reported above. The iteration has the advantage in
that it is done exclusively in the modulation plane. However, the prospect of significantly
improving diffraction efficiency further by this method is unlikely. The problem has to
do with the statistics of the intensity transmittance. These are known to be exponentially
distributed [64]. As such, there can be large excursions from the average intensity transmittance. Thus the iterated optimization of diffraction efficiency by this method corresponds
to minimizing the maximum intensity transmittance. The large deviations away from the
desired level modulation, as prescribed by Eq. (15.45), are extremely difficult to meet with
this particular approach. Akahori also noted the poor job of leveling produced by random-
phase codes [61]. Chu and Goodman [60] and Akahori [61] showed that determinsitic
codes can do a much better job of leveling the modulation if the desired reconstruction is
a binary sequence of bright and dark intensities. The simplest code is the chirped phase
function,

$$\phi_{k,l} = \frac{\pi k^2}{m} + \frac{\pi l^2}{n}, \tag{15.49}$$

for an $M = m \times n$ array of binary data sequences that is referred to as the Schroeder code.
More information on various modulation-leveling phase codes can be found in Refs. 32 and
65. These codes could provide the necessary compromise between computational complexity and improved performance in some designs.

15.4 Discussion: scenarios for fully complex representations

This chapter has reviewed several approaches to representing complex-valued images on
SLM's. The approaches taken today are substantially different from those used in the past.
Formerly the media of photographic film and pen plots were used to make static displays
whereas today SLM's can accept data at real-time rates. The former media differ from the
more recent in that there is less spatial resolution and the modulation characteristics are
significantly different. Neither medium can physically produce all complex values, at least
affordably, which leads to the various approaches for representing complex values.

A major goal in our considerations has been to identify methods that can instantly encode
the desired complex-valued modulation to the SLM. In this way it would be possible to make
the most flexible use of the SLM. Certainly many useful optoelectronic processors can be
developed by use of precomputed, globally optimal encodings. However, prebriefing is not
always possible or practical, especially if the processor is required to adapt to a changing
environment in which there is limited prior knowledge. Thus to achieve this degree of
flexibility the encoding process must not be a computational bottleneck. Of course, the
design of certain fully complex functions (e.g., some SDF filters) may often by themselves
take too much time to be implemented at real-time rates. Nonetheless, the speedups provided
by not simultaneously optimizing the complex function and the encoding to the SLM bring
closer the possibility of incorporating these design procedures on real-time processors. The
designs would be adaptive, in the sense that the digital calculations would be performed
over several frames of the SLM. To give some concreteness to this general discussion we
present some illustrative examples of useful processors that would be possible as a result
of being able to encode complex functions in real time.

15.4.1 Optical security

Recently proposed correlator-based optical security systems are an example in which custom information, such as fingerprints, would be encoded onto a fixed-pattern reference filter that is fabricated on an access key [66, 67]. If this method is adopted by credit card companies, then the design and the fabrication of a custom filter could be required for each new credit card. Today, holograms are mass produced by credit card companies to reduce forgery. However, individually customized computer-designed holograms would require the development of both rapid design and rapid fabrication procedures to meet demand. This need for a real-time system can be appreciated by considering the task of providing an optically secure credit card to each person in the United States. Suppose that 250 million cards are to be made in 1 year and there are 8 manufacturing systems. This corresponds to ∼1 card per machine per second. At these rates, numerically intensive solutions are unlikely to be fast enough, but suboptimal filter designs followed by a fast encoding algorithm seem probable. One likely fabrication procedure would be direct gray-scale recording of an intensity image from a display of SLM into a phase-relief profile.

15.4.2 Arbitrary multispot scanning and beam shaping

SLM's are optically equivalent to phased-array attennas. Phase-only SLM's are especially desirable because the far-field pattern can be formed on the optical axis or steered anywhere within the usable bandwidth of B. Arbitrary multispot patterns can also be formed, and the spots can be arbitrarily repositioned each frame. Thus scanning operations that are much more general than the usual raster scanning of electromechanical scanners can be envisioned. However, performing arbitrary scanning and beam shaping requires that functions of the form of expression (15.44) would be calculated and encoded in real time. The maximum number of frequencies to be scanned would determine whether the calculation could be computed at a reasonable rate. Two-dimensional beam shaping can be easily accomplished when the complex function is multiplied by appropriate apodizing window functions [50]. This can be used in applications of either single-spot or multispot diffraction patterns. The window functions are especially desirable because they can also significantly reduce those sidelobes that arise from the finite spatial extent of the SLM. Multispot scanning operations could be used to illuminate multiple objects or multiple track points on three-dimensional objects. Used together with video sensing, the spot positions could be actively adapted to the motions of the objects.

15.4.3 Hybrid optoelectronic correlators for autonomous recognition and tracking

The combination of optical and electronic computation has been an active area of research for over 20 years [68, 69]. The continuing increase in the speed of digital systems has not yet overtaken the speed of optical computing, in general and optical correlation, specifically. Consider in Fig. 15.1 the somewhat general architecture of a hybrid correlator that is possible today with available SLM's, video electronics, and computers. The optical correlator is no more than a specialized (albeit high computation rate) coprocessor that supports the general-purpose computer or system supervisor with numerically intensive full-frame correlation and filtering operations. In a fully adaptive system the supervisor controls all activities based on inferences about the information collected from the visual environment. However, rather than obviating the need for the correlator, in vision-based automatic control, the increasing speed of computers is actually allowing greater and more flexible use of the correlator. Some examples suggestive of this flexibility are briefly presented.

One recent example is the pattern recognition system proposed by Casasent *et al.*, in which a suite of up to 33 filtering operations would be performed by the correlator per frame of live video [70]. Procedures for designing banks of composite filters have also been developed [71–73], and these filters can be run sequentially through the hybrid system in order to improve recognition probabilities.

We can enhance speed by performing decision-directed searches rather than by running every filter in the filter bank through the correlator. A specific example of decision-guided processing was shown by Hassebrook *et al.* in which filters from a composite filter bank are correlated with the scene until the object is identified and its position is found [73]. Then a second set of composite filters is correlated with the scene region of interest to estimate object orientation. This method could be used to identify several different objects. Based on the identification, the appropriate filters for estimating the pose of the object would be recalled and used to program the filter SLM.

A related application of decision-directed searches is in acquiring and tracking a moving three-dimensional object. We would like to associate the object with the closest matching noncomposite filter in the filter-plane memory. Then models of the dynamics of the object can be used to recall the appropriate noncomposite filter for each frame. A set of composite filters designed to allow a hierarchical tree search [31] of memory could be used to eliminate the need to test each noncomposite view in sequence.

The region of interest determination or prescreening is important for cases that have a large field of regard. This corresponds to performing a coarse search to locate object roughly, followed by a refined analysis on the region by itself. Lhamon *et al.* demonstrated a method for which a complex-valued composite function is designed for (rather than the filter SLM) the input SLM [74]. The object is assumed to be smaller than a window of a given size. To explain the method, consider the case of recognizing an $M \times M$ pixel image in an $N \times N$ pixel image. The filter is constructed by shifting the image to all $N \times N$ locations and adding all the shifted versions together. This is a periodic function, and thus we can most economically calculate it by circularly shifting the image within the $M \times M$ window, adding the replicas together, and then tiling this subimage to form the $N \times N$ image. To perform correlation of the $N \times N$ scene with the $M \times M$ filter, the $N \times N$ periodic image and the scene image are first (electronically) multiplied together. Then this preprocessed image is convolved with an $M \times M$ pixel aperture function (by programming the filter plane SLM with a sinc function). This hybrid processing produces a blurred correlation peak that roughly locates the position of the object of interest. Distortion invariance can be designed into the prescreener if the periodic mask is built with complex-valued combinations of views of the object.

The use of composite functions in the above examples usually leads to complex-valued functions and encodings. Many of the SDF designs involve too many calculations to be performed on line in a reasonable amount of time. Thus the encodings, whether real time or not, could be done off-line to reduce the computing load on the supervisor and to reduce the memory required. The prescreener, though, does require real-time encoding because the input scene is multiplied by the periodic mask before the encoding can be applied. It would be desirable if adequate composite filters could be designed in real time by simple superpositions of noncomposite filters stored in memory. Such algorithms were not a focus of this review. However, the next example is interesting from the standpoint that it might be

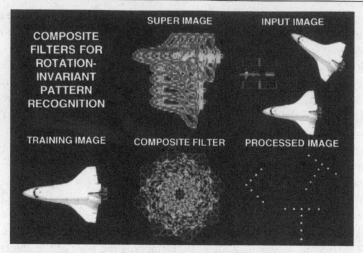

Fig. 15.12. Concept for using composite recognition filters to preprocess and transform imagery into patterns that can be recognized by real-time digital and neural processors.

possible to build adequate quality filters and encode them on line in a reasonable amount of time and also from the standpoint of the increasing amount of digital image processing that the supervisory computer can be expected to provide.

The operation is suggested by Fig. 15.12, in which composite filters are designed to transform objects of interest in the input image into other images. The resulting images form simple characters that would require far fewer electronic calculations to identify than with the original image. The superimage contains the edge-enhanced version of the training image shifted to nine locations. The images are added together with different phase weights to construct the superimage. Then the composite filter is similarly constructed with rotated versions of the superimage. The composite filter is similar in a sense to the prescreening filter. In this case, rather than producing a blurred image, the processed image produces nine correlation peaks for the Space Shuttle and seven peaks for the Hubble telescope. In doing this type of processing, it becomes possible to reduce the visual complexity of the input image without totally eliminating information on the object and its orientation. This preprocessing reduces the image complexity to a degree that the remaining task of digital image processing is greatly simplified. The simplications could allow one to consider using a low-order digitally implemented neural network to perform the remaining tasks of recognition, orientation estimation, and tracking. The actual results shown in Fig. 15.11 were simulated and are presented in Ref. 75. There only the image of the Space Shuttle was used. The Hubble telescope has been included to suggest novel approaches to using composite filters in hybrid correlators. The results were simulated with fully complex values, and so the practical limitations set by realistic SLM's have not yet been evaluated for this potentially valuable approach.

The examples described so far have focused primarily on algorithms for the hybrid correlator in Fig. 15.1. Additional flexibility is possible with other realizations of the correlator. The remainder of this section reviews alternative architectures.

In order for a hybrid correlator to be truly adaptive it must be able to acquire new and unanticipated information from the environment and develop new filters at a rate commensurate with environmental changes. The architecture, as shown in Fig. 15.1, is not adaptive as

in most instances we would assume that the supervisory computer cannot perform the necessary Fourier-transform operations fast enough to update the filter-plane memory in near-
real-time. However, this limitation is due to the specific choice of the $4f$ optical correlator.

Other optical architectures can overcome this limitation. The best known is the joint
transform correlator (JTC) [76]. Although hybrid systems have been demonstrated for
some time [64], the hybrid single-SLM JTC [77] is especially important in its compactness,
robustness, and simplicity. The hybrid JTC is easy to adapt because the reference is applied
in the image plane rather than in the filter plane. Thus reference images can be replaced and
updated at real-time rates. One issue to consider in using the JTC is that it does not use the
full-bandwidth B of the SLM because the reference and the signal are placed side by side
on the SLM and also because the patterns are holographically recorded at the filter-plane
detector. A second issue is that most research on JTC's is focused on real-valued images. It
would be an interesting extension of hybrid JTC's to consider encoding composite functions
on the reference SLM and complex-valued images on the scene SLM, especially in light of
the great speed at which the reference can be updated. Extensive discussions of applications
of JTC's can be found in Ref. 69.

An extremely flexible approach is the time-integrating interferometric optical architecture
described above in which the system synthesizes the complex Fourier transform directly
and the supervisory computer performs all other numerical operations (e.g., multiplication
of the spectra). Another interferometric architecture has been proposed by Cohn [78], in
which the complex field of the optical Fourier transform is directly measured by phase-shift
interferometry [79]. A flow chart describing the mathematics of the correlator is shown in
Fig. 15.13. In succession interference patterns of the signal spectrum with three reference
wave fronts of known phase are formed. Each interferogram is recorded by the CCD camera.
The identical pixel of each interferogram is processed to determine the phase of the spectrum.
This phase spectrum is added together with the phase spectrum of the reference, and then
the resulting spectrum is used to modulate a phase-only SLM. The phase modulation is
Fourier transformed by a lens to produce the desired correlation. A compact, single-SLM
version of the hybrid architecture is shown in Fig. 15.14, in which the scene, reference,
and correlation output are processed by the same optoelectronics in sequence. Note that
phase-shift interferometers can be used to measure not just phase, but amplitude as well.
Also note that the phase-only SLM is used to phase shift the reference wave front. In the
particular implementation a small obscuration is permitted from the reference mirror (on

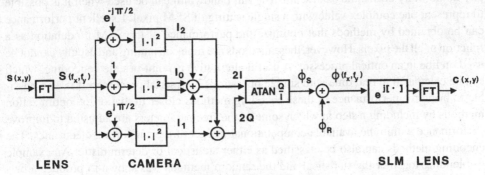

Fig. 15.13. Flow chart of the phase-only correlation algorithm. This illustrates the
algorithm for the specific case of reference phase shifts of $-\pi/2$, 0, and $\pi/2$.

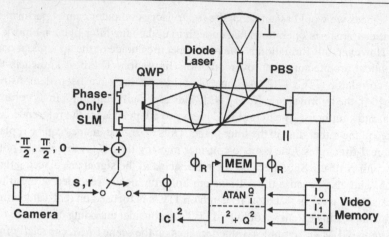

Fig. 15.14. Compact implementation of an optoelectronic correlator. The system performs Fourier transforms of the scene, the reference, and the product of the reference and scene spectra with the same hardware. The phase of the reference spectrum is stored in memory (MEM) and then recalled for subtraction from the phase of the scene spectrum. The quarter-wave plate (QWP) has a small-area mirror deposited on its front surface. The quarter-wave plate and the polarized beam splitter are used together to direct light efficiently from the laser illuminator to the video camera. The phase-only SLM serves the dual function of a signal modulator and a source of the reference phase shifts.

the quarter-wave plate) and the laser diode. In this arrangement the hybrid correlator may process as many as nine video frames for one scene frame. This hybrid correlator can also encode complex-valued functions to the SLM by the various methods presented here. An important unresolved issue is the degree to which noise from the encoding process can be minimized in a real-time environment.

15.5 Summary and conclusion

The way we choose to represent complex-valued modulations is heavily influenced by the moderately small number of pixels, resolution, or spatial bandwidth of current SLM's. The early methods of encoding from the fields of holography and computer-generated holography do not fully utilize the bandwidth. The full bandwidth can be used when it is possible to represent one complex value with a single setting of SLM pixel. Excellent performance can be obtained by methods that optimize the performance of the SLM modulation as a function of all the pixels. However, these methods are time consuming and usually cannot be used on line in an optical processor. A useful alternative is point-oriented encoding, which can be calculated in real time with simple operations by a serial processor. It is also possible to bring the performance of these encoding methods closer to that of the optimization methods by including a step in which some of the free parameters are adjusted to improve performance within the available computational budget of the supporting electronics. The encoding methods can also be classified as either statistical or deterministic. An example of blending together the statistical and the random methods was shown to produce a better performance than either method alone. The upper bound of diffraction efficiency was discussed. It provides a metric to judge the quality of the point-oriented encoding, and it may provide insight into ways that the performance of the encoding algorithms might be enhanced.

Point-oriented-encoding methods are important because they allow the manipulation and processing of complex-valued functions in real time with modulators that usually do not produce fully complex modulation. The greatest benefits will be realized in real-time systems that adapt to and incorporate new information from a rapidly changing environment. The issues in developing a hybrid optical–electronic system in these situations involve the ability to encode and also the ability to process complex-valued data. The composite functions provide extremely sophisticated functions that can greatly enhance the robustness and autonomy of a hybrid system; however, their adaptability is also limited by the computational budget of the supervisory computer and also the specific optical architecture.

We anticipate that continued focus and further emphasis on representations of complex-valued functions on SLM's will lead to significant improvements in the functionality, performance, and usefulness of hybrid optoelectronic processors.

References

[1] A. VanderLugt, "Signal detection by complex spatial filtering," IEEE Trans. Inf. Theory **IT-10**, 139–145 (1964).

[2] R. D. Juday and B. J. Daiuto, "Relaxation method of compensation in an optical correlator," Opt. Eng. **26**, 1094–1101 (1987).

[3] R. D. Juday, "Optical correlation with a cross-coupled spatial light modulator," in *Spatial Light Modulators and Applications*, Vol. 8 of 1988 OSA Technical Digest Series (Optical Society of America, Washington, DC, 1988), pp. 238–241.

[4] R. D. Juday, "Correlation with a spatial light modulator having phase and amplitude cross coupling," Appl. Opt. **28**, 4865–4869 (1989).

[5] J. W. Goodman, *Introduction to Fourier Optics*, 2nd ed. (McGraw-Hill, New York, 1996), Chaps. 8 and 9.

[6] B. R. Brown and A. W. Lohmann, "Computer-generated binary holograms," IBM J. Res. Develop. **13**, 160–168 (1969).

[7] W. H. Lee, "Sampled Fourier transform hologram generated by computer," Appl. Opt. **9**, 639–643 (1970).

[8] C. B. Burckhardt, "A simplification of Lee's method of generating holograms by computer," Appl. Opt. **9**, 1949 (1970).

[9] J. M. Florence and R. D. Juday, "Full complex spatial filtering with a phase-mostly dmd," in *Wave Propagation and Scattering in Varied Media II*, V. K. Varadan, ed., Proc. SPIE **1558**, 487–498 (1991).

[10] R. D. Juday and J. M. Florence, "Full complex modulation with two one-parameter SLMs," in *Wave Propagation and Scattering in Varied Media II*, V. K. Varadan, ed., Proc. SPIE **1558**, 499–504 (1991).

[11] D. A. Gregory, J. C. Kirsch, and E. C. Tam, "Full complex modulation using liquid-crystal televisions," Appl. Opt. **31**, 163–165 (1992).

[12] L. G. Neto, D. Roberge, and Y. Sheng, "Full-range, continuous, complex modulation by the use of two coupled-mode liquid-crystal televisions," Appl. Opt. **35**, 4567–4576 (1996).

[13] A. VanderLugt, *Optical Signal Processing* (Wiley-Interscience, New York, 1992), pp. 327–346.

[14] A. VanderLugt, "The theory of acousto-optic spectrum analyzers," in *Acousto-Optic Signal Processing Theory and Implementation*, 2nd ed., N. J. Berg and J. M. Pellegrino, eds. (Marcell Dekker, New York, 1996), Chap. 4, pp. 101–138.

[15] W. R. Franklin, "Optical signal processing at Essex Corporation," Opt. Eng. **35**, 401–414 (1996)

[16] J. L. Lafuse, P. J. Roth, T. M. Turpin, and R. E. Feinleib, "Performance modeling and measurements in optical image synthesis systems," in *Advances in Optical Information Processing VI*, D. R. Pape, ed., Proc. SPIE **2240**, 256–264 (1994).

[17] R. R. Kallman, P. Roth, and T. Turpin, "Correlation results using invariant filters in an acousto-optic correlator," in *Optical Pattern Recognition V*, D. P. Casasent and T.-H. Chao, eds., Proc. SPIE **2237**, 62–73 (1994).

[18] NEOS Acousto-optic Products Catalog, NEOS Technologies, Inc., 4300-C Fortune Place, Melbourne, FL 32904 (1995).

[19] Optical Scanning Products Brochures, General Scanning Inc., 500 Arsenal Street, Watertown, MA 02272 (1995).

[20] Refer to Ref. 5, p. 209, for a derivation of the phase-modulating properties of Bragg cells.

[21] Dalsa Camera Catalog, Dalsa Corporation, Waterloo, Ontario, Canada (1995).

[22] S. Kirpatrick, C. D. Gelatt Jr., and M. P. Vecchi, "Optimization by stimulated annealing," Science **220**, 671–679 (1983).

[23] D. Lawrence, *Genetic Algorithm & Simulated Annealing* (Kaufmann, Los Altos, CA, 1987).

[24] E. G. Johnson and M. A. Abushagur, "Microgenetic-algorithm optimization methods applied to dielectric gratings," J. Opt. Soc. Am. A **12**, 1152–1160 (1995).

[25] M. P. Dames, R. J. Dowling, P. McKee, and D. Wood, "Efficient optical elements to generate intensity weighted spot arrays: design and fabrication," Appl. Opt. **30**, 2685–2691 (1991).

[26] N. C. Gallagher and B. Liu, "Method for computing kinoforms that reduces image reconstruction error," Appl. Opt. **12**, 2328–2335 (1973).

[27] F. B. McCormick, "Generation of large spot arrays from a single laser beam by multiple imaging with binary phase gratings," Opt. Eng. **28**, 299–304 (1989).

[28] J. Bengtsson, "Kinoform design with an optimal-rotation-angle method," Appl. Opt. **33**, 6879–6884 (1994).

[29] *Diffractive Optics and Micro-Optics*, Vol. 5 of 1996 OSA Technical Digest Series (Optical Society of America, Washington, DC, 1996).

[30] D. Jared and D. Ennis, "Inclusion of filter modulation in synthetic discriminant function construction," Appl. Opt. **28**, 232–239 (1989)

[31] B. V. K. Vijaya Kumar, "Tutorial survey of composite filter designs for optical correlators," Appl. Opt. **31**, 4773–4801 (1992).

[32] W. J. Dallas, "Computer-generated holograms," in *The Computer in Optical Research*, B. R. Frieden, ed. (Springer, Berlin, 1980), Chap. 6, pp. 291–366.

[33] J. P. Kirk and A. L. Jones, "Phase-only complex-valued spatial filter," J. Opt. Soc. Am. **61**, 1023–1028 (1971).

[34] J. L. Horner and P. D. Gianino, "Phase-only matched filtering," Appl. Opt. **23**, 812–816 (1984).

[35] D. Casasent and W. A. Rozzi, "Computer-generated and phase-only synthetic discriminant function filters," Appl. Opt. **25**, 3767–3772 (1986).

[36] J. A. Davis and D. M. Cottrell, "Random mask encoding of multiplexed phase-only and binary phase-only filters," Opt. Lett. **19**, 496–498 (1994).

[37] M. W. Farn and J. W. Goodman, "Optimal maximum correlation filter for arbitrarily constrained devices," Appl. Opt. **28**, 3362–3366 (1989).

[38] R. D. Juday, "Optimal realizable filters and the minimum Euclidean distance principle," Appl. Opt. **32**, 5100–5111 (1993).

[39] M. Montes-Usategui, J. Campos, and I. Juvells, "Computation of arbitrarily constrained synthetic discriminant functions," Appl. Opt. **34**, 3904–3914 (1995).

[40] B. V. K. Vijaya Kumar and D. W. Carlson, "Optimal trade-off synthetic discriminant function filters for arbitrary devices," Opt. Lett. **19**, 1556–1558 (1994).

[41] L. G. Hassebrook, M. E. Lhamon, R. C. Daley, R. W. Cohn, and M. Liang, "Random-phase-encoding of composite fully complex filters," Opt. Lett. **21**, 272–274 (1996).

[42] R. W. Cohn and W. Liu, "Pseudorandom-encoding of fully complex modulation to bi-amplitude phase modulators," in Ref. 29, pp. 237–240.

[43] R. D. Juday and J. Knopp, "HOLOMED – An algorithm for computer generated holograms," in *Optical Pattern Recognition VII*, D. P. Casasent and T.-H. Chao, eds., Proc. SPIE **2752**, 162–172 (1996).

[44] Z. Bahri and B. V. K. Vijaya Kumar, "Algorithms for designing phase-only synthetic discriminant functions," in *Optical Information Processing Systems and Architectures*, B. Javidi, ed., Proc. SPIE **1151**, 138–147 (1989).

[45] D. C. Youla and H. Webb, "Image restoration by the method of convex projections: part 1 – theory," IEEE Trans. Med. Imag. **MI-1**, 81–94 (1982).

[46] H. Stark, W. C. Catino, and J. L. LoCicero, "Design of phase gratings by generalized projections," J. Opt. Soc. Am. A **8**, 566–571 (1991).

[47] R. W. Gerchberg and W. O. Saxton, "Practical algorithm for the determination of phase from image and diffraction plane pictures," Optik **35**, 237–250 (1972).

[48] C. F. Hester and D. Casasent, "Multivariant technique for multiclass pattern recognition," Appl. Opt. **19**, 1758–1761 (1980).

[49] Z. Bahri and B. V. K. Vijaya Kumar, "Generalized synthetic discriminant functions," J. Opt. Soc. Am. A **5**, 562–571 (1988).

[50] R. W. Cohn and M. Liang, "Approximating fully complex spatial modulation with pseudorandom phase-only modulation," Appl. Opt. **33**, 4406–4415 (1994).

[51] R. W. Cohn and M. Liang, "Pseudorandom phase-only encoding of real-time spatial light modulators," Appl. Opt. **35**, 2488–2498 (1996).

[52] M. T. Gale, M. Rossi, H. Schutz, P. Ehbets, H. P. Herzig, and D. Prongue, "Continuous-relief diffractive optical elements for two-dimensional array generation," Appl. Opt. **34**, 2526–2533 (1993).

[53] A. Papoulis, *Probability, Random Variable and Stochastic Process*, 3rd ed. (McGraw-Hill, New York, 1991).

[54] Ref. 53, pp. 101–102 and 226–229.

[55] C. Soutar, S. E. Monroe Jr., and J. Knopp, "Measurement of the complex transmittance of the Epson liquid crystal television," Opt. Eng. **33**, 1061–1068 (1994).

[56] R. W. Cohn and J. L. Horner, "Effects of systematic phase errors on phase-only correlation," Appl. Opt. **33**, 5432–5439 (1994).

[57] R. W. Cohn, "Performance models of correlators with random and systematic phase errors," Opt. Eng. **34**, 1673–1679 (1995).

[58] J. N. Mait, "Understanding diffractive optic design in the scalar domain," J. Opt. Soc. Am. A **12**, 2145–2158 (1995).

[59] F. Wyrowski, "Upper bound of the diffraction efficiency of diffractive phase elements," Opt. Lett. **16**, 1915–1917 (1991).

[60] D. C. Chu and J. W. Goodman, "Spectrum shaping with parity sequences," Appl. Opt. **11**, 1716–1724 (1970).

[61] H. Akahori, "Comparison of deterministic phase coding with random-phase coding in terms of dynamic range," Appl. Opt. **12**, 2336–2343 (1973).

[62] U. Krackhardt, J. N. Mait, and N. Streibl, "Upper bound on the diffraction efficiency of phase-only fan-out elements," Appl. Opt. **31**, 27–37 (1992).

[63] C. B. Burckhardt, "Use of random-phase masks for the recording of Fourier-transform holograms of data masks," Appl. Opt. **9**, 695–700 (1970).

[64] J. C. Dainty, ed., *Laser Speckle and Related Phenomena*, 2nd ed. (Springer, Berlin, 1984).

[65] O. Bryngdahl and F. Wyrowski, "Digital holography – computer-generated holograms," in *Progress in Optics*, E. Wolf, ed. (Elsevier, Amsterdam, 1990), Vol. XXVIII, pp. 1–86.

[66] B. Javidi and J. L. Horner, "Optical pattern recognition for validation and security verification," Opt. Eng. **33**, 1752–1756 (1994).

[67] E. G. Johnson and J. D. Brasher, "Phase encryption of biometrics in diffractive optical elements," Opt. Lett. **21**, 1271–1273 (1996).

[68] D. P. Casasent, "Hybrid processors," in *Optical Information Processing: Fundamentals* (Springer-Verlag, Berlin, 1981), Chap. 5, pp. 181–233.

[69] F. T. S. Yu and S. Jutamulia, "Hybrid-optical signal processing," in *Optical Signal Processing, Computing, and Neural Networks* (Wiley-Interscience, New York, 1992), Chap. 6.

[70] D. P. Casasent, A. Ye, J. Smokelin, and R. H. Schaefer, "Optical correlation filter fusion for object detection," Opt. Eng. **33**, 1757–1766 (1994).

[71] L. G. Hassebrook, B. V. K. Vijaya Kumar, and L. Hostetler, "Linear phase coefficient composite filter banks for distortion-invariant optical pattern recognition," Opt. Eng. **29**, 1033–1043 (1990).

[72] L. G. Hassebrook, M. Rahmati, and B. V. K. Vijaya Kumar, "Hybrid composite filter banks for distortion-invariant optical pattern recognition," Opt. Eng. **31**, 923–933 (1992).

[73] L. G. Hassebrook, M. E. Lhamon, and M. Wang "Distortion parameter estimation using complex distortion-invariant correlation filter bank responses," in *Optical Pattern Recognition VI*, D. P. Casasent and T.-H. Chao, eds., Proc. SPIE **2490**, 64–76 (1995).

[74] M. E. Lhamon, L. G. Hassebrook, and R. C. Daley, "Translation-invariant optical pattern recognition without correlation," Opt. Eng. **35**, 2700–2709 (1996).

[75] M. E. Lhamon, L. G. Hassebrook, and J. P. Chatterjee, "Complex spatial images for multi-parameter distortion-invariant optical pattern recognition and high-level morphological transformations," in *Optical Pattern Recognition VII*, D. P. Casasent and T.-H. Chao, eds., Proc. SPIE **2752**, 22–30 (1996).

[76] C. S. Weaver and J. W. Goodman, "A technique for optically convolving two functions," Appl. Opt. **5**, 1248–1249 (1966).

[77] F. T. S. Yu, S. Jutamulia, T. W. Lin, and D. A. Gregory, "Adaptive real-time pattern recognition using a liquid crystal TV based joint transform correlator," Appl. Opt. **26**, 1370–1372 (1987).

[78] R. W. Cohn, "Adaptive real-time architectures for phase-only correlation," Appl. Opt. **32**, 718–725 (1993).

[79] J. H. Bruning, D. R. Herriott, J. E. Gallagher, D. P. Rosenfeld, A. D. White, and D. J. Brangaccio, "Digital wavefront measuring interferometer for testing optical surfaces and lenses," Appl. Opt. **13**, 2693–2703 (1974).

Index